Optoelectronic Devices

Joachim Piprek (Editor)

Optoelectronic Devices
Advanced Simulation and Analysis

With 238 Illustrations

Springer

Joachim Piprek
Department of Electrical and Computer Engineering
University of California
Santa Barbara, CA 93106
USA
piprek@ece.ucsb.edu

Library of Congress Cataloging-in-Publication Data is available.

ISBN 0-387-22659-1   Printed on acid-free paper.

© 2005 Springer Science+Business Media, Inc.
All rights reserved. This work may not be translated or copied in whole or in part without the written permission of the publisher (Springer Science+Business Media, Inc., 233 Spring Street, New York, NY 10013, USA), except for brief excerpts in connection with reviews or scholarly analysis. Use in connection with any form of information storage and retrieval, electronic adaptation, computer software, or by similar or dissimilar methodology now known or hereafter developed is forbidden.
The use in this publication of trade names, trademarks, service marks, and similar terms, even if they are not identified as such, is not to be taken as an expression of opinion as to whether or not they are subject to proprietary rights.

Typesetting: Pages created by the authors using a Springer TeX macro package.

Printed in the United States of America.   (BS/DH)

9 8 7 6 5 4 3 2 1   SPIN 10946619

springeronline.com

Preface

Optoelectronic devices have received great attention in recent years, as they are key components of the Internet and other optical communication systems. Other breakthrough developments, for instance with GaN-based light emitters, also contribute to the increased interest in optoelectronics. The complexity of physical mechanisms within such devices makes computer simulation an essential tool for performance analysis and design optimization. Advanced software tools have been developed for optoelectronic devices and several commercial software providers have emerged. These tools enable engineers and scientists to design and understand ever-more sophisticated nanostructure devices.

The specific challenge of optoelectronic device simulation lies in the combination of electronics and photonics, including the sophisticated interaction of electrons and light. The large variety of materials, devices, physical mechanisms, and modeling approaches often makes it difficult to select appropriate theoretical models or software packages. This book presents a review of devices and advanced simulation approaches written by leading researchers and software developers. The intended audience is scientists and device engineers in optoelectronics, who are interested in using advanced software tools. Each chapter describes the theoretical background as well as practical simulation results that help to better understand internal device physics. The software packages used are available to the public, on a commercial or noncommercial basis, so that interested readers can perform similar simulations on their own. Software providers and Internet addresses are given in each chapter.

The book starts with a chapter on electron–photon interaction, followed by several chapters on laser diodes, including Fabry–Perot lasers, distributed feedback lasers, multisection and tunable lasers, vertical-cavity lasers, as well as mode locking. Following GaN-based light-emitting diodes, three prominent types of light receivers are addressed (solar cells, image sensors, and infrared optical detectors). The book concludes with chapters on optoelectronic device integration and active photonic circuits. Various semiconductor material systems are involved, including structures grown on InP (Chaps. 1–7, 14, 15), GaAs (Chaps. 8–9), GaN (Chap. 10), Si (Chaps. 11–12), and CdTe (Chap. 13). A wide wavelength spectrum is covered, from the ultraviolet (340 nm) to the far infrared (10 μm), with a main focus on communication wavelengths from 850 nm to 1.55 μm.

I would like to sincerely thank all contributors to this book for their chapters. To further support and connect researchers in this rapidly developing field, I have started the website http://www.nusod.org as well as the annual international conference *Numerical Simulation of Optoelectronic Devices (NUSOD)*, which presents the latest research results and software tools. I encourage readers to contact me with questions and suggestions at piprek@nusod.org.

Santa Barbara *Joachim Piprek*
March 2004

Contents

1 Gain and Absorption: Many-Body Effects
S. W. Koch, J. Hader, A. Thränhardt, J. V. Moloney 1
1.1 Introduction 1
1.2 Theory.. 2
1.3 Simplified Models 7
    1.3.1 General Features; Single-Particle Gain and Absorption.. 7
    1.3.2 Fair Approximations 10
    1.3.3 Poor Approximations............................... 13
1.4 Commercial Applications.................................. 18
    1.4.1 Gain Tables 18
    1.4.2 On-Wafer Device Testing 19
1.5 Carrier Dynamics 23
References 24

2 Fabry–Perot Lasers: Temperature and Many-Body Effects
B. Grote, E. K. Heller, R. Scarmozzino, J. Hader, J. V. Moloney, S. W. Koch 27
2.1 Introduction 27
2.2 Theory.. 31
    2.2.1 Transport 31
    2.2.2 Optics 38
    2.2.3 Gain 39
2.3 Temperature Sensitivity of InGaAsP Semiconductor
    Multi-Quantum Well Lasers 41
    2.3.1 Laser Structure 42
    2.3.2 Sample Characterization 42
    2.3.3 Gain Spectra 44
    2.3.4 Light-Current Characteristics and Model Calibration ... 47
    2.3.5 Self-Heating 53
2.4 Summary 58
References 59

3 Fabry–Perot Lasers: Thermodynamics-Based Modeling
U. Bandelow, H. Gajewski, and R. Hünlich....................... 63
3.1 Introduction 63

3.2	Basic Equations		64
	3.2.1	Poisson Equation	64
	3.2.2	Transport Equations	64
	3.2.3	State Equations	65
	3.2.4	Optics	65
3.3	Heating		67
	3.3.1	Free Energy, Entropy, Energy	67
	3.3.2	Current Densities	69
	3.3.3	Heat Equation	70
	3.3.4	Entropy Balance	72
3.4	Boundary Conditions		74
3.5	Discretization		75
	3.5.1	Time Discretization	75
	3.5.2	Space Discretization	75
	3.5.3	Discretization of the Currents	76
3.6	Solution of the Discretized Equations		78
	3.6.1	Decoupling, Linearization	78
	3.6.2	Solution of Linear Algebraic Equations	78
3.7	Example		78
	3.7.1	Stationary Characteristics	79
	3.7.2	Modulation Response	82
3.8	Conclusion		83
A	Temperature Dependence of Model Parameters		84
References			85

4 Distributed Feedback Lasers: Quasi-3D Static and Dynamic Model
X. Li 87

4.1	Introduction		87
4.2	Governing Equations		89
	4.2.1	Optical Wave Equations	89
	4.2.2	Carrier Transport Equations	93
	4.2.3	Optical Gain Model	95
	4.2.4	Thermal Diffusion Equation	97
4.3	Implementation		98
	4.3.1	General Approach	98
	4.3.2	Solver for Optical Wave Equations	103
	4.3.3	Solver for Carrier Transport Equations	104
	4.3.4	Solver for Optical Gain Model	104
	4.3.5	Solver for Thermal Diffusion Equation	104
4.4	Model Validation		106
4.5	Model Comparison and Application		107
	4.5.1	Comparison among Different Models	107
	4.5.2	1.3-μm InAlGaAs/InP BH SL-MQW DFB Laser Diode	108
	4.5.3	1.55-μm InGaAsP/InP RW SL-MQW DFB Laser Diode	110

4.6	Summary	117
References		117

5 Multisection Lasers: Longitudinal Modes and their Dynamics

M. Radziunas, H.-J. Wünsche 121

5.1	Introduction	121
5.2	Traveling Wave Model	122
5.3	Model Details and Parameters	123
	5.3.1 Model Details	123
	5.3.2 Parameters	125
5.4	Simulation of a Passive Dispersive Reflector Laser	126
5.5	The Concept of Instantaneous Optical Modes	129
5.6	Mode Expansion of the Optical Field	131
5.7	Driving Forces of Mode Dynamics	133
5.8	Mode-Beating Pulsations in a PhaseCOMB Laser	134
	5.8.1 Simulation	135
	5.8.2 Mode Decomposition	135
	5.8.3 Spatio-temporal Properties of Mode-beating Self-pulsations	137
5.9	Phase Control of Mode-beating Pulsations	139
	5.9.1 Simulation of Phase Tuning	139
	5.9.2 Mode Analysis	139
	5.9.3 Regimes of Operation	140
	5.9.4 Bifurcations	140
5.10	Conclusion	142
A	Numerical Methods	142
	A.1 Numerical Integration of Model Equations	142
	A.2 Computation of Modes	144
	A.3 Mode Decomposition	147
References		149

6 Wavelength Tunable Lasers: Time-Domain Model for SG-DBR Lasers

D. F. G. Gallagher 151

6.1	The Time-Domain Traveling Wave Model	151
	6.1.1 Gain Spectrum	153
	6.1.2 Noise Spectrum	155
	6.1.3 Carrier Equation	155
	6.1.4 Carrier Acceleration	156
	6.1.5 Extension to Two and Three Dimensions	156
	6.1.6 Advantages of the TDTW Method	158
	6.1.7 Limitations of the TDTW Method	159
6.2	The Sampled-Grating DBR Laser	159
	6.2.1 Principles	159

	6.2.2	Reflection Coefficient	163
	6.2.3	The Three-section SG-DBR Laser	164
	6.2.4	The Four-section SG-DBR Laser	166
	6.2.5	Results	169
6.3	The Digital-Supermode DBR Laser		178
	6.3.1	Principle of Operation	178
	6.3.2	Simulations	179
6.4	Conclusions		182
References			184

7 Monolithic Mode-Locked Semiconductor Lasers
E. A. Avrutin, V. Nikolaev, D. Gallagher 185

7.1	Background and General Considerations		185
7.2	Modeling Requirements for Specific Laser Designs and Applications		187
7.3	Overview of Dynamic Modeling Approaches		189
	7.3.1	Time-Domain Lumped Models	189
	7.3.2	Distributed Time-Domain Models	192
	7.3.3	Static or Dynamic Modal Analysis	198
7.4	Example: Mode-Locked Lasers for WDM and OTDM Applications		200
	7.4.1	Background	200
	7.4.2	Choice of Modeling Approach	200
	7.4.3	Parameter Ranges of Dynamic Regimes: The Background	200
	7.4.4	Choice of Cavity Design: All-Active and Active/Passive, Fabry–Perot and DBR Lasers	202
	7.4.5	Passive Mode Locking	203
	7.4.6	Hybrid Mode Locking	207
7.5	Modeling Semiconductor Parameters: The Absorber Relaxation Time		210
7.6	Directions for Future Work		213
7.7	Summary		214
References			214

8 Vertical-Cavity Surface-Emitting Lasers: Single-Mode Control and Self-Heating Effects
M. Streiff, W. Fichtner, A. Witzig 217

8.1	VCSEL Device Structure		217
8.2	Device Simulator		221
	8.2.1	Optical Model	221
	8.2.2	Electrothermal Model	222
	8.2.3	Optical Gain and Loss	226
	8.2.4	Simulator Implementation	227
8.3	Design Tutorial		230
	8.3.1	Single-Mode Control in VCSEL Devices	231
	8.3.2	VCSEL Optical Modes	232

	8.3.3	Coupled Electrothermo-Optical Simulation	238
	8.3.4	Single-Mode Optimization Using Metallic Absorbers and Anti-Resonant Structures	242

8.4 Conclusions 245
References 246

9 Vertical-Cavity Surface-Emitting Lasers: High-Speed Performance and Analysis

J. S. Gustavsson, J. Bengtsson, A. Larsson 249

9.1 Introduction to VCSELs 249
9.2 Important Characteristics of VCSELs 251
    9.2.1 Resonance and Damping: Modulation Bandwidth 251
    9.2.2 Nonlinearity 253
    9.2.3 Noise 254
9.3 VCSEL Model 255
    9.3.1 Current Transport 255
    9.3.2 Heat Transport 259
    9.3.3 Optical Fields 261
    9.3.4 Material Gain 265
    9.3.5 Noise 265
    9.3.6 Iterative Procedures 268
9.4 Simulation Example: Fundamental-Mode-Stabilized VCSELs.... 270
    9.4.1 Surface Relief Technique 271
    9.4.2 Device Structure 275
    9.4.3 Simulation Results 276
9.5 Conclusion 290
References 291

10 GaN-based Light-Emitting Diodes

J. Piprek, S. Li 293

10.1 Introduction 293
10.2 Device Structure 293
10.3 Models and Parameters 295
    10.3.1 Wurtzite Energy Band Structure 295
    10.3.2 Carrier Transport 298
    10.3.3 Heat Generation and Dissipation 302
    10.3.4 Spontaneous Photon Emission 303
    10.3.5 Ray Tracing 304
10.4 Results and Discussion 306
    10.4.1 Internal Device Analysis 306
    10.4.2 External Device Characteristics 308
10.5 Summary 311
References 311

11 Silicon Solar Cells
P. P. Altermatt 313
11.1 Operating Principles of Solar Cells 313
11.2 Basic Modeling Technique 315
11.3 Techniques for Full-Scale Modeling 318
11.4 Derivation of Silicon Material Parameters 319
11.5 Evaluating Recombination Losses 327
11.6 Modeling the Internal Operation of Cells 330
11.7 Deriving Design Rules for Minimizing Resistive Losses 334
References 339

12 Charge-Coupled Devices
C. J. Wordelman, E. K. Banghart 343
12.1 Introduction 343
12.2 Background 344
    12.2.1 Principles of Operation of CCDs 344
    12.2.2 CCD Architectures 345
12.3 Models and Methods 348
    12.3.1 Process Models 349
    12.3.2 Device Models 349
    12.3.3 Solution Methods 352
12.4 Charge Capacity 353
12.5 Charge Transfer 357
    12.5.1 Charge Transport Mechanisms 359
12.6 Charge Blooming 364
12.7 Dark Current 370
12.8 Charge Trapping 373
12.9 Summary 377
A   Example Distribution 377
References 378

13 Infrared HgCdTe Optical Detectors
G. R. Jones, R. J. Jones, W. French 381
13.1 Introduction 381
13.2 Photon Detection 381
13.3 Summary of Simulation Tools 383
    13.3.1 Introduction 383
    13.3.2 Fundamentals of Device Simulation 384
    13.3.3 Carrier Generation and Recombination Mechanisms 387
    13.3.4 Shockley–Read–Hall Recombination 387
    13.3.5 Auger Recombination 388
    13.3.6 Recombination Through Photon Emission 388
13.4 Optoelectronic Simulation 389
    13.4.1 Optical Beam Characteristics 389
    13.4.2 Light Absorption and Photogeneration 390

| 13.5 | Device Simulation | 391 |

- 13.5.1 Material Parameters ... 391
- 13.5.2 Device Structure ... 393
- 13.5.3 Cross Talk Considerations ... 395
- 13.5.4 Photogeneration and Spectral Response ... 396
- 13.5.5 Recombination Studies ... 398
- 13.6 Temperature Studies ... 400
- 13.7 Variation of Composition ... 402
- 13.8 Conclusion ... 402
- References ... 403

14 Monolithic Wavelength Converters: Many-Body Effects and Saturation Analysis
J. Piprek, S. Li, P. Mensz, J. Hader ... 405

- 14.1 Introduction ... 405
- 14.2 Device Structure ... 405
- 14.3 General Device Physics ... 406
  - 14.3.1 Optical Waveguiding ... 406
  - 14.3.2 Quantum Well Active Region ... 410
  - 14.3.3 Carrier Transport ... 414
- 14.4 Simulation Results ... 415
  - 14.4.1 Amplifier ... 416
  - 14.4.2 Photodetector ... 420
  - 14.4.3 Sampled-Grating DBR Laser ... 422
- 14.5 Summary ... 425
- References ... 425

15 Active Photonic Integrated Circuits
A. J. Lowery ... 427

- 15.1 Introduction ... 427
- 15.2 Fundamental Requirements of a Simulator ... 428
  - 15.2.1 Single-Mode Interfaces ... 428
  - 15.2.2 Backward-Propagating Waves ... 428
  - 15.2.3 Nonlinearities ... 429
  - 15.2.4 Optical Time Delays ... 430
  - 15.2.5 Time Domain versus Frequency Domain ... 430
  - 15.2.6 Transmission Line Laser Models ... 431
- 15.3 The Simulation Environment ... 433
- 15.4 Simulation Example ... 434
  - 15.4.1 Phase Discriminator ... 435
  - 15.4.2 Internal Clock Source ... 437
  - 15.4.3 External Clock Source ... 438
  - 15.4.4 Phase Locking the Clock Sources ... 440
  - 15.4.5 Optical AND Gate ... 443
  - 15.4.6 Open Design Issues ... 446

15.5 Conclusions 446
References 447

Index 449

1 Gain and Absorption: Many-Body Effects

S. W. Koch[1], J. Hader[2], A. Thränhardt[1], and J. V. Moloney[2]

[1] Department of Physics and Material Sciences Center, Philipps Universität Marburg, Renthof 6, 35032 Marburg, Germany,
stephan.w.koch@physik.uni-marburg.de
[2] Arizona Center for Mathematical Sciences and Optical Sciences Center, University of Arizona, Tucson, AZ 85721, U.S.A., jhader@acms.arizona.edu

1.1 Introduction

For the theoretical analysis and modeling of semiconductor lasers, it is desirable to have access to reliable absorption/gain and refractive index calculations. Whereas some generic device features may be understood on the basis of rate equations with empirical gain coefficients or using free-carrier theory, most quantitative predictions require realistic gain models that are based on microscopic calculations.

Many applications critically depend on fine details of the optical material properties. For example, in most vertical-cavity surface-emitting lasers (VCSELs), the Bragg-mirrors select a well-defined operating wavelength. For optimal use of the laser medium, this cavity wavelength should be at the position of the gain maximum, which shifts as a function of carrier density and temperature.

Other examples are devices that are designed to be polarization insensitive. Here, different regions of the device are usually made of different materials. Part of them influence predominantly transverse electric (TE) polarized light—i.e. have stronger TE gain or absorption—others influence transverse magnetic (TM) polarized light stronger. For true polarization insensitivity, the TE and TM gains or absorptions have to be as similar as possible, which can only be achieved if the material composition is chosen properly. Furthermore, if one wants to use an optical amplifier to amplify several wavelengths at the same time, one would like the gain to be almost wavelength insensitive over a given spectral region. Thus, one has to know the spectral lineshape of the gain very precisely.

Generally, not only for the design and optimization of such advanced devices, but also for more conventional structures, the optical material properties have to be known very precisely if one wants to avoid the expensive and time-consuming process of fabricational trial and error.

In this chapter, we review our approach to compute optical material properties of semiconductor-based optoelectronic devices. The theory is based on a fully microscopic model in which scattering and dephasing processes are calculated explicitly using generalized quantum Boltzmann equations. As input to the calculations, we need material parameters such as the Luttinger parameters, bulk band gaps, bulk dipole matrix elements, band offsets for heterostructure interfaces, and bulk material constants like the background refractive index or phonon energies. As illustrative examples, we show some theory-experiment comparisons and demonstrate that the calculations reliably predict the optical properties of good quality systems.

1.2 Theory

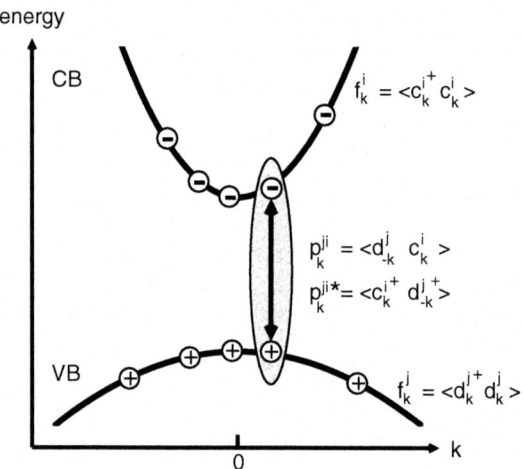

Fig. 1.1. Schematic of the energetically lowest conduction subband (CB) and highest hole subband (VB) and the processes light can induce in the system.

The absorption of light in a semiconductor excites an electron from a valence band into a conduction band, see Fig. 1.1. Thereby, it creates a positively charged hole in the valence band and an electron in the conduction band. In second quantization, this process is represented by the product of a creation operator for an electron in momentum state $\mathbf{k}$ and subband i, $c_{\mathbf{k}}^{i\dagger}$ and that of a creation operator for a hole in momentum state $-\mathbf{k}$ and subband j, $d_{-\mathbf{k}}^{j\dagger}$. Here, we employed the dipole approximation where the momentum of the photon is neglected. The statistical average of this operator pair is a microscopic polarization, $P_{\mathbf{k}}^{ji\star}$.

Aside from absorption, light interacting with a pre-excited system can also induce electron–hole recombination. This process is described by the statistical average of the product of an annihilation operator for an electron (c) and a hole (d), $P_{\mathbf{k}}^{ji} = <d^j_{-\mathbf{k}} c^i_{\mathbf{k}}>$.

As electrons and holes are fermions, the creation of an electron hole pair is only possible if the corresponding electron and hole states have initially been empty. Hence, one has to monitor the occupation probabilities for electrons, $f^i_{\mathbf{k}} = <c^{i\dagger}_{\mathbf{k}} c^i_{\mathbf{k}}>$, and holes, $f^j_{\mathbf{k}} = <d^{j\dagger}_{\mathbf{k}} d^j_{\mathbf{k}}>$. The values of these functions vary between zero and one.

The optical polarization entering into Maxwell's equation is obtained as:

$$P(t) = \frac{1}{V} \sum_{i,j,\mathbf{k}} P^{ji}_{\mathbf{k}}(t) \mu^{ij}_{\mathbf{k}}, \tag{1.1}$$

where V is volume of the active region. The dipole matrix elements $\mu^{ij}_{\mathbf{k}}$ are given by the matrix element of the dipole operator $\hat{e} \cdot \mathbf{p}$ between the corresponding electron and hole states $\psi^{i/j}_{\mathbf{k}}(\mathbf{r})$. $\hat{e}$ is the polarization vector of the light field.

In many applications, the light field is relatively weak such that it does not cause significant changes in the carrier distributions. In that case, the macroscopic polarization can be linearized in the amplitude of the light field, allowing us to calculate the linear optical susceptibility χ by a Fourier transform:

$$\chi(\omega) = \frac{P(\omega)}{4\pi\epsilon_0 \epsilon E(\omega)}. \tag{1.2}$$

Here, ϵ is the background dielectric constant and ω is the angular frequency. From this susceptibility, the linear (intensity-) absorption/gain $\alpha(\omega)$ and the carrier-induced refractive index $\delta n(\omega)$ can be derived:

$$\alpha(\omega) = -\frac{\omega}{n_b c} \mathcal{I}m\{\chi(\omega)\} \tag{1.3}$$

$$\delta n(\omega) = \frac{1}{2n_b} \mathcal{R}e\{\chi(\omega)\}, \tag{1.4}$$

where $\mathcal{I}m$ and $\mathcal{R}e$ denote the imaginary and real parts, n_b is the background refractive index, ϵ_0 is the permittivity, and c is the vacuum speed of light.

If the system is in thermodynamic quasi-equilibrium, the carrier distribution functions assume the form of Fermi–Dirac distributions. Under this condition, one can obtain the spontaneous emission (photoluminescence) from the gain/absorption spectra [1]:

$$S(\omega) = -\frac{1}{\hbar}\left(\frac{n_b \omega}{\pi c}\right)^2 \alpha(\omega) \left[exp\left(\frac{\hbar\omega - \mu}{k_B T}\right) - 1\right]^{-1}, \tag{1.5}$$

where k_B is Boltzmann's constant, T is the temperature, and μ is the interband chemical potential.

Another useful quantity is the linewidth-enhancement factor (alpha factor) α_L:

$$\alpha_L(\omega) = \left(\frac{d\mathcal{R}e\{\chi(\omega)\}}{dN}\right)\left(\frac{d\mathcal{I}m\{\chi(\omega)\}}{dN}\right)^{-1}, \quad (1.6)$$

where N is the carrier density and α_L is a density- and temperature-dependent function that determines the stability of a laser and its tendency to exhibit nonlinearities like pulse chirping, filamentation, and the resulting far-field broadening.

In order to calculate the macroscopic polarization, we use the Heisenberg equations for the electron–hole creation and annihilation operators. From these, we obtain the equation of motion for the single-particle density matrix ρ:

$$\rho_{\mathbf{k}} = \begin{pmatrix} 1 - f_{\mathbf{k}}^{j} & P_{\mathbf{k}}^{ji} \\ P_{\mathbf{k}}^{ji\star} & f_{\mathbf{k}}^{i} \end{pmatrix}. \quad (1.7)$$

The diagonal elements are the distribution functions f, and the off-diagonal elements are the polarizations P.

To evaluate the Heisenberg equations, we have to specify the system Hamiltonian H:

$$\begin{aligned}
H = & \sum_{i,\mathbf{k}} \varepsilon_{\mathbf{k}}^{i} c_{\mathbf{k}}^{i\dagger} c_{\mathbf{k}}^{i} + \sum_{j,\mathbf{k}} \varepsilon_{\mathbf{k}}^{j} d_{\mathbf{k}}^{j\dagger} d_{\mathbf{k}}^{j} + \sum_{\mathbf{q}} \hbar\omega_{\mathbf{q}} b_{\mathbf{q}}^{\dagger} b_{\mathbf{q}} \\
& - \sum_{ij,\mathbf{k}} \left[\mu_{\mathbf{k}}^{ij} c_{\mathbf{k}}^{i\dagger} d_{-\mathbf{k}}^{j\dagger} E^{+}(t) + \mu_{\mathbf{k}}^{ij\star} d_{-\mathbf{k}}^{j} c_{\mathbf{k}}^{i} E^{-}(t) \right] \\
& + \frac{1}{2} \left[\sum_{\substack{i_1 i_2 i_3 i_4 \\ \mathbf{q},\mathbf{k},\mathbf{k}'}} V_{\mathbf{q}}^{i_1 i_2 i_3 i_4} c_{\mathbf{k}+\mathbf{q}}^{i_1\dagger} c_{\mathbf{k}'-\mathbf{q}}^{i_2\dagger} c_{\mathbf{k}'}^{i_3} c_{\mathbf{k}}^{i_4} + \sum_{\substack{j_1 j_2 j_3 j_4 \\ \mathbf{q},\mathbf{k},\mathbf{k}'}} V_{\mathbf{q}}^{j_4 j_3 j_2 j_1} d_{\mathbf{k}+\mathbf{q}}^{j_1\dagger} d_{\mathbf{k}'-\mathbf{q}}^{j_2\dagger} d_{\mathbf{k}'}^{j_3} d_{\mathbf{k}}^{j_4} \right] \\
& - \sum_{\substack{i_1 i_2 j_1 j_2 \\ \mathbf{q},\mathbf{k},\mathbf{k}'}} V_{\mathbf{q}}^{i_1 j_2 j_1 i_2} c_{\mathbf{k}+\mathbf{q}}^{i_1\dagger} d_{\mathbf{k}'-\mathbf{q}}^{j_1\dagger} d_{\mathbf{k}'}^{j_2} c_{\mathbf{k}}^{i_2} \\
& + \sum_{\substack{i_1 i_2 \\ \mathbf{k},\mathbf{q}}} \left[g_{\mathbf{q}}^{i_1 i_2} c_{\mathbf{k}}^{i_1\dagger} b_{\mathbf{q}} c_{\mathbf{k}-\mathbf{q}}^{i_2} + g_{\mathbf{q}}^{i_1 i_2 \star} c_{\mathbf{k}-\mathbf{q}}^{i_2\dagger} b_{\mathbf{q}}^{\dagger} c_{\mathbf{k}}^{i_1} \right] \\
& + \sum_{\substack{j_1 j_2 \\ \mathbf{k},\mathbf{q}}} \left[g_{\mathbf{q}}^{j_2 j_1} d_{\mathbf{k}}^{j_1\dagger} b_{\mathbf{q}} d_{\mathbf{k}-\mathbf{q}}^{j_2} + g_{\mathbf{q}}^{j_2 j_1 \star} d_{\mathbf{k}-\mathbf{q}}^{j_2\dagger} b_{\mathbf{q}}^{\dagger} d_{\mathbf{k}}^{j_1} \right]. \quad (1.8)
\end{aligned}$$

The first line contains the kinetic energies of the carriers in the heterostructure potential and the phonon modes. The second line gives the dipole interaction between carriers and the light field $E(t) = E^{+}(t) + E^{-}(t) = E_0(t)e^{i\omega_L t} + E_0(t)e^{-i\omega_L t}$, for which the rotating wave approximation is used. ω_L is the central frequency of the light. The third and fourth lines describe the Coulomb interaction between the carriers. The last two lines are the electron–phonon interaction. Here, ε are the single-particle energies (subband energies), $\hbar\omega_{\mathbf{q}}$ are the phonon energies, $b/b^{\dagger}$ are phonon creation/annihilation operators, V is the Coulomb potential, and g is the electron–phonon interaction potential. For the electron–phonon interaction in direct semiconductors, it is usually sufficient to take into account just the Fröhlich interaction with one dispersionless branch of longitudinal optical phonons ($\hbar\omega_\mathbf{q} = \hbar\omega_0$ for all $\mathbf{q}$).

With this Hamiltonian, the equations of motion for the microscopic interband polarizations $P_\mathbf{k}^{ji}$ are obtained in the form (see [2] for details of the derivation):

$$\frac{d}{dt}P_\mathbf{k}^{ji} = \frac{1}{i\hbar}\left\{\sum_{i',j'}\left[\mathcal{E}_\mathbf{k}^{jj'}\delta_{ii'} + \mathcal{E}_\mathbf{k}^{ii'}\delta_{jj'}\right]P_\mathbf{k}^{j'i'} + \left[1 - f_\mathbf{k}^i - f_\mathbf{k}^j\right]\mathcal{U}_\mathbf{k}^{ij}\right\} + \left.\frac{d}{dt}P_\mathbf{k}^{ji}\right|_{corr}, \tag{1.9}$$

where

$$\mathcal{E}_\mathbf{k}^{ii'} = \varepsilon_\mathbf{k}^i \delta_{ii'} - \sum_{i'',\mathbf{q}} V_{\mathbf{k-q}}^{ii''i'i''} f_\mathbf{q}^{i''} \tag{1.10}$$

$$\mathcal{E}_\mathbf{k}^{jj'} = \varepsilon_\mathbf{k}^j \delta_{jj'} - \sum_{j'',\mathbf{q}} V_{\mathbf{k-q}}^{j'j''jj''} f_\mathbf{q}^{j''} \tag{1.11}$$

$$\mathcal{U}_\mathbf{k}^{ij} = -\mu_{ij,\mathbf{k}}E(t) - \sum_{i',j',\mathbf{q}} V_{\mathbf{k-q}}^{ij'ji'} P_\mathbf{q}^{j'i'}. \tag{1.12}$$

These are the so-called Semiconductor Bloch equations. Here, $\mathcal{E}$ are the renormalized energies. The indices i, i', i'' (j, j', j'') label the subbands (SBs) for electrons (holes) and n runs over all electron and hole SBs. $\mathcal{U}$ is the effective field that includes the influence of the attractive Coulomb interaction between electrons and holes. As a consequence, the absorption is increased ("Coulomb enhancement") and quasi-bound states below the subband edges, excitonic resonances, appear.

In (1.9), the term $d/dt P_\mathbf{k}^{ji}|_{corr}$ comprises all higher order correlations as a result of the Coulomb and electron–phonon interactions. These are responsible for the screening of the Coulomb interaction and the dephasing of the polarization because of electron–electron scattering and electron–phonon scattering. In our approach, we include the effects of screening at the random-phase approximation (RPA) level (Lindhardt approximation) and the carrier scattering in the second Born–Markov limit. The explicit form of the resulting equations can be found in [3].

The single-particle energies $\varepsilon_\mathbf{k}^{i/j}$ and the wavefunctions used to calculate the Coulomb-, dipole-, and electron–phonon coupling matrix elements are obtained using a 10×10 $\mathbf{k} \cdot \mathbf{p}$ model. The $\mathbf{k} \cdot \mathbf{p}$ matrix takes the general form:

$$\begin{matrix} |N> & |cb> & |vb's> \\ \begin{pmatrix} E_N(z) & V_{NM}(z) & 0 \\ V_{NM}(z) & E_c(\mathbf{k},z) & A(\mathbf{k},z) \\ 0 & A^\dagger(\mathbf{k},z) & E_v(\mathbf{k}) \end{pmatrix} \end{matrix}. \tag{1.13}$$

$E_{c/v}$ are the parabolic conduction band/valence bands in the absence of band coupling. A describes the coupling between these bands and includes the Luttinger submatrix for the coupling in between the valence bands. $A^\dagger$ is the hermitian of the matrix-subblock A.

The lower right 2×2 block of the matrix (1.13) describes the coupling between the energetically lowest bulk conduction band $|cb>$ and the energetically highest three valence bands $|vb's>$. In a zincblende crystal, the latter are the heavy hole, light hole, and spin-orbit split-off hole bands. In a wurtzite crystal, the split-off hole band is replaced by the crystal-field split-hole band. All bands are twice spin degenerate. The explicit forms for this part of the matrix can be found for zincblende structures in [4] and for wurtzite structures in [5].

For dilute nitride systems like $GaInNAs$, the analysis shows that the conduction band is coupled to a dispersionless band E_N originating from the states of the locally, well-separated nitrogen atoms. This band is degenerate with the conduction band and couples with it. This coupling and the resulting bands can be described well by an anti-crossing Hamiltonian with coupling matrix elements V_{NM}. Within the uncertainties of the bandstructure parameters and the experiment, the coupling between the nitrogen band and the valence bands can be neglected. Details of this part of the Hamiltonian can be found in [6, 7, 8]. This part of the Hamiltonian is absent in the absence of dilute nitrogen.

Although the functional form of the matrix (1.13) is the same in all layers of a heterostructure, the bandstructure parameters entering the matrix are dependent on the growth position z. The confinement potential appears through the z dependence of the bulk bandedge energies. The strain induced by mismatching lattice constants in different layers also modifies the bandedge energies and lifts, e.g., the degeneracy between the heavy and light hole band in zincblende materials. This effect can be taken into account by adding strain-dependent shifts to the bandedge energies, i.e., the diagonal terms of the matrix (see [9] for details). In wurtzite structures, the strain induces piezoelectric fields and spontaneous polarization fields [5]. They, as well as other modifications of the confinement potential, like, e.g., external electric potentials or potentials due to local charges stemming from ionized dopants, are added to the diagonal matrix elements.

As the heterostructure potential in growth direction is inhomogeneous, the eigenstates are no longer Bloch functions. In order to calculate the resulting subbands and confined states, k_z in the $\mathbf{k} \cdot \mathbf{p}$ Hamiltonian has to be replaced by its real-space analog $-i\partial_z$ in a hermitian way as demonstrated, e.g., in [10]. An elegant approach to deal with the resulting system of second-order differential equations is to take its Fourier transform. The resulting eigenvalue matrix problem can then be solved with standard diagonalization software. This method has the advantage that boundary conditions at heterostructure interfaces are taken into account implicitly. Also, unphysical spurious solutions corresponding to energies within the band gap and remote bands at very high energies do not appear.

Once the eigenenergies (subbands) are known, carriers can be filled into the states according to Fermi statistics. In general, the potential due to charges from these carriers will modify the confinement potential. These modifications of the total potential are determined by solving the corresponding Poisson equation. The calculation of the subbands, wavefunctions, and charge potentials has to be done in a self-consistency cycle (see [11]), which usually converges within a few iterations.

The parameters entering the $\mathbf{k}\cdot\mathbf{p}$ matrix are bulk bandstructure parameters for the material at the growth position z. These are the bandedge energies, Luttinger parameters, the bulk electron effective mass and momentum matrix elements, as well as strain-related and lattice constants. For the anti-crossing part of dilute nitrides, one needs the energy of the nitrogen band and the coupling matrix element V_{NM}. For most commonly used materials, all these parameters are known and can be found in the standard literature, like [12, 13, 14, 15]. For the dilute nitrides, we use the parameters given in [3]. For some less-investigated material systems, the accuracy of the simulations is limited by the knowledge of these parameters. However, in most cases, this uncertainty mostly influences only the fundamental transition energy. This can be corrected, e.g., by determining the actual bandedge from a comparison with experimental photoluminescence spectra, as discussed in [16].

1.3 Simplified Models

1.3.1 General Features; Single-Particle Gain and Absorption

In simplified approaches, the correlation contribution in (1.9) is often replaced by $P_{\mathbf{k}}^{ji}/T_2$, introducing the phenomenological dephasing time T_2. If one additionally neglects the Coulomb interaction, (1.9) can be solved analytically and one obtains the well-known expression for the single-particle absorption:

$$\alpha(\omega) \propto \sum_{i,j,\mathbf{k}} \left|\mu_{\mathbf{k}}^{ij}\right|^2 \left[1 - f_{\mathbf{k}}^i - f_{\mathbf{k}}^j\right] L(\varepsilon_{\mathbf{k}}^i - \varepsilon_{\mathbf{k}}^j - \hbar\omega)\omega. \tag{1.14}$$

Here:

$$L(\varepsilon - \hbar\omega) = \frac{\gamma^2}{\gamma^2 + (\varepsilon - \hbar\omega)^2}, \tag{1.15}$$

is the Lorentzian lineshape function and $\gamma = 2\pi\hbar/T$. The term in the square brackets is the so-called inversion factor whose sign determines the presence of absorption or gain. In the ground state of the system there are no excited carriers, $f_{\mathbf{k}}^i = f_{\mathbf{k}}^j = 0$ for all i, j, $\mathbf{k}$. All electron and hole states are empty and carriers can be excited into them. Thus, absorption is possible but no stimulated emission. The inversion factor is one for all states and the absorption is maximal. If carriers are filled into the states, the distributions assume values between zero and one. Under strong excitation conditions, $f_\mathbf{k}^i = f_\mathbf{k}^j = 1$ for a range of k values. The inversion factor for these k values is -1. The absorption becomes negative, i.e., one has gain. Under quasi-equilibrium conditions, the carriers in the system occupy the energetically lowest states. At the Fermi level, the (Fermi) distribution functions are one-half and the inversion is zero. At this energy, the semiconductor is transparent, for lower energies one has gain and for higher energies there is absorption.

If one, furthermore, neglects the **k**-dependence of the dipole matrix element, the absorption is proportional to the interband density of states times the inversion factor. In the absence of carriers and for parabolic bands, this is a steplike function for quantum-well systems, with an additional step at each intersubband bandedge. The absorption is multiplied by the square of the corresponding dipole matrix element. This gives rise to selection rules representing the symmetry of the electron and hole wavefunctions involved.

Figure 1.2(a) compares the linear (zero carrier density) absorption as calculated by the full model with that calculated by a model that neglects the Coulomb interaction and uses a dephasing time of 100 fs. The structure has been investigated in [17]. It nominally consists of three 6-nm dilute nitride $Ga_{0.62}In_{0.38}N_{0.019}As_{0.981}$ wells between 20-nm $Ga_{0.95}In_{0.05}N_{0.015}As_{0.985}$ barriers. Comparisons between experimental gain spectra at various excitation densities and calculations of the full model are shown in Fig. 1.2(b) using an inhomogeneous broadening of 32 meV full-width half-maximum (FWHM). The experimental current densities in Figs. 1.2(b)–(d) are 0.56, 0.65, 0.74, 0.84, and 0.93 I_{Thr}, where I_{Thr} is the threshold current density. The theoretical carrier densities are 1.0, 1.7, 2.0, 2.4, 2.6, 2.85, 3.1, 3.5, and $4.0 \times 10^{12}/\text{cm}^2$.

In the absence of Coulomb interaction, the excitonic resonance at the subband edge is missing and the absorption is too small in comparison with experimental values. This is a consequence of the Coulomb enhancement of the absorption. The Coulomb enhancement factor is a frequency and bandstructure-dependent function, that has a maximum value of two at the bandedge for ideal parabolic bands. However, one sees in Fig. 1.2 that the enhancement in the continuum of the subband absorption, e.g., between about 1160 nm and 1220 nm, is significantly less than two. For some structures, even Coulomb "enhancement" less than one has been demonstrated for parts of the spectra [18].

As shown in Fig. 1.2(c), even with such a strongly simplified model, a reasonably good agreement with experimental data can be achieved if the correct parameters are not known and one uses fitting parameters instead. Starting from the parameters used for Figs. 1.2(a) and (b), the theoretical gain spectra had to be shifted by 33 nm to longer wavelengths and had to be scaled by a factor of about 1.5. Using the correct parameters would give the disagreement shown in Fig. 1.2(d). If one would not know the correct

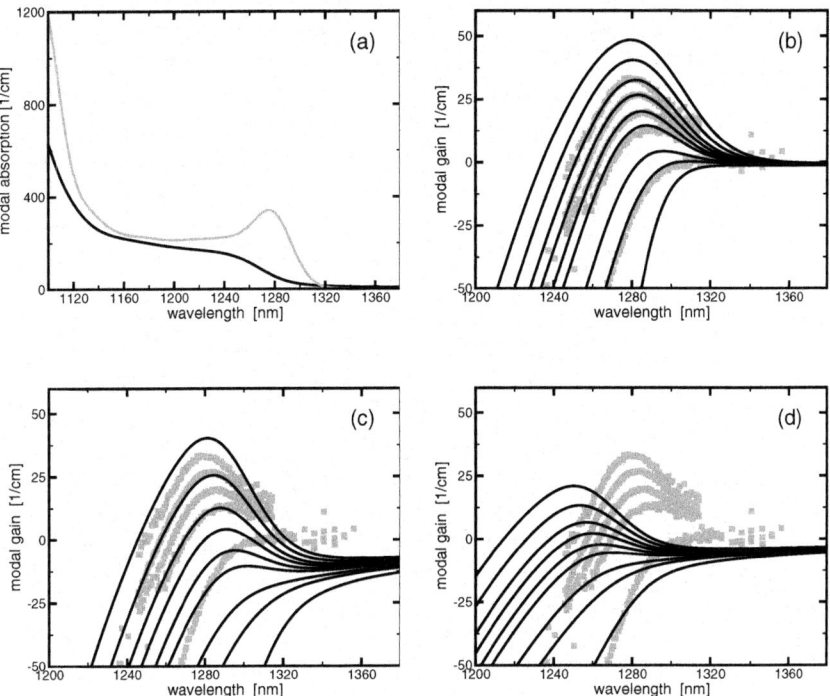

Fig. 1.2. (a): Modal zero-density absorption. Black: calculated with the single-particle model and dephasing time of 100 fs. Gray: calculated with the full model. (b): Experimental gain spectra (gray symbols) and results of the full model (black lines) for various carrier densities (taken from [17]). (c): Best possible fit between experimental gain spectra and spectra calculated with the single-particle model and a dephasing time of 100 fs. (d): Comparison between the spectra of (c) when the same parameters are used, as in (b).

bandedge at zero carrier density, the shift could be (incorrectly) attributed to differences between the actual and nominal structural parameters.

When using such a simplified model, even the best fit can only lead to good agreement with the experiment for limited parameter ranges. We see in Fig. 1.2 that, in the present example for small carrier densities, the agreement becomes worse, and, for high densities, the gain maximum shifts not as it is observed in the experiment. Also, the density dependence of the gain is different from that obtained with the full model. Thus, predictions such as for output versus pump power based on such a model are not correct.

Another problem results from using a lineshape function like (1.15) for the broadening of the transitions because the low-energy tails of these functions extend far below the band gap. In the case of inversion/gain at the band gap, the tails of noninverted transitions at higher energies are larger than the negative tails from inverted states. This leads to the prediction of unphysical absorption energetically below the band gap and gain. This absorption indicates that, at other energies, the spectral amplitudes and lineshapes obtained with this approximation cannot be correct. One can eliminate the absorption below the band gap using an empirical broadening function other than the Lorentzian (1.15). However, this has little or no microscopic justification and the overall predictive capability of the results is not improved.

1.3.2 Fair Approximations

1.3.2.1 Axial Approximation for the Bandstructure

A commonly applied approximation is the so-called axial approximation for the bandstructure [19]. In this approximation, the in-plane angular dependence of the bandstructure is neglected and one uses an angular averaged bandstructure instead. It is an advantage of this approximation that the microscopic polarization has to be calculated only for one angle, ϕ_0, and the modulus of the two-dimensional in-plane momentum, k. The contributions for other angles, ϕ, can be obtained from [18]:

$$P_{k,\phi}^{ji} = \sum_l e^{il(\phi-\phi_0)} P_{l,k,\phi_0}^{ji}. \quad (1.16)$$

In this approximation, all potentials entering the equations of motion for the polarizations, including the Coulomb interaction, have axial symmetry around the growth direction. Therefore, the z component of the angular momentum is conserved and the equations for polarizations with different angular momentum quantum numbers l decouple. Although the single-particle energies are angular independent, the wavefunctions ψ are not. However, they can be derived from the ones for a specific angle by using:

$$\psi_{k,\phi}^n(\mathbf{r}) = \sum_j e^{i\mathbf{k}\cdot\rho} e^{-iM_j^n(\phi-\phi_0)} \zeta_{k,\phi_0}^{n,j}(z) u^j(\mathbf{r}), \quad (1.17)$$

where $\mathbf{k}$ is (k,ϕ), ρ is the in-plane part of $\mathbf{r}$, $\zeta^{n,j}$ is the j'th spinor component of the envelope function, and M_j^n is the angular momentum of that spinor. u^j is the atomic part of the Bloch function (at $\mathbf{k} = \mathbf{0}$).

Using (1.17) and a Fourier expansion, one finds:

$$\mu_{l,k}^{nm} = \int d\phi \mu_{l,k,\phi}^{nm} \propto \sum_{M_i^n, M_j^m} \delta(l - M_i^n + M_j^m) \mathbf{P}^{M_i^n M_j^m} \cdot \hat{e}. \quad (1.18)$$

Here, $\mathbf{P}^{M_i^n M_j^m}$ is the momentum matrix element between the atomic parts of the wavefunctions, $< u^i|\mathbf{p}|u^j >$. The symmetry of the atomic wavefunctions leads to selection rules for the matrix elements. For a zincblende crystal, the nonvanishing matrix elements are listed in [4]. One obtains the following selection rules for TE-polarized light $[\hat{e} = (a, \sqrt{1-|a|^2}, 0)]$ and TM-polarized light $[\hat{e} = (0,0,1)]$:

$$\mu_l \propto \begin{cases} \delta(l, \pm 1) & \text{for TE-polarized light} \\ \delta(l, 0) & \text{for TM-polarized light} \end{cases}. \qquad (1.19)$$

One important conclusion is that a cross-coupling susceptibility, which has been suggested to couple TE- and TM-polarized light [20], is exactly zero for a system with this symmetry. This follows from:

$$P^{TE/TM} = \frac{1}{V} \sum_{i,j,l,k} P_{l,k}^{TE,ji} \mu_{l,k}^{TM,ij} \propto \sum_l \delta(l, \pm 1)\delta(l, 0) = 0. \qquad (1.20)$$

1.3.2.2 Simplified Coulomb Model

The result of (1.19) indicates that a correct treatment of the angular dependence of the wavefunctions is important to obtain correct selection rules. However, a complete treatment of the Coulomb interaction as presented in [18] leads to very involved numerics that reach the limits of todays workstation-level computers. The main problem is that the overlaps of the four confinement wavefunctions $\zeta_{\mathbf{k}}^{n,j}$ in the matrix elements have to be calculated for all $\mathbf{k}$ and all spinor components j. However, the calculations simplify significantly (typically by more than two orders of magnitude in CPU-time) if one neglects the $\mathbf{k}$-dependence of the wavefunctions in the calculation of the Coulomb interaction (and only there) and uses instead only the dominant spinor component at $\mathbf{k} = 0$. On the same level of inaccuracy, one can neglect the angular dependence of the wavefunctions everywhere and simply pick one direction ϕ_0 for the wavefunctions. Figure 1.3 compares some results obtained using this approximation with those of the full model that uses the axial approximation, but takes the angular dependence of the wavefunctions fully into account. The structure is an 8-nm-wide $In_{0.05}Ga_{0.95}As$ well between $Al_{0.2}Ga_{0.8}As$ barriers.

For the zincblende $\mathbf{k} \cdot \mathbf{p}$ Hamiltonian, one can show that if one uses the wavefunctions calculated along the [110] direction, the dipole matrix elements are the same as the nonzero ones (here, for TE: $l = \pm 1$) in the full calculation. Thus, for this case, the continuum absorption agrees with the one of the full model. For other directions, the continuum absorption has a wrong amplitude and slope.

As one sees in Fig. 1.3, the differences are mostly within the uncertainties of a typical experiment. The only significant difference is the break in the selection rules that leads to a double peak around 1.45 eV instead of just one resonance in the full model. In the full model, one has only the $e1 - lh1$ transition at this energy, whereas in the simplified model, a resonance due to

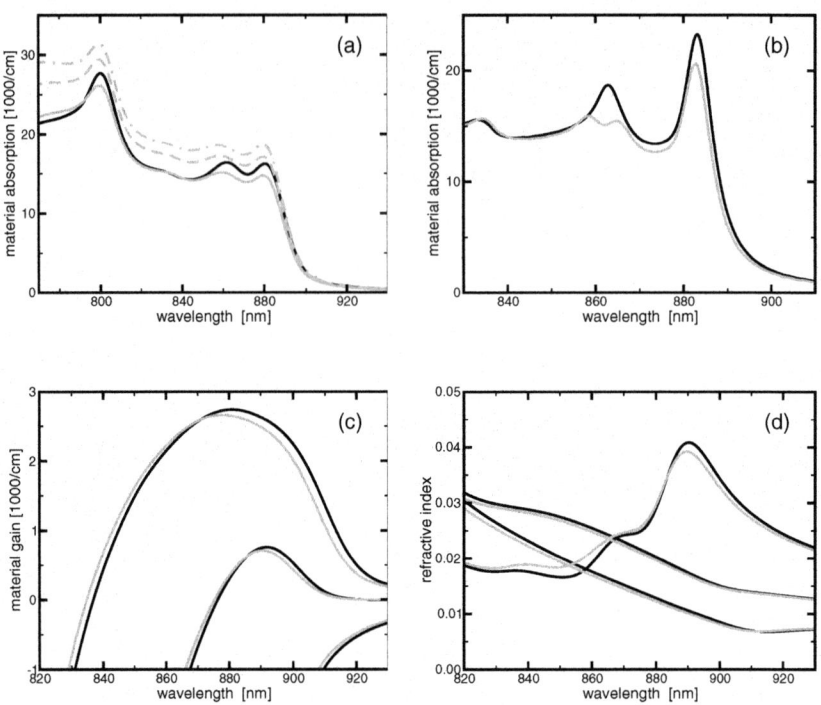

Fig. 1.3. Black lines: spectra calculated taking the angular dependence of the confinement wavefunctions into account and calculating the Coulomb interaction using the **k**-dependent wavefunctions. Gray lines: spectra when only the dominant spinor component at **k** = **0** is used for the Coulomb interaction and the axial approximation is used for the bandstructure calculation, but the angular dependence of the wavefunctions is neglected in the calculation of the dipole matrix elements. (a): Zero density absorption. Solid gray line: wavefunctions calculated for the [110] direction; Dashed lines: [210]; Dash-dotted lines: [100]. An inhomogeneous broadening of 20 meV (FWHM) has been used. (b): As in (a); here, for an inhomogeneous broadening of only 8 meV. (c): As in (a); here, the gain for densities of 0.0, 2.5, and $5.0 \times 10^{12}/\text{cm}^2$. (d): As (c); here, the carrier-induced refractive index δn. For increasing carrier density, δn decreases at the long wavelength side.

the $e1-hh2$ transition appears. For realistic inhomogeneous broadenings, this double-peak structure is washed out, it only becomes visible if one reduces the inhomogeneous broadening as shown in Fig. 1.3(b). Here, the [110] direction has been used. Thus, the dipole matrix elements are the same as in the full model, showing that the break in the selection rules is a direct consequence of the simplified Coulomb model.

In the gain region where the Coulomb interaction is strongly screened, the fine details of its calculation become less relevant. As can be seen in Fig. 1.3 (c), in this density regime, the results of the simplified model agree very well with the full model. Gain amplitudes and lineshapes are virtually the same in both models. The only difference is a very small wavelength shift. It would be impossible to notice this shift in a comparison with the experiment. It is almost completely density independent and, therefore, not distinguishable from a possible deviation between nominal and actual structural parameters.

1.3.3 Poor Approximations

1.3.3.1 Dephasing-Time Approach

As already indicated in Sect. 1.3.1, the replacement of the correlation / scattering terms by a simple dephasing time leads to several problems. First, of course, one needs a reasonable value for this time; i.e., one needs experimental input. Also, the results will only agree well with the experiment in a limited range of parameters, i.e., carrier densities, temperatures, and spectral positions.

For the structure discussed in Sect. 1.3.1, Fig. 1.4(a) compares absorption spectra calculated with the full model with those of the same model, however, using a dephasing time of 100 fs instead of calculating the correlation/scattering terms explicitly. For zero carrier density, the absorptions agree quite well when the same inhomogeneous broadening is used for both cases. This simply is a consequence of the inhomogeneous broadening being much stronger than the homogeneous one. Thus, the details of the homogeneous broadening are rather insignificant. However, as soon as the carrier density is increased, the scattering becomes more efficient. Figure 1.4(c) compares gain spectra for the same parameters with experimental ones. As had been shown in Fig. 1.2(b), the full model gives excellent agreement with these gain spectra for all wavelengths and carrier densities. When using the dephasing time, one can obtain a good fit to the experimental spectra [see Fig. 1.4(c)]. However, to obtain this fit, the spectra have to be shifted by about 9 nm toward shorter wavelengths and they have to be amplified by a factor of 1.9. The spectral deviation is due to the neglect of the shifts induced by the imaginary parts of the correlation/scattering contributions and the inaccurate treatment of the screening of the Coulomb interaction in the coherent parts of the Semiconductor Bloch equations, (1.9). The incorrect amplitudes are partly a consequence of the incorrect screening used. In addition, as indicated by the unphysical nonzero absorption below the band gap, the use of the (Lorentzian) lineshapes leads to a background absorption that reduces the gain. At higher densities than shown here, this background absorption becomes less important. Thus, the problem in the amplitudes becomes less evident.

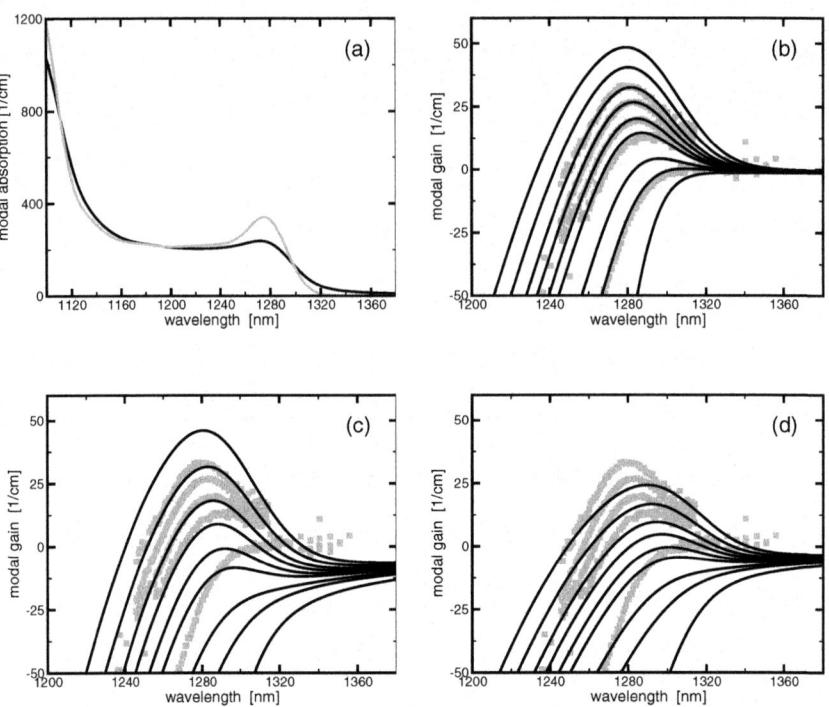

Fig. 1.4. (a): Modal zero-density absorption. Gray: calculated with the full model. Black: calculated with a dephasing time of 100 fs. (b): Experimental gain spectra (gray symbols) and results of the full model (black lines) for various carrier densities (taken from [17]). (c): Best possible fit between experimental gain spectra and spectra calculated using a dephasing time of 100 fs. (d): Comparison between the spectra of (c) when the same parameters are used, as in (b).

Even in the case of an "optimal fit," as shown in Fig. 1.4(c), the agreement is only good for a rather small range of densities. Here, it becomes very bad for low densities, as well as transition energies below the band gap. The density-dependent shift of the gain maxima and the density dependence of the gain amplitudes is not correct, as can be seen by comparing the results with those in the full model, Fig. 1.4(b).

1.3.3.2 Neglect of Conduction-to-Valence Band Coupling

As discussed in [21], a rather poor approximation results if one neglects the coupling between conduction and valence bands in the bandstructure calculation. Figure 1.5(a) shows conduction and valence bands for a structure consisting of a 6-nm-wide $In_{0.60}Ga_{0.40}As_{0.77}P_{0.23}$ well between InP barriers.

In one case, a **k · p** Hamiltonian is used, in which only the coupling between the heavy holes and the light holes is taken into account via a 4 × 4 Luttinger Hamiltonian. The coupling between those and the split-off hole and the conduction bands is neglected. In the other case, the fully coupled 8 × 8 Hamiltonian is used.

The coupling between the hole and conduction bands leads to a nonparabolicity in the conduction bands and also modifies the hole bands. This effect leads to a higher density of states above the band gap, which, in turn, leads to increased absorption, especially at higher transition energies [see Fig. 1.5(b)]. This increased density of states also leads to higher gain amplitudes [see Fig. 1.5(c)]. As the density of states is smaller in the uncoupled model, a given carrier density fills states up to a larger energy. Thus, the Fermi level is higher in this model, and consequently, the gain bandwidth is larger.

Even more than the absorption or gain, the refractive index is strongly influenced by contributions from transitions energetically high above the band gap. As the band nonparabolicities become more significant for elevated energies, the refractive index is more sensitive to the correct band model than the gain and absorption. As shown in Fig. 1.5(d), the neglect of the conduction-to-valence band coupling leads to much too small values for the carrier-induced refractive index contribution.

Generally, the coupling between conduction and valence bands becomes increasingly more important for materials with smaller band gaps. Thus, it is particularly important, e.g., for narrow-band-gap materials used for applications in telecommunication or for antimonide-based materials emitting in the far infrared.

1.3.3.3 Neglect of Coulomb-Induced Intersubband Coupling

The numerics for solving the semiconductor Bloch equations (1.9) can be simplified by about one order of magnitude if one neglects the Coulomb-induced intersubband coupling [21]. This coupling is mediated by Coulomb matrix elements $V^{n_1 n_2 n_3 n_4}$ that are off-diagonal with respect to the subband indices, i.e., $n_1 \neq n_4$ and $n_2 \neq n_3$. If this coupling is neglected, the coherent part of the equations of motion for the polarizations P^{ji} decouples for all i and j. Then, the absorption/gain spectrum is simply the sum of the spectra for each individual subband pair. Each subband spectrum has its individual excitonic resonance according to the two-dimensional density of states.

The Coulomb-induced subband coupling allows for the transfer of excitonic oscillator strength from energetically higher subband transitions toward the bandedge. In the limit of very wide wells, i.e., when the situation of a bulk semiconductor is approached, the excitonic resonances of the individual subbands disappear and their oscillator strength combines into that of the bulk exciton resonance energetically below the fundamental band gap. This bulk limit cannot be reproduced in calculations where the subband coupling is neglected. We thus conclude that this coupling is especially important for the

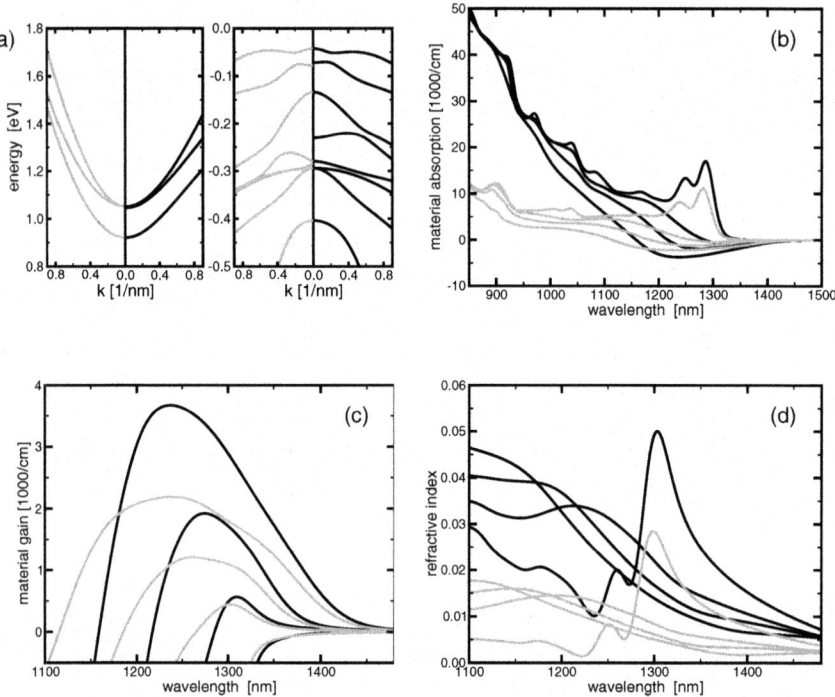

Fig. 1.5. (a): Left: conduction bands with/without conduction-to-valence band coupling (right/left). The energetically higher band is the bulk conduction band of the barrier material, the other bands are confined in the well. Right: valence bands with/without conduction-to-valence band coupling (right/left). The energetically lowest three bands are the bulk heavy-, light-, and split-off hole bands of the barrier material. The other bands are confined in the well. (b): Absorption for carrier densities of 0.05, 2.5, 5.0, and $10.0 \times 10^{12}/\mathrm{cm}^2$. Black: with conduction-to-valence band coupling; Gray: without. (c): As (b); here an enlarged view on the gain. (d): As (b); here: the refractive index δn. For increasing carrier density the index decreases at the long wavelength side.

correct absorption in structures with wide wells where the energetic subband separation becomes similar to the exciton binding energy.

Figure 1.6(a) compares the absorption computed with and without the Coulomb induced subband coupling for a 10-nm-wide and a 50-nm-wide $In_{0.08}Ga_{0.92}As$ well between $GaAs$ barriers. Without the coupling, the absorption in the wide well shows the step-like variation resulting from the summation of individual subband transitions. With the coupling, this step-like shape disappears. The shift of oscillator strength to lower energies leads

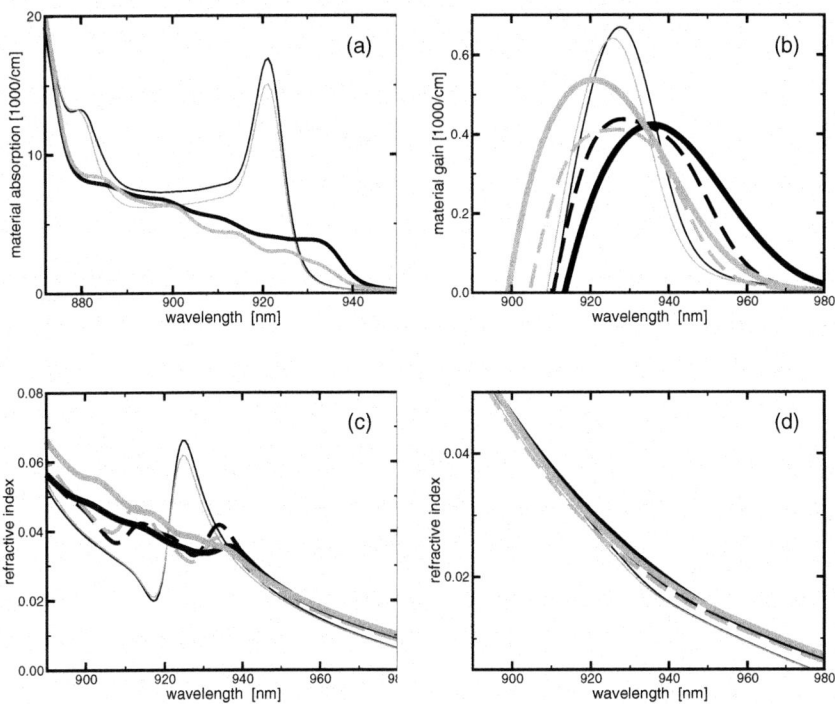

Fig. 1.6. Black lines: spectra calculated with the Coulomb-induced subband coupling; Gray lines: spectra calculated without the coupling. (a): Zero carrier density absorption. Thin lines: for a 10-nm-wide well; thick lines: for a 50-nm-wide well. (b): Thin/medium/thick lines: gain for well widths of 10 nm/20 nm/50 nm and carrier densities of $7 \times 10^{12}\,\mathrm{cm}^{-2}$, $10 \times 10^{12}\,\mathrm{cm}^{-2}$, and $20 \times 10^{12}\,\mathrm{cm}^{-2}$, respectively. (c): Carrier-induced refractive index change δn at zero carrier densities. Notation as in (b). (d): δn at elevated carrier densities. Densities and labeling, as in (b).

to a bulk-like absorption spectrum with a single bulk-band exciton. For the narrow well, the influence of the coupling is more or less negligible.

In the gain regime, the omission of the Coulomb-induced subband coupling mainly leads to a wrong density-dependent wavelength shift of the spectra [see Fig. 1.6(b)]. Typically, in a two-dimensional system, the position of the gain maximum shifts to shorter wavelengths with increasing density, whereas in bulk, this shift is much smaller or even a shift to longer wavelengths. Thus, neglecting the subband coupling leads to an overestimation of the shifts toward shorter wavelengths. Again, the error increases with the well width.

The carrier-induced refractive index δn at a given wavelength is strongly influenced by contributions from energetically detuned transitions. The refractive index is influenced by contributions from a wider energy range than the absorption/gain. Therefore, it is not very sensitive to the details of the excitonic interaction in a specific energetic region. As can be seen from Figs. 1.6(c) and (d), the refractive index spectra, with or without the Coulomb-induced subband coupling, are far more similar than the corresponding gain spectra. This is true particularly in the high-density (gain-) regime in which the Coulomb interaction is strongly screened.

1.4 Commercial Applications

1.4.1 Gain Tables

The numerical effort in calculating optical spectra at the level of the full microscopic approach is quite substantial. Even with the approximations discussed in Sect. 1.3.2, it typically takes somewhere between 10 and 100 minutes on a state-of-the-art workstation computer to calculate one spectrum. The calculation time increases with the third power of the number of confined subbands. Thus, it is especially long for structures with wide or deep wells or for structures with electronically coupled wells. On the other hand, when using a dephasing time model or even a single-particle model, the calculation takes only seconds or less.

Usually, in an operating electrooptical device, parameters like the carrier density or the temperature vary locally and temporally. Thus, a simulation of the overall performance of such a device requires the optical material properties for many situations. It would take prohibitively long if one would have to calculate the optical spectra with the fully microscopic approach each time a parameter changes. However, changes in parameters like density or temperature usually happen on a much longer timescale than the dephasing time of the polarizations (some tens of femtoseconds) and the time it takes carriers to relax into a quasi-thermal equilibrium (of the order of one picosecond). Thus, it is not necessary to solve the dynamic problem for the microscopic polarizations and distribution functions parallel to the time dependence of macroscopic quantities like the electric current, heat dissipation, or light field propagation. Instead, precalculated data bases ("gain-tables") can be used. Here, the optical spectra are calculated on a discrete grid for a range of temperatures or densities using the quasi-equilibrium approximation (time-independent Fermi-distribution functions).

A simulator of the macroscopic properties of the device then just has to look up the spectra corresponding to the local conditions, eventually interpolating between neighboring precalculated values. This requires only fractions of seconds independent of the model used to calculate the optical spectra.

Precalculated gain tables are commercially available through Nonlinear Control Strategies Inc., Tucson/AZ [22]. Figure 1.7 shows some examples for rather extensive gain-tables that include many well widths and material

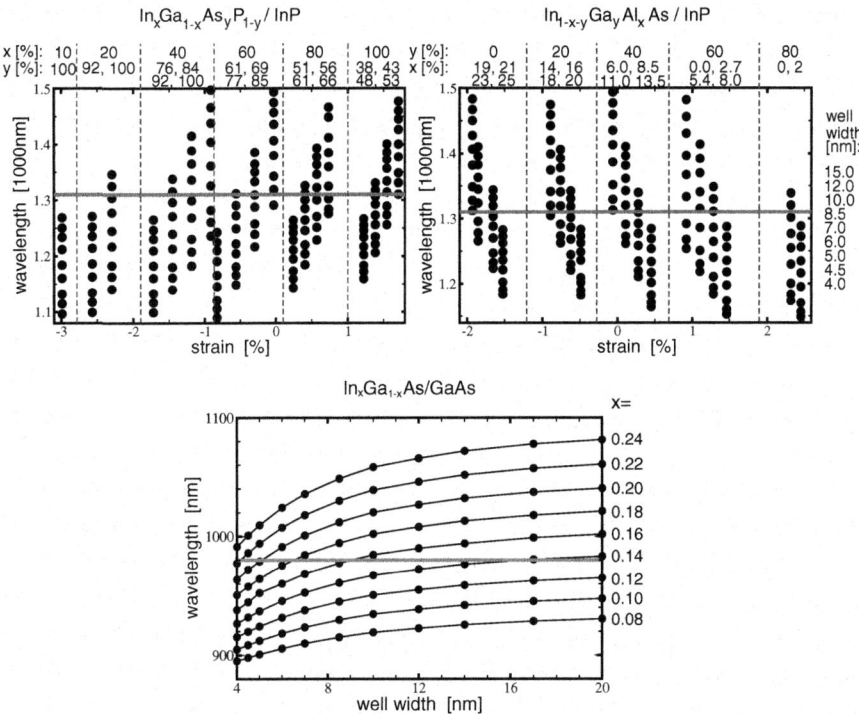

Fig. 1.7. Wavelengths of the energetically lowest subband transition at 300 K for all well widths and material compositions in Nonlinear Control Strategies' $InGaAsP$ 1310-nm gain table (left), $InGaAlAs$ 1310-nm gain table (right), and $InGaAs$ 980-nm gain table (bottom).

compositions for material systems, like $In_xGa_{1-x}As$ wells with $GaAs$ barriers, $In_xGa_{1-x}As_yP_{1-y}$ wells with InP barriers, or $In_{1-x-y}Ga_yAl_xAs$ wells with InP barriers and a specific target wavelength. Also, customized tables for more specialized material systems or including specific doping profiles are available.

Commercially available simulator programs, like LaserMOD from the Rsoft Design Group [23], are already offering to use these gain-tables as a basis for the simulations of electrooptical devices. In addition, such a data base can be used in the search of optimized active regions as, e.g., in the search for a well design with a maximum gain amplitude at a given wavelength, density, or temperature.

1.4.2 On-Wafer Device Testing

The gain-tables and the corresponding photoluminescence spectra obtained through (1.5) can be used very effectively to obtain crucial information about the quality of semiconductor laser structures while the devices are still unprocessed and on the wafer [16].

Very often, the only experimentally obtainable optical information, while the device is still on the wafer, is the photoluminescence. This is typically measured by illuminating the sample with a low-intensity pump source and collecting the resulting photoluminescence vertically to the surface. However, if the experimental results are not compared with advanced theoretical models, as discussed here, the information obtainable through these measurements is rather limited. The wavelength of the photoluminescence often differs from that of the peak gain in the processed and running device by several tens of nanometers. Part of this is due to the different excitation conditions: For the luminescence, low-intensity optical pumping is used, whereas in running semiconductor lasers typically strong electrical pumping is used. With high-power electrical pumping, possible doping-related electrical fields are almost completely screened. However, they are present if weak optical pumping is used. These fields lead to density-dependent spectral shifts in addition to those due to many-body effects. Also, other parameters, like the temperature, might be different in the luminescence measurement and the running device; however, all these differences are automatically included in the microscopic calculations.

The linewidth of the luminescence spectra can give valuable information about the structural quality of the device. However, if one does not know the intrinsic density- and temperature-dependent homogeneous linewidth due to electron–electron and electron–phonon scattering, it is impossible to extract the contribution of the inhomogeneous broadening due to deviations from the ideal crystal structure.

The microscopic calculations outlined in this chapter can help to solve all these problems. They can predict, for the nominally ideal structure, the low-intensity luminescence as well as the high-density gain spectra. From deviations between the computed spectra and experimental luminescence measurements, possible deviations from the nominal structural parameters, like well widths or material compositions, and the inhomogeneous broadening, can be determined. Then, the optical properties of the operating device can be predicted using the actual structural parameters.

This idea is illustrated for a realistic example in Fig. 1.8. Shown are experimental and theoretical spectra for a structure nominally consisting of three 5-nm-wide $In_{0.2}Ga_{0.8}As$ wells between 10-nm-wide $GaAs$ barriers. In a $p-i-n$ doping configuration the undoped active layers are surrounded by p^{++}- and n-doped layers (see [16] for details). In the absence of electric pumping, partial ionization of the dopants leads to an electric field across the active region of $37.6\,\text{kV/cm}$.

In the first step, the low-density absorption spectra are calculated for the ideal nominal structure [Fig. 1.8(a)]. Then, these spectra are compared with experimental luminescence spectra obtained for some low-excitation condi-

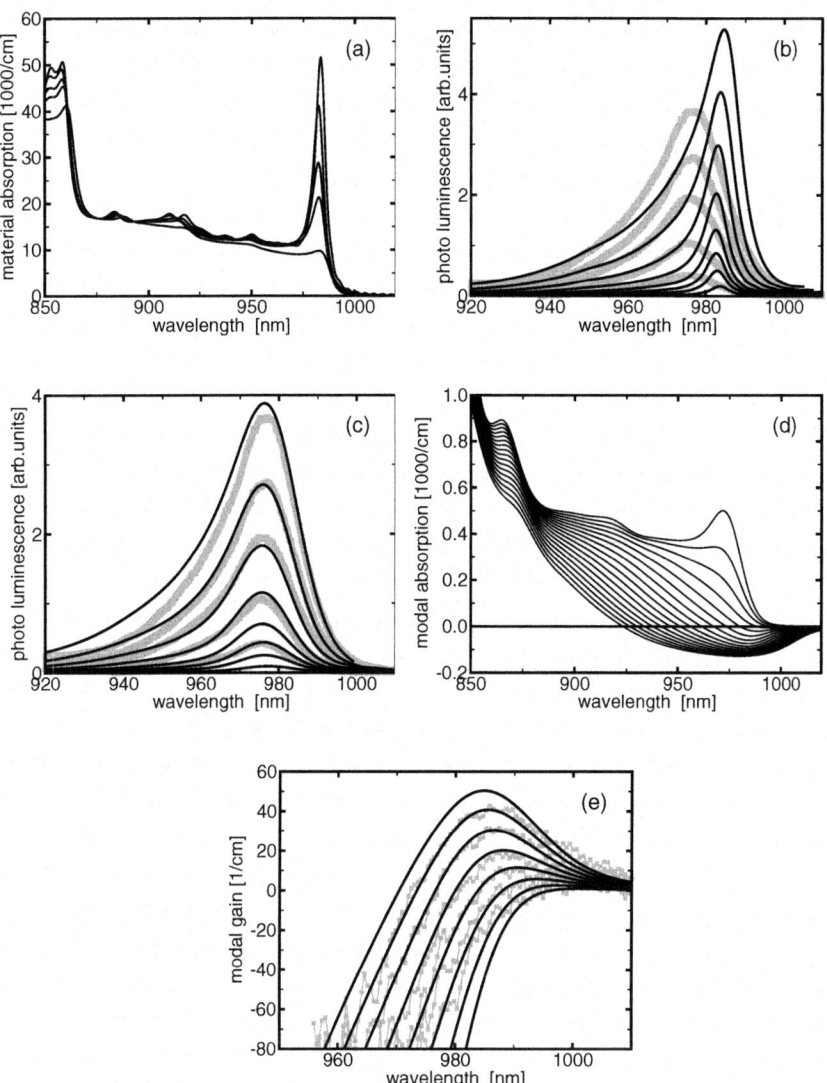

Fig. 1.8. (a): Theoretical absorption for densities of 1.0, 2.3, 4.0, 5.3, and 6.8 × 10^{11}cm^{-2} and no inhomogeneous broadening. (b): Black: theoretical photoluminescence (PL) for 1.0, 1.7, 2.3, 3.0, 4.0, 5.3, 6.8, and 8.5 × 10^{11}cm^{-2} and no inhomogeneous broadening. Gray: experimental PL for pump powers of 12, 16, 18, 21, and 24 mW. (c): As (b); here, the theoretical spectra have been inhomogeneously broadened by 16 meV (FWHM) and shifted by −6 nm. (d): Theoretical absorption for densities from 0.5 to 4.5 × 10^{12}cm^{-2} and increments of 0.25 × 10^{12}cm^{-2} and broadening and shift as in (c). (e): Black: Theoretical gain for densities from 1.0 to 1.875 × 10^{12}cm^{-2} and increments of 0.125 × 10^{12}cm^{-2} and broadening and shift as in (c). Gray: experimental gain for pump currents of 6.0, 6.5, 7.0, 7.5, 8.3, and 9.0 mA.

tions [Fig. 1.8(b)]. The theoretical spectra are then shifted in order to get agreement of the peak positions. The wavelength shift needed in this example was only 6 nm. This spectral mismatch is most likely due to an actual Indium concentration in the well of 19% instead of the nominal 20%. The other explanation would be a rather unlikely deviation from the targeted well width by two monolayers. After broadening the theoretical spectra according to an inhomogeneous broadening of 16 meV the almost perfect agreement shown in Fig. 1.8(c) is obtained.

Here, it is important to notice that it is not necessary to know the actual carrier densities in the experiment. These are usually very hard to determine. Instead, theoretical spectra are calculated for the nominal structure and a density range that covers the expected experimental one. Although the luminescence spectra look rather featureless, usually only one match between theoretical and experimental spectra for a set of excitation conditions can be found. In other words, only for one inhomogeneous broadening and one association between theoretical and experimental densities can a very good match be achieved. The experimental densities can then be deduced from the comparison with the theory.

Once the actual material parameters and the inhomogeneous broadening are known, the theory has no free parameters left. Then, the optical spectra for the running device can be predicted [Fig. 1.8(d)]. On this basis, variations of other parameters, like the temperature, polarization of the light field, or external electric fields, can then be studied realistically.

As a test of the approach for the structure discussed here, the gain spectra were measured independently. Figure 1.8(e) shows the experimental results plotted on top of the theoretically predicted spectra without adjustments like amplitude scaling or spectral shifts. It is very satisfying to see that the agreement between experiment and theoretical prediction is very good.

The procedure described here could be automated and integrated into software used in diagnostic tools that measure the luminescence of devices while still on the wafer. The ideal spectra could be precalculated and tabulated for several inhomogeneous broadenings. The comparison between experimental and theoretical luminescence spectra would then be done by scanning through the data sets for different broadenings while shifting the spectra according to the mismatch in the peak position. Then, the spectra for the operating device could be read from the data base of precalculated spectra. The whole process would only take seconds or less per device. Based on this information, one could decide already on this stage whether one wants to process or discard the individual structure. This could also be used to give the customer precise information about the operating characteristics, e.g., temperature and density dependence of each individual device, without having to measure them.

1.5 Carrier Dynamics

We now consider the case of a more intensive excitation. The carrier distributions are then no longer described by Fermi functions but are governed by the following equations of motion [3]:

$$\frac{\partial}{\partial t} f_{\mathbf{k}}^i = -2\mathrm{Im}(\mathcal{U}_{\mathbf{k}}^{ij} P_{\mathbf{k}}^{ij*}) + \left.\frac{d}{dt} f_{\mathbf{k}}^i\right|_{corr}. \qquad (1.21)$$

Again, the term $d/dt f_{\mathbf{k}}^i|_{corr}$ comprises higher order correlations due to Coulomb interaction and carrier–phonon scattering treated in second Born-Markov approximation.

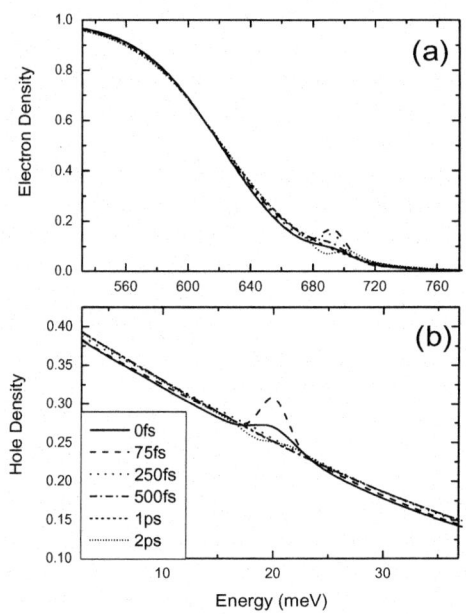

Fig. 1.9. (a) Electron and (b) hole distributions at $T = 300$ K for a 4.5-nm $In_{0.75}Ga_{0.25}As/InP$ quantum well after excitation by a 50-fs pulse at time $t = 0$.

Here, we consider a time-dependent optical excitation $E(t)$. Since optical transitions are vertical in momentum space, the excitation induces a deviation in the electron and hole distributions at equal momentum values, see Fig. 1.1. The specific momentum value depends on the transition energy.

Figure 1.9 shows the (a) electron and (b) hole distributions at $T = 300$ K for a 4.5-nm $In_{0.75}Ga_{0.25}As/InP$ quantum well excited by a 50-fs pulse at time

$t = 0$. The excitation energy was 0.7 meV. Before the optical excitation, the quantum well system was populated by a Fermi distribution of electrons/holes with a density of $2 * 10^{12}/\text{cm}^2$ at room temperature ($T = 300$ K). The lines show the distributions at different times after excitation. The optical excitation thus induces a peak in both electron and hole distribution. We observe that the maximum deviation occurs not at the time of the pulse maximum ($t = 0$) but several tens of femtoseconds delayed. The carrier density is driven by the product of the renormalized electromagnetic field and polarization, see (1.21). Only after this source term has completely switched off, pump-induced population change decreases by carrier–phonon and carrier–carrier scattering. As carrier–carrier scattering times in a laser are about an order of magnitude faster than carrier–phonon scattering times at room temperature, they dominate the relaxation process. From Fig. 1.9, we see that hole scattering is faster than electron scattering and the hole distribution returns to equilibrium more quickly, relaxation of the peaked distribution taking about 70 fs for the holes and 350 fs for the electrons. This faster relaxation is a consequence of the higher hole mass.

The carrier distributions return to a Fermi distribution on the time scale of hundreds of femtoseconds. However, because carrier–carrier scattering entails energy conservation of the carrier distribution, the energy brought into the system by the optical excitation cannot dissipate on the femtosecond timescale but on the longer timescale of carrier–phonon scattering (on the order of picoseconds at room temperature). We thus observe heating effects in our system. Deviations from thermal equilibrium and heating, as explained above, have to be taken into account, e.g., to realistically model lasers under fast modulation conditions or under strong optical pumping. Scattering rates for a specific experimental situation may be derived from a fit to experiment, their prediction for a wide range of densities, temperatures, and excitation energies, however, is only possible with a microscopic theory, as introduced in this chapter.

References

1. W. W. Chow, M. Kira, and S. W. Koch: Phys. Rev. B **60**, 1947 (1999)
2. W. W. Chow and S. W. Koch: *Semiconductor-Laser Fundamentals; Physics of the Gain Materials* (Springer, Berlin 1999)
3. J. Hader, S. W. Koch, and J. V. Moloney: Sol. Stat. Electron. **47**, 513 (2003)
4. J. Hader, N. Linder, and G. H. Döhler: Phys. Rev. B **55**, 6960 (1997)
5. S-H. Park and S-L. Chuang: J. Appl. Phys. **87**, 353 (2000)
6. W. Shan, W. Walukiewicz, J. W. Ager, et al.: Phys. Rev. Lett. **82**, 1221 (1999)
7. E. P. O'Reilly and A. Lindsay: phys. stat. sol. (b) **216**, 131 (1999)
8. J. Hader, S. W. Koch, J. V. Moloney, and E. P. O'Reilly: Appl. Phys. Lett. **76**, 3685 (2000)
9. S-L. Chuang: Phys. Rev. B **43**, 9649 (1991)
10. R. Winkler and U. Rössler: Phys. Rev. B **48**, 8918 (1993)

11. D. Ahn and S-L. Chuang: J. Appl. Phys. **64**, 6143 (1988)
12. O. Madelung: *Landolt-Börnstein, New Series III, Vol. 17a: Semiconductors* (Springer, Berlin 1982)
13. O. Madelung: *Landolt-Börnstein, New Series III, Vol. 22a: Semiconductors* (Springer, Berlin 1987)
14. E. H. Li: Physica E **5**, 215 (2000)
15. I. Vurgaftman, J. R. Meyer, and L. R. Ram-Mohan: J. Appl. Phys. **89**, 5815 (2001)
16. J. Hader, A. R. Zakharian, J. V. Moloney, et al.: IEEE Photon. Technol. Lett. **14**, 762 (2002)
17. M. R. Hofmann, N. Gerhardt, A. M. Wagner, et al.: IEEE J. Quantum Electron. **38**, 213 (2002)
18. R. Winkler: Phys. Rev. **B 51**, 14395 (1995)
19. B. Zhu and K. Huang: Phys. Rev. **B 34**, 3917 (1987)
20. R. Paiella, G. Hunziker, U. Koren, and K. J. Vahala: IEEE J. Sel. Topics Quantum Electron. **3**, 529 (1997)
21. J. Hader, J. V. Moloney, and S. W. Koch: IEEE J. Quantum Electron. **40**, 330 (2004)
22. Nonlinear Control Strategies Inc., Tucson, AZ (http://www.nlcstr.com)
23. Rsoft Design Group, Ossining, NY (http://www.rsoftdesign.com)

2 Fabry–Perot Lasers: Temperature and Many-Body Effects

B. Grote[1], E. K. Heller[1], R. Scarmozzino[1], J. Hader[2], J. V. Moloney[2], and S. W. Koch[3]

[1] RSoft Design Group, 200 Executive Blvd, Ossining, New York 10562, bernhard_grote@rsoftdesign.com
[2] Arizona Center for Mathematical Sciences, University of Arizona, Tucson, Arizona 85721, jhader@dinha.acms.arizona.edu
[3] Physics Department and Material Sciences Center, University of Marburg, Renthof 5, 35037 Marburg, Germany

2.1 Introduction

In this chapter, we demonstrate the integration of microscopic gain modeling into the laser design tool LaserMOD, which is derived from the Minilase II simulator developed at the University of Illinois [1]. Multidimensional carrier transport, interaction with the optical field via stimulated and spontaneous emission, as well as the optical field are computed self-consistently in our full-scale laser simulations. Giving additional details with respect to our previous work [2], we demonstrate the effectiveness of this approach by investigating the temperature sensitivity of a broad-ridge Fabry–Perot laser structure with InGaAsP multi-quantum wells for 1.55 μm emission wavelength.

Monochromatic light sources are key components in optical telecommunication systems. Predominantly, this need has been filled by semiconductor lasers due to their narrow linewidth. However, increasingly stringent requirements for bandwidth, tunability, power dissipation, temperature stability, and noise are being placed on these devices to meet network demands for higher capacity and lower bit error rates. As in the semiconductor industry, where electronic design automation assists in designing multimillion-gate integrated circuits, it is becoming common practice to employ simulation tools for designing and optimizing telecommunication networks and components. As a consequence of predictive modeling, the time to market as well as development cost of telecommunication infrastructure can be reduced, as fewer cycles between design and experimental verification are necessary.

Different levels of model abstraction are used to describe the behavior of devices, depending on whether they are being simulated alone or with other components in an optical system. At the lowest level, simulations treat the fundamental device physics rigorously, whereas behavioral modeling, which allows for an increased number of elements to be treated, is applied at the system or network level. This chapter will focus on the former approach.

To predict the performance of a new design, a successful commercial laser simulator must account for the many complex physical processes that contribute to the device operation. Mainly, the optical field and its interactions with the carrier populations must be described self-consistently, as they are strongly coupled via radiative recombination.

Classic approaches for solving the carrier and energy transport equations, such as drift-diffusion, heat flow, and energy balance equations, have been well established by the silicon device simulation industry [3–10]. Historically, the drift-diffusion model constitutes the first approach developed for semiconductor device simulation [3]. Within a momentum expansion of Boltzmann's transport equations, the charge conservation represents only the lowest order contribution. The hydrodynamic and energy balance models include differential equations describing the conservation of momentum and energy [5, 6]. From this system of equations, the drift-diffusion equations can be obtained by assuming constant and equal temperatures for electrons, holes, and the crystal lattice. The thermodynamic model applies principles of irreversible thermodynamics and linear transport theory to derive a system of equations describing carrier concentrations as well as carrier and lattice temperatures [9, 11]. It can be shown that the thermodynamic and the hydrodynamic approaches result in equivalent equations for the thermal transport [12, 13].

However, aside from the simulation of electronic transport within a classic framework, lasers also require quantum mechanical methods to treat light emission and amplification. These theoretical modeling techniques are far less mature. Specifically, bound quantum well states that give rise to lasing transitions must be modeled quantum mechanically; yet they must also be coupled to the classically modeled propagating states that describe the electronic transport. Progress has been made in this area by employing rate equations to describe carrier capture and scattering between classic propagating states and bound quantum well states [1, 14].

Optical gain/absorption and spontaneous emission or photoluminescence couple transport and optics in optoelectronic devices, as they correlate the complex refractive index with radiative recombination and generation of carriers. In lasers, strong coupling occurs due to stimulated emission. Methodologies for treating the optical aspect of the coupled problem are based on different proven approaches of solving Maxwell's equations. Gain/absorption and photoluminescence are determined by the energetic position and oscillator strength of optical transitions. The optical transitions in a semiconductor quantum well depend on the detailed bandstructure, which has to be calculated for the given geometry and material composition of quantum well and barrier. $\mathbf{k} \cdot \mathbf{p}$-perturbation theory has been widely adopted to compute the electron and hole energy dispersion, the optical transition matrix elements, the subband levels, and the confinement wavefunctions based on the knowledge of the bulk bandstructure [15–22].

In free-carrier or oscillator model approaches, interactions between the carriers are neglected and analytic expressions for the gain can be obtained based on Fermi's Golden Rule or by deriving equations of motion from the corresponding system Hamiltonian of a noninteracting electron gas or two-level system, respectively. Carrier collision effects are introduced phenomenologically in terms of a lineshape function that has a width determined by an effective decay rate [23, 24]. These kinds of models are often employed in laser simulation tools, as their simplicity allows us to avoid the excessive computational effort imposed by the full treatment of carrier interactions, which requires making the Hamiltonian diagonal with respect to the quasi-particle interactions.

However, it has been shown that accurate modeling of gain and absorption spectra requires the accounting of many-body interactions, leading to effects such as Coulomb enhancement, excitonic correlation, and band gap renormalization. In Chapt. 1, "Gain and Absorption: Many-Body Effects" by S. W. Koch et al., a discussion of many-body effects and the underlying theory can be found. In Sect. 3.1 of Chapt.1 and Fig. 2 within that section, the free-carrier gain model is compared with the full many-body calculation and experimental gain spectra. It can be seen that, for a series of measurements, such as the density-dependent set of gain spectra shown in Fig. 2 of Chapt. 1, satisfying agreement between theory and experiment can only be achieved by taking into account many-body effects. For low carrier densities and temperatures, the absorption of semiconductors around the bandedge is dominated by a pronounced exciton resonance. These electron–hole correlations persist even for higher densities and temperatures and influence the spectra beyond the effects due to band gap renormalizations alone. Furthermore, dephasing by carrier–carrier and carrier–phonon scattering can lead to a broadening energy of the spectra. It is clear that reliable prediction of the density and temperature dependence of such spectral characteristics as the energetic position, broadening, and oscillator strength of optical transitions requires that all significant interactions are taken into account (see, e.g., [25–27]). In the many-body gain model employed here, real and imaginary parts of eigenenergy renormalizations describe band gap and excitonic shifts as well as collision broadening. These effects are not treated phenomenologically, but within the framework of a quantum kinetic theory that rigorously treats the Coulomb interaction in system Hamiltonian for electrons and holes [25, 26].

In the phenomenological treatment, a gain model would require the user to specify energetic shift and broadening parameters. This prohibits the prediction of absolute magnitude and shape of the gain over a range of temperatures and densities for a specific material system. Although several experimental behaviors of laser devices can be described by the phenomenological effective decay rate treatment, it cannot reproduce certain experimental features in the gain spectra that are important for advanced laser structures. It has been shown that the accurate description of gain and absorption spectra, in the neighborhood of the transparency carrier density and over a variety of carrier densities and temperatures, requires the full treatment of carrier interactions. Section 3.3 of Chapt. 1 discusses the dephasing-time approach and its limitations. Approximate results are compared to gain spectra obtained employing the full many-body calculation including scattering terms and to experimental findings. Without the predictive knowledge, calibration of gain parameters might imply additional iterations between simulation and experimental verification during the laser design process, increasing effort in cost and time.

A completely microscopic treatment, alternatively, does not require any experimentally measured fitting parameters such as lineshape broadenings or spectral shifts. It can quantitatively predict the absorption, gain, and photoluminescence spectra of ideal structures using only basic bandstructure parameters, which are independent of carrier density, temperature, and the design of the quantum well. Real samples are usually affected by a certain amount of disorder. In semiconductor quantum wells, crystal inhomogeneities, such as local well width fluctuations or local fluctuations of the material composition, lead to an inhomogeneous broadening of spectra in addition to the homogeneous broadening due to electron–electron and electron–phonon scattering. Moreover, magnitude of fluctuations, as well as average material composition and sample geometry, can vary across the wafer. The results of a simple low-excitation photoluminescence measurement performed directly on the wafer for the nonideal sample may be compared with predictions of the microscopic theory for the ideal structure. This allows for the determination of the inhomogeneous broadening, as well as possible deviations between nominal and actual structural parameters, which are all independent of carrier density and temperature. This method has recently received attention for application as an on-wafer testing tool [28, 29]. Once the amount of disorder has been characterized, the theory is completely parameter free and can predict the optical properties for the device under high excitation operating conditions.

In a rigorous laser simulation, the overall complexity of treating the coupled transport and optoelectronic problem leads to significant computational effort. In practice, a number of approximations have to be made in order to produce a more tractable simulation. Complexity is reduced with respect to the full microscopic description, at the expense of some predictability, by relying on careful calibration of model parameters introduced through a phenomenological treatment. A good phenomenological model requires only a few parameters that need calibration, while preserving the basic functional dependencies given by the microscopic theory. This concept is very common in semiconductor device and laser simulation where many processes involved in the carrier transport have to be described. However, it is important to identify the most critical processes, where enhancements in the level of accuracy can improve the quality of the result of the overall simulation with acceptable increase of computational effort.

The microscopic many-body theory of the gain employed by our simulator allows us to re-examine the underlying approximation of noninteracting particles leading to the simpler free-carrier gain model with phenomenological broadening parameter as well as the trade-off between computational effort and predictability. The rigorous microscopic many-body theory of the semiconductor, which is based on the semiconductor Bloch equations, allows for the accurate modeling of the spectral characteristics of the material gain. With such a model, the energetic position of the gain peak, the broadening due to quasi-particle collisions, and therefore, the absolute magnitude of the gain can be predicted based solely on fundamental bulk material parameters. The properties of the gain are found to be the most critical contributions to the overall slope efficiency, threshold current, and emission wavelength of the laser. By comparing computations with experimental results, we will show, that the use of the advanced gain model can improve the overall predictability of the simulator.

We note that significant computational effort is associated with the quantum many-body gain calculation. In order to avoid the gain calculation during full-scale laser simulations, which can be numerous in design optimization cycles, the gain and related quantities, such as refractive index change and photoluminescence, are precomputed and stored as a data base. For different common material systems associated with specific telecommunication laser emission wavelength, the creation of libraries, which are parameterized by material composition and quantum well geometry, might be attractive. During a full-scale laser simulation the gain, refractive index, and photoluminescence data can be retrieved from the precomputed spectra for the current operating condition, to perform self-consistent computations of optical field, carrier transport, and their interactions to obtain steady-state, transient, or frequency responses for a particular laser geometry. Taking advantage of this methodology, computation times for the full-scale laser simulation using the many-body gain model are comparable with or even faster than those using a run-time gain calculation based on the free-carrier approach.

In Sect. 2.2 we describe the theoretical background of the transport, optical, and optoelectronic modeling applied in our simulator. The temperature sensitivity of an InGaAsP multi-quantum well laser is analyzed in Sect. 2.3 using the many-body gain theory in comparison to the free-carrier gain model. Our findings are summarized in Sect. 2.4.

2.2 Theory

2.2.1 Transport

A methodology for the carrier transport has been developed and established for silicon device simulation in multiple dimensions, which we adapt for material systems common to semiconductor lasers, to describe electronic transport through bulk regions, in which active layers may be embedded [10, 11]. The injection current into the active quantum well region determines the

Table 2.1. Nomenclature.

Symbol	Definition
B	Einstein coefficient
$C_{e/h}^{Auger}$	electron/hole Auger recombination coefficient
c	vacuum speed of light
c_L	crystal lattice heat capacity
$E_{e/h}^{Auger,act}$	Auger recombination activation energy
$F_{e/h}$	electron/hole Fermi level
$f_{e/h}$	electron/hole distribution functions
$f_{e/h}^{2D/3D}$	quantum well electron/hole distribution function for propagating/bound states
$G(\omega)$	gain spectrum
$G_{thermal}$	lumped thermal conductivity
$g_{e/h}^{3D/2D}$	density of states for quantum well propagating/bound states
H	total heat generation rate
H_{Joule}	Joule heat source
H_{rec}	recombination heat source
H_{trans}	heat source due to transient modulation of carrier concentrations
$H_{Peltier+Thomson}$	sum of Peltier and Thomson heat sources
$\mathbf{J}_{e/h}$	electron/hole current
k_B	Boltzmann constant
k_0	wavenumber for Helmholtz equation
$k_{e/h}^{fca}$	free-carrier absorption coefficients for electrons/holes
$\mathcal{L}_{i,j}$	Lorentzian broadening around transition frequency $\omega_{i,j}$
q	elementary charge
$N_{D/A}^{\pm}$	ionized donors/acceptors
$n_{e/h}$, $n_{e/h}^{3D}$, $n_{e/h}^{2D}$, $n_{e/h}^{2D,i}$	electron/hole concentrations, total propagating ($3D$) and bound ($2D$) quantum well densities, individual subband (i) contributions for bound states
n_i	intrinsic carrier density
$n_{e/h}^t$	electron/hole trapped carrier density
n_{ph}	phonon density
$n_{eff,\nu}$	effective index as given by eigenvalue ν of the Helmholtz equation
$P_{e/h}$	electron/hole thermoelectric power
R_{dark}, R_{Auger}, R_{SRH}	total nonradiative, Auger and Shockley–Read–Hall recombination
R_{stim}, $R_{spon,bound/bulk}$	stimulated and spontaneous recombination for bound/bulk states
$R_{capture}^{e/h}$, $R_{capture,cc/ph}^{e/h}$	electron/hole quantum well net capture rate, contribution due to carrier–carrier/carrier–phonon scattering
$s_{e/h}^{capture/escape,cc/ph}$	electron/hole capture/escape coefficient for carrier–carrier/carrier–phonon scattering
$S_{\nu,\omega}$	photon occupation of mode (ν,ω)
$\mathbf{S}_{e/h}$	electron/hole energy flux
T, T_C	temperature, contact temperature
t	time
$U(\omega)$	spontaneous emission spectrum
$Z(\omega)$	photon density of states
α_0	background absorptive loss (carrier independent)
ϵ	static dielectric permittivity
Φ	electric potential
γ_0, γ_{inh}	homogeneous, inhomogeneous broadening
$\kappa_{L/e/h}$	crystal lattice/electron gas/hole gas heat conductivity
$\tau_{e/h}^{SRH}$	Shockley–Read–Hall electron/hole lifetime
$\tau_{\nu,\omega}^{photon}$, τ_{mirror}, $\tau_{scatter}$	total photon losses, mirror losses, scatter losses
ω_{LO}	longitudinal optical phonon frequency
$\psi_i^{e/h}$	quantum well confinement wavefunction of subband i
ζ_ν	νth Helmholtz eigenmode

carrier densities within bound quantum well states and, therefore, the degree of inversion. For carrier transport through bulk semiconductor regions, the drift-diffusion system of equations is applied. The electric potential ϕ is determined by Poisson's equation,

$$\nabla \cdot \epsilon \nabla \phi = q(n_e - n_h - N_D^+ + N_A^-), \qquad (2.1)$$

where the charges are given by the densities of electrons n_e, holes n_h, ionized donors N_D^+, and acceptors N_A^- (q is the elementary charge and ϵ is the static dielectric permittivity of the respective material). A list of symbols used in this chapter and their definition can by found in Table 2.1. For quantum wells, we distinguish between carriers in propagating ($n_{e/h}^{3D}$) and bound states ($n_{e/h}^{2D}$). Although all states are considered propagating in bulk regions ($n_{e/h}^{2D} = 0$), the confinement potential leads to bound states localized within the quantum well. Within an envelope function approximation, the spatial distribution of carriers in the confined direction of the quantum well is described by wavefunctions, which are obtained by solving the Schroedinger equation for the potential given by the solution of Poisson's equation and the band gap offsets. With wavefunctions $\psi_i^{e/h}$ for subband i, the spatial distribution of bound electrons and holes (indicated by indices e/h) in the direction perpendicular to the quantum well plane is given by:

$$n_{e/h}^{2D}(\mathbf{r}) = \sum_i |\psi_i^{e/h}(y)|^2 n_{e/h}^{2D,i}(\mathbf{r}_\parallel). \qquad (2.2)$$

Here, y denotes the growth direction and coordinates perpendicular to the y-axis and parallel to the quantum well plane are indicated by $\parallel$. Light propagation within the Fabry–Perot cavity is assumed to be along the z-axis. The carrier density $n_{e/h}^{2D,i}$ in subband i is related to the distribution function $f_{e/h}^{2D}$ via the density of states $g_{e/h}^{2D,i}$,

$$n_{e/h}^{2D,i} = \int_{E_i}^{\infty} g_{e/h}^{2D,i}(E) f_{e/h}^{2D,i}(E), \qquad (2.3)$$

which results from our eight band $\mathbf{k} \cdot \mathbf{p}$ calculation. The total charge due to electrons is given by the sum over bound and propagating contributions, $n_{e/h} = n_{e/h}^{3D} + n_{e/h}^{2D}$.

Figure 2.1 illustrates our transport model applied to quantum well regions. Capture due to carrier–carrier and carrier–phonon scattering couples the classic propagating bulk and quantum well continuum states to the quantum confined bound states. The corresponding rates enter the continuity equations for carriers in bound and continuum states [1, 30]. Carriers entering the quantum well in continuum states can transit through or get captured into bound states as a result of scattering events involving other carriers or phonons. For sufficiently fast inter-subband scattering leading to thermalization between

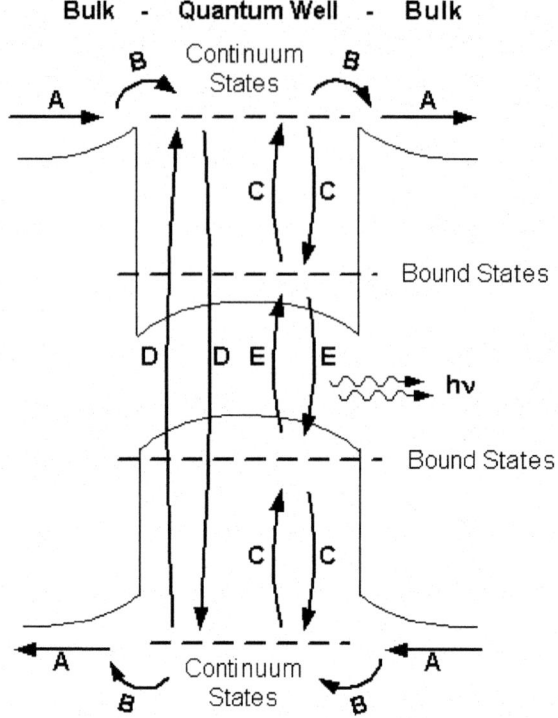

Fig. 2.1. Schematic illustration of the transport model applied to quantum well regions including drift-diffusion currents in bulk regions (A), transport across interfaces (B), carrier capture from continuum into bound states (C), radiative and nonradiative recombination from bound states (E), and nonradiative recombination from continuum states or for bulk carriers (D) (from [2]).

the subbands, the occupation of bound states can be described by a single Fermi level. This reduces the set of rate equations in the quantum wells to a four-level system defined by effective rates and density of states, which are the sum over individual subband contributions $f_{e/h}^{2D,i} = f_{e/h}^{2D}$, $g_{e/h}^{2D} = \sum_i g_{e/h}^{2D,i}$. The total bound carrier concentration is described by a continuity equation:

$$\frac{\partial n_{e/h}^{2D}}{\partial t} = \pm \frac{1}{q} \nabla_\| \cdot \mathbf{J}_{e/h\,\|} - R_{dark} - R_{stim} - R_{spon,bound} + R_{capture}^{e/h}. \quad (2.4)$$

Via the capture rate $R_{capture}^{e/h}$, the bound carriers couple to the continuum concentrations, where it enters the continuity equation as a loss:

$$\frac{\partial n_{e/h}^{3D}}{\partial t} = \pm \frac{1}{q} \nabla \cdot \mathbf{J}_{e/h} - R_{dark} - R_{spon,bulk} - R_{capture}^{e/h}. \quad (2.5)$$

Capture into and escape from bound states is modeled by Master equation type rates for in- and out-scattering due to carrier–carrier and carrier–phonon interaction. The net capture rate $R_{capture}^{e/h} = R_{capture,cc}^{e/h} + R_{capture,ph}^{e/h}$ is given by:

$$R_{capture,cc}^{e/h} = \int_{E_{e/h}^0}^{\infty} dE \int_{E_{e/h}^0}^{\infty} dE' g_{e/h}^{3D}(E) g_{e/h}^{2D}(E') \times$$
$$\times \left(s_{e/h}^{cc,capture}(E, E') f_{e/h}^{3D}(E) (1 - f_{e/h}^{2D}(E')) \right.$$
$$\left. - s_{e/h}^{cc,escape}(E, E') (1 - f_{e/h}^{3D}(E)) f_{e/h}^{2D}(E') \right), \quad (2.6)$$

describing carrier–carrier scattering, and:

$$R_{capture,ph}^{e/h} = \int_{E_{e/h}^0}^{\infty} dE \int_{E_{e/h}^0}^{\infty} dE' g_{e/h}^{3D}(E) g_{e/h}^{2D}(E') \times$$
$$\times \left(s_{e/h}^{ph,capture}(E, E') (n_{ph} + 1) f_{e/h}^{3D}(E) (1 - f_{e/h}^{2D}(E')) \right.$$
$$\left. - s_{e/h}^{ph,escape}(E, E') n_{ph} (1 - f_{e/h}^{3D}(E)) f_{e/h}^{2D}(E') \right), \quad (2.7)$$

modeling scattering of carriers with longitudinal optical phonons. In general, the scattering coefficients $s_{e/h}^{cc/ph,capture/escape}$ would have to be assumed to be dependent on the occupation of the involved states to emulate the dynamics of the quantum Boltzmann equation used for the microscopic description of the scattering process. Following [1], the scattering coefficients are given by constant rates normalized by the final density of states. In (2.7), absorption or emission of longitudinal optical phonons allows for interaction of energetically nonresonant states by transfer of a phonon energy. The phonon occupation is assumed to be [1]:

$$n_{ph} = \frac{1}{\exp(\frac{\hbar \omega_{LO}}{k_B T}) - 1}, \quad (2.8)$$

with longitudinal optical phonon energy $\hbar \omega_{LO}$, Boltzmann constant k_B, and temperature T. We assume energy conservation within the electron gas under elastic carrier–carrier scattering described by (2.6).

The net recombination rate $R_{dark} = R_{SRH} + R_{Auger}$ balances recombination and generation due to nonradiative processes, such as Auger recombination:

$$R_{Auger} = (C_h^{Auger} n_h + C_e^{Auger} n_e)(n_e n_h - n_i^2), \quad (2.9)$$

and Shockley–Read–Hall recombination:

$$R_{SRH} = \frac{n_e n_h - n_i^2}{\tau_h^{SRH}(n_e + n_e^t) + \tau_e^{SRH}(n_h + n_h^t)}. \quad (2.10)$$

Here, n_i is the intrinsic carrier density, $n_{e/h}^t$ are the respective trapped carrier densities, $\tau_{e/h}^{SRH}$ are the lifetimes of electrons and holes for trap-assisted recombination, and $C_{e/h}^{Auger}$ in (2.9) are the coefficients for the electron and hole Auger process. The temperature dependence of the Auger recombination can be modeled by [31, 32]:

$$C_{e/h}^{Auger}(T) = C_{e/h}^{Auger}(300 \text{ K}) \exp\left(-E_{e/h}^{Auger,act}\left(\frac{1}{k_B T} - \frac{1}{k_B 300 \text{ K}}\right)\right), \quad (2.11)$$

with activation energies $E_{e/h}^{Auger,act}$. Nonradiative processes, as well as spontaneous emission into modes other than the lasing modes, decrease the carrier densities without contributing to the laser output power.

Aside from nonradiative recombination, stimulated and spontaneous emission lead to a decrease of the carrier population within bound quantum well states. The spontaneous emission is given by:

$$R_{spon,bound} = \int d\omega Z(\omega) U(\omega), \quad (2.12)$$

where ω is the angular frequency, $Z(\omega)$ is the spectral density of photon states, and U is the photoluminescence spectrum. The recombination due to stimulated emission,

$$R_{stim} = \sum_{\nu,\omega} S_{\nu,\omega} |\zeta_\nu|^2 \frac{c}{n_{eff,\nu}} G(\omega), \quad (2.13)$$

is the sum over light emission into all Fabry–Perot modes denoted by ω and eigenmodes ζ_ν of the plane perpendicular to the light propagation. G is the gain spectrum as given by the interband transitions of the quantum well. Eigenmodes and effective refractive indices $n_{eff,\nu}$ are determined by the solution of a Helmholtz eigenvalue problem. c is the speed of light. The photon occupation number $S_{\nu,\omega}$ obeys a rate equation (see Sect. 2.2.2). In bulk regions and for continuum states, a simplified model is applied to account for recombination due to spontaneous emission:

$$R_{spon,bulk} = B(n_e n_h - n_i^2), \quad (2.14)$$

where B is the spontaneous recombination coefficient.

The current densities $\mathbf{J}_{e/h}$ are calculated within the framework of the drift-diffusion theory. For Fermi statistics, the diffusivity is related to the carrier mobility by a generalized Einstein relation. Transport across material interfaces constituting a bandedge discontinuity is described in terms of thermionic emission. Within the quantum well plane, indicated by $\parallel$, continuum and bound carriers can drift leading to injection losses due to lateral leakage currents, see (2.4). For bulk regions, (2.5) reduces to the well-known continuity equations for electrons and holes, as all states are considered propagating.

Due to carrier–phonon scattering, part of the electronic energy is transferred to the crystal lattice resulting in an increase of the lattice temperature.

Instead of deriving quantum mechanical equations of motion for the phonon population interacting with electrons and holes, classic thermal transport equations, which describe the average energy of the respective subsystems of electrons, holes, and host lattice, are typically employed in device simulation drastically reducing the problem complexity [5–11, 33]. In semiconductor laser diodes, self-heating under continuous-wave (CW) operation is known to affect the characteristics significantly and often limits the performance, as recombination and transport processes explicitly or implicitly depend on the temperature. In order to model these effects, a lattice heat flow equation has to be solved to determine the temperature profile within the device [9, 34]:

$$\left(c_L + \frac{3}{2}k_B(n_e + n_h)\right)\frac{\partial T}{\partial t} = \nabla \cdot \left(\kappa_L \nabla T - \mathbf{S}_e - \mathbf{S}_h\right) + H. \tag{2.15}$$

Here, c_L is the heat capacity of the lattice and κ_L is the lattice thermal conductivity. The energy fluxes $\mathbf{S}_{e/h}$ are given by:

$$\mathbf{S}_{e/h} = \mp P_{e/h} T \mathbf{J}_{e/h} - \kappa_{e/h} \nabla T, \tag{2.16}$$

with thermoelectric powers $P_{e/h}$ and thermal conductivity mediated by electrons and holes $\kappa_{e/h}$. Including contributions due to temperature gradients, the current densities are given by:

$$\mathbf{J}_{e/h} = n_{e/h}\mu_{e/h}\left(\nabla F_{e/h} \pm qP_{e/h}\nabla T\right), \tag{2.17}$$

where $\mu_{e/h}$ are the mobilities and $F_{e/h}$ are the Fermi levels of electrons and holes, respectively. For the derivation of (2.15), the electron and hole temperature were assumed to be equal to the lattice temperature. The heat capacity and conductivities of the electron and hole gas add to the thermal properties of the lattice, as can be seen from the heat capacity term on the left-hand side of (2.15) as well as the heat flux term on the right-hand side. Heat generation $H = H_{Joule} + H_{rec} + H_{trans}$ in (2.15) is due to Joule heat:

$$H_{Joule} = -\frac{1}{q}\left(\mathbf{J}_e \cdot \nabla F_e + \mathbf{J}_h \cdot \nabla F_h\right), \tag{2.18}$$

recombination heat expressed by:

$$H_{rec} = (F_e - F_h)R_{dark}, \tag{2.19}$$

and an additional heat production rate originating from the transient modulation of the carrier concentrations:

$$H_{trans} = -T\frac{\partial F_e}{\partial T}\frac{\partial n_e}{\partial t} + T\frac{\partial F_h}{\partial T}\frac{\partial n_h}{\partial t}. \tag{2.20}$$

The sum of Peltier and Thomson heat:

$$H_{Peltier+Thomson} = -\mathbf{J}_e \cdot T\nabla P_e - \mathbf{J}_h \cdot T\nabla P_h, \tag{2.21}$$

is included in the convective part of the energy fluxes (see 2.16 and 2.15). The recombination heat source accounts for the energy of the order of the band gap dissipated to the lattice by nonradiative recombination. Note that radiative processes transfer this energy to the light field and, therefore, should not be included here.

Two different thermal contact models are common in physical device simulation. By assuming an isothermal contact with given temperature T_C, a Dirichlet boundary condition is imposed:

$$T = T_C. \tag{2.22}$$

The second model is of Cauchy type. It associates a finite thermal conductance $G_{thermal}$ with the contact to determine the heat flux through the contact area [35]:

$$-\kappa_L \mathbf{n} \cdot \nabla T = \frac{G_{thermal}}{A}(T - T_C), \tag{2.23}$$

where $\mathbf{n}$ is the surface normal vector at the contact and A is the contact surface area.

2.2.2 Optics

The light propagation within a waveguide structure is determined by the solution of Maxwell's equations. A set of approximations has been developed that reduces the computational effort with respect to the full solution for specific cavity structures. For Fabry–Perot lasers, the lateral modes can be described by a Helmholtz equation, whereas the resonator modes determine the spectrum of the axial direction. Photon rate equations describe the light intensity within the individual eigenmodes. Modal gain and spontaneous emission compensate losses due to different scattering and absorption mechanisms.

The waveguide properties of the semiconductor lasers are expressed by a Helmholtz eigenvalue equation describing stationary solutions of Maxwell's equations:

$$\left(\nabla_{x,y}^2 + k_0^2(n_b^2 - n_{eff,\nu}^2)\right)\zeta_\nu(x,y) = 0. \tag{2.24}$$

Here, n_b is the background refractive index, and k_0 is equal to ω/c with the angular frequency ω. The eigenvalues $n_{eff,\nu}$ determine the propagation along the optical z axis of the Fabry–Perot cavity:

$$\zeta_\nu(\mathbf{r}) = \zeta_\nu(x,y) \exp(i n_{eff,\nu} k_0 z). \tag{2.25}$$

The mode profile is updated by solving (2.24) during the simulation to reflect index changes in the active layer. The light intensity within the different modes is determined by a set of photon rate equations:

$$\frac{\partial}{\partial t}\mathcal{S}_{\nu,\omega} = (G_{\nu,\omega} - \frac{1}{\tau_{\nu,\omega}^{photon}})\mathcal{S}_{\nu,\omega} + U_{\nu,\omega}. \tag{2.26}$$

Via the modal gain,
$$G_{\nu,\omega} = \int dV |\zeta_\nu|^2 \frac{c}{n_{eff,\nu}} G, \qquad (2.27)$$
and the spontaneous emission,
$$U_{\nu,\omega} = \int dV |\zeta_\nu|^2 U, \qquad (2.28)$$
electronic transport and optical properties are coupled. The losses,
$$\frac{1}{\tau^{photon}_{\nu,\omega}} = \frac{1}{\tau_{mirror}} + \frac{1}{\tau_{scatter}} + \frac{c}{n_{eff,\nu}} \alpha_b, \qquad (2.29)$$
entering (2.26) are the sum of losses due to light leaving the cavity through the facets (τ_{mirror}), light scattered out of the waveguide ($\tau_{scatter}$), and absorptive losses (α_b). For narrow ridge waveguides, scatter losses can become considerable due to surface imperfections. Substantial losses due to intervalence band absorption in InGaAsP materials have been reported [32]. The absorptive losses α_b include free-carrier absorption mechanisms such as intervalence band absorption, which are modeled as being proportional to the carrier densities:
$$\alpha_b = \int dV |\zeta_\nu|^2 \left(k_e^{fca} n_e + k_h^{fca} n_h + \alpha_0 \right), \qquad (2.30)$$
with free-carrier absorption coefficients $k_{e/h}^{fca}$ (α_0 accounts for carrier-independent background absorptive losses). Usually, barrier and other layer materials in semiconductor quantum well lasers are designed to be transparent at the laser wavelength given by the effective quantum well band gap. However, although band-to-band absorption may vanish, absorption spectra usually exhibit an Urbach tail due to phonon-assisted absorption and disorder-induced localized states below the band gap, contributing to background losses.

2.2.3 Gain

By coupling the electronic and optical subsystems, the spontaneous and, especially, the stimulated emission play a crucial role in the laser simulation. The gain computation for semiconductor quantum well lasers involves determining the bandstructure via methods such as $\mathbf{k} \cdot \mathbf{p}$ calculation and computing the subband spectrum to obtain optical matrix elements and density of states. Within the framework of a free-carrier approach, a noninteracting electron gas is assumed, allowing for the derivation of analytical expressions for gain and spontaneous emission. Broadening mechanisms are introduced via a phenomenological linewidth broadening function. In contrast, the many-body theory microscopically accounts for the effects of carrier–carrier interaction, including excitonic correlations as well as carrier–phonon scattering. The predictive capabilities of this method in determining the material gain spectra for a variety of materials has been proven in several publications [25, 27, 29].

2.2.3.1 Free-Carrier Gain Model

Neglecting interactions of carriers with other quasi-particles of the semiconductor, the gain G and spontaneous emission spectra U can be expressed as:

$$G = \int_0^\infty dE \sum_{i,j} \langle \psi_i^e | \psi_j^h \rangle (Bg_{red})_{i,j} (f_e^{2D} + f_h^{2D} - 1) \mathcal{L}_{i,j} \quad (2.31)$$

$$U = \int_0^\infty dE \sum_{i,j} \langle \psi_i^e | \psi_j^h \rangle (Bg_{red})_{i,j} f_e^{2D} f_{2D}^h \mathcal{L}_{i,j}. \quad (2.32)$$

Here, $\langle \psi_i^e | \psi_j^h \rangle$ is the overlap integral for the optical transition involving subbands i,j. The optical matrix element $(Bg_{red})_{i,j}$ is the product of Einstein coefficient and the reduced density of states, which results from our eight-band $\mathbf{k} \cdot \mathbf{p}$ bandstructure calculation. The range of integration includes all relevant transitions. The collision broadening term is modeled by a Lorentzian:

$$\mathcal{L}_{i,j}(E) = \frac{\gamma_0/2\pi}{(\hbar\omega_{i,j} - E)^2 + (\gamma_0/2)^2} \quad (2.33)$$

around the transition frequency $\omega_{i,j}$ and with the full-width half-maximum of γ_0.

2.2.3.2 Many-Body Gain Theory

The microscopic calculation of gain/absorption, refractive index, and photoluminescence spectra is described in detail in Chapt. 1, [25, 36], and references therein. It is based on solving the semiconductor Bloch equations, i.e., the equations of motion for the reduced density matrix, which are derived from the system Hamiltonian, including Coulomb interaction between carriers and carrier–phonon interaction. Consequently, Coulomb-induced effects like band gap renormalization, Coulomb enhancement of the absorption, and excitonic resonances are taken into account self-consistently. The electron–electron and electron–phonon scattering processes that lead to the dephasing of the polarizations and, therefore, spectral broadening and spectral shifts are calculated in second Born approximation. The resulting scattering equations take the form of generalized quantum-Boltzmann equations.

The diagonal elements of the reduced density matrix are the distribution functions, whereas the off-diagonal elements are the microscopic polarizations. The resulting polarizations are added up to the total macroscopic optical polarization, from which the optical susceptibility can be obtained through a simple Fourier transform. The real part of the susceptibility gives the carrier-induced change of the refractive index, and the imaginary part gives the gain/absorption. The photoluminescence spectra are derived from the absorption/gain spectra using the Kubo–Martin–Schwinger relation [37].

Spectra of gain, spontaneous emission, and carrier-induced change of the refractive index are precalculated and stored for a sufficiently dense grid of carrier density and temperature points. The full-scale laser simulation accesses this data base and interpolates for the required bound carrier density, temperature, and wavelength. Derivatives with respect to internal variables used in the Newton–Raphson method, which is applied to the solution of the system of coupled differential equations, are evaluated numerically.

2.3 Temperature Sensitivity of InGaAsP Semiconductor Multi-Quantum Well Lasers

Semiconductor lasers based on the InGaAsP-InP material system have gained great attention because the band gap allows tuning into the 1.3 μm – 1.6 μm range of emission wavelength suitable for optical fibers. Their performance is known to be strongly temperature dependent [38]. However, from an application point of view, it is desirable to achieve high-temperature stability of laser diodes. The necessity of cost-intensive cooling can be avoided, while gaining tolerance with respect to thermal interaction and, therefore, increasing flexibility in terms of packaging and integration. In this context, the performance of InGaAsP lasers at elevated temperatures is of interest, especially with regard to threshold current and slope efficiency.

Different physical processes have been discussed with respect to their role in determining the temperature sensitivity of these lasers: Auger recombination [39, 40], intervalence band absorption [41], carrier leakage out of the active region [42], lateral current spreading [43], barrier absorption and spontaneous recombination [44], as well as gain reduction [45]. The self-consistent modeling of all of these mechanisms is required for a theoretical investigation of the temperature sensitivity. In the past, most theoretical studies have been performed based on simple gain models, such as the free-carrier model or the oscillator model, which describe collision effects phenomenologically in terms of a gain broadening parameter. We will show that temperature and carrier-density-induced collision broadening of the gain spectrum is a dominant factor influencing the temperature sensitivity [2]. Carrier–carrier and carrier–phonon scattering-induced gain broadening can only be accounted for correctly by employing a theory that takes these many-body interactions into account self-consistently. We will point out shortcomings in phenomenological models for the gain broadening and identify the need for enhancements. Oversimplified models for the gain might lead to misinterpretations concerning the relative importance of physical processes being involved, as adjustment of carrier transport parameters can have a similar impact on the laser characteristics as modifications of gain parameters.

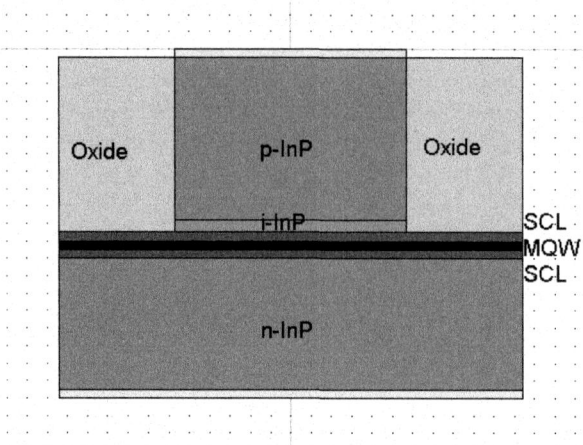

Fig. 2.2. Cross section of the ridge waveguide Fabry–Perot laser structure studied in our simulations.

2.3.1 Laser Structure

We study a multi-quantum well (MQW) ridge waveguide laser diode with a Fabry–Perot cavity, as described in detail in [11, 32]. Figure 2.2 depicts a cross section of the device. In the active region, six compressively strained $In_xGa_{1-x}As_yP_{1-y}$ quantum wells with nominal composition $x = 0.76$ and $y = 0.79$ and a thickness of 6.4 nm are embedded in 5-nm-thick barriers made of $In_{0.71}Ga_{0.29}As_{0.55}P_{0.45}$. The multi-quantum well region is sandwiched between two undoped $In_{0.171}Ga_{0.829}As_{0.374}P_{0.626}$ separate confinement layers (SCLs). The cladding layers are made of InP. On the p-side of the structure the first 0.14 μm are undoped to avoid diffusion of acceptors into the active layers. The acceptor concentration is $4 \times 10^{17} cm^{-3}$ within the p-InP cladding layer. In the n-type cladding layer, the donor concentration is $8 \times 10^{17} cm^{-3}$. The p-type cladding layer is etched down to the separate confinement to leave a 57 μm-wide ridge forming the lateral mode confinement. The cleaved facets have a reflectivity of $R = 0.28$ and the cavity length is 269 μm. Figure 2.3 shows conduction and valence band within the regions surrounding the active layer for a vertical cut along the symmetry plane.

2.3.2 Sample Characterization

Due to the predictive capabilities of the many-body theory, comparison of measured and calculated photoluminescence spectra allows for the characterization of the sample. A methodology has been described that makes it possible to use this procedure for on-wafer testing [28, 29]. Static disorder in the quantum wells leads to an inhomogeneous broadening of optical resonances due to local fluctuations of the confinement potential. By applying

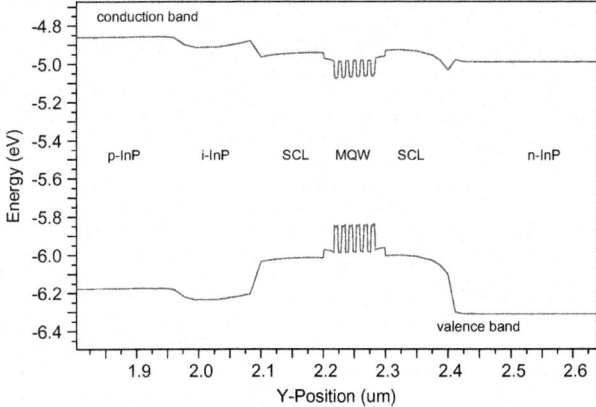

Fig. 2.3. Conduction and valence band above threshold for $T = 333$ K.

a Gaussian broadening convolution to the computed spectra, the sample quality can be assessed by fitting the experimental findings. Furthermore, comparison of the spectral position of the photoluminescence spectrum can indicate deviations of the sample geometry and material composition from the nominal specification. Temperature- and density-dependent experiments allow us to distinguish between the homogeneous collision broadening and the disorder-induced inhomogeneous broadening, because in contrast to the particle collisions, the influence of static disorder does not exhibit any dependence on temperature or carrier density.

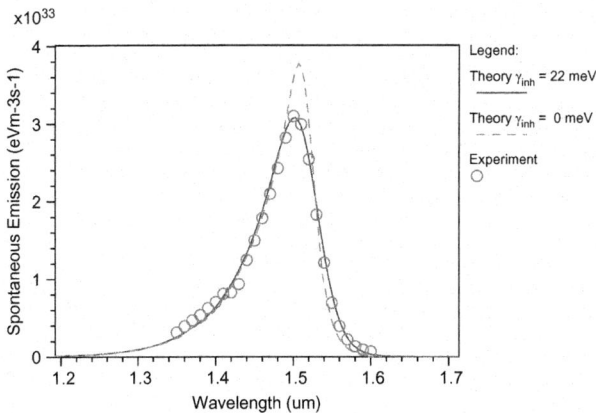

Fig. 2.4. Comparison of measured photoluminescence with calculations based on the many-body gain theory for specified inhomogeneous broadening γ_{inh} for density $N = 8.5 \times 10^{11} \text{cm}^{-2}$ and temperature $T = 300$ K.

In Fig. 2.4, experimental photoluminescence is reproduced by computations based on the many-body gain theory. In order to achieve agreement, a blue shift of 23 meV with respect to the measurement had to be assumed. This indicates a slight deviation of the material composition from the nominal values. A 3% reduction of the Indium or Arsenic concentration would lead to such a shift. Assuming the nominal concentrations are correct, a deviation of the well width by 1 nm from the nominal value would also explain the shift. Such a deviation seems very unlikely because the growth process can be controlled to a good degree in this regard. Therefore, we proceed assuming $x = 0.745$ and $y = 0.775$. The nominal material composition, as specified in [11, 32], is based on an analysis of the photoluminescence and strain using a simpler theory and bowing parameters for the bulk material band gap different from those used here, which explains the deviations in the resulting material composition [46].

For determining the inhomogeneous broadening, a series of density-dependent photoluminescence spectra would be desirable to distinguish carrier-collision-induced homogeneous broadening from disorder-induced inhomogeneous broadening. However, the carrier density dependence of the lineshape helps to determine the inhomogeneous broadening to be approximately 22 meV using only one experimental photoluminescence spectrum at low carrier density. Note that this sample characterization should not be regarded solely as a parameter calibration step, as in return, it allows for the identification of deviations from the nominal geometry and composition, and it helps to assess the quality of the growth process.

2.3.3 Gain Spectra

Figure 2.5 shows gain spectra for different temperatures and carrier densities obtained using a free-carrier model [left column, Fig. 2.5(a) and (b)] in comparison with the results of the microscopic many-body theory [right column, Fig. 2.5(c) and (d)] [48]. The upper row [Fig. 2.5(a) and (c)] depicts gain spectra for increasing carrier density, while the lower row [Fig. 2.5(b) and (d)] shows gain spectra for increasing temperature. For further illustration, Fig. 2.6 depicts the peak gain obtained using the free-carrier model in comparison with computations based on the many-body theory as a function of the carrier density for different temperatures. The gain spectra obtained using the two models exhibit significant differences in amplitude and overall spectral shape as a consequence of interactions, which are neglected in the simpler free-carrier approach but included in the many-body theory.

In the absorptive regime, the attractive part of the Coulomb interaction leads to formation of excitons leading to a distinct resonance below the bandedge [Fig. 2.5(c)], which is missing in the free-carrier model spectra. Above the bandedge, Coulomb enhancement effects influence the spectral shape of the absorption due to excitonic continuum states. For higher density, the excitonic resonance disappears. However, excitonic correlations persist leading

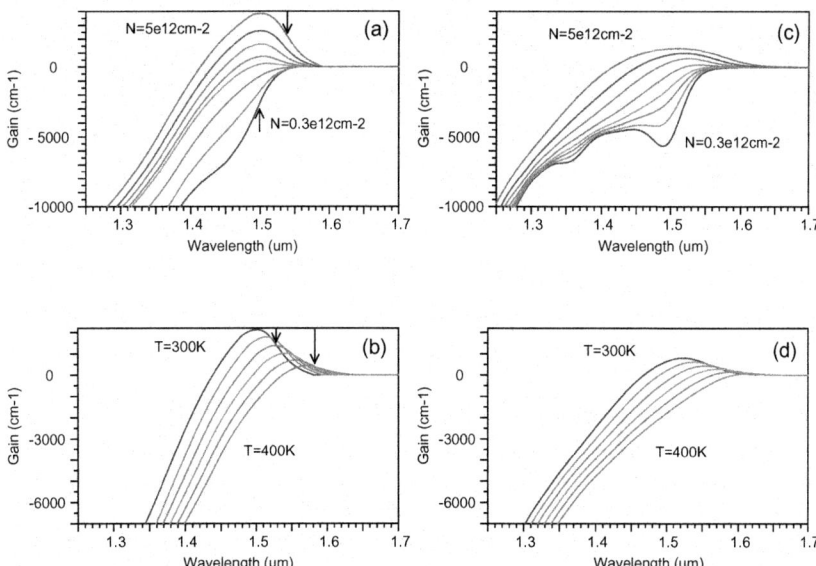

Fig. 2.5. Gain spectra computed using the free-carrier model are shown in (a) and (b) and gain spectra based on the many-body gain theory are shown in (c) and (d). (a,c): for increasing carrier density $N = 0.3/0.6/0.9/1.2/1.7/2.5/3.5/5.0 \times 10^{12} \text{cm}^{-2}$ at $T = 300$ K. (b,d): for increasing temperature $T = 300/320/340/360/380/400$ K at $N = 3.0 \times 10^{12} \text{cm}^{-2}$. The arrows in (a) and (b) indicate the energetic position of the lowest optical transition for the lowest and highest density in (a) and for the lowest and highest temperature in (b).

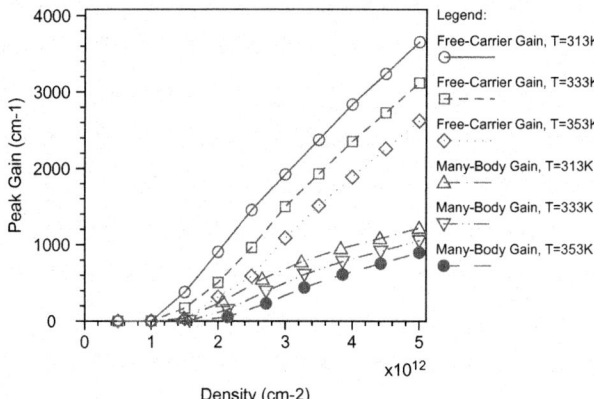

Fig. 2.6. Peak material gain as a function of quantum well carrier density for different temperatures. Results obtained using the free-carrier gain model are compared with calculations based on the many-body theory.

to a red-shift of the gain/absorption with respect to the material band gap in addition to carrier-induced band gap renormalizations. By adjusting the optical band gap, an error of the order of an excitonic binding energy can be corrected in the simpler free-carrier model for the gain regime. However, it is still unable to predict the position accurately and requires experimental data to calibrate the spectrum. Although band gap renormalizations are a consequence of microscopically including the Coulomb interaction in the many-body theory, the free-carrier approach accounts for the decrease of transition energy with increasing carrier density by applying a phenomenological band gap reduction, which is based on the local density approximation for the carrier Coulomb self-energies [47]. The resulting energetic position of the lowest subband transition is indicated by arrows in Fig. 2.5 for the lowest and the highest carrier density. In spite of the density-induced band gap reduction, both models predict a blue-shift of the gain peak for increasing density due to higher subband contributions overcompensating band gap renormalizations. This blue-shift is stronger in the spectra computed based on the many-body theory than in the results of the free-carrier approach.

In a Fabry–Perot cavity of common resonator length of several hundred micron, a quasi-continuum of longitudinal modes is available. Spectral shifts might lead to inaccuracies in predicting the laser frequency, but they have limited impact on the output characteristics, as a resonator mode close to the gain peak can always be found. Although spectral details have minor influence on the characteristics of a Fabry–Perot laser, it becomes more significant in cavities with a strong optical confinement such as in vertical-cavity surface-emitting lasers (VCSELs) An error on the order of an excitonic binding energy can shift the spectral region of maximum gain into the reflection band of the cavity. Furthermore, in such structures, spectral details around the transparency point can be important. In our example, the macroscopic results, such as light-current characteristics, depend on the gain peak, but they are insensitive to the rest of these spectral details. The position of the gain maximum determines the lasing modes and, therefore, the emission wavelength. The threshold current density and the slope efficiency are affected by the magnitude of the gain and its density and temperature dependence. Carrier–phonon and carrier–carrier scattering induce broadening, and subsequently reduce the peak amplitude of the gain. As these interactions are included in the microscopic theory within the Markovian limit, the temperature- and density-dependent broadening and, therefore, the total magnitude of the gain can be predicted as a function of the injection current.

Due to carrier–carrier scattering-induced broadening, the increase of gain with increasing density shows a stronger saturation behavior than the gain spectra obtained by neglecting these effects in the free-carrier approach, as can be seen from Fig. 2.5(a) and (c) and Fig. 2.6. Lower differential gain and stronger saturation behavior of the peak gain for increasing carrier concentration can be observed for the many-body gain compared with the free-carrier model. In the free-carrier model, a linewidth broadening parameter has to be specified to phenomenologically account for carrier–carrier and carrier–phonon interactions [see (2.33)], which prohibits the model from predicting the absolute gain amplitude and requires calibration using experimental data. We use a linewidth broadening of 26 meV for the gain spectra shown in Fig. 2.5. Comparing the gain amplitude of the free-carrier model with the many-body gain spectra (see Fig. 2.6), the broadening parameter would have to be increased to reduce the gain peak and differential gain to the correct values as given by the many-body gain. However, for increased broadening, the deviation of the free-carrier gain from the many-body gain around the transparency density would increase.

Naturally, the complex details of the carrier–carrier and carrier–phonon scattering processes cannot be included in one number. Therefore, the linewidth broadening can be adjusted to reproduce the gain amplitude of experiments only in a narrow window of carrier densities. The effective decay time approach fails to describe the behavior over a wider range of densities or temperatures, as discussed in Chap. 1. Improving the model for more complex lineshapes or introducing a density and temperature dependence would add fit parameters. Although experimental results can be reproduced, additional effort in model calibration has to be invested.

For a constant linewidth broadening parameter, we observe a weaker temperature dependence of the free-carrier gain [Fig. 2.5(b)] compared with the results obtained by the many-body theory [Fig. 2.5(d)]. Although a quantitative change of the gain amplitude is observed for the free-carrier model, the gain computed based on the many-body theory is almost completely suppressed by increasing the temperature from 300 K to 400 K for constant carrier density. The temperature dependence of the free-carrier gain is implicitly determined by the temperature dependence of the carrier distribution functions [see (2.31)]. In addition to the inversion factor, the many-body gain shows the influence of the increased phonon scattering for higher temperatures. A red-shift of the gain with increasing temperature is predicted by both models and is caused by band gap narrowing. The energetic position of the lowest subband transition is indicated by arrows in Fig. 2.5(b) for the lowest and highest temperature.

2.3.4 Light-Current Characteristics and Model Calibration

In Figs. 2.7 and 2.8, light-current curves obtained by the free-carrier model and the many-body model (lines) are compared with measured data (denoted by symbols) for the structure described in Sect. 2.3.1. The laser structure and experimental data have been taken from [32]. The experiments were performed under pulsed conditions to avoid self-heating. Therefore, a uniform temperature distribution, as given by the stage temperature, can be assumed. Both models can reproduce the experimental findings after model calibration.

With known gain spectrum calculated using the many-body theory, the calibration of remaining transport model parameters is greatly simplified due to a significant reduction of the parameter space. Aside from the gain, the spontaneous recombination is completely determined, as well as refractive index changes within the active layer. The use of simpler gain models requires simultaneous adjustments of transport and gain model parameters to account for density and temperature dependence of experimental findings.

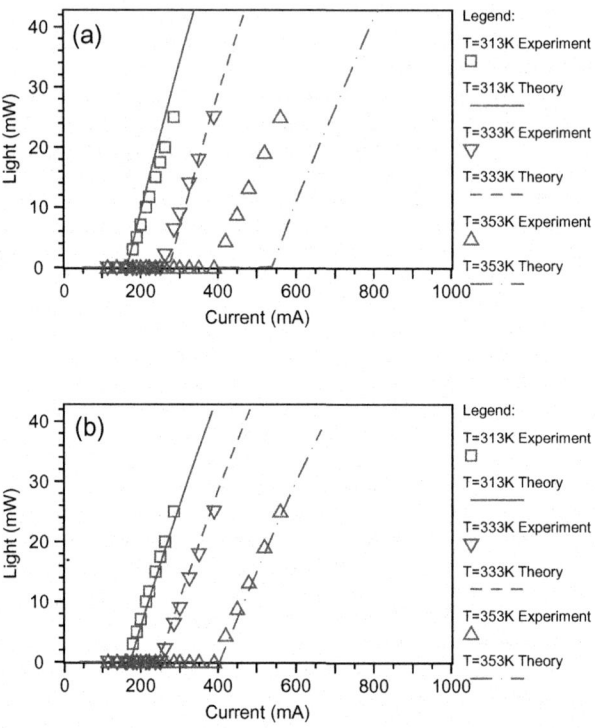

Fig. 2.7. Light-current characteristics for increasing temperature using the manybody gain model in comparison with experimental data. (a): Prediction based on default parameters of the simulator. (b): Calibrated result.

The uncalibrated simulation results obtained for each model, using default parameters from the material library of our simulator, show that the prediction based on the many-body theory is much closer to the experimental findings, concerning threshold current and slope efficiency, than the prediction based on the free-carrier approach [denoted by "default parameters" in Fig. 2.8(a)]. The free-carrier model predicts a lower threshold current and a much smaller shift of the threshold as a function of temperature. The slope efficiency is higher for the free-carrier approach than the microscopic many-body calculation, which agrees well with the experimental data. The weaker temperature sensitivity of the free-carrier model-based prediction is based on the weaker temperature dependence of the gain due to neglecting the electron–phonon interaction, as observed in Fig. 2.5. Furthermore, as a consequence of not accounting for carrier–carrier and carrier–phonon collision effects, the total amplitude of the gain as well as the differential gain is higher for the free-carrier model compared with the many-body gain theory (see Fig. 2.5), leading to lower threshold of the light-current characteristic and influencing the slope efficiency.

Besides the gain, nonradiative recombination affects the threshold current density and slope efficiency. Auger recombination, Shockley–Read–Hall recombination, and spontaneous emission into nonlasing modes decrease the carrier densities in the active layers without contributing to light amplification. Waveguide and background absorptive losses, intervalence band absorption, and lateral leakage are other processes affecting the device behavior. The Auger recombination, in particular, has been pointed out as one of the dominating processes [32, 39]. Within the nonradiative recombination mechanisms, it has the strongest density dependence, being cubic, whereas other recombination mechanisms such as spontaneous recombination and Shockley–Read–Hall processes have a weaker, quadratic and linear density dependence, respectively [see (2.9), (2.14), (2.10)]. As a decrease of gain with increasing temperature has to be compensated by an increase of carrier density within the active layers, the threshold injection current increases with temperature. The efficiency of the injection is affected by the dark recombination, which acts to reduce the carrier density. Due to the nonlinearity with respect to the carrier density dependence, the Auger recombination tends to strongly amplify the effects of the temperature-dependent gain reduction, leading to further reduction of slope efficiency and increase of threshold currents. This mechanism scales with the Auger coefficients. The same argument applies to a lesser degree to other carrier-dependent processes involved in the coupled problem of carrier transport and radiative interaction, as an increase in temperature will indirectly increase the carrier density in the quantum wells. Note that the strong decrease of the differential gain with increasing carrier density observed for the many-body gain tends to nonlinearly increase the carrier densities required to compensate a temperature-induced reduction of the gain, thus enhancing the temperature sensitivity.

As the gain is completely determined by the microscopic calculation, only transport parameters have to be adjusted to fit the measurement. Due to the prediction overestimating the temperature dependence in Fig. 2.7(a), we reduce the Auger coefficients to $C_e^{Auger} = 1 \times 10^{-31} \text{cm}^6 \text{s}^{-1}$, $C_h^{Auger} = 5 \times 10^{-31} \text{cm}^6 \text{s}^{-1}$ and compensate with a decrease of the carrier lifetime in the Shockley–Read–Hall recombination within the quantum wells: $\tau_e^{SRH} = 5 \times 10^{-9}$s, $\tau_h^{SRH} = 5 \times 10^{-8}$s. We slightly reduce the carrier-dependent loss and increase carrier-independent contributions. Based on the photoluminescence comparison, an inhomogeneous broadening of 22 meV is used. A good match of the experimental curves can be obtained using one set of transport model parameters for all temperatures (see Table 2.2).

Table 2.2. Material and Model Parameters Used for the Active Layers. Default Parameters of the Simulator, Parameters Calibrated Using the Many-body Gain Model, and Parameters Published in [32] are Listed.

Parameter	Default	Calibration	[32]
τ_e^{SRH}/s	2.44e-8	5.e-9	2.e-8
τ_h^{SRH}/s	2.88e-6	5.e-8	2.e-8
$B/(\text{cm}^3\text{s}^{-1})$	2.26e-10	1.2e-10	1.2e-10
$C_e^{Auger}/(\text{cm}^6\text{s}^{-1})$	4.43e-31	1.0e-31	0.0
$C_h^{Auger}/(\text{cm}^6\text{s}^{-1})$	3.95e-30	5.0e-31	1.6e-28
$E_e^{Auger,act}/\text{meV}$	0.0	0.0	60
$E_h^{Auger,act}/\text{meV}$	0.0	0.0	0.0
$\alpha_0/(\text{cm}^{-1})$	0.0	16.0	0.0
k_e^{fca}/cm^2	1.e-18	1.0e-18	1.0e-18
k_h^{fca}/cm^2	2.e-17	1.8e-17	8.2e-17
γ_0/meV (only free-carrier gain)	26	120/190/265	41
γ_{inh}/meV (only many-body gain)	0.0	22.0	–

Based on our advanced treatment of the gain, we find that the influence of the Auger recombination on the temperature dependence of InGaAsP quantum well lasers is secondary. Our Auger parameters used in our calibrated simulations are about one to two orders of magnitude lower than parameters published in the literature for InGaAsP [11]. The large spread of values in the literature indicates that the Auger process in quantum wells deserves further scientific investigation. Experimentally, Auger coefficients are often determined as the coefficient of the cubic term in a polynomial fit of a density-dependent measurement. Other processes could contribute to the measured coefficient leading to an overestimation of the Auger recombination. Theoretical Auger parameters found in [49, 50] for unstrained InGaAsP quantum wells are in good qualitative agreement with our findings. The influence of 1% strain present in our quantum well structure should be negligible compared with the general uncertainty associated with Auger recombination for this material system. It has been suggested that strain could be used to reduce the Auger recombination in InGaAsP quantum wells [40, 51]. Note that parameters for quantum well Auger recombination can deviate significantly from coefficients known for bulk material. Moreover, for narrow quantum wells, thresholdless and quasi-threshold Auger processes with weak temperature dependence dominate over threshold Auger processes present in the bulk limit, which are characterized by a higher activation energy [50]. Our simulations indicate that the gain reduction, due to carrier–phonon and carrier–carrier scattering-induced dephasing, primarily determines the temperature sensitivity of this material system for the temperature range reasonable for laser applications investigated here. For even higher temperatures, it can be expected that the cubic density dependence of the Auger process will increase its relative importance due to the implicit increase of density with temperature. Furthermore, vertical carrier leakage out of the active region becomes more important [32, 52].

The role of nonradiative recombination and gain in this context has been discussed in the literature [39, 40, 45]. In particular, the influence of the reduction of differential gain with increasing temperature has been discussed in experimental investigations of the temperature sensitivity [40, 45]. The decrease of differential gain with temperature is consistent with our findings, as, in order to sustain lasing operation, the effects of the increased temperature will be compensated by an increase of carrier density, implying a collision-induced decrease in differential gain, as can be seen from the saturation behavior of the gain spectra as a function of carrier density in Fig. 2.5(c).

Pointing out the significance of the gain in determining the temperature sensitivity, Piprek et al. required higher Auger parameters than used here to reproduce their experimental data for constant gain broadening [32]. Following their calibration procedure, we obtain a good fit to the experimental findings using the free-carrier gain approach with constant broadening parameter in connection with increased Auger recombination for similar model parameters, as is shown in Fig. 2.8(a) (denoted by "published calibrated parameters"). We use Auger parameters of $C_e^{Auger} = 0$, $C_h^{Auger} = 1.1 \times 10^{-28}$ and $E_h^{Auger,act} = 80$ meV. Comparison of this fit with the result computed using our lower default parameters demonstrates the influence of the Auger recombination on the temperature sensitivity. Our low default Auger parameters ($C_e^{Auger} = 4.43 \times 10^{-31} \text{cm}^6 \text{s}^{-1}$, $C_h^{Auger} = 3.95 \times 10^{-30} \text{cm}^6 \text{s}^{-1}$) explain the weak temperature dependence of the results obtained in the uncalibrated prediction using the free-carrier gain model. Moreover, temperature independent Auger coefficients were used [$E_{e/h}^{Auger,act} = 0$ in (2.11)]. The temperature sensitivity observed in the light-current characteristic for the low default Auger parameters is more directly related to the weak dependence observed in the gain spectra in Fig. 2.5, because the amplification effect described above is reduced with respect to the parameter set with high Auger recombination.

As our advanced model indicates the domination of collision broadening in determining the temperature sensitivity, we keep the calibrated transport model used within the computations based on the many-body theory and only switch to the free-carrier gain model. In order to obtain the fit in Fig. 2.8, we use the linewidth broadening parameter to fit the curves individually for each temperature, while keeping all other parameters unchanged. The linewidth broadening can be understood as a simplified model for dephasing due to quasi-particle scattering. However, it fails to describe the temperature and density dependence of these processes. Using increased

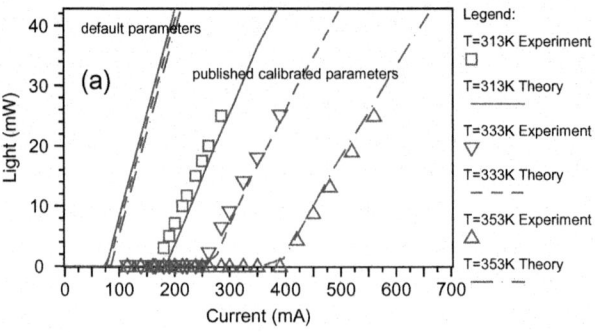

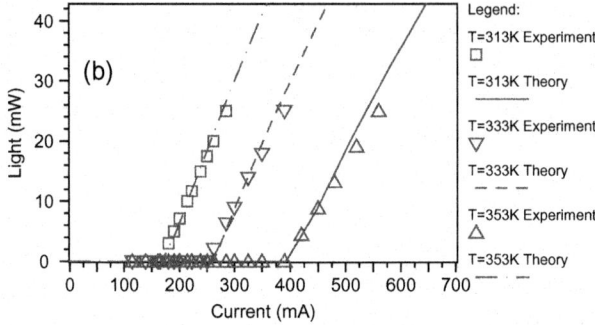

Fig. 2.8. Light-current characteristics for increasing temperature using the free-carrier gain model in comparison with experimental data. (a): Prediction based on default parameters and calibration based on Auger recombination parameters from [32]. (b): Calibration based on parameters obtained using the many-body gain model.

linewidth broadening parameters with increasing temperature, the threshold current can be adjusted to match the experimental data as shown in Fig. 2.8. The slope efficiency of the fitted simulation results tends to be slightly higher in the simulation compared with the experiment, especially for higher temperatures. As the simple broadening model cannot capture the complex dynamics of carrier–carrier scattering, the density dependence is not taken into account correctly. As the density will increase to compensate for the effects of the increased temperature, the more linear differential gain of the free-carrier model will cause stronger discrepancies with respect to the carrier-collision-induced decrease of differential gain. Furthermore, the broadening is uniformly applied to the full range of subband transitions instead of having an energy dependence according to the spectrally resolved carrier densities. As a significant reduction of gain requires a strong increase of the uniform broadening, unphysically high fit parameters are obtained (120/190/265 meV for 313/333/353 K), which should not be interpreted in terms of a directly associated scattering mechanism. Introducing proper dependencies with respect to density, temperature, and energy, the linewidth broadening model could be improved. However, new model parameters would be increasing the effort for calibration. In spite of the shortcomings of the linewidth broadening model, we have demonstrated that it provides an alternative method to the increase of Auger recombination for reproducing the experimental data using a free-carrier approach, which is in agreement with our findings from the advanced treatment of the gain.

Without detailed knowledge of the gain spectrum of a given material, ambiguities between different processes determining the temperature and density dependence of the light output characteristics can exist. A good fit can be obtained by calibrating transport parameters only, indicating a correct description of the underlying physics. Auger recombination has been regarded as the dominant mechanism determining the temperature sensitivity of InGaAsP lasers in the past. As our investigation indicates, this conclusion was based on the use of oversimplified gain models leading to an overestimation of the Auger recombination. In order to avoid these ambiguities, a calibration procedure should attempt to calibrate the gain spectrum first. Density- and temperature-dependent photoluminescence measurements could help in determining broadening parameters describing density and temperature dependence of the gain. Using gain and photoluminescence spectra based on the microscopic many-body theory, this step is already completed. Additional benefit is provided by sample characterization. After the gain and spontaneous emission are calibrated, the remaining transport parameters can be adjusted by performing full-scale laser simulations.

2.3.5 Self-Heating

In Fig. 2.9, light-current characteristics (a) and the corresponding maximum lattice temperature within the device (b) are shown for calculations involving the heat flow equation (2.15) in comparison with results based on a uniform temperature distribution. The many-body gain model was used for these computations. Different boundary conditions can be imposed at the metal contacts. At the top contact on the p-type side of the device, a reflective boundary is assumed leading to vanishing heat flux. A lumped thermal resistor that describes the heat conductance of the substrate and the heat sink is associated with the bottom electrode. For ideal cooling of the laser, corresponding to vanishing lumped thermal resistance in our example, the temperature at the thermal contact is forced to the external stage temperature described by imposing a Dirichlet boundary condition. The more realistic boundary condition (2.23) with finite thermal conductance associated with the contact accounts for the thermal properties of regions that are not included in the simulation domain, such as substrate, mounting, or packaging [35].

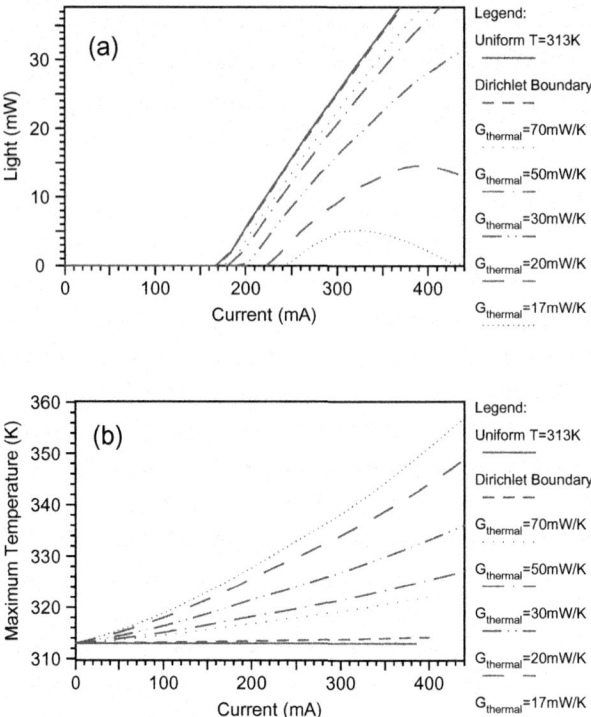

Fig. 2.9. Light-current characteristics for increasing temperature using the many-body gain model for different boundary conditions for the heat flow equation: Uniform temperature distribution with $T = 313$ K, Dirichlet boundary condition forcing $T = 313$ K at the contacts, lumped thermal conductance associated with the bottom contact with given thermal conductance $G_{thermal}$.

For the Dirichlet boundary condition, we observe negligible effect on the light-current characteristic. As can be seen from Fig. 2.9(b), the maximum temperature increases by only 2 K with respect to the stage temperature of $T_C = 313$ K. By applying a finite thermal conductance, the local temperature at the thermal contact, and therefore within the device, is allowed to rise. Higher temperatures are reached within the device as shown in Fig. 2.9(b) for decreasing thermal conductance $G_{thermal}$ associated with the bottom electrode, describing less-efficient cooling of the laser. As a consequence, an increase in threshold current and a decrease in slope efficiency with respect to the isothermal case can be observed. Furthermore, the light-current curves show the characteristic thermal roll-off. For thermal conductivities $G_{thermal} < 15$ mW/K, lasing is completely suppressed at stage temperature $T_C = 313$ K due to self-heating effects. As our findings from Sect. 2.3.4 indicate, the temperature-induced reduction of the gain is the primary cause of these thermal effects. As can be seen from Fig. 2.10, the maximum temperature occurs in the p-type ridge and decreases across the active layers toward the heat sink at the bottom contact. The heat is generated dominantly due to Joule heat around the quantum wells and the p-type layers as well as recombination heat around the active region. The lattice heat equation was deactivated in the oxide regions (see Fig. 2.2), thus keeping the temperature equal to the ambient temperature in those regions.

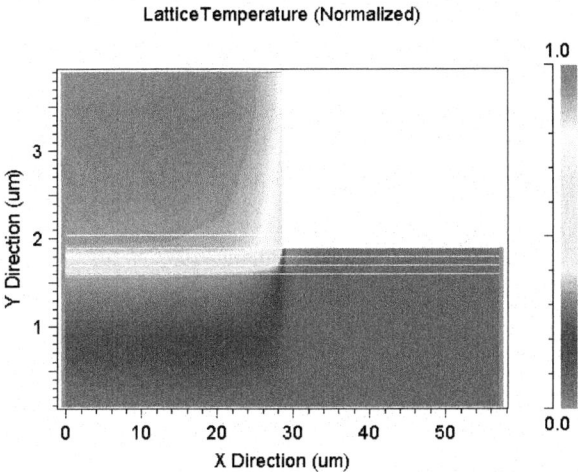

Fig. 2.10. Temperature profile of the half-domain ridge-waveguide laser structure as obtained by solving the heat flow equation.

Figure 2.11 shows a comparison of the effects of self-heating obtained using the many-body gain compared with the free-carrier gain model. For the many-body gain, strong self-heating effects can be observed in the light-current characteristic caused by quasi-particle interaction-induced gain broadening. Calculations based on the free-carrier model with the same parameter settings, in particular, the low Auger parameters obtained by our calibration procedure, show very weak influence of the temperature increase (indicated by "Free-Carrier Gain A" in Fig. 2.11). In contrast to the temperature and carrier density dependence of the collision broadening included in the many-body theory, the gain broadening of the free-carrier model is constant. Weaker effects of the self-heating-induced temperature increase are observed compared with the computations using the advanced gain model. This result was expected from the weak temperature sensitivity observed for the uniform temperature simulations for our low default Auger parameters [see Fig. 2.8(a)]. Moreover, our calibrated Auger coefficients are even lower (see Table

2.2). Although it is possible to adjust the constant gain broadening for the isothermal case, the self-heating simulation demonstrates the shortcoming of a missing temperature and carrier density dependence of the broadening because the temperature varies spatially and with the bias conditions. Using the higher Auger parameters from [32] for this structure, the influence of self-heating leads to considerable increase of threshold and decrease of slope efficiency with respect to the uniform temperature case, as the strong carrier density dependence of the Auger recombination tends to amplify the effects of the temperature-induced gain reduction aside from its explicit exponential temperature dependence (see Fig. 2.11(a) "Free-Carrier Gain B").

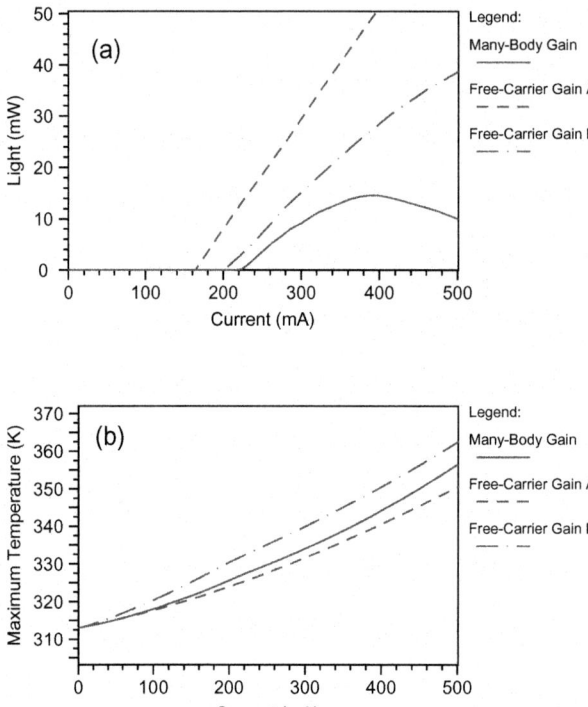

Fig. 2.11. Light-current characteristics for stage temperature $T = 313$ K using the many-body gain compared with results obtained for the free-carrier model for parameters calibrated based on the many-body calculations (Free-Carrier Gain A) and for calibrated parameters taken from [32] (Free-Carrier Gain B).

Figure 2.11(b) depicts the maximum temperature within the device occurring during the respective simulations. As can be seen from Fig. 2.10, the maximum occurs in the vicinity of the active layer; thus, it roughly indicates the local temperature influencing the optical transitions. For all of the three model settings, similar temperatures occur in the device. Slightly higher temperatures are reached for the high Auger recombination parameter settings ("Free-Carrier Model B") compared with the lower Auger parameters obtained by the calibration based on the many-body gain models ("Many-Body Gain", "Free-Carrier Gain A") as the higher Auger recombination implies higher contributions to the recombination heat source.

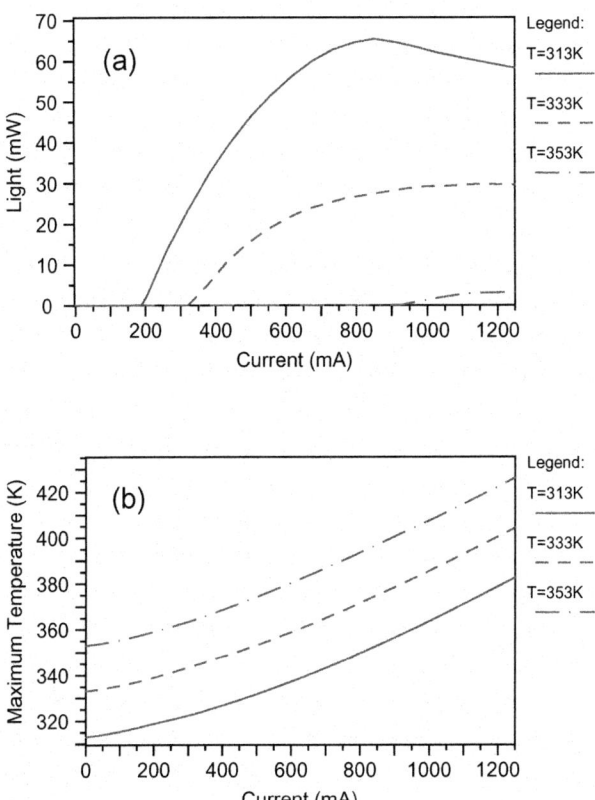

Fig. 2.12. (a): Light-current characteristics for increasing stage temperature T using the many-body gain model and including self-heating, (b): corresponding maximum temperature within the device.

In self-heating simulations, modeling the correct temperature dependence of the gain broadening cannot be avoided in order to obtain reliable results, as can be seen from Fig. 2.11. The effective decay time approximation can reproduce the gain spectra for a narrow window of carrier densities and temperatures as discussed in Sect. 2.3.3 and in Chapt. 1. For isothermal full-scale laser simulations, experimental light-current characteristics can be reproduced once parameters are calibrated, as the increase of carrier density above the lasing threshold is weak and the temperature is kept constant.

However, in self-heating simulations, the temperature varies spatially and as a function of the injection current. Temperature-induced gain reductions imply a compensating increase of the carrier concentrations. The increased range of variation in carrier density and temperature cannot be covered by the constant linewidth broadening. Therefore, the self-heating results obtained using the free-carrier model differ strongly from the advanced many-body computations (see Fig. 2.11), in spite of the agreement between the light-current characteristics computed using the respective models with the same parameters for isothermal conditions (see Figs. 2.7(b), 2.8(b) and 2.8(a) "published calibrated parameters").

In Fig. 2.12, simulation results are shown for increasing stage temperature. The many-body gain model was used for these calculations. The thermal conductance associated with the heat sink at the bottom electrode was set to $G_{thermal} = 45$ mW/K. As can be seen by comparing these light-current characteristics [Fig. 2.12(a)] with Fig. 2.7(b), temperature effects are stronger under self-heating conditions. A uniform temperature distribution can be prepared experimentally by pulsed operation, where the bias is applied only for a short period of time during one duty cycle. In this mode of operation, the heat generated within the device can be dissipated to the environment efficiently, whereas during the short lasing period, the heat capacity of the device prevents significant increase of the temperature [see left-hand side of (2.15)]. This situation is desirable from an experimental point of view, because it eliminates self-heating effects and eases analysis of the dominant physical mechanisms influencing the performance of the device. However, in most applications, diode lasers are operated under CW conditions. In this mode of operation, a steady state between heating within the device due to transport and recombination processes and external cooling will occur. A significant increase of the temperature with respect to the stage temperature will occur in the active layer, as can be seen from Fig. 2.12(b). The theoretical description of such a situation requires the coupled solution of carrier and thermal transport as well as radiative interaction with the light field.

2.4 Summary

We have demonstrated the integration of advanced gain modeling based on a microscopic many-body theory into full-scale laser simulations. Our approach has been applied to the investigation of the temperature sensitivity of InGaAsP quantum well lasers. It has been shown that the gain broadening due to carrier–carrier and carrier–phonon scattering-induced dephasing dominantly determines the temperature sensitivity of these laser structures rather than nonradiative recombination.

Our microscopic gain model allows for an accurate prediction of the gain spectrum for a specific material system based solely on material parameters. The energetic position and the collision broadening of the gain maximum have a significant impact on the optical properties of Fabry–Perot laser diodes, in particular, emission wavelength, threshold current, and slope efficiency, as discussed here. The detailed spectral behavior of the gain can be expected to be of more importance for advanced structures like VCSELs, which exhibit a strong optical confinement in the longitudinal direction.

The advanced many-body gain theory has been compared with the free-carrier gain model, which is a common approach in commercial laser simulators. The advantage of the predictive modeling of the gain by the microscopic many-body theory with respect to simpler models carries over to the full-scale laser simulation. Calibration effort can be reduced while improving the overall predictive capabilities of the simulation. In order to improve the free-carrier approach, density, temperature, and energy dependences would have to be added to the gain broadening model to describe effects of carrier–carrier and carrier–phonon scattering phenomenologically. We suggest a calibration procedure that determines the gain model parameters first by performing optical experiments in order to avoid ambiguities between transport and gain models in describing temperature and density dependence of the overall laser performance. Using the microscopic many-body theory, this additional calibration step can be avoided.

Acknowledgment

We thank J. Piprek for valuable discussions and for providing experimental data. The Tucson effort is supported by a grant from the U.S. Air Force Office for Scientific Research under grant number: AFOSR F49620-01-1-0380. The Marburg research is supported by the Max Planck Research Prize.

References

1. M. Grupen and K. Hess: IEEE J. Quantum Electron. **34**, 120 (1998)
2. B. Grote, E. K. Heller, R. Scarmozzino, J. Hader, J. V. Moloney, and S. W. Koch: Integration of microscopic gain modeling into a commercial laser simulation environment. In: *Physics and Simulation of Optoelectronic Devices XI* Proc. SPIE, vol. 4986, ed by M. Osinski, H. Amano, and P. Blood, pp. 413-422 (2003)
3. W. Van Roosbroeck: Bell System Tech. J. **29**, 560 (1950)
4. D. L. Scharfetter and H. K. Gummel: IEEE Trans. Electron Devices **ED-16**, 64 (1969)
5. K. Blotekjaer: IEEE Trans. Electron Devices **ED-17**, 38 (1970)
6. R. K. Cook and J. Frey: COMPEL **1**, 65 (1982)
7. D. Chen, E. C. Kan, U. Ravaioli, C. W. Shu, and R. W. Dutton: IEEE Electron Device Lette. **EDL-13**, 26 (1992)
8. D. Chen, Z. Yu, K. C. Wu, R. Goossens, and R. W. Dutton: Dual Energy Transport Model with Coupled Lattice and Carrier Temperatures. In: Proceedings 5th SISDEP Conference Vienna, Austria, pp. 157-160 (1993)

9. G. Wachutka: IEEE Trans. on Computer-Aided Design **CAD-9**, 1141 (1990)
10. S. Selberherr: *Analysis and Simulation of Semiconductor Devices* (Springer Verlag, New York 1984)
11. J. Piprek: *Semiconductor Optoelectronic Devices* (Academic Press, San Diego 2003)
12. G. Wachutka: COMPEL **10**, 311 (1991)
13. H. Brand: Thermoelektrizität und Hydrodynamik. Dissertation, Technical University Vienna, Austria (1994) (ohttp://www.iue.tuwien.ac.at/phd/brand/diss.html)
14. M. A. Alam, M. S. Hybertsen, R. K. Smith, and G. A. Baraff: IEEE Trans. Electron Devices **47**, 1917 (2000)
15. J. Luttinger and W. Kohn: Phys. Rev. **97**, 869 (1955)
16. G. L. Bir and G. E. Pikus, eds.: *Symmetry and Strain-induced Effects in Semiconductors* (Wiley, New York 1974)
17. C. Hermann and C. Weisbuch,: Phys. Rev. B **15**, 823 (1977)
18. E. O. Kane: Energy band theory. In: *Handbook on Semiconductors* ed. by T. S. Moss (North-Holland, New York 1982) PP. 193–217
19. S. L. Chuang: Phys. Rev. B **43**, 9649 (1991)
20. P. von Allmen: Phys. Rev. B **46**, 15382 (1992)
21. M. S. Hybertsen, G. A. Baraff, S. K. Sputz, D. A. Ackermann, G. E. Shtengel, J. M. Vandenberg, and R. Lum: Modeling of optical spectra for characterization of multi-quantum well InGaAsP-based lasers. In: *Physics and Simulation of Optoelectronic Devices IV* Proc. SPIE, vol. 2693, ed. by W. W. Chow and M. Osinski, pp. 430-441 (1996)
22. F. Oyafuso, P. von Allmen, M. Grupen, and K. Hess: Gain calculation in a quantum well simulator using an eight band $\mathbf{k} \cdot \mathbf{p}$ model. In: *Proc. 4th Int. Workshop Computat. Electron.* (Tempe, AZ 1995)
23. B. Zee: IEEE J. Quantum Electron. **QE-14**, 727 (1978)
24. M. Yamada and Y. Suematsu: J. Appl. Phys. **52**, 2653 (1981)
25. W. W. Chow and S. W. Koch: *Semiconductor-Laser Fundamentals: Physics of the Gain Materials* (Springer, Berlin 1999)
26. H. Haug and S. W. Koch: *Quantum Theory of the Optical and Electronic Properties of Semiconductors*, 3rd edn (World Scientific, Singapore 1994)
27. J. Hader, J. V. Moloney, S. W. Koch, and W. W. Chow: IEEE J. Sel. Topics in Quantum Electron. **9**, 688 (2003)
28. J. Hader, A. R. Zakharian, J. V. Moloney, T. R. Nelson, W. J. Siskaninetz, J. F. Ehret, K. Hantke, S. W. Koch, and M. Hofmann: Optics and Photonics News **13 (12)**, 22 (2002)
29. J. Hader, A. R. Zakharian, J. V. Moloney, T. R. Nelson, W. J. Siskaninetz, J. F. Ehret, K. Hantke, M. Hofmann, and S. W. Koch: IEEE Photon. Technol. Lett., **14**, 762 (2002)
30. M. Grupen, K. Hess, and G. H. Song: Simulation of transport over heterojunctions. In: *Proc. 4th Int. Conf. Simul. Semicon. Dev. Process.*, vol 4, pp. 303-311 (Zurich, Switzerland 1991)
31. S. Seki, W. W. Lui, and K. Yokoyama: Appl. Phys. Lett. **66**, 3093 (1995)
32. J. Piprek, P. Abraham, and J. E. Bowers: IEEE Journal of Quantum Electron. **36**, 366 (2000)
33. Z. Yu, D. Chen, L. So, and R. W. Dutton: *PISCES-2ET and Its Application Subsystems*, Manual, Stanford University, Stanford, California (1994)

34. R. Thalhammer: Internal Laser Probing Techniques for Power Devices: Analysis, Modeling, and Simulation. Dissertation, Technical University Munich, Germany (2000) (http://tumb1.biblio.tu-muenchen.de/publ/diss/ei/2000/thalhammer.pdf)
35. T. Grasser and S. Selberherr: IEEE Trans. on Electron Devices **48**, 1421 (2001)
36. J. Hader, S. W. Koch, and J. V. Moloney: Sol. Stat. Electron. **47**, 513 (2003)
37. W. Chow, M. Kira, and S. W. Koch: Phys. Rev. B **60**, 1947 (1999)
38. G. P. Agrawal and N. K. Dutta: *Semiconductor Lasers* (Van Nostrand Reinhold, New York 1993)
39. J. Braithwaite, M. Silver, V. A. Wilkinson, E. P. O'Reilly, and A. R. Adams: Appl. Phys. Lett. **67**, 3546 (1995)
40. Y. Zou, J. S. Osinski, P. Grodzinski, P. D. Dapkus, W. C. Rideout, W. F. Sharfin, J. Schlafer, and F. D. Crawford: IEEE J. Quantum Electron. **29**, 1565 (1993)
41. J. Piprek, D. Babic, and J. E. Bowers: J. Appl. Phys. **81**, 3382 (1997)
42. L. J. P. Ketelsen and R. F. Kazarinov: IEEE J. Quantum Electron. **34**, 811 (1995)
43. Y. Yoshida, H. Watanabe, K. Shibata, A. Takemoto, and H. Higuchi: IEEE J. Quantum Electron. **34**, 1257 (1998)
44. A. A. Bernussi, H. Temkin, D. L. Coblentz, and R. A. Logan: Appl. Phys. Lett. **66**, 67 (1995)
45. D. A. Ackerman, G. E. Shtengel, M. S. Hybertsen, P. A. Morgan, R. F. Kazarinov, T. Tanbun-Ek, and R. A. Logan, IEEE J. Select. Topics Quantum Electron. **1**, 250 (1995)
46. J. Piprek, private communication
47. L. Hedin and B. Lundquist, *J. Phys. C* **4**, 2064 (1971)
48. Gain / refractive index / photoluminescence tables courtesy of Nonlinear Control Strategies, 1001 East Rudasill Rd., Tucson, AZ 85718
49. *New Semiconductor Materials. Characteristics and Properties*, Ioffe Physico-Technical Institute, St. Petersburg, Russia, (http://www.ioffe.ru/SVA/NSM/Semicond/)
50. A. S. Polkovnikov and G. G. Zegrya: Phys. Rev. B **58 (7)**, 4039 (1998)
51. O. Gilard, F. Lozes-Dupuy, G. Vassilieff, S. Bonnefont, P. Arguel, J. Barrau, and P. Le Jeune: J. Appl. Phys. **86(11)**, 6425 (1999)
52. J. Piprek, J. K. White, and A. J. Spring Thorpe: IEEE J. Quantum. Electron. **38**, 1253 (2002)

3 Fabry–Perot Lasers: Thermodynamics-Based Modeling

U. Bandelow, H. Gajewski, and R. Hünlich

Weierstrass Institute for Applied Analysis and Stochastics
Mohrenstr. 39, 10117 Berlin, Germany, bandelow@wias-berlin.de

Summary. This chapter describes the modeling and the simulation of edge-emitting quantum well (QW) lasers, based on the drift-diffusion equations and equations for the optical field. By applying fundamental thermodynamic principles such as the maximum entropy principle and the principle of local thermal equilibrium, we derive a self-consistent energy transport model that can be proven to meet the thermodynamic requirements. Its numerical solution is discussed explicitly, by starting from the discretization procedure and by ending up with the iteration scheme. As an example, we demonstrate the simulation of a long-wavelength ridge-waveguide multi-quantum well laser.

3.1 Introduction

In modern semiconductor devices such as high-power transistors or lasers, thermal effects caused by strong electric and optical fields and by strong recombination play an important role and have to be included in the mathematical models.

Indeed, there is a large variety of energy models for semiconductor devices. Typically, these models base on the usual state equations and continuity equations for the carrier densities and on the balance of the total energy expressed by the equation:

$$\partial_t u + \nabla \cdot j_u = -\gamma \qquad (3.1)$$

for the density u and the current density j_u of the total energy, where γ counts for the radiation that is emitted from the device. Furthermore, differential relations for u and general thermodynamic relations for j_u are used to transform the energy balance equation (3.1) into a heat flow equation [1]:

$$c_h \partial_t T - \nabla \cdot (\kappa_L \nabla T) = H, \qquad (3.2)$$

where c_h is the heat capacity and κ_L is the heat conductivity. Although the heat flow equation (3.2) with the description of the source term H is well established, the discussion about its relation to the conservation law of energy is still ongoing.

In this contribution, based on an expression for the density of the free energy, we derive a thermodynamics-based system of evolution equations for edge-emitting quantum well lasers in a deductive way. Thereby we only apply first principles like the entropy maximum principle and the principle of partial local equilibrium [2]. Moreover, we assume that the free energy is the sum of the internal free energy, of the electrostatic field energy, and of the energy of the optical field. For the simulation of semiconductor lasers, the energy transport model is coupled to the evolution equations for the optical field in a self-consistent manner. The resulting energy transport model is, in this explicit form, new to our knowledge and can be shown to meet the thermodynamic requirements. Boundary conditions as well as proper discretization schemes will be discussed too. Furthermore, the numerical solution procedure will be discussed. The complete energy transport model has been implemented in WIAS-TeSCA [3], a numerical code for simulation of semiconductor devices. On this basis, we demonstrate the simulation of long-wavelength edge-emitting QW lasers, with a special focus on the self-heating of the device and the modulation response.

3.2 Basic Equations

Let $\Omega \subset \mathbb{R}^2$ be the transverse cross section of the device under consideration. By ∇, we denote the transverse part of the Nabla operator, i.e., the Nabla operator with respect to $r = (x, y)$ in Ω. We denote by n and p the densities of the mobile charge carriers, electrons, and holes, respectively, and by C the net doping profile. For shorter notation, these densities as well as the band-edge densities of states N_c and N_v are scaled by a reference density N_{ref}. The lattice temperature T is scaled by a reference temperature T_s. Moreover, the electrostatic potential φ as well as the quasi-Fermi potentials f_n and f_p are scaled by $u_{T_s} = k_B T_s / q$ and the conduction and valence band edges e_c and e_v by $k_B T_s$. Here k_B and q are Boltzmann's constant and the elementary charge, respectively.

3.2.1 Poisson Equation

The electrostatic potential φ satisfies Poisson's equation:

$$-\nabla \cdot (\varepsilon \nabla \varphi) = C + p - n \tag{3.3}$$

in the transverse cross section Ω of the laser. Here $\varepsilon = \varepsilon(r)$ is the static dielectric constant in the possibly heterogeneous semiconductor material.

3.2.2 Transport Equations

The charge carrier densities n and p have to fulfill the continuity equations:

3 Fabry–Perot Lasers: Thermodynamics-Based Modeling

$$\frac{\partial n}{\partial t} - \nabla \cdot \boldsymbol{j}_n = -R \tag{3.4}$$

$$\frac{\partial p}{\partial t} + \nabla \cdot \boldsymbol{j}_p = -R \tag{3.5}$$

for $(t, \boldsymbol{r}) \in \mathbb{R}_+^1 \times \Omega$, i.e., for all times $t > 0$ in the transverse cross section. $\boldsymbol{j}_n$ denotes the electron current density and $\boldsymbol{j}_p$ the hole current density. The recombination rate R in (3.4) and (3.5) involves all nonradiative and radiative recombination processes, as in particular the Shockley–Read–Hall recombination rate R^{SRH}, the Auger recombination rate R^{AUG}, and the spontaneous radiative recombination rate R^{sp}. The recombination processes stimulated by the optical field are included by the stimulated recombination rate R^{stim}

$$R^{stim} = v_g g |\chi|^2 N_s, \tag{3.6}$$

where v_g denotes the group velocity and g denotes the material gain. The modal intensity distribution $|\chi|^2$ and the photon number N_s will be discussed in Sect. 3.2.4.

As a consequence of the Poisson equation (3.3) and the continuity equations (3.4) and (3.5), we infer the current conservation equation:

$$\nabla \cdot \boldsymbol{j} = 0, \quad \boldsymbol{j} = -\varepsilon \nabla \frac{\partial \varphi}{\partial t} + \boldsymbol{j}_n + \boldsymbol{j}_p \tag{3.7}$$

to hold in $\mathbb{R}_+^1 \times \Omega$.

3.2.3 State Equations

The quasi-Fermi potentials are linked with the carrier concentrations by means of Fermi–Dirac statistics:

$$n = N_c \mathcal{F}_{1/2}\left(\frac{\varphi - f_n - e_c}{T}\right) \tag{3.8}$$

$$p = N_v \mathcal{F}_{1/2}\left(\frac{e_v + f_p - \varphi}{T}\right). \tag{3.9}$$

The size quantization by the quantum wells essentially induces a modified density of states. We simulate this by multiband kp-models to obtain at least net coefficients that model the QW's, like classic materials with specific material parameters. This has been extensively described recently in [4].

3.2.4 Optics

Assuming stable transverse waveguiding allows us to express the main component of the optical field vector $E(\boldsymbol{r}, z, t)$:

$$E(\boldsymbol{r}, z, t) = e^{i\omega t}\left[\Psi^+(z, t)e^{-ikz} + \Psi^-(z, t)e^{ikz}\right]\chi(\boldsymbol{r}) \tag{3.10}$$

in terms of the transverse main mode χ. Transverse modes χ are eigensolutions of the waveguide equation:

$$\left[\nabla^2 + \frac{\omega^2}{c^2}\varepsilon_{opt}(\omega, \boldsymbol{r}) - \beta^2\right]\chi(\boldsymbol{r}) = 0, \qquad (3.11)$$

corresponding to their respective (complex) eigenvalues β.[1] Here:

$$\varepsilon_{opt}(\omega, \boldsymbol{r}) = (n_r(\omega, \boldsymbol{r}) + ic\left[g(\omega, \boldsymbol{r}) - \alpha_{bg}(\boldsymbol{r})\right]/2\omega)^2$$

denotes the complex dielectric function of the pumped laser averaged over one section in longitudinal direction. Via the gain $g(\omega, \boldsymbol{r})$, the background absorption $\alpha_{bg}(\boldsymbol{r})$, and the refractive index $n_r(\omega, \boldsymbol{r})$, the dielectric function ε_{opt} depends on almost all properties of the device and its operating state, as well as on properties of the optical field, as its polarization and its frequency. As a consequence, the eigenvalues β as well as the eigenfunctions χ will parametrically depend on the carrier density distribution and on the temperature profile as well, both of which can change in time. By modeling this, we allow for the corresponding changes of the eigenvalues and of the optical field profile, but we do not allow for explicit time derivatives of β and χ throughout this chapter. This is at least due to the very different time scales of the optical and the electrothermal processes, the latter of which slave the optical processes.

In two-dimensional simulations with WIAS-TeSCA, longitudinal properties are only considered by assuming a longitudinally homogeneous power distribution, which is approximately met in Fabry–Perot lasers or in edge-emitting lasers with properly designed Bragg gratings [5]. In our calculations, the fundamental mode has been involved, the number of photons N_s of which is balanced by a corresponding photon rate equation:

$$\dot{N}_s = v_g(2\Im m\beta - \alpha_0 - \alpha_m)N_s + r^{sp}. \qquad (3.12)$$

In (3.12), $\dot{N}_s$ is a short notation for the time derivative of N_s, r^{sp} gives the spontaneous emission into the mode, α_0 are the longitudinal scattering losses, and:

$$\alpha_m = \frac{1}{2L}\ln\frac{1}{R_0 R_L} \qquad (3.13)$$

are the output losses for a Fabry–Perot laser with facet reflectivities R_0 at $z = 0$ and R_L at $z = L$. The energy density loss:

$$\gamma = \hbar\omega v_g(\alpha_0 + \alpha_m)|\chi|^2 N_s \qquad (3.14)$$

counts the radiation energy that is emitted from the device per time, either by the lasing mode through the end facets (α_m) or by the excitation of other modes (α_0).

---

[1] Equation (3.11) is formulated here for transverse electric (TE) polarization. Transverse magnetic (TM) polarization can be accounted for as well, without restrictions for the considerations here, but with a different version of (3.11).

The modal gain $2\Im m\beta$ is the imaginary part of the corresponding eigenvalue subject to (3.11). For the modal gain $2\Im m(\beta)$, an alternative expression can be obtained from (3.11) in first order of perturbation theory:

$$2\Im m(\beta) = \int (g - \alpha_{bg})|\chi|^2 d\Omega. \tag{3.15}$$

The photon rate equation (3.12) is valid for Fabry–Perot and Bragg resonators with a homogeneous longitudinal field distribution. The intensity distribution of the lasing mode along the transverse plane is readily obtained by multiplying N_s with the (suitable normalized) transverse intensity distribution $|\chi|^2$ of the lasing mode. The photon flow over the boundary of the transverse domain Ω is negligible as long as the lasing mode is a guided mode, which we will assume throughout this chapter.

3.3 Heating

Recently, the drift-diffusion model (3.3)–(3.5) has been coupled to a heat transport equation [2]. In this section, we want to motivate the heat transport equation and the resulting energy model by considering fundamental thermodynamic requirements. For the sake of simplicity, we restrict us here to the case of Boltzmann statistics.

3.3.1 Free Energy, Entropy, Energy

We define densities of free energy f, entropy s and energy u by:

$$f = \frac{\varepsilon}{2}|\nabla\varphi|^2 + c_L T(1 - \log T) + u_{rad} - Ts_{rad}$$
$$+ n[T(\log \frac{n}{N_c} - 1) + e_c] + p[T(\log \frac{p}{N_v} - 1) - e_v] \tag{3.16}$$

$$s = -\frac{\partial f}{\partial T} = nP_n + pP_p + c_L \log T + s_{rad} \tag{3.17}$$

$$u = f + Ts = \frac{\varepsilon}{2}|\nabla\varphi|^2 + c_L T + u_n n + u_p p + u_{rad}. \tag{3.18}$$

Here c_L is the lattice heat capacity, the prime ' means differentiation with respect to temperature T, and:

$$P_n = 1 + R_n - \log\left(\frac{n}{N_c}\right) - e'_c, \quad R_n = T(\log N_c)' \tag{3.19}$$

$$P_p = 1 + R_p - \log\left(\frac{p}{N_v}\right) + e'_v, \quad R_p = T(\log N_v)' \tag{3.20}$$

are the entropies per electron and hole, respectively. The energies per particle u_n and u_p are defined by:

$$u_n = T(R_n - 1) + e_c - Te'_c \qquad (3.21)$$
$$u_p = T(R_p - 1) - e_v + Te'_v \qquad (3.22)$$

for electrons and holes, respectively. For completeness, we note ($e_g = e_c - e_v$):

$$u_n + u_p = T(R_n + R_p) + e_g - Te'_g = T(P_n - 1) + T(P_p - 1) + (f_p - f_n). \quad (3.23)$$

Throughout this chapter, we assume R_n and R_p to be independent on T. The energy density u_{rad} and the entropy density s_{rad} of the optical field are given by (c.f. [6]):

$$u_{rad} = \int \rho_{rad} \hbar\omega |\chi|^2 N_s d\hbar\omega \qquad (3.24)$$

$$s_{rad} = k \int \rho_{rad} |\chi|^2 \left[(N_s + 1)\log(N_s + 1) - N_s \log(N_s)\right] d\hbar\omega, \quad (3.25)$$

where $\rho_{rad} = \frac{8\pi n_r^3 (\hbar\omega)^2}{h^3 c^3}$ is the density of states for the photons [7].

Then, we define free energy F, entropy S and energy U by:

$$F = \int_\Omega f \, d\Omega, \quad S = \int_\Omega s \, d\Omega, \quad U = \int_\Omega u \, d\Omega. \qquad (3.26)$$

To find equilibrium values for n, p, and T, we maximize the entropy S under the constraints:

$$Q = \int_\Omega (C + p - n) \, d\Omega = const., \quad U = const. \qquad (3.27)$$

Following Lagrange's method, this can be done by maximizing the augmented entropy:

$$S_\lambda = S + \lambda Q + \lambda_3 U. \qquad (3.28)$$

The resulting Euler–Lagrange equations read as:

$$\lambda_3 = -\frac{1}{T}, \quad q_{n,p}\lambda = \partial_{n,p}\left[S - \frac{U}{T}\right], \quad q_p = 1 = -q_n, \quad \log\left(\frac{N_s + 1}{N_s}\right) = \frac{\hbar\omega}{T},$$

with constant Lagrange multipliers λ and λ_3. Solving these equations for n, p, and N_s, we arrive at the Bose distribution for the photons[2]:

---

[2] In [8], a generalization of (3.29) can be found. There, n_s is the average optical-mode occupation factor of photons that are in (a hypothetical) equilibrium with a biased semiconductor, where the carriers separated by $f_p - f_n$. Such a hypothetical equilibrium can only occur if the photons are associated with an optical cavity that is closed and loss-free, except for optical transitions between conduction and valence band [8].

$$n_s = \frac{1}{1 - \exp\left(\frac{\hbar\omega}{T}\right)} \quad (3.29)$$

and the state equations (3.8) and (3.9) but according to Boltzmann statistics:

$$n = N_c \exp\left(\frac{\varphi - f_n - e_c}{T}\right), \quad p = N_v \exp\left(\frac{e_v + f_p - \varphi}{T}\right) \quad (3.30)$$

and with coinciding constant equilibrium quasi-Fermi potentials defined by [c.f. (3.8) and (3.9)]:

$$\frac{f_n}{T} = \frac{f_p}{T} = \lambda. \quad (3.31)$$

3.3.2 Current Densities

According to the principle of local thermal equilibrium, we assume the state equations (3.8) and (3.9), respectively, (3.30) to hold also in the case of different, nonconstant quasi-Fermi potentials f_n, f_p and nonhomogeneous temperature T. Moreover, we suppose the vector $\nabla\boldsymbol{\lambda} = (\nabla\lambda_1, \nabla\lambda_2, \nabla\lambda_3)$ of gradients of Lagrange multipliers:

$$\lambda_1 = \frac{f_n}{T}, \quad \lambda_2 = \frac{f_p}{T}, \quad \lambda_3 = -\frac{1}{T} \quad (3.32)$$

to be the driving force toward equilibrium; i.e., we make the ansatz:

$$\boldsymbol{j} = (0, 0, \varphi \dot{\boldsymbol{D}}) - L\,\nabla\boldsymbol{\lambda}, \quad \boldsymbol{j} = (\boldsymbol{j}_n, \boldsymbol{j}_p, \boldsymbol{j}_u), \quad (3.33)$$

with a (3×3) conductivity-matrix L, which has to be positively definite and symmetric in view of the second law of thermodynamics and Onsager's reciprocity relations. $\dot{\boldsymbol{D}} = -\varepsilon\nabla\varphi_t$ is the electric displacement current density.[3] We specify L such that (3.33) becomes:

$$\boldsymbol{j}_n = T\Big[-(\sigma_n + \sigma_{np})\nabla\lambda_1 + \sigma_{np}\nabla\lambda_2 + [\sigma_n(P_n - \lambda_1) - \sigma_{np}(\lambda_1 - \lambda_2)]\nabla\lambda_3\Big]$$

$$\boldsymbol{j}_p = T\Big[-(\sigma_p + \sigma_{np})\nabla\lambda_2 + \sigma_{np}\nabla\lambda_1 - [\sigma_p(P_p + \lambda_2) + \sigma_{np}(\lambda_1 - \lambda_2)]\nabla\lambda_3\Big]$$

$$\boldsymbol{j}_u = -\varphi\varepsilon\nabla\varphi_t - T^2[\kappa_L\nabla T + (P_n - \lambda_1)^2 + (P_p + \lambda_2)^2 + \sigma_{np}(\lambda_1 - \lambda_2)^2]\nabla\lambda_3$$
$$+ T[\sigma_n(P_n - \lambda_1) - \sigma_{np}(\lambda_1 - \lambda_2)]\nabla\lambda_1$$
$$- T[\sigma_p(P_p + \lambda_2) - \sigma_{np}(\lambda_2 - \lambda_1)]\nabla\lambda_2, \quad (3.34)$$

where:

---

[3] Note that due to perfect optical waveguiding, the optical energy flow is only along the z axis and hence does not occur in the flow (3.33) along the transverse plane.

$$\sigma_n = a\, n\mu_n, \quad \sigma_p = a\, p\mu_p, \quad \sigma_{np} = \sigma_{pn} = a\, np\mu_n\mu_p\, b \qquad (3.35)$$

$$b = \frac{c_1[1 + c_2(n+p)/2]}{1 + c_3(n+p)/2}, \quad a = \frac{1}{1 + (p\mu_n + n\mu_p)b}, \qquad (3.36)$$

and $\mu_n, \mu_p, c_1, c_2, c_3$, and κ_L are material parameters (see [9] and [10]). It is worth noting, that the appearance of $\varphi\varepsilon\nabla\varphi_t$ in $\boldsymbol{j}_u$ is a unique feature of our model. Replacing $\nabla\lambda$ by the more familiar vector $(\nabla f_n, \nabla f_p, \nabla T)$, we can rewrite the current densities as:

$$\boldsymbol{j}_n = -\sigma_n(\nabla f_n - P_n\nabla T) + \sigma_{np}[\nabla(f_p - f_n) + (P_n + P_p)\nabla T] \qquad (3.37)$$
$$\boldsymbol{j}_p = -\sigma_p(\nabla f_p + P_p\nabla T) + \sigma_{pn}[\nabla(f_n - f_p) - (P_n + P_p)\nabla T] \qquad (3.38)$$
$$\boldsymbol{j}_u = -\varphi\varepsilon\nabla\varphi_t - \kappa_L\nabla T - (P_n T - f_n)\boldsymbol{j}_n + (P_p T + f_p)\boldsymbol{j}_p. \qquad (3.39)$$

For later use, we define here also the temperature current density $\boldsymbol{j}_T$ and the heat current density $\boldsymbol{j}_q$ by:

$$\boldsymbol{j}_T = -\kappa_L\nabla T - T(P_n\boldsymbol{j}_n - P_p\boldsymbol{j}_p) \qquad (3.40)$$
$$\boldsymbol{j}_q = -\kappa_L\nabla T. \qquad (3.41)$$

3.3.3 Heat Equation

Now we can transform the energy balance equation (3.1) into a heat flow equation. With the heat capacity:

$$c_h = c_L + nu'_n + pu'_p, \qquad (3.42)$$

we find by using the current conservation equation (3.7) and the definition of the energy u:

$$\begin{aligned}\frac{\partial u}{\partial t} &= \varepsilon\nabla\varphi\cdot\nabla\varphi_t + c_h T_t + n_t u_n + p_t u_p + \partial_t u_{rad} \\ &= \varepsilon\nabla\varphi\cdot\nabla\varphi_t + \varphi(\nabla\cdot(\varepsilon\nabla\varphi_t) + p_t - n_t) + c_h T_t + n_t u_n + p_t u_p + \partial_t u_{rad} \\ &= \nabla\cdot(\varphi\varepsilon\nabla\varphi_t) + c_h T_t + n_t(u_n - \varphi) + p_t(u_p + \varphi) + \partial_t u_{rad}.\end{aligned} \qquad (3.43)$$

We note:

$$u_n - \varphi = T(P_n - 1) - f_n \qquad (3.44)$$
$$u_p + \varphi = T(P_p - 1) + f_p. \qquad (3.45)$$

We insert now (3.43) and (3.39) in the energy balance equation (3.1):

$$c_h T_t + n_t(u_n - \varphi) + p_t(u_p + \varphi) + \partial_t u_{rad} + \nabla\cdot\boldsymbol{j}_T + \nabla\cdot(f_n\boldsymbol{j}_n + f_p\boldsymbol{j}_p) = -\gamma. \qquad (3.46)$$

Hence:

$$\begin{aligned}
c_h T_t + \nabla \cdot \boldsymbol{j}_T &= -n_t(u_n - \varphi) - p_t(u_p + \varphi) - \partial_t u_{rad} - \nabla \cdot (f_n \boldsymbol{j}_n + f_p \boldsymbol{j}_p) - \gamma \\
&= -(u_n - \varphi)(\nabla \cdot \boldsymbol{j}_n + R) + (u_p + \varphi)(\nabla \cdot \boldsymbol{j}_p + R) \\
&\quad -\partial_t u_{rad} - \nabla \cdot (f_n \boldsymbol{j}_n + f_p \boldsymbol{j}_p) - \gamma \\
&= (u_n + u_p)R - \partial_t u_{rad} - \gamma - TP_n \nabla \cdot \boldsymbol{j}_n + TP_p \nabla \cdot \boldsymbol{j}_p \\
&\quad + T(\nabla \cdot \boldsymbol{j}_n - \nabla \cdot \boldsymbol{j}_p) - \boldsymbol{j}_n \cdot \nabla f_n - \boldsymbol{j}_p \cdot \nabla f_p \\
&= (u_n + u_p)R - \partial_t u_{rad} - \gamma - TP_n \nabla \cdot \boldsymbol{j}_n + TP_p \nabla \cdot \boldsymbol{j}_p \\
&\quad + T \nabla \cdot (\boldsymbol{j}_n - \boldsymbol{j}_p) - \boldsymbol{j}_n \cdot (\nabla f_n - P_n \nabla T) \\
&\quad - \boldsymbol{j}_p \cdot (\nabla f_p + P_p \nabla T) - P_n \nabla T \cdot \boldsymbol{j}_n - P_p \nabla T \cdot \boldsymbol{j}_p \, .
\end{aligned}$$

Replacing now $\boldsymbol{j}_T$ by $\boldsymbol{j}_q$ (3.41), we find:

$$\begin{aligned}
c_h T_t + \nabla \cdot \boldsymbol{j}_q &= \nabla \cdot T(P_n \boldsymbol{j}_n - P_p \boldsymbol{j}_p) - TP_n \nabla \cdot \boldsymbol{j}_n - P_n \nabla T \cdot \boldsymbol{j}_n \\
&\quad + TP_p \nabla \cdot \boldsymbol{j}_p + P_p \nabla T \cdot \boldsymbol{j}_p + T \nabla \cdot (\boldsymbol{j}_n - \boldsymbol{j}_p) \\
&\quad + (u_n + u_p) R - \partial_t u_{rad} - \gamma \\
&\quad - \boldsymbol{j}_n \cdot (\nabla f_n - P_n \nabla T) - \boldsymbol{j}_p \cdot (\nabla f_p + P_p \nabla T)
\end{aligned}$$

and by straightforward calculation finally the desired heat flow equation (3.2):

$$c_h T_t - \nabla \cdot \kappa_L \nabla T = H \tag{3.47}$$

with the heat source term:

$$\begin{aligned}
H &= T \nabla P_n \cdot \boldsymbol{j}_n - T \nabla P_p \cdot \boldsymbol{j}_p + T \nabla \cdot (\boldsymbol{j}_n - \boldsymbol{j}_p) \\
&\quad - \boldsymbol{j}_n \cdot (\nabla f_n - P_n \nabla T) - \boldsymbol{j}_p \cdot (\nabla f_p + P_p \nabla T) \\
&\quad + (u_n + u_p) R - \partial_t u_{rad} - \gamma.
\end{aligned} \tag{3.48}$$

The first two terms on the right-hand-side of (3.48) represent the Thomson–Peltier heat, which can be positive and negative as well. The term in the second row of (3.48) is the Joule heating which is strictly positive. The first term in the last row of (3.48) is the recombination heat. Each recombining electron-hole pair sets free its energy $u_n + u_p$, which is immediately transferred to the lattice, unless it is transferred to the radiation field. This latter radiative part does not heat the device and has therefore to be subtracted from the heat source, as indicated by the appearance of $-\partial_t u_{rad}$ in (3.48). Similarly, γ counts the total energy loss from our system [see (3.1)], which cools down the device too. So far, in the stationary case ($\partial_t D = 0$), our model does not differ from [1] and [11]. Concerning the nonstationary case, there are differences to [1] and [11], with respect to the electrostatic potential φ and with respect to the definition of u and $\boldsymbol{j}_u$, which is required by the conservation of energy.

In particular, H contains contributions H_{rad} from radiative processes:

$$H_{rad} = [R^{sp} + R^{stim}](u_n + u_p) - \partial_t u_{rad} - \gamma \, , \tag{3.49}$$

which have to be modeled in the following. Thereby we will restrict our consideration to the lasing mode, because it gives the main contribution above threshold. For this, we note:

$$T\partial_t s_{rad} = \partial_t u_{rad} = \hbar\omega|\chi|^2 \dot{N}_s. \tag{3.50}$$

For the lasing mode, the energy balance (3.50) is governed by the global equation (3.12). To get the local heat sources, we have to model the localization of $\dot{N}_s$ in the r-plane. For this purpose, we assume that the carriers generated by absorption of the lasing field transfer their energy to the lattice directly, without traveling. We use (3.15) and remove the integral $\int d\Omega$ from all terms to find in this approximation

$$\partial_t u_{rad} = \hbar\omega v_g(g - \alpha_{bg})|\chi|^2 N_s - \gamma + \hbar\omega|\chi|^2 r^{sp}, \tag{3.51}$$

for the energy density averaged along the laser axis.[4] Then, we obtain:

$$H_{rad} = \hbar\omega|\chi|^2 v_g \alpha_{bg} N_s - v_g g|\chi|^2 N_s \left[\hbar\omega - (u_p + u_n)\right] \tag{3.52}$$
$$- [\hbar\omega r^{sp} - (u_p + u_n)R^{sp}],$$

which is the net heat source caused by the lasing mode. The first terms in (3.52) describe the background absorption of radiation. It is a strictly positive contribution that dominates H_{rad} above the laser threshold. The second term in (3.52) is caused by a possible incomplete energy transfer from the carrier ensemble to the radiation field during stimulated processes, which heats the device. The last term deals with the spontaneous emission and can only be discussed reasonably together with the complete radiation field, i.e., including the incoherent radiation field too, which we cannot calculate at this time. In our simulations, we have dropped the two last terms of (3.52) for simplicity.[5]

3.3.4 Entropy Balance

We first note the following relations between the entropies of the electron- and hole subsystem $s_n = nP_p$, $s_p = pP_p$ and the specific entropies per particle P_n, P_p:

---

[4] The change $\partial_t u_{rad}$ of the energy density of the total radiation field might be the sum over the balances (3.51) for all modes, at least in the case when they are uncorrelated.

[5] This corresponds to the assumption that all radiative recombination processes directly create/annihilate photons, without energy loss. However, also the absorption of the incoherent field would heat the device, but its calculation remains an open problem here.

3 Fabry–Perot Lasers: Thermodynamics-Based Modeling

$$\frac{\partial s_n}{\partial n} = P_n + n\frac{\partial P_n}{\partial n} = P_n - n\frac{1}{n} = P_n - 1 \quad (3.53)$$

$$\frac{\partial s_p}{\partial p} = P_p + p\frac{\partial P_p}{\partial p} = P_p - p\frac{1}{p} = P_p - 1 \quad (3.54)$$

$$\frac{\partial (s_n + s_p)}{\partial T} = n\frac{\partial P_n}{\partial T} + p\frac{\partial P_p}{\partial T} = n(\frac{R_n}{T} - e_c'') + p(\frac{R_p}{T} + e_v'')$$

$$= \frac{n}{T}\frac{\partial u_n}{\partial T} + \frac{p}{T}\frac{\partial u_p}{\partial T} \,. \quad (3.55)$$

Equipped with that, we differentiate now the entropy density (3.17) with respect to time:

$$\frac{\partial s}{\partial t} = c_L T_t/T + n_t(P_n - 1) + p_t(P_p - 1) + \left(\frac{n}{T}\frac{\partial u_n}{\partial T} + \frac{p}{T}\frac{\partial u_p}{\partial T}\right) T_t + \partial_t s_{rad}$$

$$= n_t(P_n - 1) + p_t(P_p - 1) + c_h T_t/T + \partial_t s_{rad}$$

$$= n_t(P_n - 1) + p_t(P_p - 1) + \frac{1}{T}\nabla \cdot \kappa_L \nabla T + H/T + \partial_t s_{rad}$$

$$= (P_n - 1)(\nabla \cdot \boldsymbol{j}_n - R) - (P_p - 1)(\nabla \cdot \boldsymbol{j}_p + R) + \frac{1}{T}\nabla \cdot \kappa_L \nabla T$$
$$+ \nabla P_n \cdot \boldsymbol{j}_n - \nabla P_p \cdot \boldsymbol{j}_p + \nabla \cdot (\boldsymbol{j}_n - \boldsymbol{j}_p)$$
$$- \frac{1}{T}\left(\boldsymbol{j}_n \cdot (\nabla f_n - P_n \nabla T) - \boldsymbol{j}_p \cdot (\nabla f_p + P_p \nabla T)\right)$$
$$+ \frac{1}{T}(u_n + u_p)R + \partial_t s_{rad} - \frac{1}{T}\partial_t u_{rad} - \gamma/T$$

$$= \nabla \cdot (P_n \boldsymbol{j}_n - P_p \boldsymbol{j}_p) + \frac{1}{T}\nabla \cdot \kappa_L \nabla T$$
$$- \frac{1}{T}\left(\boldsymbol{j}_n \cdot (\nabla f_n - P_n \nabla T) - \boldsymbol{j}_p \cdot (\nabla f_p + P_p \nabla T)\right)$$
$$+ \left(\frac{u_n + u_p}{T} - P_n + 1 - P_p + 1\right) R + \partial_t s_{rad} - \frac{1}{T}\partial_t u_{rad} - \gamma/T$$

$$= \nabla \cdot \left(\frac{\kappa_L}{T}\nabla T + P_n \boldsymbol{j}_n - P_p \boldsymbol{j}_p\right) + \frac{1}{T^2} \cdot \kappa_L |\nabla T|^2$$
$$- \frac{1}{T}\left(\boldsymbol{j}_n \cdot (\nabla f_n - P_n \nabla T) + \boldsymbol{j}_p \cdot (\nabla f_p + P_p \nabla T)\right)$$
$$+ (f_p - f_n)R + \partial_t s_{rad} - \frac{1}{T}\partial_t u_{rad} - \gamma/T, \quad (3.56)$$

where we have used (3.23) and the identity:

$$\frac{1}{T}\nabla \cdot \kappa_L \nabla T = \nabla \cdot \frac{\kappa_L}{T}\nabla T + \frac{1}{T^2} \cdot \kappa_L |\nabla T|^2. \quad (3.57)$$

Equation (3.56) is a continuity equation for the entropy density s:

$$\frac{\partial s}{\partial t} + \nabla \cdot \boldsymbol{j}_s = d/T, \qquad (3.58)$$

with the entropy current density $\boldsymbol{j}_s = \boldsymbol{j}_T/T$ defined by (3.40) and the dissipation rate d:

$$d = \frac{\kappa_L}{T}|\nabla T|^2 + \sigma_n|\nabla f_n - P_n \nabla T|^2 + \sigma_p|\nabla f_p + P_p \nabla T|^2$$
$$+ \sigma_{np}|\nabla(f_n - f_p) - (P_n + P_p)\nabla T|^2 + (f_p - f_n)R - \gamma. \qquad (3.59)$$

The dissipation rate d in (3.59) is always positive for a device that is isolated from the outside world ($\gamma = 0$). Therefore, by partial integration of (3.58) and supposing no-flux boundary conditions and $\gamma = 0$, it follows, according to the second law of thermodynamics,

$$\frac{dS}{dt} = \int_\Omega \frac{ds}{dt}\, d\Omega = \int_\Omega \frac{d}{T}\, d\Omega \geq 0. \qquad (3.60)$$

In conclusion, it is the main achievement of this work that we are able to proof the thermodynamic correctness of our model in view of the second law of thermodynamics (3.60).

3.4 Boundary Conditions

Let $\Gamma = \partial \Omega$ be the boundary of the transverse cross section Ω of the device under consideration, and let $\boldsymbol{\nu}$ be the normal unit vector at any point $r_\Gamma \in \Gamma$. In order to describe nonequilibrium situations, we have to complete the system (3.3), (3.4), (3.5), (3.47), and (3.11) of coupled nonlinear partial differential equations by boundary conditions. To include homogeneous and nonhomogeneous as well Dirichlet- as Neumann-conditions, we choose the form:

$$\begin{aligned}
\boldsymbol{\nu} \cdot (\varepsilon \nabla \varphi) + \alpha_\varphi(\varphi - \varphi_\Gamma) &= 0 \\
\boldsymbol{\nu} \cdot \boldsymbol{j}_n + \alpha_n(n - n_\Gamma) &= 0 \\
\boldsymbol{\nu} \cdot \boldsymbol{j}_p - \alpha_p(p - p_\Gamma) &= 0 \qquad (3.61) \\
\boldsymbol{\nu} \cdot \boldsymbol{j}_q + \alpha_T(T - T_\Gamma) &= 0 \\
\boldsymbol{\nu} \cdot \nabla \chi + \alpha_\chi(\chi - \chi_\Gamma) &= 0,
\end{aligned}$$

where $\alpha_\varphi, \alpha_n, \alpha_p, \alpha_T, \alpha_\chi \in (\Gamma \to \mathbb{R}^1)$, and $\varphi_\Gamma, n_\Gamma, p_\Gamma, T_\Gamma, \chi_\Gamma \in (\Gamma \to \mathbb{R}^1)$ are given coefficients and boundary values, respectively. By choosing α very small, one can manage homogeneous Neumann-conditions, which means that there will be no flux over the boundary of the domain. This applies, e.g., for the temperature at the surface-to-the-air contacts of a device. On the other hand, by choosing α very large, one can manage Dirichlet-conditions, which fix the quantity to the value given at the boundary. The latter applies, e.g., for the temperature at the heat-sink contact of the device. Finite values of α account for finite penetration depth's of the quantity into the outside world. In some situations, for instance, to describe the interaction of the device with outer circuits, it is appropriate to replace the boundary condition for the Poisson equation by a (possibly nonlocal) boundary condition for the current conservation equation (3.7).

3.5 Discretization

3.5.1 Time Discretization

In order to maintain a Clausius–Duhem relation like (3.60) and to get numerically stable algorithms, we use an Euler-backward time discretization scheme. Accordingly, we replace the time derivatives in (3.4), (3.5), and (3.47) by backward difference quotients, which means:

$$w_\tau(t, \boldsymbol{r}) = \frac{1}{\tau}(w(t, \boldsymbol{r}) - w(t - \tau, \boldsymbol{r})), \qquad (3.62)$$

where t is the time, $\tau > 0$ is the (backward) timestep, and $\boldsymbol{r} = (x, y) \in \Omega$ is a space point. The remaining terms in these equations as well as in (3.3) and (3.11) are taken at the time level t.

3.5.2 Space Discretization

For the spatial discretization of the system (3.3), (3.4), (3.5), (3.47), (3.11) completed by the boundary conditions (3.61), we apply the finite volume method. Accordingly to that method, we suppose to have triangulation $\{E_l, \ l = 1, ..., n_e\}$, such that:

$$\Omega = \cup_{l=1}^{n_e} E_l \ .$$

Let n_v be the number of vertices of that triangulation, i.e., the number of elements of the set $\{v_i \ : \ E_l = (v_i, v_j, v_k), \ 1 \leq l \leq n_e\}$. We assign to each vertex $v_i = (x_i, y_i)$ its Voronoi volume V_i and its Voronoi surface ∂V_i defined by:

$$V_i = \{\boldsymbol{r} \in \mathbb{R}^2 \ : \ |\boldsymbol{r} - v_i| < |\boldsymbol{r} - v_j| \, , v_j \in \Omega\}, \qquad \partial V_i = \bar{V}_i \setminus V_i \ .$$

Further, we denote by $\boldsymbol{\nu}_i$ the normal unit vector with respect to ∂V_i and by $|V_i|$ the measure of V_i.
Now we integrate the time-discretized versions of (3.3), (3.4), (3.5), (3.47), and (3.11). Lumping terms w without derivatives, i.e.:

$$\int_{V_i} w \, d\Omega = w_i \, |V_i| \, , \qquad w_i = w(v_i) \, , \qquad (3.63)$$

and applying the Gauss-theorem:

$$\int_{V_i} \nabla \cdot \boldsymbol{w}\, d\Omega = \int_{\partial V_i} \boldsymbol{\nu}_i \cdot \boldsymbol{w}\, d\Gamma \tag{3.64}$$

to divergence terms, we get for $1 \leq i \leq n_e$:

$$-\int_{\partial V_i} \boldsymbol{\nu}_i \cdot (\varepsilon \nabla \varphi)\, d\Gamma + \int_{\partial V_i \cap \Gamma} \alpha_\varphi(\varphi - \varphi_\Gamma)\, d\Gamma = [C + p - n]_i |V_i| \tag{3.65}$$

$$n_{\tau i} |V_i| - \int_{\partial V_i} \boldsymbol{\nu}_i \cdot \boldsymbol{j}_n\, d\Gamma + \int_{\partial V_i \cap \Gamma} \alpha_n(n - n_\Gamma)\, d\Gamma = -[R]_i |V_i| \tag{3.66}$$

$$p_{\tau i} |V_i| + \int_{\partial V_i} \boldsymbol{\nu}_i \cdot \boldsymbol{j}_p\, d\Gamma + \int_{\partial V_i \cap \Gamma} \alpha_p(p - p_\Gamma)\, d\Gamma = -[R]_i |V_i| \tag{3.67}$$

$$[c_h T_\tau]_i |V_i| + \int_{\partial V_i} \boldsymbol{\nu}_i \cdot \boldsymbol{j}_q\, d\Gamma + \int_{\partial V_i \cap \Gamma} \alpha_T(T - T_\Gamma)\, d\Gamma = H_i |V_i| \tag{3.68}$$

$$\int_{\partial V_i} \boldsymbol{\nu}_i \cdot \nabla \chi\, d\Gamma - \int_{\partial V_i \cap \Gamma} \alpha_\chi(\chi - \chi_\Gamma)\, d\Gamma$$
$$+[\frac{\omega^2}{c^2}\varepsilon_{opt}(\omega, v_i) - \beta^2]\chi_i |V_i| = 0 . \tag{3.69}$$

In (3.68), the term H_i on the right-hand-side contains the divergence expression $T \nabla \cdot (\boldsymbol{j}_n - \boldsymbol{j}_p)$, which we shift to the left-hand-side by partial integration. We obtain:

$$[c_h T_\tau]_i |V_i| - \int_{\Gamma_i} \boldsymbol{\nu}_i \cdot [\kappa_L \nabla T + T(\boldsymbol{j}_n - \boldsymbol{j}_p)]\, d\Gamma +$$
$$\int_{\partial V_i \cap \Gamma} [\alpha_T(T - T_\Gamma) - T\alpha_n(n - n_\Gamma) - T\alpha_p(p - p_\Gamma)]\, d\Gamma = H_{T_i} |V_i|, \tag{3.70}$$

with:

$$H_T = H - \nabla \cdot \left(T(\boldsymbol{j}_n - \boldsymbol{j}_p) \right) \tag{3.71}$$

where H is defined by (3.48).

3.5.3 Discretization of the Currents

In order to transform the system (3.65), (3.66), (3.67), (3.70), (3.69) into a time- and space-discretized form, it remains to discretize the currents $\boldsymbol{j}_n, \boldsymbol{j}_p, \boldsymbol{j}_T$ and the gradient operator. We can restrict us to demonstrate our approach for $\boldsymbol{j}_n$ with $\sigma_{np} = 0$. We consider three cases:

(1) We start with the case of Boltzmann statistics and homogeneous temperature $T = T_0$. Then the state, respective current equation for electrons can be rewritten as:

$$n = N_0 \exp \frac{\varphi + \psi_0 - f_n}{T_0}; \quad \boldsymbol{j}_n = T_0 \mu_n N_0 \left[\nabla \frac{n}{N_0} - \frac{n}{N_0} \nabla \frac{\varphi + \psi_0}{T_0} \right] \quad (3.72)$$

where:

$$N_0 = \sqrt{N_c N_v} \exp \frac{e_v - e_c}{2T_0}, \quad \psi_0 = \frac{\log \frac{N_c}{N_v} - e_c - e_v}{2}.$$

For this situation, the well-tried discretization schema due to Scharfetter–Gummel [12]:

$$\int_{\partial V_i} \boldsymbol{\nu}_i \cdot \boldsymbol{j}_n \, d\Gamma =$$

$$\sum_{l \,:\, v_i \in E_l} \left[a_j \left(\frac{n_j}{N_{0_j}} B(s_j) - \frac{n_i}{N_{0_i}} B(-s_j) \right) + a_k \left(\frac{n_k}{N_{0_k}} B(s_k) - \frac{n_i}{N_{0_i}} B(-s_k) \right) \right]$$

can be used. Here $B(s) = \frac{s}{e^s + 1}$ is the Bernoulli-function, (v_i, v_j, v_k) are the vertices of the element E_l, and the quantities:

$$a_i = \frac{1}{4|E_l|}(|v_i - v_j|^2 + |v_i - v_k|^2 - |v_j - v_k|^2), \quad s_i = \frac{(\varphi + \psi_0)_k - (\varphi + \psi_0)_j}{T_0},$$

are to be cyclically exchanged.

(2) Next we consider the case of Fermi–Dirac statistics and again homogeneous temperature T_0. In order to reduce (2) to (1) we replace N_0, ψ_0 in (3.72) by:

$$N = N_0 \sqrt{\gamma_n \gamma_p}, \quad \psi = \psi_0 + 0.5 \log \frac{\gamma_n}{\gamma_p},$$

where:

$$\gamma_n = \frac{\mathcal{F}_{1/2}(\frac{c_n}{T_0})}{\exp \frac{c_n}{T_0}}, \quad c_n = \varphi - f_n - e_c, \quad \gamma_p = \frac{\mathcal{F}_{1/2}(\frac{c_p}{T_0})}{\exp \frac{c_p}{T_0}}, \quad c_p = e_v + f_p - \varphi.$$

The correction factors γ_n and γ_p from Boltzmann to Fermi–Dirac statistics have to be updated during the iteration procedure, which is needed for solving the nonlinear Euler-backward system.

(3) Finally, we consider the most involved case of Fermi–Dirac statistics and nonhomogeneous temperature T. We define now:

$$N = N_0 \sqrt{\gamma_n \gamma_p \exp\left[(e_v - e_c)(1/T_0 - 1)\right]}$$
$$\psi = \psi_0 + 0.5[\log \frac{\gamma_n}{\gamma_p} + (e_c + e_v)(1 - 1/T_0)], \quad (3.73)$$

where:

$$T_0 = \min_{1 \leq i \leq n_v} T_i$$

$$\gamma_n = \frac{\mathcal{F}_{1/2}(\frac{c_n}{T})}{\exp\frac{c_n}{T_0}} \qquad c_n = \varphi - f_n - e_c$$

$$\gamma_p = \frac{\mathcal{F}_{1/2}(\frac{c_p}{T})}{\exp\frac{c_p}{T_0}} \qquad c_p = e_v + f_p - \varphi \,.$$

Replacing N_0, ψ_0 in (3.72) by N, ψ from (3.73), we can proceed as in (1). However, as in (2), the correction factors γ_n and γ_p have to be updated iteratively.

3.6 Solution of the Discretized Equations

3.6.1 Decoupling, Linearization

The equations (3.65), (3.66), (3.67), (3.70), and (3.69) form in view of the current discretization approach, described in the last subsection, a coupled system of nonlinear algebraic equations. We solve this system timestep by timestep and start with initial values:

$$\varphi(0, v_i),\ n(0, v_i),\ p(0, v_i),\ T(0, v_i)\,, \chi(0, v_i), \qquad i = 1, n_v \,,$$

such that the discrete Poisson-equation (3.65) and the waveguide equation (3.69) are satisfied. To calculate a first iterate of:

$$\varphi(\tau, v_i),\ n(\tau, v_i),\ p(\tau, v_i),\ T(\tau, v_i)\,, \chi(0, v_i), \qquad i = 1, n_v \,, \qquad \tau > 0 \,,$$

we start solving the continuity equations (3.66) and (3.67) to update the quasi-Fermi potentials f_n and f_p. Next we calculate the potential φ by using the discretized, with respect to φ linearized current conservation equation. We repeat this iteration until the numerical defects of the equations are sufficiently small. Then we solve (3.69) and (3.70) and pass over to the next timestep. By "freezing" the material coefficients depending on the solution at the foregoing timestep, we solve during that iteration process only decoupled linear algebraic equations.

3.6.2 Solution of Linear Algebraic Equations

For solving the decoupled linear algebraic equations, we apply sparse matrix techniques. Hereby new factorizations are made as rare as possible.

3.7 Example

In the modeling of semiconductor lasers, combinations of several effects need to be considered to explain experimental results [13], [14]. Moreover, experimental findings have to be used for the calibration of simulation parameters, which are often not well known. To decrease the number of uncertain simulation parameters, we have separately simulated the bandstructure of the strained multi-quantum well (MQW) active region by eight-band kp calculations [4]. In particular, we have computed the optical response function as well as other properties of the strained MQW active region of III-V semiconductor lasers by kp calculations with WIAS-QW [15]. Results from these calculations entered the device simulations with WIAS-TeSCA in terms of net coefficients, which allowed us to treat the quantum wells like classic materials with specific material parameters [4].

We consider a structure similar to that described in [16]. It is a long-wavelength InP-based (substrate) ridge waveguide semiconductor laser. The active region is a $Ga_xIn_{1-x}As_yP_{1-y}$ strained MQW structure that is designed for emission at 1.55 μm. The structure consists of six 1% compressively strained 7-nm-thick quantum wells ($x = 0.239, y = 0.826$), which are separated by 10-nm-thick 0.3% tensile-strained barriers ($x = 0.291, y = 0.539$).

The active region is sandwiched between two undoped $Ga_xIn_{1-x}As_yP_{1-y}$ ($x = 0.1732, y = 0.3781$) waveguide layers. The lower waveguide layer (n-side) has a thickness of 250 nm, whereas the upper waveguide layer (p-side) has a thickness of 180 nm.

A part of the transverse cross section is shown in Fig. 3.1. Note that the lateral coordinate x and the vertical coordinate y are not equally scaled. The quantum well structure is centered in $y = 0$ (dark gray), and the upper and lower waveguide layers are highlighted (some lighter gray).

The ridge has a width of 2.4 μm and is highly p-doped ($\approx 10^{18}/cm^3$), where the p-doping increases up to $\approx 10^{20}/cm^3$ close to the p-contact. The layers below the lower waveguide are n-doped with nominally $10^{18}/cm^3$.

At the contacts, we allowed for a heat-flow (see Fig. 3.1), no thermal isolation of the device has been assumed. The thermal resistance of the top contact is low. Besides the top of the ridge, the p-contact has been passivated by a 200 nm SiN_x layer that is also shown in Fig. 3.1.

The length L of the device was 400 μm, and a longitudinal scattering loss $\alpha = 10/cm$ has been assumed as well as reflectivities of $R_0 = R_L = 0.3$ at the end facets of the laser.

3.7.1 Stationary Characteristics

Throughout all of the calculations for this structure, single-mode emission has been observed. The second transverse mode shown in Fig.3.1 has been calculated for control but it never reached lasing threshold. Therefore, its power could be neglected compared with that of the lasing mode.

Figure 3.2 displays the calculated P-I characteristics of the device. The influence of self-heating can be detected by the curvature of the P-I curves, which becomes significant in the region above 100 mA. The ambient temperature influences the device characteristics too, and its impact is indicated by the different lines in Fig. 3.2, where a significant thermal rollover for high

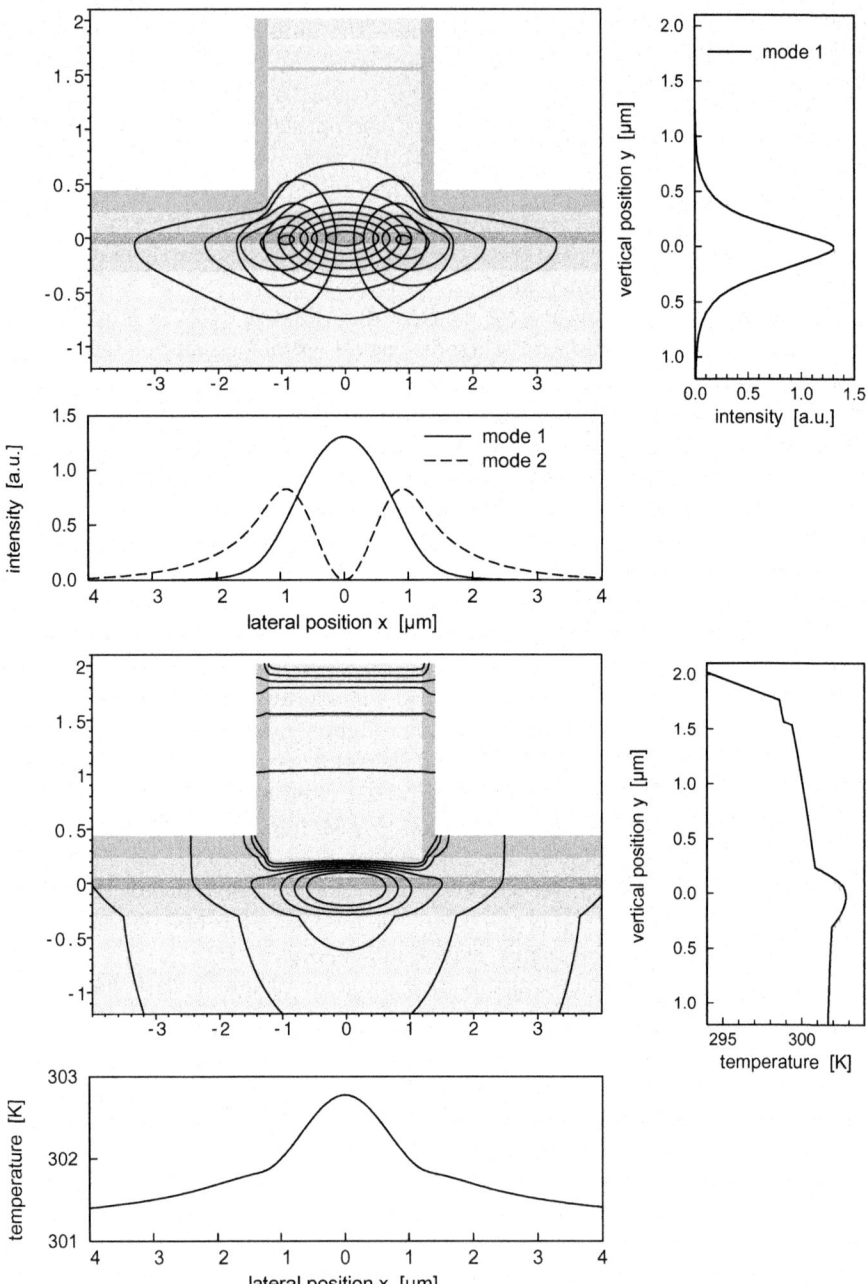

Fig. 3.1. Simulation results for given injection current 150 mA and ambient temperature 293 K. We show isolines in a part of the cross section, the lateral distribution for $y = 0$ (in the middle of the QW structure) and the vertical distribution for $x = 0$ (in the middle of the ridge) for the intensity pattern of the first and second optical mode (top) as well as for the device temperature (bottom).

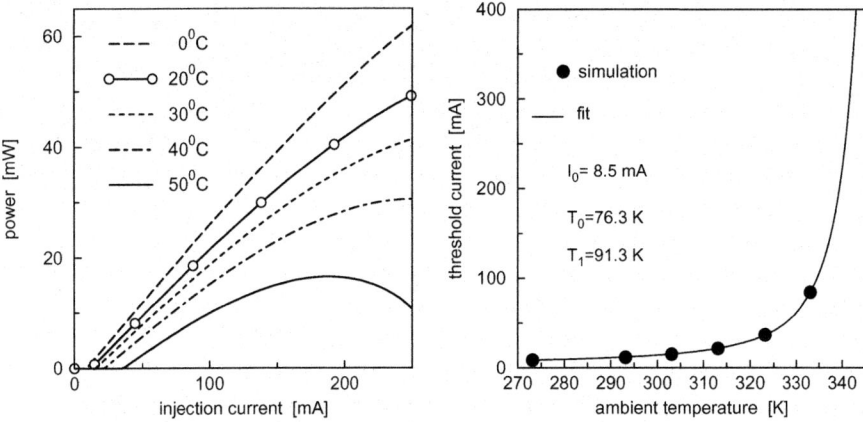

Fig. 3.2. Left: PI-characteristics of the device for different ambient temperatures. Above 50°C, a significant thermal rollover occurs. Right: Dependence of the threshold current on the ambient temperature for the above device. Points: Calculation with WIAS-TeSCA, line: fit according to the modified T_0-rule (3.74).

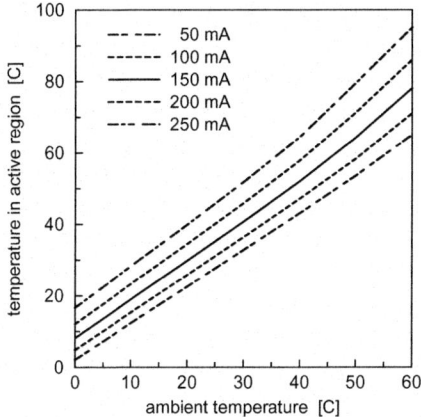

Fig. 3.3. Maximum temperature (in the active region) vs. ambient temperature for different injection currents.

ambient temperatures occurs. The threshold current grows roughly exponentially with the ambient temperature T, for which we have found a fit by:

$$I_{th}(T) = I_0 \cdot \exp\left((T - T^*)/(T_0(1 - (T - T^*)/T_1))\right), \quad (3.74)$$

with $I_0 = 8.5$ mA, $T_0 = 76.3$ K, $T_1 = 91.3$ K, and $T^* = 273.15$ K, which is displayed in Fig. 3.2 right. This fit very precisely reproduces the simulation results, whereas a fit to the usual exponential T_0-law had been much worse. The latter coincides with (3.74) at $T = T^*$ to yield a $T_0 = 76.3$ K at 0°C, but this usual T_0 would decrease with rising temperature, according to (3.74). The formula (3.74) correctly displays that the threshold becomes infinite large at $T = T_1 + T^*$. In practice, there is no lasing threshold above 345 K in our example.

Throughout all calculations above threshold, at least in the thermal regime, the maximum temperature was located in the active region close to the peak position of the lasing mode (c.f. Fig. 3.1). The evolution of the maximum MQW temperature with the ambient temperature is displayed in Fig. 3.3 for different injection currents. A superlinear increase can be observed there. Once the temperature rises, the optical gain decreases, but the threshold gain is maintained by a higher quantum well carrier density. The required compensation can only be realized by a higher carrier density in the active region. The higher carrier density in turn increases the recombination processes (especially Auger) and hence the recombination heating. Moreover, the recombination coefficients and the background absorption itself depend on the temperature (see the Appendix). Altogether, this causes the superlinear increase of the maximum temperature with the ambient temperature as well as with the injection current as displayed in Fig. 3.3.

3.7.2 Modulation Response

Finally, let us note that WIAS-TeSCA allows us to study the small-signal modulation response of the device by performing the AC analysis (linearization of all equations in some fixed operating point characterized by the injection current and the ambient temperature; transformation into the frequency domain). The result of such a simulation is the complex normalized frequency response function $H_n(f)$, where f denotes the frequency of the small modulation of the injection current. Some results are given in Fig. 3.4 for fixed ambient temperature and fixed injection current, respectively.

We found that the function $H_n(f)$ can be very well fitted by the expression:

$$H_n(f) = \frac{1}{1 + j\tau_1 f} \frac{1}{1 + j\tau_2 f} \frac{f_r^2}{f_r^2 - f^2 + j\gamma_r f}, \qquad (3.75)$$

where the parameters τ_1, τ_2, f_r, γ_r depend on the injection current and on the ambient temperature. The last factor in (3.75) is well known from rate equation models whereas the other factors are low-pass filters due to parasitic effects. The order of magnitude of τ_1 and τ_2 is about 10 ns and 60 ns, respectively. Figure 3.5 shows the values of f_r and γ_r, which correspond to the curves drawn in Fig. 3.4.

In the figure on the left, f_r^2 and γ_r are linear functions of the injection current for $I < 125$ mA as one would expect for the "nonthermal" regime. Above 125 mA, heating effects come into play and yield a sublinear dependence of f_r^2 and γ_r on I. In the right figure, especially the dependence of f_r on the ambient temperature is complicated and cannot be expressed by a power law.

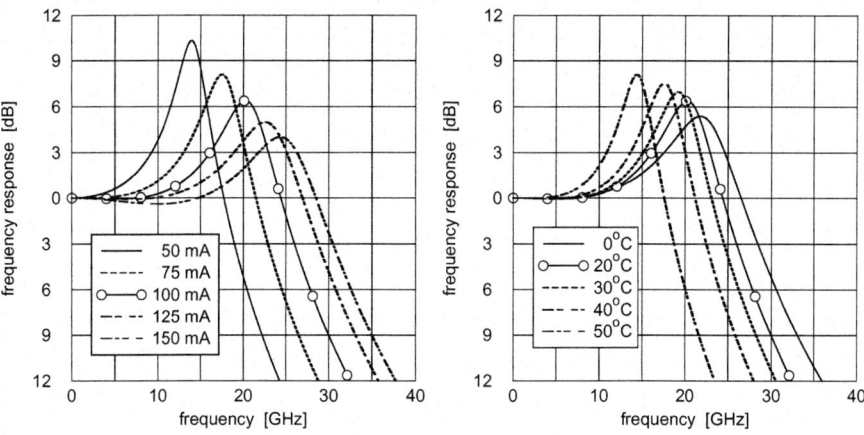

Fig. 3.4. Frequency response functions $10\lg|H_n(f)|^2$ for given ambient temperature 20° C and different injection currents (left) as well as for given injection current 100 mA and different ambient temperatures (right).

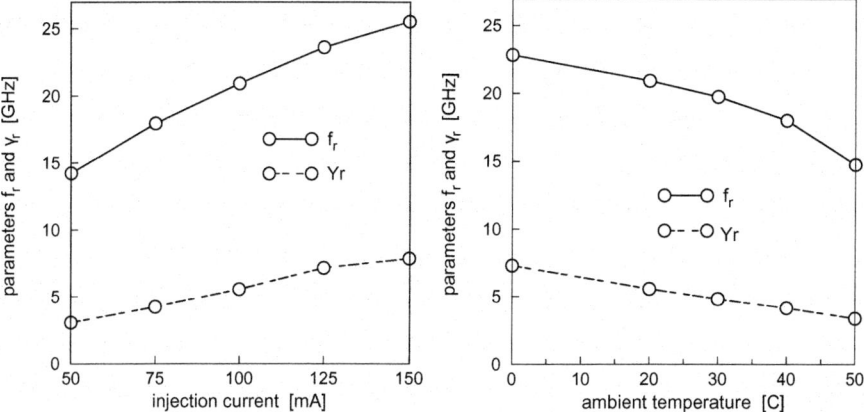

Fig. 3.5. Extracted parameters f_r and γ_r for given ambient temperature 20° C and different injection currents (left) as well as for given injection current 100 mA and different ambient temperatures (right).

3.8 Conclusion

We have derived a thermodynamics-based model for edge-emitting quantum well lasers. The model comprises the drift-diffusion equations, the Poisson equation, as well as equations for the optical field, which have been discussed in detail. Following basic thermodynamic principles, we have successfully derived the heat flow equation (3.47). In deriving this equation, we started with an energy balance equation and general expressions for the densities of the energy and the entropy. Following Lagangre's method and the maximum entropy principle we first defined the thermodynamic equilibrium. Then, by applying the principle of local thermal equilibrium and taking into account the second law of thermodynamics as well as the Onsager symmetry relations, we obtained expressions for the current densities that guided us to the desired heat flow equation (3.47). Boundary conditions as well as proper discretization schemes have been given explicitly and discussed with respect to the solution procedure. The self-consistent numerical solution of the full problem can be obtained by our code WIAS-TeSCA, which is a software for the numerical simulation of semiconductor devices. Using WIAS-TeSCA, we have demonstrated the simulation of a long-wavelength ridge-waveguide multi-quantum well laser, with a special focus on the self-heating of the device and its modulation response.

A Temperature Dependence of Model Parameters

Spontaneous radiative recombination:

$$R^{sp} = B \cdot (n \cdot p - n_i^2) \tag{3.76}$$

$$B(T) = B_0(T/T_0) \tag{3.77}$$

$B_0 = 4 \cdot 10^{-10}/\text{cm}^3\text{s}$, $T_0 = 300$ K.
Auger Recombination:

$$R^{AUG} = (C_n n + C_p p) \cdot (n \cdot p - n_i^2) \tag{3.78}$$

$$C(T) = C_0 \left(\frac{kT}{E_a}\right)^\gamma \exp\left(\frac{E_a}{kT_0} - \frac{E_a}{kT}\right) \tag{3.79}$$

$C_0 = 4 \cdot 10^{-28}/\text{cm}^6\text{s}$, $E_a = 0.5\text{eV}$, $\gamma = 0.5$, $T_0 = 300$ K.
Background Absorption $\alpha_{bg} = \alpha_{IVB} + f_{cn} \cdot n + f_{cp} \cdot p$.
Intervalence Band Absorption:

$$\alpha_{IVB}(T) = \alpha_0 \exp\left(\frac{E_a}{kT_0} - \frac{E_a}{kT}\right) \tag{3.80}$$

$\alpha_0 = 80/\text{cm}$ (in quantum wells), $E_a = 0.1\text{eV}$, $T_0 = 300$ K.
Free Carrier Absorption: $f_{cn} = f_{cp} = 10^{18}\text{cm}^2$.
Refractive Index:

$$n(T) = n_0 + n' \cdot \left(\frac{T - T_0}{kT_0}\right) \tag{3.81}$$

$n' = 6.8 \cdot 10^{-4}$, $T_0 = 300$ K. For the above calculations we have used a very simple gain model that approximates the gain in the vicinity of its spectral maximum under the constraint $E_g \approx \hbar\omega$:

$$g = \kappa \cdot \left[\exp\left(\frac{f_p - f_n - \hbar\omega}{kT}\right) - 1\right] \cdot \frac{np}{n_i^2} \exp\left(\frac{f_n - f_p}{kT}\right). \tag{3.82}$$

The factor:

$$\frac{np}{n_i^2} \exp\left(\frac{f_n - f_p}{kT}\right)$$

is $= 1$ in the Boltzmann case and ≈ 1 in the Fermi case. Therefore, the prefactor determines the gain maximum and has been adjusted to $\kappa = 3000/\text{cm}$ throughout our calculations.

References

1. G. K. Wachutka: IEEE Transactions CAD **9**, 1141 (1990)
2. G. Albinus, H. Gajewski, and R. Hünlich: Nonlinearity **15**, 367–383 (2002)
3. WIAS-TeSCA (2003). (http://www.wias-berlin.de/software/tesca)
4. T. Koprucki U. Bandelow, R. Hünlich: IEEE J. Sel. Topics Quantum Electron. **9**, 798–806 (2003)
5. H. J. Wünsche, U. Bandelow, and H. Wenzel: IEEE J. Quantum Electron. **29**, 1751–1761 (1993)
6. L. D. Landau and E. M. Lifschitz: *Course of Theoretical Physics, Vol. V: "Statistical Physics"* (Pergamon Press, London 1971)
7. S. L. Chuang: *Physics of Optoelectronic Devices* (Wiley, New York 1995)
8. C. H. Henry and R. F. Kazarinov: Reviews of Modern Physics **68**, 801–853 (1996)
9. T. T. Mnatsakanov: physica status solidi (b) **143**, 225 (1987)
10. D. E. Kane and R. M. Swanson: IEEE Trans. Electron Devices **40**, 1496 (1993)
11. J. Piprek. *Semiconductor Optoelectronic Devices: Introduction to Physics and Simulation* (Academic Press, San Diego 2003)
12. D. L. Scharfetter and H. K. Gummel: IEEE Trans. Electron Devices **16**, 64 (1969)
13. M. Grupen and K. Hess: IEEE J. Quantum Electron. **34**, 120–140 (1998)
14. J. Piprek, P. Abraham, and J. E. Bowers: IEEE J. Quantum Electron. **36**, 366–374 (2000)
15. WIAS-QW (2003). (http://www.wias-berlin.de/software/qw)
16. M. Möhrle, A. Sigmund, J. Kreissl, F. Reier, R. Steingrüber, W. Rehbein, and H. Röhle: Integratable high-power small-linewidth $\lambda/4$ phase-shifted 1.55 µm InGaAsP-InP-Ridge-Waveguide DFB Lasers. In: *Workbook of 26th Int. Conf. on Compound Semiconductors (ISCS'99), Berlin, Germany*, Paper Th A3.1 (1999)

4 Distributed Feedback Lasers: Quasi-3D Static and Dynamic Model

X. Li

Department of Electrical and Computer Engineering, McMaster University, 1280 Main Street, Hamilton, Ontario, L8S 4k1, Canada, lixun@mcmaster.ca

4.1 Introduction

As the wafer growth, E-beam writing, and holography technologies are getting more mature and standardized, less uncertainties are left in the device manufacturing. The device design, therefore, becomes the major issue in the development of new components for system and network applications as well as in the optimization of existing components for cost reduction. Computer-aided design, modeling, and simulation are highly desirable, particularly for those semiconductor optoelectronic devices with complicated structures such as strained-layer multi-quantum well (SL-MQW) distributed feedback (DFB) laser diodes (LDs). On the other hand, progresses on both numerical techniques and the computing hardware over the recent decades have provided a powerful platform. This made sophisticated computer-aided simulation tools for SL-MQW DFB LDs possible.

Between the two major approaches in device modeling and simulation, we will focus on the physics modeling approach, i.e., a direct approach based on the physics-based first principles. Unlike the other approach, known as behavior modeling, developed for the device users and system level researchers, this model gives the physical description of what exactly happens inside the device. Thus, it is capable of generating the device performance in almost every aspect for the given material constants and geometrical sizes under specified operating conditions. For this reason, this approach is usually adopted by those device designers who work on the development of the device itself.

Illustrated by the schematic diagram Fig. 4.1, a typical DFB LD consists of an optical waveguide that guides the optical waves along a designated spatial direction, a grating that couples the forward and backward traveling optical waves to form the oscillation, an active region (SL-MQW) that provides optical gain, a forward-biased diode junction near the active region, plus a pair of electrodes that introduce both types of carriers, i.e., electrons and holes, for optical gain generation through recombination in the active region.

Shown in the block diagram Fig. 4.2, there are four coupled physical processes in the semiconductor laser diodes. Through the diode junction, the injected current introduces both holes and electrons in the active region. The holes and electrons recombine and generate optical gain accompanied by

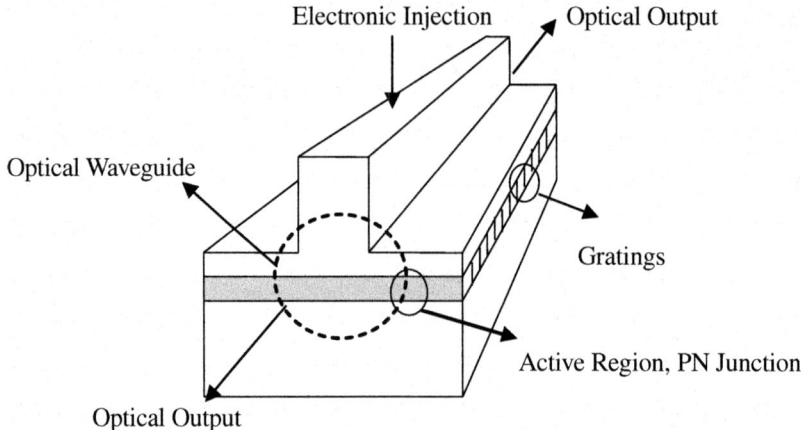

Fig. 4.1. Schematic semiconductor DFB laser diode structure.

material refractive index change. The optical waves originated from the spontaneous recombination of the carriers propagate along the optical waveguide and get coherently amplified by the optical gain through stimulated recombination and the selective feedback through the grating (in DFB LDs) or the end facets (in Fabry–Perot or FP LDs). Finally, the thermal diffusion process is also involved as the temperature varies with the heating generated

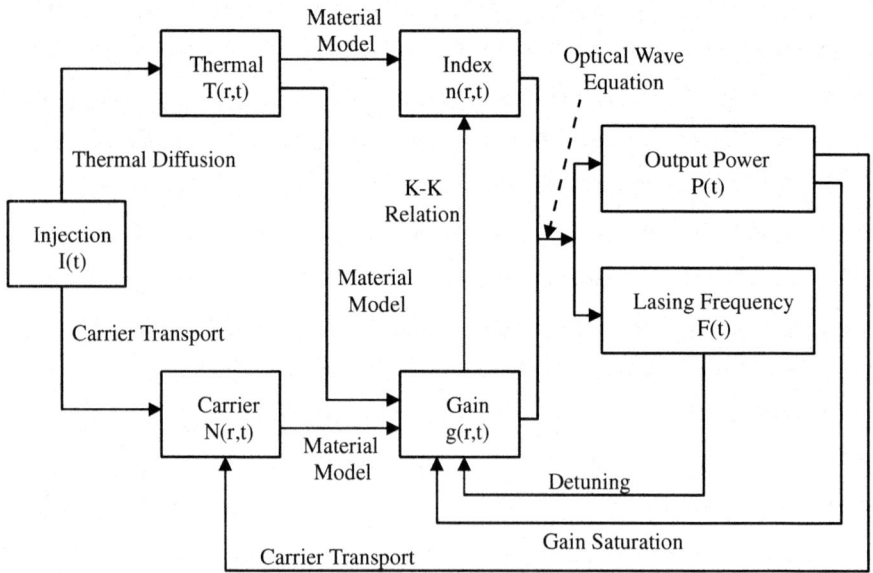

Fig. 4.2. Physical processes and governing equations in semiconductor laser modeling.

through various nonradiative carrier recombination and noncoherent photon reabsorption processes, which in turn changes the material properties.

To capture these physical processes, the following coupled physics-based governing equations are therefore required:

1. Equation that describes the wave propagation along the optical waveguide from electromagnetic wave theory.
2. Equation that describes the carrier transport from quasi-electrostatic and charge continuity theory.
3. Equation that describes the material optical properties from semiconductor physics.
4. Equation that describes the material thermal properties from thermal diffusion theory.

In this chapter, the governing equations on all the above four aspects are given by the first principles, such as the Maxwell's equations for wave propagation and carrier transport, the Schrödinger equation for material optical properties, and the thermal diffusion equation for material thermal properties. Numerical methods for solving these coupled equations are given, and some critical issues in the implementation are discussed. To validate the modeling and simulation, the calculated results are compared with the experimental data. Simulated static and dynamic performance of typical SL-MQW DFB LDs are also given as application examples, which shows the capability of this model.

4.2 Governing Equations

4.2.1 Optical Wave Equations

Derived from the Maxwell equations, the optical wave equation can be expressed as [1]:

$$\nabla^2 \boldsymbol{E}(\boldsymbol{r},t) = \frac{1}{c^2}\frac{\partial^2}{\partial t^2}\boldsymbol{E}(\boldsymbol{r},t) + \frac{1}{c^2}\frac{\partial^2}{\partial t^2}\boldsymbol{P}(\boldsymbol{r},t) + \mu_0 \frac{\partial}{\partial t}\boldsymbol{J}_{\mathrm{sp}}(\boldsymbol{r},t) \qquad (4.1)$$

$$\boldsymbol{P}(\boldsymbol{r},t) \equiv \int_{-\infty}^{t} \chi(\boldsymbol{r},t-\tau)\,\boldsymbol{E}(\boldsymbol{r},\tau)\,\mathrm{d}\tau\,, \qquad (4.2)$$

where $\boldsymbol{E}$ denotes the electric field vector [V/m], $\boldsymbol{P}$ denotes the polarization vector [V/m], and $\boldsymbol{J}_{\mathrm{sp}}$ denotes the spontaneous emission current density [A/m^2]; $c = 1/(\varepsilon_0\mu_0)^{1/2}$ [m/s] is the speed of the light in vacuum, and χ is the dimensionless material susceptibility; ε_0 [F/m] and μ_0 [H/m] are the electrical permittivity and magnetic permeability in vacuum, respectively.

Under the following assumptions:

1. Wave propagates along the waveguide direction only (one directional)

2. Wave is confined in the waveguide cross-sectional area
3. Wave has discrete fast oscillation frequencies (in optical domain) with relatively slow-varying envelopes,

we may write the electric field as:

$$\boldsymbol{E}(x,y,z,t) = \frac{1}{2}\boldsymbol{s}\sum_{i=1}^{M}\phi_i(x,y)[e_i^{\mathrm{f}}(z,t)\,e^{j\beta_i z} + e_i^{\mathrm{b}}(z,t)\,e^{-j\beta_i z}]e^{-j\omega_i t} + \mathrm{c.c.},\tag{4.3}$$

with x, y denoting the cross-sectional vertical and horizontal dimensions and z denoting the propagation direction; ϕ_i is the ith cross-sectional field distribution [1/m] known as the ith transverse mode or the ith eigenfunction of the optical waveguide, a real function without change along the propagation direction; $e_i^{\mathrm{f,b}}$ are the forward- and backward-traveling slow-varying envelopes [V] of the ith transverse mode, a complex function that shows relatively slow change with time and along the propagation direction; $\exp[j(\omega_i t \pm \beta_i z)]$ represents the forward- and backward-propagating harmonic plane waves with frequency ω_i [rad/s] and propagation constant β_i [1/m], known as the traveling wave factors moving along $\pm z$ at the speed of ω_i/β_I; $\boldsymbol{s}$ is a unit vector within the cross-section (x,y) plane and c.c. denotes the complex conjugate of the first term on the right-hand side (RHS) of (4.3).

By assuming the linearity of (4.2) (χ must be independent of $\boldsymbol{E}$ in this case), we can simply ignore the complex conjugate term in (4.3) in further derivations, as we know that the field will be obtained by taking two times the real part of the final solution according to the linear superposition theory. For the same reason, we will only consider a single transverse mode with single polarization along x [$\boldsymbol{s}=\boldsymbol{x}$ or transverse electric (TE)] in the following derivations: It is trivial to consider the summation in a linear system when multiple transverse modes with both polarizations are involved. Even when the material shows insignificant nonlinearity, such as nonlinear gain suppression, the above conclusion still holds as the nonlinear term can be treated by the perturbation method.

Under these assumptions, the following coupled equations can be obtained for the slow-varying envelopes by plugging (4.3) into the original wave equation (4.1) and then ignoring all the second-order derivatives:

$$\frac{1}{v_g}\frac{\partial}{\partial t}e^{\mathrm{f}} + \frac{\partial}{\partial z}e^{\mathrm{f}} = \varGamma\left[j\delta + jk_0\varDelta n(N,P,T,\omega_0)\right.$$
$$\left.+\frac{1}{2}g(N,P,T,\omega_0) - \frac{1}{2}\alpha\right]e^{\mathrm{f}} + j\kappa e^{\mathrm{b}} + \tilde{s}$$
$$\frac{1}{v_g}\frac{\partial}{\partial t}e^{\mathrm{b}} - \frac{\partial}{\partial z}e^{\mathrm{b}} = \varGamma\left[j\delta + jk_0\varDelta n(N,P,T,\omega_0)\right.$$
$$\left.+\frac{1}{2}g(N,P,T,\omega_0) - \frac{1}{2}\alpha\right]e^{\mathrm{b}} + j\kappa^* e^{\mathrm{f}} + \tilde{s}\,.$$

(4.4)

In these equations, $e^{f,b}$ are the forward- and backward-traveling wave slow-varying envelopes [V], respectively, and $v_g = c/n_g$ [m/s] is the group velocity with n_g defined as the group refractive index.

In the derivation of the coupled traveling wave equations (4.4), the bandwidth of the material susceptibility χ is assumed much larger than that of the slow-varying envelope e. This assumption means that the system (χ) bandwidth must be much broader than the signal (e) baseband in frequency domain, or the system (χ) responds much faster than the signal (e) varying in time domain. This is true particularly for semiconductor laser diodes under single-frequency operation, as the FWHM bandwidth of χ is usually around $5 \sim 10$ THz ($50 \sim 100$ nm in C-band), whereas the lasing mode linewidth can hardly go beyond 100 GHz (0.8 nm) even under fast modulation with frequency chirping. Therefore, we find from (4.2):

$$\boldsymbol{P}(x,y,z,t) = \frac{1}{2}\boldsymbol{x} \int_{-\infty}^{t} \chi(x,y,z,t-\tau)\,\phi(x,y)\,e^{f,b}(z,\tau)\,e^{-j(\omega_0\tau \mp \beta_0 z)}\,d\tau$$

$$\approx \frac{1}{2}\boldsymbol{x}\chi(x,y,z,\omega_0)\,\phi(x,y)\,e^{f,b}(z,t)\,e^{-j(\omega_0 t \mp \beta_0 z)}\,. \tag{4.5}$$

The material susceptibility χ at the reference frequency is further assumed as:

$$\chi(x,y,z,\omega_0) = n^2(x,y,\omega_0) - 1 + 2n(x,y,\omega_0)\,\Delta n(N,P,T,\omega_0)$$
$$- j\frac{1}{k_0}n(x,y,\omega_0)g(N,P,T,\omega_0)\,, \tag{4.6}$$

where n denotes the cross-sectional refractive index profile of the optical waveguide; g is the stimulated emission gain [1/m], a function of the carrier densities (N, P) and the temperature (T); Δn is the material refractive index change due to the optical gain change linked through the Kramers–Kronig relationship; $k_0 = \omega_0/c$ [1/m] is the reference propagation constant. In (4.4), the detuning factor, the grating coupling coefficient, and the waveguide confinement factor are given by:

$$\delta = \beta - \beta_0 = k_0 n_{\text{eff}} - \frac{m\pi}{\Lambda} = \frac{\omega_0 n_{\text{eff}}}{c} - \frac{m\pi}{\Lambda} = \frac{2\pi n_{\text{eff}}}{\lambda_0} - \frac{m\pi}{\Lambda} \tag{4.7}$$

$$\kappa = \frac{k_0}{2n_{\text{eff}}} \int_{\Sigma} [\chi_H(x,y,z,\omega_0) - \chi_L(x,y,z,\omega_0)]\phi^2(x,y)\,dx\,dy \tag{4.8}$$

$$\Gamma = \int_{AR} \phi^2(x,y)\,dx\,dy\,, \tag{4.9}$$

with $\beta \equiv k_0 n_{\text{eff}}$ and $\beta_0 \equiv m\pi/\Lambda$ (both in [1/m]) defined as the wave propagation and the Bragg constants, respectively. n_{eff} is known as the effective index. Integer m denotes the grating order and Λ the grating period [1/m]. $\chi_{H,L}$ are the susceptibilities in the high and low refractive index or gain areas given by (4.6), respectively. α [1/m] counts in all the optical losses except the interband absorption that is already included in the imaginary part of the material susceptibility.

Moreover, the cross-sectional field distribution (eigenfunction) and the wave propagation constant β (eigenvalue) are solved by the following eigenvalue equation in (x, y) plane:

$$\nabla_t^2 \phi(x,y) + k_0^2 n^2(x,y,\omega_0)\, \phi(x,y) = \beta^2 \phi(x,y) \tag{4.10}$$

subject to the waveguide boundary condition and with the field distribution normalized by $\int_\Sigma \phi^2(x,y)\,dx\,dy = 1$. The two-dimensional (2D) integrations $\int_{\Sigma,\mathrm{AR}}$ go over the entire cross-sectional area and the active region only, respectively.

The electric field (TE mode) is thus given by:

$$\boldsymbol{E}(x,y,z,t) = \boldsymbol{x}\phi(x,y)\mathrm{Re}\{[e^{\mathrm{f}}(z,t)\,e^{j\beta_0 z} + e^{\mathrm{b}}(z,t)\,e^{-j\beta_0 z}]\,e^{-j\omega_0 t}\}. \tag{4.11}$$

Although the spontaneous emission noise contribution in the governing equation (4.4) is given in the form of:

$$\tilde{s}(z,t) = -j\frac{e^{j(\omega_0 t - \beta z)}}{n_{\mathrm{eff}}\omega_0}\sqrt{\frac{\mu_0}{\varepsilon_0}}\int_\Sigma \frac{\partial}{\partial t}\boldsymbol{J}_{\mathrm{sp}}\phi(x,y)\,dx\,dy, \tag{4.12}$$

we do not use this expression for the calculation of the driving noise $\boldsymbol{s}$ [V/m], as $\boldsymbol{J}_{\mathrm{sp}}$ cannot easily be evaluated. Noting the fact that the noise power can usually be evaluated in an easier way, we take a small step d_z [m] along the wave propagation direction to integrate equation (4.4) at the steady state and assume there is a forward traveling wave only:

$$e^{\mathrm{f}}(z+d_z) = e^{\Gamma[j\delta + jk_0\Delta n(N,P,T,\omega_0) + \frac{1}{2}g(N,P,T,\omega_0) - \frac{1}{2}\alpha]\,d_z} e^{\mathrm{f}}(z) + \tilde{s}(z)\,d_z. \tag{4.13}$$

Multiplying (4.13) by its complex conjugate and factor $n_{\mathrm{eff}}(\varepsilon_0/\mu_0)^{1/2}/2\,[1/\Omega]$ on both sides, we obtain:

$$P^{\mathrm{f}}(z+d_z) = e^{\Gamma[g(N,P,T,\omega_0) - \alpha]\,d_z} P^{\mathrm{f}}(z) + \frac{n_{\mathrm{eff}}d_z^2}{2}\sqrt{\frac{\varepsilon_0}{\mu_0}}|\tilde{s}(z)|^2, \tag{4.14}$$

where the cross terms all disappear as the noise term has a random phase and their averaged contribution after the integration gives zero. Equation (4.14) clearly shows the power balance with the second term on the RHS indicating the spontaneous noise power contribution in the local section d_z; the latter, in turn, can be evaluated as [2]:

$$\frac{n_{\mathrm{eff}}d_z^2}{2}\sqrt{\frac{\varepsilon_0}{\mu_0}}|\tilde{s}(z)|^2 = \gamma \Gamma v_g g_{\mathrm{sp}}(N,P,T,\omega_0)\hbar\omega_0, \tag{4.15}$$

where γ denotes the coupling coefficient of the spontaneous emission emitted over the entire spatial sphere to the waveguide mode and of that spread over the entire spectrum to the lasing mode (frequency), $\hbar\omega_0$ denotes the photon energy [J] at the reference frequency with $\hbar$ defined as the Plank constant, and g_{sp} denotes the spontaneous emission gain [1/m]. Therefore, the amplitude of the noise contribution in (4.4), $|\tilde{s}|$, can be modeled by a Gaussian random process with its autocorrelation function given by:

$$\langle |\tilde{s}(z,t)| |\tilde{s}(z',t')| \rangle = 2\sqrt{\frac{\mu_0}{\varepsilon_0}} \frac{\gamma \Gamma v_g g_{\mathrm{sp}}(N,P,T,\omega_0)\hbar\omega_0}{n_{\mathrm{eff}} d_z^2} \delta(z-z')\delta(t-t') , \quad (4.16)$$

whereas the phase of the noise should be modeled by a random process with uniform distribution over $[0, 2\pi]$.

4.2.2 Carrier Transport Equations

In general, the carrier distribution is also nonuniform along the cavity, and therefore, the derivative term $\partial J/\partial z$ should be included in the continuity equations to account for the carrier transport along the cavity. However, the diffusion length of the minority carrier is much smaller than the cavity length that can be regarded as the characteristic length of carrier concentration variation due to the longitudinal spatial hole burning (LSHB) effect. Hence, for typical conditions, the derivative along the laser cavity is negligible in comparison with the recombination terms in the continuity equations. Under this condition, the carrier transport process will be considered only in the cross-sectional area and, thus, is described by the following 2D equations in (x, y) plane [3]:

$$\nabla_t \circ [\varepsilon_{\mathrm{r}}(x,y)\nabla_t \Phi(x,y,z,t)] \\ -(q/\varepsilon_0)[P(x,y,z,t) - N(x,y,z,t) - N_{\mathrm{AD}}(x,y)] \quad (4.17)$$

$$\frac{\partial}{\partial t} N(x,y,z,t) = \frac{1}{q}\nabla_t \circ \boldsymbol{J}_n - R_{\mathrm{nr}}(N,P) \\ \frac{\partial}{\partial t} P(x,y,z,t) = -\frac{1}{q}\nabla_t \circ \boldsymbol{J}_p - R_{\mathrm{nr}}(N,P) , \quad (4.18)$$

where the electron and the hole currents are further expressed by:

$$\boldsymbol{J}_n = q[-\mu_n(x,y)N\nabla_t\Phi + D_n(x,y)\nabla_t N + D_n^T(x,y)N\nabla_t T(x,y,z,t)] \\ \boldsymbol{J}_p = q[-\mu_p(x,y)P\nabla_t\Phi - D_p(x,y)\nabla_t P - D_p^T(x,y)P\nabla_t T(x,y,z,t)] . \quad (4.19)$$

The Poisson equation (4.17) is derived from the Maxwell's equation under the quasi-electrostatic field assumption, whereas (4.18) is the continuity equation. The classic drift and diffusion model is used for the current evaluation in (4.19).

In these equations, Φ is the potential distribution [V]; N and P are the electron and hole concentrations [1/m^3], respectively; $\boldsymbol{J}_{n,p}$ are the electron and hole current densities [A/m^2], respectively; T is the temperature distribution [K]. The parameter q is the electron charge [C]; ε_r is the dimensionless relative dielectric constant; N_{AD} is the net doping concentration [1/m^3]; $\mu_{n,p}$ are the electron and hole mobilities [m^2/Vs], respectively; $D_{n,p}$ are the electron and hole diffusivities [m^2/s], respectively; and $D_{n,p}^{\text{T}}$ are the electron and hole thermal diffusivities [m^2/Ks], respectively. The mobility and the diffusivities are linked through the Einstein relation $D_{n,p}/\mu_{n,p} = k_{\text{B}} T/q$ with k_{B} defined as the Boltzmann constant, whereas the thermal diffusivities are approximated by $D_{n,p}^{\text{T}} = D_{n,p}/2T$ [4].

The above classic model can be used to treat the bulk region only. In the quantum well region, part of the carriers (N_{w}, P_{w}) will be captured by the bounded energy levels inside the well and will in turn contribute to the optical gain. The rest carriers (N, P) will stay at the unbounded energy levels above the well. These two different types of carriers are coupled through a phenomenologically introduced capture and escape model [5, 6] given as follows:

$$\frac{\partial}{\partial t} N_{\text{w}}^i(y,z,t) = \frac{N(x_{\text{wb}}^i, y, z, t)}{\tau_n^{\text{c}}} - \frac{N(x_{\text{wt}}^i, y, z, t)}{\tau_n^{\text{e}}}$$
$$- R_{\text{r}}(N_{\text{w}}^i, P_{\text{w}}^i) - R_{\text{nr}}(N_{\text{w}}^i, P_{\text{w}}^i)$$
$$\frac{\partial}{\partial t} P_{\text{w}}^i(y,z,t) = \frac{P(x_{\text{wt}}^i, y, z, t)}{\tau_p^{\text{c}}} - \frac{P(x_{\text{wb}}^i, y, z, t)}{\tau_p^{\text{e}}}$$
$$- R_{\text{r}}(N_{\text{w}}^i, P_{\text{w}}^i) - R_{\text{nr}}(N_{\text{w}}^i, P_{\text{w}}^i) \, .$$
(4.20)

In (4.20), the superscript i indicates the ith quantum well, $i = 1, 2, \ldots, M$, with M as the total number of quantum wells. N_{w}^i, P_{w}^i denote the electron and hole concentrations [1/m^3] in the ith quantum well, respectively. $x_{\text{wt,wb}}^i$ are the coordinates of the top and bottom edges of the ith quantum well (by assuming the P-type cladding layers are on the top and the N-type cladding layers are on the bottom), respectively. $\tau_{n,p}^{\text{c,e}}$ represent the electron and hole capture and escape time constants, respectively.

The continuity equation (4.18) must be modified in the quantum well region accordingly:

$$\frac{\partial}{\partial t} N(x,y,z,t) = \frac{1}{q} \nabla_t \circ \boldsymbol{J}_n$$
$$- \left[R(x,y,z,t) + \frac{N(x_{\text{wb}}^i, y, z, t)}{\tau_n^{\text{c}}} - \frac{N(x_{\text{wt}}^i, y, z, t)}{\tau_n^{\text{e}}} \right]$$
$$\frac{\partial}{\partial t} P(x,y,z,t) = -\frac{1}{q} \nabla_t \circ \boldsymbol{J}_p$$
$$- \left[R(x,y,z,t) + \frac{P(x_{\text{wt}}^i, y, z, t)}{\tau_p^{\text{c}}} - \frac{P(x_{\text{wb}}^i, y, z, t)}{\tau_p^{\text{e}}} \right] .$$
(4.21)

Finally, the carrier nonradiative (R_{nr}) and radiative (R_{r}) recombination rates [1/m^{3}s] in the bulk and active regions are given as:

$$R_{\mathrm{nr}}(N, P) = (NP - N|_{\mathrm{eq}}P|_{\mathrm{eq}})$$
$$\times \left[\frac{1}{\tau_n(x,y)P + \tau_p(x,y)N} + C_n(x,y)N + C_p(x,y)P \right] \quad (4.22\mathrm{a})$$

$$R_{\mathrm{nr}}(N_{\mathrm{w}}^i, P_{\mathrm{w}}^i) = (N_{\mathrm{w}}^i P_{\mathrm{w}}^i - N_{\mathrm{w}}^i|_{\mathrm{eq}}P_{\mathrm{w}}^i|_{\mathrm{eq}})$$
$$\times \left[\frac{1}{\tau_n^i(y)P_{\mathrm{w}}^i + \tau_p^i(y)N_{\mathrm{w}}^i} + C_n^i(y)N_{\mathrm{w}}^i + C_p^i(y)P_{\mathrm{w}}^i \right] \quad (4.22\mathrm{b})$$

$$R_{\mathrm{r}}(N_{\mathrm{w}}^i, P_{\mathrm{w}}^i) = B^i(y)(N_{\mathrm{w}}^i P_{\mathrm{w}}^i - N_{\mathrm{w}}^i|_{\mathrm{eq}}P_{\mathrm{w}}^i|_{\mathrm{eq}}) \quad (4.23)$$
$$+ \frac{n_{\mathrm{eff}}}{2\hbar\omega_0 L_x} \sqrt{\frac{\varepsilon_0}{\mu_0}} g(N_{\mathrm{w}}^i, P_{\mathrm{w}}^i, T, \omega_0) \int_{x_{\mathrm{wb}}^i}^{x_{\mathrm{wt}}^i} |\phi(x,y)|^2 \, \mathrm{d}x$$
$$\times |e^{\mathrm{f}}(z,t) e^{j\beta_0 z} + e^{\mathrm{b}}(z,t) e^{-j\beta_0 z}|^2 ,$$

where $\tau_{n,p}$ [s] are the electron and hole SRH recombination time constants, $C_{n,p}$ [m^6/s] are the Auger recombination coefficients, and B [m^3/s] is the spontaneous emission recombination coefficient, respectively. Subscript $|_{\mathrm{eq}}$ indicates taking the variable values at their equilibrium state. And L_x is the well thickness [m]. The second term on the RHS of (4.23) gives the stimulated emission recombination rate, whereas the rest part on the RHS of (4.23) and (4.22) are the nonradiative and spontaneous emission recombination contributions.

Due to the nonuniformity of the optical wave intensity along the wave propagation direction z, the stimulated emission recombination becomes z-dependent. As a result, all of the variables such as the potential, the electron and hole concentrations, the electron and hole current densities, and the temperature also become z-dependent, although the carrier transport effect is treated in the cross-section (x,y) plane only. In this sense, carrier transport happens locally in each cross-section (x,y) sheet but is coupled from sheet to sheet along the wave propagation direction through the optical wave only.

4.2.3 Optical Gain Model

For the SL-MQW active region, the stimulated and spontaneous emission gains are calculated through [7]:

$$g(N,P,T,\omega_0) = \frac{q^2}{c\varepsilon_0\sqrt{\varepsilon_\mathrm{r}}m_0^2\omega_0 L_x}$$

$$\times \sum_{\eta,\sigma}\sum_{l,m}\int_0^\infty k\,dk M_{lm}^{\eta\sigma}(k)\frac{[f_\mathrm{c}^l(k)-f_{\mathrm{v},\sigma}^m(k)]\Gamma_t/(2\pi)}{[E_\mathrm{g}+E_l^\mathrm{c}(k)-E_m^{\mathrm{v},\sigma}(k)-\hbar\omega_0]^2+(\Gamma_t/2)^2} \quad (4.24)$$

$$g_\mathrm{sp}(N,P,T,\omega_0) = \frac{q^2}{c\varepsilon_0\sqrt{\varepsilon_\mathrm{r}}m_0^2\omega_0 L_x}$$

$$\times \sum_{\eta,\sigma}\sum_{l,m}\int_0^\infty k\,dk M_{lm}^{\eta\sigma}(k)\frac{f_\mathrm{c}^l(k)[1-f_{\mathrm{v},\sigma}^m(k)]\Gamma_t/(2\pi)}{[E_\mathrm{g}+E_l^\mathrm{c}(k)-E_m^{\mathrm{v},\sigma}(k)-\hbar\omega_0]^2+(\Gamma_t/2)^2}, \quad (4.25)$$

where Γ_t is the Lorentzian linewidth broadening factor [eV] introduced to describe the intraband scatterings. The integrations are carried out over the momentum k space. σ indicates the upper (U) or lower (L) blocks of the Hamiltonian, η indicates the electron spin state, and l and m indicate the subband indices. $M_{lm}^{\eta\sigma}$ denotes the optical dipole matrix element between the lth subband in the conduction band with a spin state η and the mth subband in the valence band of the 2×2 Hamiltonian H^σ. f_c^l, $f_{\mathrm{v},\sigma}^m$ are the dimensionless Fermi functions for the lth subband in the conduction band and the mth subband in the valence band of H^σ, respectively. E_l^c, $E_m^{\mathrm{v},\sigma}$ are the lth subband energy [eV] in the conduction band and the mth subband energy [eV] in the valence band of H^σ at k, respectively. m_0 is the free electron mass [kg], and E_g is the band gap energy [eV].

For any given SL-MQW structures, the bandstructure (energy) and the electron and hole wave functions are calculated as the eigenvalue and eigenfunctions by employing the multiband effective-mass theory (KP method), where the conduction band and the valence band are assumed decoupled. The isotropic parabolic band assumption is used for the conduction band, whereas a 4×4 Luttinger–Kohn Hamiltonian matrix with the subband mixing derived from the KP method is used in the calculation of the valence band [8]. The strain effect is considered by including a Pikus–Bir perturbation term in the Hamiltonians [9]. In solving the valence band structure, the 4×4 Hamiltonian is further block-diagonalized to two 2×2 matrices under a unitary transformation [7]. Through this method, we just need to solve a set of two coupled eigenvalue equations twice, instead of solving a set of four coupled eigenvalue equations. After the electron and hole wave functions are obtained by solving the eigenvalue equations, the optical dipole matrix elements can readily be calculated through various integrals given in [7]. Together with the subband energies also obtained from these equations, the dipole matrix elements are plugged into the above expressions (4.24) and (4.25) for the stimulated and spontaneous gain calculation.

Also in the gain expressions, the Fermi functions are given as:

$$f_c^l(k) = \frac{1}{1+e^{(E_g+E_i^c(k)-F_c)/k_BT}} \qquad (4.26a)$$

$$f_{v,\sigma}^m(k) = \frac{1}{1+e^{(E_m^{v,\sigma}(k)-F_v)/k_BT}}, \qquad (4.26b)$$

with $F_{c,v}$ defined as the quasi-Fermi energies [eV] in the conduction and valence bands, respectively. They are linked to the electron and hole concentrations through:

$$N = \frac{1}{\pi L_x}\sum_l \int_0^\infty k\,\mathrm{d}k\, f_c^l(k) \qquad (4.27a)$$

$$P = \frac{1}{\pi L_x}\sum_\sigma\sum_m \int_0^\infty k\,\mathrm{d}k\, f_{v,\sigma}^m(k), \qquad (4.27b)$$

where k_B denotes the Boltzmann constant [J/K].

The material refractive index change is calculated from the Kramers–Kronig transformation as long as the stimulated emission gain is obtained [10]:

$$\Delta n(N,P,T,\omega_0) = -\frac{c}{2\pi\omega_0}\int_{-\infty}^{+\infty}\frac{g(N,P,T,\omega)}{\omega-\omega_0}\,\mathrm{d}\omega. \qquad (4.28)$$

To count in the nonlinear gain suppression due to the carrier heating and/or spectral hole burning (SBH), a factor in the form of [11]

$$\frac{1}{1+\varepsilon_s|\phi(x,y)|^2|e^f(z,t)\,e^{j\beta_0 z}+e^b(z,t)\,e^{-j\beta_0 z}|^2} \qquad (4.29)$$

is further multiplied to the gains given in the forms (4.24) and (4.25), with ε_s phenomenologically introduced as the gain suppression coefficient [m^2/V^2].

4.2.4 Thermal Diffusion Equation

Restricted by the geometrical structure of the edge-emitting devices, thermal flow will mainly happen inside the 2D cross-sectional area, as the structure is uniform along the cavity if the small corrugation introduced by the DFB grating is ignored. If the optical field intensity and the carriers are all uniformly distributed along the cavity (i.e., assuming no LSHB), the temperature distribution would also be constant, or $\partial T/\partial z = 0$. With the LSHB effect considered, the temperature derivative along the cavity should be included in the thermal diffusion equation as the temperature is no longer uniform in this direction. However, $\partial T/\partial z$ is still smaller comparing with $\partial T/\partial x$ and $\partial T/\partial y$, as the temperature varying in z is in the same scale of the optical field intensity and the carriers. For the same temperature variation ΔT, Δx and

Δy are in the order of a few micron or even less than one micron, whereas Δz is in the order of several tens of microns for typical edge-emitting SL-MQW DFB laser diodes. Therefore, $\partial T/\partial z$ is ignored and we only solve the thermal diffusion equation in the 2D cross-sectional area [12]:

$$\frac{\partial}{\partial t}T(x,y,z,t) = D_T(x,y)[\nabla_t^2 T(x,y,z,t) + C_T^{-1}(x,y)H(x,y,z,t)], \quad (4.30)$$

where D_T is the thermal diffusion constant [m²/s], and C_T is the thermal conductivity [w/mK]. The heat generation rate $H(x,y,z,t)$ [W/m³] is given as [13]:

$$\begin{aligned}H = & |J_n|^2/[q\mu_n(x,y)N] + |J_p|^2/[q\mu_n(x,y)P] \\ & + R_{nr}(N,P)\{F_c - F_v + T[P_n(x,y) + P_p(x,y)]\} \\ & + \frac{n_{\text{eff}}}{2}\sqrt{\frac{\varepsilon_0}{\mu_0}}\alpha|\phi(x,y)|^2|e^f(z,t)\,e^{j\beta_0 z} + e^b(z,t)\,e^{-j\beta_0 z}|^2,\end{aligned} \quad (4.31)$$

with $P_{n,p}$ defined as the electron and hole thermoelectric powers [J/K], respectively, and counted as constant material parameters.

The first, second, and third terms on the RHS of (4.31) represent the Joule heating, recombination heating, and optical absorption heating, respectively. As the optical field varies along the propagation direction, the electron and hole current densities and the electron and hole concentrations all become z-dependent; the heating source H, therefore, is z-dependent. In this sense, the 2D thermal diffusion equation (4.30) holds locally in each cross-section (x,y) sheet along the optical wave propagation direction and couples to each other through the optical field, current densities, and carrier concentrations.

4.3 Implementation

4.3.1 General Approach

To solve all of the above equations in a self-consistent manner, the following procedures are used in the numerical implementation for the steady-state (DC) analysis.

The three-dimensional (3D) SL-MQW DFB laser structure is first subdivided into a number of sections along the wave propagation (z) direction (laser cavity). For each section, (4.10) for the optical field distribution; (4.17) through (4.21) for the potential, carrier, and current density distributions; (4.24), (4.25), and (4.28) for the stimulated and spontaneous emission gains and material refractive index change; and (4.30) for the temperature distribution are solved by switching off all time derivatives in the 2D cross-sectional (x,y) plane through the finite-difference (FD) algorithm. In order to solve these distributions efficiently, three different computation windows and meshes are adopted for the optical field; the potential, carrier, and current density; and the temperature calculations in 2D (x, y) plane, respectively; whereas a 1D mesh along x direction (vertical) is used for the material gain calculation. The neighborhood interpolation technique is used to obtain the values of the different variables at any arbitrary mesh point. Based on the calculated band structure for the given SL-MQW active region and the 2D carrier and temperature distributions in each section along the laser cavity, the material optical gains, refractive index change, and optical loss (mainly due to free carrier absorption) are calculated accordingly. Taking the 2D optical field intensity as the weighting function, we calculate all of the modal parameters that will be used for solving (4.4) along the laser cavity through weighted integrals in each section. These modal parameters obtained are usually different from section to section.

Under the steady state, the time derivatives in the coupled traveling wave (4.4) should be replaced by $\partial e^{f,b}/\partial t = -j\Delta\omega e^{f,b}$ as the laser will oscillate at frequency $\omega_0 + \text{Re}[\Delta\omega]$ with its linewidth proportional to $|\text{Im}(\Delta\omega)|$. The complex frequency deviation $\Delta\omega$ should be viewed as the eigenvalue associated with the eigenfunction $e^{f,b}$ as the solution of (4.4) in its reduced form

$$\frac{\partial}{\partial z}e^f = \Gamma\left[j\left(\delta + \frac{\Delta\omega}{\Gamma v_g}\right)\right.$$
$$\left. + jk_0\Delta n(N,T,\omega_0) + \frac{1}{2}g(N,T,\omega_0) - \frac{1}{2}\alpha\right]e^f + j\kappa e^b + \tilde{s}$$
(4.32)
$$-\frac{\partial}{\partial z}e^b = \Gamma\left[j\left(\delta + \frac{\Delta\omega}{\Gamma v_g}\right)\right.$$
$$\left. + jk_0\Delta n(N,T,\omega_0) + \frac{1}{2}g(N,T,\omega_0) - \frac{1}{2}\alpha\right]e^b + j\kappa e^f + \tilde{s},$$

subject to the boundary conditions at the two ends of the laser

$$e^f(0,t) = r_l e^b(0,t), \quad e^b(L,t) = r_r e^f(L,t), \quad (4.33)$$

with L denoting the total cavity length [m] and r_l, r_r denoting the dimensionless amplitude reflectivity at the left and right facets, respectively.

By further substituting the model parameters into (4.32), the static optical field distribution along the cavity, known as the optical standing wave pattern (longitudinal mode) or the eigenfunction, as well as the complex frequency deviation can readily be obtained under the given facet and operating (bias) conditions through the transfer matrix method [14] with each section along the cavity (in z direction) represented by a 2×2 matrix. As the eigenvalue of the device, the complex frequency takes clear physical meaning: its real part is the detuning of the lasing frequency from the reference and its imaginary part the net gain known as the difference between the modal gain provided by the injection and the total losses scaled by the optical wave group velocity. The total losses include various modal losses seen by the wave propagation along the waveguide inside the laser cavity plus the terminal loss due to the power escape on both ends of the laser cavity. It is the latter that forms the laser optical power output. As the net gain of the laser, the imaginary part of the eigenvalue is also a measure of the laser coherence: Its reciprocal simply gives the laser coherent time.

The lasing mode power, frequency, and longitudinal field distribution calculated from (4.32) are used for the re-evaluation of the 2D potential, carrier, and current density distributions through (4.17) to (4.21), and the 2D temperature distribution through (4.30), respectively, from section to section. The material gains and refractive index change are also recalculated through (4.24), (4.25), and (4.28) based on the new band structure obtained for the updated potential distribution. The cross-sectional optical field distribution is also re-evaluated through (4.10) for the updated refractive index in the active region.

This 2D cross-sectional (x, y) – 1D longitudinal (z) iteration loop continues until convergence is reached. By updating the bias, we can let this loop start again until the maximum required bias is reached. Such a scheme for the DC analysis can be further summarized as follows:

1. 3D geometrical structure input
2. Material parameter input
3. Longitudinal subdivision (in z direction)
4. Mesh setup for cross-section sheets [in (x, y) plane]
5. Solver initialization
6. Operating condition input (bias looping starts here)
7. Variable scaling (physical to numerical)
8. 2D–1D iteration loop starts
9. Call solver-D1 [for (4.10), to get the cross-sectional optical field distribution and effective index]
10. Call solver-D2 [for static equations (4.17) to (4.21), to get the potential, carrier, and current density distributions]
11. Call solver-D3 [for static equation (4.30), to get the temperature distribution]
12. Call solver-D4 [through (4.24), (4.25), and (4.28), to get the material gains and refractive index change]
13. Call solver-D5 [for (4.32), to get the lasing mode optical power, frequency, and longitudinal field distribution]
14. Go to the iteration starting point (step 8) if not converged, otherwise
15. Variable scaling (numerical to physical)
16. Post processing for required output assembly
17. Go to step 6 for bias looping until the maximum setting is reached, otherwise
18. Stop

In the subdivision along the longitudinal (cavity) direction, a section length in 50 ∼ 100 wavelength periods is normally sufficient to capture the change of the slow-varying optical field envelope caused by LSHB. This requires 10 ∼ 40 subsections in total for the simulation of a typical semiconductor laser operated in the C-band with the cavity length and normalized coupling strength (κL) no more than 1 mm and 10, respectively. The computation effort will be greatly saved if the adaptive longitudinal subdivision scheme is introduced particularly for those lasers with long cavity length. Namely, the cavity will initially be subdivided into a small number of subsections and the longitudinal field obtained by this division scheme will be stored. The cavity will then be redivided into a doubled number of subsections, and the longitudinal field will be calculated again and compared with the stored one. This process will continue until the longitudinal field converges. The corresponding subdivision scheme will then be adopted.

Small-signal dynamic (SS-AC) analysis is straightforward based on the DC analysis. By letting the time-varying bias, thus every time-dependent function, be equal to its reference DC value plus a time-dependent harmonic term with a small amplitude (comparing with the reference DC value), we can expand all time-dependent equations according to the frequency harmonic orders. By balancing the terms in the same harmonic order, we can obtain an extra set of linearized equations in addition to the original DC equations. This set of linear equations is, therefore, used for the computation of the unknown variable amplitudes for the given bias amplitude. The reference DC values can obviously be obtained by solving the corresponding set of DC equations following the same approach as previously discussed. If the harmonic frequency is taken as the variable and is looped in these calculations, the small signal frequency dependence of all variables, particularly the lasing mode optical power and frequency known as the intensity modulation (IM) and parasitic frequency modulation (FM) responses, respectively, can readily be obtained. Higher order responses, such as the intermodulation and higher order harmonic distortions, can be computed through a similar process where an extra harmonic term with a small amplitude at a different frequency will be further added to the time-varying bias. In this case, every time-dependent function is assumed to be the summation of a reference DC value and a series of harmonic terms with their frequencies positioned at $i_1 f_1 + i_2 f_2 = 0, \pm 1, \pm 2, \ldots$, where $f_{1,2}$ are the two different modulation frequencies (tones) of the bias. As only linear equations are involved, numerical SS-AC analysis is straightforward; hence, it will not be further discussed here.

Large-signal dynamic (LS-AC) analysis has to be performed directly by following the variable evolution in time domain (TD). A TD marching scheme is usually adopted for such analysis:

1. 3D geometrical structure input
2. Material parameter input
3. Longitudinal subdivision (in z direction)

4. Mesh setup for cross-section sheets [in (x,y) plane]
5. Solver initialization
6. Variable scaling (physical to numerical)
7. TD marching starts (by setting $\Delta t = \Delta z / v_g$)
8. Operating condition input (read in bias as function of time)
9. Call solver-A1 [same as solver-D1 for (4.10), to get the cross-sectional optical field distribution and effective index]
10. Call solver-A2 [for dynamic equations (4.17) to (4.21), to get the potential, carrier, and current density distributions]
11. Call solver-A3 [for dynamic equations (4.30), to get the temperature distribution]
12. Call solver-A4 [same as solver-D4, to get the material gains and refractive index change through (4.24), (4.25), and (4.28)]
13. Call solver-A5 [for (4.4), to get the lasing mode optical power and longitudinal field distribution]
14. Go to the TD marching starting point (step 7) if the maximum time is not reached, otherwise
15. Variable scaling (numerical to physical)
16. Post processing for required output assembly (optical power spectrum computed by taking FFT of a power data stream cut in TD)
17. Stop

Ideally, the cavity subdivision and TD marching schemes, i.e., the longitudinal section length (Δz) and timestep (Δt), should be selected from different considerations. The former should be selected to catch up the spatial nonuniformity of the functions along the cavity due to LSHB, whereas the latter should be chosen to follow the varying of the functions due to the time-dependent bias change. Therefore, two possibilities exist if the two steps are chosen from different considerations: the one that makes the marching (numerical propagation) at a faster speed $\Delta z/\Delta t > v_g$ and the other that makes the marching at a slower speed $\Delta z/\Delta t < v_g$, both in comparing with the physical wave propagation speed. The former scheme is obviously non-causal and, thus, should be avoid in any case. The latter requires 2D–1D looping within each timestep until certain convergence is reached; otherwise, the timestep will be too big to give an accurate evaluation on the variable changes from section to section. As the 2D–1D looping will take tremendous computation time, in LS-AC analysis, we therefore cannot introduce such looping at every timestep. This leaves us the only choice to keep the marching speed exactly the same as the wave propagation speed. To do this, we first select Δt to be sufficiently small in order to capture the fastest possible varying in the bias and then select Δz to satisfy $\Delta z = v_g \Delta t$. If Δt is smaller than 0.25 ps, Δz calculated by this new rule will be smaller than 25 μm or $\sim 50\times$ wavelength inside the semiconductor laser cavity. Such longitudinal section length is therefore sufficient to treat the LSHB effect. For DFB structures with moderate grating coupling coefficient, although Δz can be made larger in DC and SS-AC analyses due to the weak LSHB effect, we have to stay with this selection rule in LS-AC analysis; otherwise this numerical marching algorithm will generate incorrect results. Another extreme happens when the given time-varying bias changes slowly. Δt selected by the varying rate of the bias could be significantly larger than 0.25 ps, and the related Δz will therefore be too large to capture the LSHB effect. In order to treat the LSHB accurately, smaller Δt still has to be used even if it is not necessary in handling the time-dependent bias. To satisfy the requirement from both aspects, the time-varying bias and the LSHB effect, we usually have to select smaller step sizes. However, this marching algorithm still saves the total computation effort as the 2D–1D iteration is no longer necessary as the time advances step by step. The accuracy is retained as the numerical marching is at the same speed of the physical wave propagation.

A careful examination on the above algorithms indicates that (1) the convergence failure mainly comes from the incorrect root-searching result and (2) the computation time is mainly spent on the carrier transport and optical gain calculations. To solve the convergence problem and to enhance the computation efficiency, several interrelated techniques are used in our solvers and will be briefly introduced in the following discussions.

4.3.2 Solver for Optical Wave Equations

The coupled traveling wave equations (solver-A5) can easily be solved through time-domain iterations if the time-derivatives are replaced by a forward leap finite difference scheme as the equations are explicit. However, both real and complex variable root-searching problems are encountered in dealing with the eigenvalue problems in the cross-sectional area (solver-D1 or A1) and along the cavity direction (solver-D5). Depending on the complexity of the DFB laser structure, sometimes we find that the searching fails or hits the wrong root, even when a sophisticated Muller search algorithm is adopted. To solve this problem, we have developed a successive scanning method that can guarantee the searching success.

Actually, we let the searching always start from a simple grating structure by assuming, e.g., zero facet reflections from both ends. In this case, the searching will not fail. Altering the parameters gradually and successively toward the given structure, we can always find the right root as long as we keep updating the initial guess by taking the previous exact solution. A similar idea also applies to those solvers in dealing with the arbitrary material parameters and operating conditions, as we can always reach the given condition by gradually and successively altering the parameters from a well-handled (known) initial condition.

4.3.3 Solver for Carrier Transport Equations

For any new bias, the rigorous carrier transport solver (D2) is called first. In the following 2D–1D iterations at the same bias, a simplified carrier balance equation [15] is used within the active region until quasi-convergence is reached. The rigorous solver is called again to check if this quasi-convergence is true. If not, the above process starts again until a real convergence is reached. As the simplified carrier balance equation solver takes significantly less amount of the execution time, this arrangement always saves computation effort.

4.3.4 Solver for Optical Gain Model

For the given geometrical structure and material compositions in the SL-MQW active region, both stimulated and spontaneous emission gain profiles and the associated refractive index change profile are first calculated based on the KP theory, as described in Sect. 4.2.3, under different injection levels and for different temperatures prior to any iteration process. Using the data collected in this preprocessing stage, analytical formulas taking the form of (4.34) are extracted for the gains and refractive index change by fitting all unknown coefficients to the precalculated numerical results [16]:

$$\left[\sum_{i=0}^{I} a_i e^{\frac{T-T_{\text{ref}}}{T_i^a}} N^i + \sum_{j=0}^{J} b_j e^{\frac{T-T_{\text{ref}}}{T_j^b}} P^j\right] \frac{\sum_{l=0}^{L} c_l e^{\frac{T-T_{\text{ref}}}{T_l^c}} \omega^l}{\sum_{m=0}^{M} d_m e^{\frac{T-T_{\text{ref}}}{T_m^d}} \omega^m}. \qquad (4.34)$$

Given in rational functions with their orders also determined by the best fitting, these analytical formulas are general enough to reproduce the dependence of the gains and refractive index change on the carrier, temperature, and frequency within the laser operating range. In those 2D–1D iterations in DC and SS-AC analyses or in the TD marching in LS-AC analysis, we actually use the extracted analytical formulas for the evaluation of the material gains and refractive index change instead of calling the numerical solver (D4 or A4) repeatedly.

4.3.5 Solver for Thermal Diffusion Equation

Noting the fact that the temperature cannot vary rapidly, in LS-AC analysis, we update the temperature only after the accumulated marching time exceeds $0.1\,\mu\text{s}$. In DC and SS-AC analyses, although a semi-analytical treatment [17] can still be introduced in a similar way as we treat the carrier solver in Sect. 4.3.3, we do not adopt it as we have found that the convergence of the 2D–1D iteration is not sensitive to the result from the numerical thermal solver (D3 or A3).

4 Distributed Feedback Lasers   105

Table 4.1. List of Parameters Used for the Simulation of 1.55-μm InGaAsP/InP Partially Gain-coupled RW SL-MQW DFB Lasers Made by Nortel Networks.

Layer	Thickness [nm]	band gap Wavelength [nm]	Composition (x, y) of $\text{In}_{1-x}\text{Ga}_x\text{As}_y\text{P}_{1-y}$	Doping $[10^{18}\text{cm}^{-3}]$	Compressive Strain
Cross-sectional Structure					
Substrate	100×10^3	918.6	0.000, 0.000	N 3.0	0
N-Buffer	1.5×10^3	918.6	0.000, 0.000	N 1.0	0
GRINSCH-1	25	1050.0	0.108, 0.236	Undoped	0
GRINSCH-2	25	1100.0	0.145, 0.317	Undoped	0
GRINSCH-3	25	1150.0	0.181, 0.395	Undoped	0
GRINSCH-4	25	1250.0	0.249, 0.539	Undoped	0
Well $\times$ 5	5.5	1570.0	0.122, 0.728	Undoped	1.5%
Barrier $\times$ 4	10	1250.0	0.249, 0.539	Undoped	0
GRINSCH	70	1000.0	0.068, 0.150	Undoped	0
Grating		Period: 242.6 nm, Depth: 110 nm, Triangular-shape with duty-cycle: 60%			
P-Buffer	200	918.6	0.000, 0.000	P 0.2	0
Etching Stop	3	1050.0	0.108, 0.236	P 0.2	0
P-Cladding	1.5×10^3	918.6	0.000, 0.000	P 0.5–1.0	0
P$^+$-Cap	200	1654.0	0.468, 1.000	P 10.0	0
Ridge Width			2.0 μm		
Longitudinal (Cavity) Structure					
Cavity Length			250 μm		
Rear/Front Facet Reflectivity			75%/0.05% (power)		

Table 4.2. List of Calculated and Extracted Modal Parameters for 1.55-μm InGaAsP/InP Partially Gain-coupled RW SL-MQW DFB Lasers.

Modal Parameter	Self-consistently Calculated	After Calibration	Final Adjustment
Series resistance [Ω]	3.26	No change	No change
Effective index	3.18	No change	No change
Group index	3.64	No change	No change
Confinement factor	2.89%	No change	No change
Normalized coupling coefficient [$\|\kappa L\|$]	6.57	No change	No change
Gain coupling ratio $[(\chi_H - \chi_L)/(\chi_H + \chi_L)]$	24.4%	No change	No change
Modal loss [cm^{-1}]	7.16	**13**	No Change
Characteristic temperature [K]	–	–	**90**
Gain coefficient [cm^{-1}]	1192	No change	No change
Characteristic temperature [K]	−347	**−150**	No change
Transparency carrier density [10^{18}cm^{-3}]	0.51	**0.7**	No change
Characteristic temperature [K]	260	**100**	No change
Nonlinear gain saturation factor [10^{-17}cm^3]	–	**3**	No change
Linewidth enhancement factor	−3.15	No change	No change
Thermal induced index change rate [10^{-3}K^{-1}]	0.97	No change	No change
SRH recombination time constants τ_n, τ_p [ns]	0.76, 0.55	No change	No change
Characteristic temperature [K]	–	–	**120**
Spontaneous emission rec. coef. [10^{-10}cm^3/s]	1.74	No change	No change
Characteristic temperature [K]	–	–	**120**
Auger rec. coef. C_n, C_p [10^{-29}cm^6/s]	3.21, 1.85	No change	No change
Characteristic temperature [K]	–	–	**150**
Rear/Front facet phase [degree]	0/0	**135/0**	No change

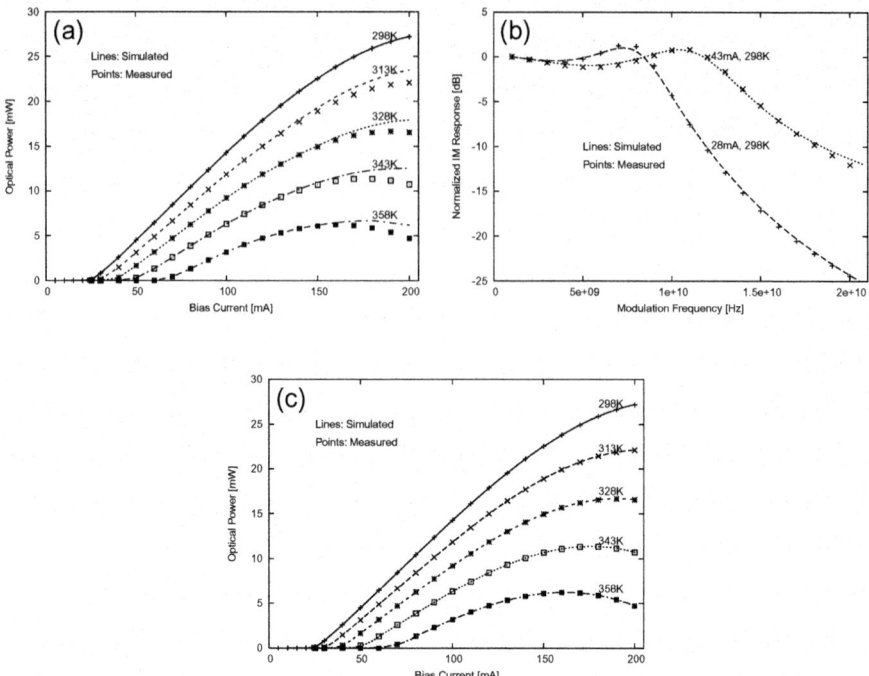

Fig. 4.3. (a) Optical power and bias current dependence (*L-I* curve) under different heat sink temperatures (without parameter fitting at high temperature; *Lines:* calculated data; *Points:* measured data), (b) Normalized small-signal intensity modulation response under different DC bias currents (*Lines:* calculated data, *Points:* measured data), (b) Optical power and bias current dependence (*L-I* curve) under different heat sink temperatures (after parameter fitting at high temperature; *Lines:* calculated data; *Points:* measured data).

4.4 Model Validation

In order to validate this model, we have compared the simulated results with the experimentally measured data for 1.55-μm InGaAsP/InP partially gain-coupled ridge waveguide (RW) SL-MQW DFB lasers made by Nortel Networks [18, 19]. The structure design parameters are provided in Table 4.1. Some of the self-consistently calculated modal parameters are summarized in Table 4.2, column 2. A few of the modal parameters are manually adjusted by fitting the simulated optical power and bias current (*L-I*) dependence to the measured data at room temperature (25°C). The altered values are listed in Table 4.2, column 3. This calibration must be introduced before any meaningful comparison, as the design parameters in Table 4.1 are just nominal values; hence, they may not be accurate after the fabrication. After the calibration, all parameters are fixed and the *L-I* dependence under different heat sink temperatures and the small-signal intensity modulation response at different bias currents are calculated as shown in Fig. 4.3(a) and (b), respectively. The corresponding measured data are also plotted in those figures. From these comparisons, we find that excellent agreement between the simulated and measured results is achieved near room temperature, which indicates the consistency of this model. However, discrepancies in L-I curves under high temperatures are also observed. This can be attributed to the inaccuracy in the evaluation of parameters, such as the nonradiative recombination coefficients and optical losses at high temperatures, as they are assumed to be constants but are actually temperature dependent. By letting these parameters (X) take exponential dependence on temperature (i.e., $X = X_0 \exp[(T - T_{\text{ref}})/T_X]$) and extracting their characteristic temperatures (T_X) through L-I curve fitting at a higher temperature (55°C), we can again achieve excellent agreement between the simulated and measured results for the entire operating temperature region, as shown in Fig. 4.3(c). The newly fitted characteristic temperatures are given in Table 4.2, column 4.

4.5 Model Comparison and Application

4.5.1 Comparison among Different Models

Among numerous models for the simulation of semiconductor laser diodes in the literature, we will limit our scope of discussion to those physics-based models only. We can generally classify those existing physics models into two categories. The models in the first category [20–24] are mainly established for the simulation of the carrier transport process in the cross-sectional area with optical mode distribution, material gain, and thermal diffusion considered from the first principles. However, the optical wave propagation along the cavity is normally treated as uniform. Therefore, these models are 2D in cross-sectional area, hence, suitable for studying devices with complex cross-sectional but simple longitudinal structures such as RW or buried hetero-junction (BH) SL-MQW FP laser diodes, where the coupled physical processes, such as the carrier transport, optical mode distribution, material gain, and thermal diffusion, need to be accurately described. The models in the second category [25–29] are mainly derived for the simulation of the optical wave propagation along the cavity where the carrier transport, optical mode distribution, material gain and thermal diffusion in the cross-sectional area are either treated through simplified equations and analytical formulas or lumped into effective and modal parameters. Therefore, these models are 1D along the cavity, hence, suitable for handling devices with simple cross-sectional but complex longitudinal structures such as DFB laser diodes with bulk active regions, where accurate treatment of the optical wave propagation along the nonuniform cavity is required, as the LSHB effect is much pronounced, which leads to a complex lasing process.

Our quasi-3D model combines the conventional 2D and 1D models in the two categories in a self-consistent manner. It is, therefore, suitable for handling those devices with combined complexity in cross section and along the cavity such as RW or BH SL-MQW DFB laser diodes.

The following examples show how this model is applied to the design and simulation of low distortion, coolerless 1.3-μm InAlGaAs/InP BH SL-MQW DFB laser diodes for analog optical communication systems and directly modulated 1.55-μm InGaAsP/InP RW SL-MQW DFB laser diodes for 10-Gbps digital optical communication systems. In both cases, low-threshold and high-temperature performance must be achieved, which requires optimized design on the cross-sectional structure. On the other hand, the LSHB effect will affect the *L-I* linearity that determines the cross-talk level among different cable television (CATV) channels under subcarrier multiplexed (SCM) scheme in the first example [30], and cause the lasing mode to be unstable and introduce frequency chirp that jointly determine the quality of the optical signal waveform in the second example [31–33]. Therefore, the optimized DFB grating design is also necessary in order to reduce the LSHB effect. For such problems where both cross-sectional and longitudinal structures are crucial, this model provides users the best design and optimization tool as the whole 3D structure is incorporated and the device performance can be efficiently simulated for the user-provided 3D structure and operating condition.

4.5.2 1.3-μm InAlGaAs/InP BH SL-MQW DFB Laser Diode

Low threshold current, high quantum efficiency, and high characteristic temperature are crucial for coolerless operation. To raise the characteristic temperature, InAlGaAs is first chosen as the active region material to replace the conventional InGaAsP quaternary [34]. In the laser structure design, the transparency carrier density, current leakage, and modal loss all need to be minimized, whereas the gain coefficient and confinement factor need to be maximized. To further reduce the intermodulation distortion introduced by the LSHB effect, the coupling coefficient of the grating, the cavity length, and the facet coating must be optimized as well. By following the approach summarized below, a set of optimized design parameters are provided in Table 4.3.

1. Well design: larger compressive strain (1.5%) is preferred in order to reduce the threshold and increase the gain coefficient [35, 36]; 6 × 5-nm wells are used to increase the confinement factor and to improve the optical mode confinement; the well composition is therefore determined by setting the gain peak at 1300 nm.
2. Barrier design: for the given wells, the barrier thickness is selected as 10 nm to avoid any bounded state coupling among different wells; the barrier composition is optimized in terms of the transparency carrier

Table 4.3. List of Optimized Parameters in the Design and Simulation of Low-distortion, Coolerless 1.3-μm InAlGaAs/InP BH SL-MQW DFB Lasers for Analog Optical Communication Systems.

Layer	Thickness [nm]	band gap Wavelength [nm]	Composition (x, y) of $In_{1-x-y}Al_x Ga_y As$	Doping $[10^{18}cm^{-3}]$	Compressive Strain
Cross-sectional Structure					
Substrate	100×10^3	918.6	InP	N 3.0	0
N-Buffer	1.5×10^3	918.6	InP	N 1.0	0
SCH-1	80	936.3	0.393, 0.081	Undoped	0
SCH-2	100	949.5	0.383, 0.092	Undoped	0
SCH-3	10	1000.0	0.347, 0.127	Undoped	0
Well $\times$ 6	5	1359.0	0.190, 0.064	Undoped	1.5%
Barrier $\times$ 5	10	1000.0	0.347, 0.127	Undoped	0
SCH-3	10	1000.0	0.347, 0.127	Undoped	0
SCH-2	100	949.5	0.383, 0.092	Undoped	0
SCH-1	120	843.9	0.480, 0.000	Undoped	0
Grating		Period: 202.0nm, Depth: 40nm, Rectangular-shape with duty-cycle: 50%			
P-Cladding	1.7×10^3	918.6	InP	P 0.5–2.0	0
P$^+$-Cap	250	1654.0	0.000, 0.468	P 30.0	0
Mesa Shape		Stripe width: 2.5 μm, Mesa height: 3.0 μm			
BH Block		N-InP block layer: 1.0 μm, P-InP block layer: 2.0 μm			
Longitudinal (Cavity) Structure					
Cavity Length		300 μm			
Rear/Front Facet Reflectivity		75%/1% (power)			

Table 4.4. List of Calculated Modal Parameters for 1.3-μm InAlGaAs/InP BH SL-MQW DFB Lasers.

Modal Parameter	Self-consistently Calculated
Series resistance [Ω]	1.89
Effective index	3.22
Group index	3.54
Confinement factor	3.33%
Normalized coupling coefficient [$|\kappa L|$]	0.71
Gain coupling ratio [$(\chi_H - \chi_L)/(\chi_H + \chi_L)$]	0
Modal loss [cm^{-1}]	5.01
Gain coefficient [cm^{-1}]	1966
Transparency carrier density [10^{18}cm^{-3}]	0.3
Linewidth enhancement factor	-1.1
SRH recombination time constants τ_n, τ_p [ns]	0.75, 0.54
Spontaneous emission rec. coef. [10^{-10}cm^3/s]	2.03
Auger rec. coef. C_n, C_p [10^{-29}cm^6/s]	2.59, 1.11

density and the mode confinement, as lower barrier (longer band gap wavelength) will result in multiple bounded states in the conduction band, whereas higher barrier (shorter band gap wavelength) may cause poorer optical mode confinement.
3. SCH and the stripe width design: for the given wells and barriers, the SCH layer thickness, composition, and stripe width are optimized to form an almost rounded far-field pattern.
4. Only one InAlAs layer is inserted at the P-side to block the electron leakage along the injection path. The N-side InAlAs layer is replaced by an InAlGaAs layer with its band gap energy smaller than that of InP to avoid any blocking of the electron injection to the quantum wells, as the N-side InAlAs layer cannot effectively block the hole leakage anyway due to the small band offset in the valence band at the InAlGaAs/InP heterojunction [34, 37].
5. Standard BH structure is adopted to block the carrier lateral leakage.
6. Index-coupled grating is designed in such a way that the normalized coupling coefficient will be 0.7 as it minimizes the LSHB effect if combined with HR/AR facet coating [38].

Some of the self-consistently calculated modal parameters are summarized in Table 4.4. The calculated L-I curve, quantum well valence band structure, material gain profile, optical mode distribution, far-field pattern, longitudinal field and carrier density distributions, and optical spectrum are given in Fig. 4.4(a)–(g). We find that the device performs well at 85°C with threshold < 20 mA and almost no degradation on slope efficiency. The calculated second- and third-order intermodulation distortion spectra are also shown in Fig. 4.5(a) and (b). Theoretical results show that the composite second-order (CSO) and the composite triple beat (CTB) harmonic distortions ($f_1 - f_2$ and $f_1 - 2f_2$ or $2f_1 - f_2$) are below -60 and -65 dB, respectively, over a frequency range up to 2 GHz. Hence, a such designed device satisfies the requirement in SCM optical transmission systems for CATV and cellular mobile communications [39, 40].

4.5.3 1.55-μm InGaAsP/InP RW SL-MQW DFB Laser Diode

Carrier transport through the undoped SCH and barrier layers becomes the major limit for high-speed direct modulation of SL-MQW DFB laser diodes up to 10 Gbps [41]. However, the SCH and barrier layers are essential and the reduction of their total thickness is limited by mode confinement and number of quantum well requirements. Structures with reduced SCH thickness and less number of quantum wells will likely lead to the reduction on the confinement factor and gain coefficient, and hence cause a reduced 3 dB modulation bandwidth [42]. To solve this problem, we will dope the SCH layers on both N- and P-sides and introduce P-type doping in the barrier layers [43]. In the device structure design, our goal is to maximize the relaxation oscillation frequency as the device must have broad 3-dB modulation bandwidth. This can be realized through maximizing the composite factor $\eta_i(\mathrm{d}g/\mathrm{d}N)\Gamma/V_{\mathrm{ar}}$ and reducing the threshold current [42], where η_i, $\mathrm{d}g/\mathrm{d}N$, Γ and V_{ar} are the internal quantum efficiency, differential gain, optical confinement factor, and active region volume, respectively. As Γ is roughly scaled by the active region area where gain appears over the entire optical field distributed area, the composite factor can be rewritten as $\eta_i(\mathrm{d}g/\mathrm{d}N)/(L\Sigma)$, where L and Σ denote the cavity length and the entire optical field distributed area, respectively. By introducing large compressive strain, we will be able to increase the internal quantum efficiency and differential gain [44], as well as to reduce the threshold current. The major task in the cross-sectional structure design, therefore, becomes to reduce the optical mode spot. In the DFB grating design, we need a large coupling coefficient (κ), as the laser cavity L must be short and yet the normalized coupling coefficient (κL) should be large enough to retain a sufficient side-mode suppression ratio (SMSR). On the other hand, the coupling cannot be made too strong, as a large κL reduces the relaxation oscillation frequency [45]. Moreover, the LSHB effect must be minimized as

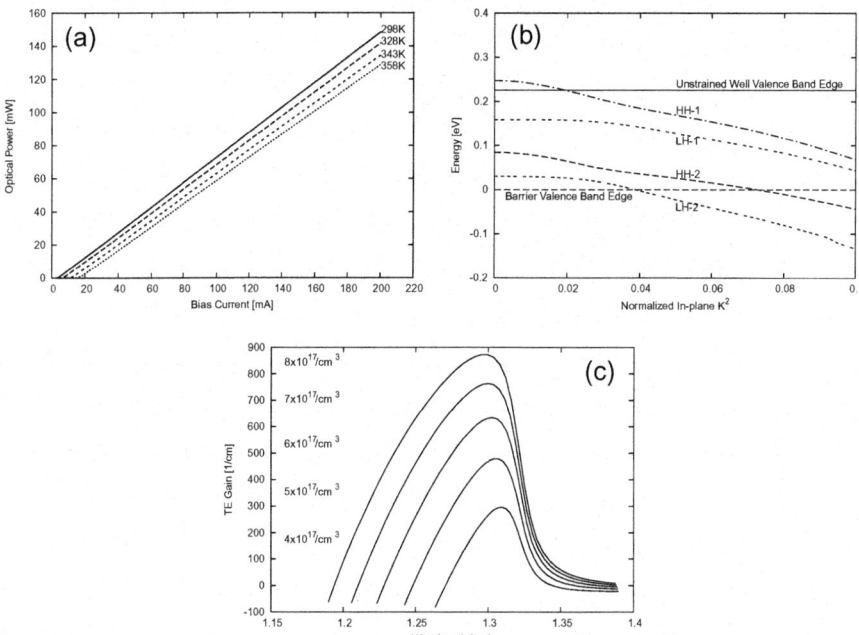

Fig. 4.4. (a) Simulated optical power and bias current dependence (*L-I* curve) under different heat sink temperatures for a coolerless 1.3-μm InAlGaAs/InP BH SL-MQW DFB laser diode. (b) Valence band energy and momentum dependence (dispersion curve) of $\mathrm{In}_{0.746}\mathrm{Al}_{0.190}\mathrm{Ga}_{0.064}\mathrm{As}$ (well) – $\mathrm{In}_{0.526}\mathrm{Al}_{0.347}\mathrm{Ga}_{0.127}\mathrm{As}$ (barrier) structure. (c) Material gain profile under different carrier injection levels.

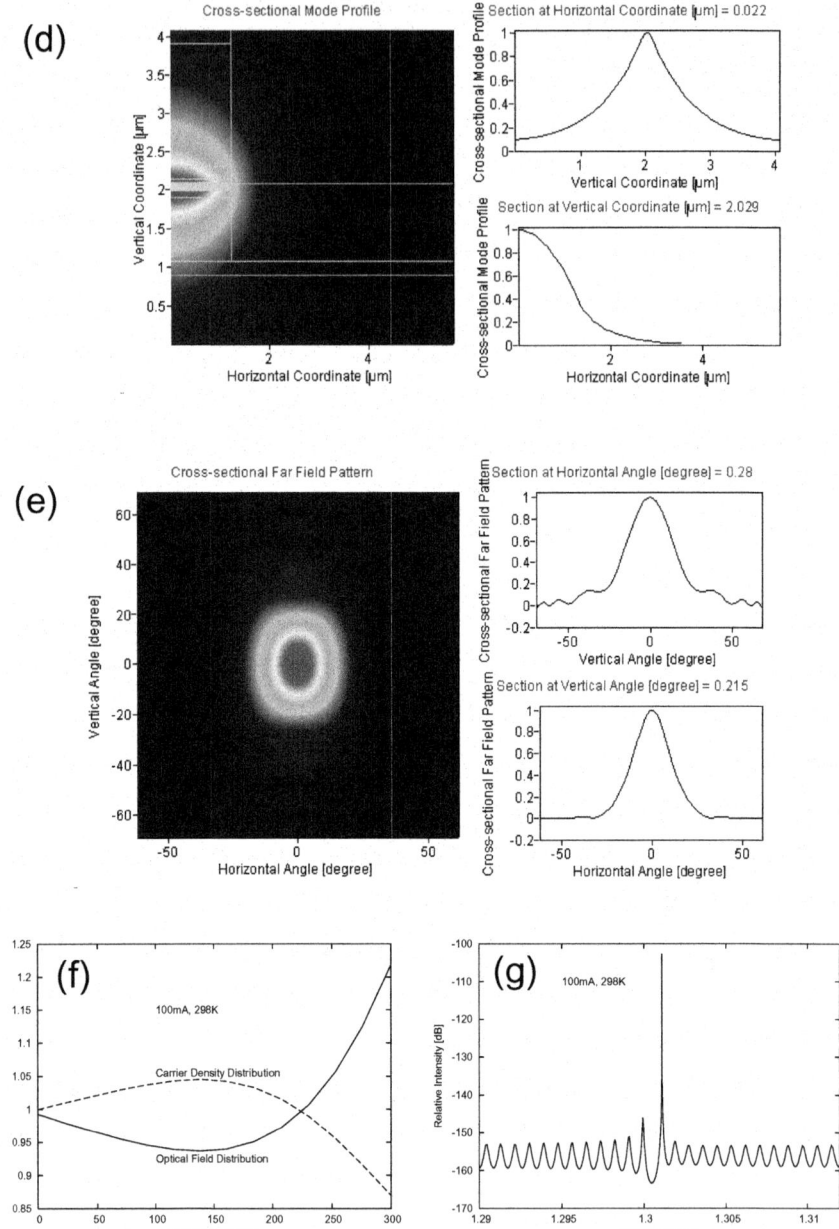

Fig. 4.4. (d) Cross-sectional optical field distribution (left half structure only). (e) Far-field pattern. (f) Normalized optical field and carrier density distributions along the laser cavity under a bias current of 100 mA and at room temperature (25°C). (g) Optical spectrum measured from the front facet under a bias current of 100 mA and at room temperature (25°C).

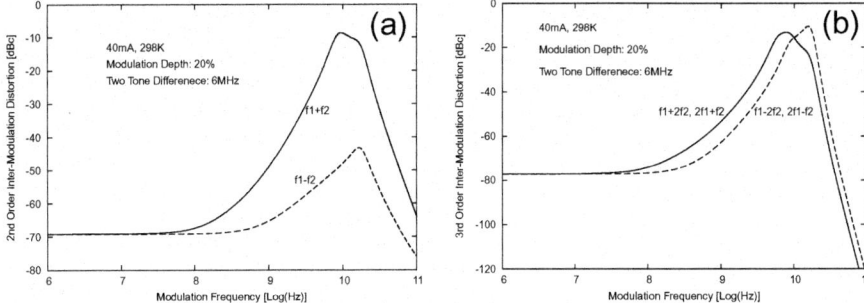

Fig. 4.5. (a) Small-signal second order intermodulation distortion spectrum when the two tones are separated by 6 MHz with 20% modulation depth under a bias current of 40 mA and at room temperature (25°C). (b) Small-signal third order intermodulation distortion spectrum when the two tones are separated by 6 MHz with 20% modulation depth under a bias current of 40 mA and at room temperature (25°C).

it will cause lasing mode instability and introduce frequency chirp when the laser is directly modulated at very high speed. Therefore, the index-coupled uniform grating DFB with relatively short cavity is adopted rather than the quarter-wavelength ($\lambda/4$) shifted DFB where LSHB is significant especially when it is designed for strong coupling. The DFB mode degeneracy problem associated with the index-coupled uniform grating is solved by introducing HR/AR facet coating.

Based on all of these considerations, a set of optimized design parameters are given in Table 4.5, where the well and barrier Ga compositions are intentionally designed the same to release the stringent requirement on the wafer growth as the barriers must be P-type doped. Some of the self-consistently calculated modal parameters are summarized in Table 4.6. The calculated optical power, SMSR, relaxation oscillation frequency, and the damping K-factor as functions of the bias current for structures with different number of quantum wells are given in Fig. 4.6(a)–(d). The calculated results show that the six-well structure gives the best DC performance, whereas the eight-well structure provides the highest relaxation oscillation frequency. At room temperature, a relaxation oscillation frequency over 20 GHz can be reached for both six-well and eight-well structures. This can be attributed to (1) P-type doping of the barriers; (2) improved $\eta_i(dg/dN)/\Sigma$; and (3) short cavity and strong coupling. We also find that a SMSR over 35 dB can be obtained for these structures for the entire operating range due to the low-LSHB design, which indicates that the lasing mode stability is acceptable. However, none of the structure seems to have sufficient 3-dB modulation bandwidth when the heat sink temperature is lifted to 85°C because the relaxation oscillation frequency drops rapidly as temperature increases, as shown in Fig. 4.6(c). We also find that the optical modal loss increases rapidly with the num-

Table 4.5. List of Optimized Parameters in the Design and Simulation of Directly Modulated 1.55-μm InGaAsP/InP RW SL-MQW DFB Lasers for 10-Gbps Digital Optical Communication Systems.

		Cross-sectional Structure			
Layer	Thickness [nm]	band gap Wavelength [nm]	Composition (x, y) of $\text{In}_{1-x}\text{Ga}_x \text{As}_y\text{P}_{1-y}$	Doping $[10^{18}\text{cm}^{-3}]$	Compressive Strain
Substrate	100×10^3	918.6	0.000, 0.000	N 3.0	0
N-Buffer	1.5×10^3	918.6	0.000, 0.000	N 1.0	0
GRINSCH-1	20	1100.0	0.145, 0.317	N 0.5	0
GRINSCH-2	20	1150.0	0.181, 0.395	N 0.5	0
GRINSCH-3	20	1200.0	0.216, 0.469	N 0.5	0
GRINSCH-4	20	1250.0	0.249, 0.539	N 0.5	0
Well × **N**	5	1650.0	0.249, 0.850	Undoped	1.0%
Barrier × **N–1**	8.5	1250.0	0.249, 0.539	P 0.3	0
GRINSCH-4	40	1250.0	0.249, 0.539	P 0.3	0
GRINSCH-3	40	1200.0	0.216, 0.469	P 0.3	0
GRINSCH-2	40	1150.0	0.181, 0.395	P 0.3	0
GRINSCH-1	40	1100.0	0.145, 0.317	P 0.3	0
Grating	Period: 242.0 nm, Depth: 160 nm, Rectangular-shape with duty-cycle: 50%				
P-Buffer	100	918.6	0.000, 0.000	P 0.5	0
Etching Stop	5	1100.0	0.145, 0.317	P 0.5	0
P-Cladding	1.5×10^3	918.6	0.000, 0.000	P 0.5–3.0	0
P$^+$-Cap	250	1654.0	0.468, 1.000	P 30.0	0
Ridge Width		1.8 μm			
		Longitudinal (Cavity) Structure			
Cavity Length		200 μm			
Rear/Front Facet Reflectivity		90%/1% (power)			

Table 4.6. List of Calculated Modal Parameters for 1.55-μm InGaAsP/InP RW SL-MQW DFB Lasers.

Modal Parameter	Self-consistently Calculated
Series resistance [Ω]	3.42
Effective index	3.20
Group index	3.53
Confinement factor	3.54%(4 wells)–8.44%(8 wells)
Normalized coupling coefficient [$|\kappa L|$]	4.24
Gain coupling ratio [$(\chi_H - \chi_L)/(\chi_H + \chi_L)$]	0
Modal loss [cm^{-1}]	20.6(4 wells)–34.6(8 wells)
Gain coefficient [cm^{-1}]	1484(4 wells)–2104(8 wells)
Transparency carrier density [10^{18}cm^{-3}]	1.10(4 wells)–1.79(8 wells)
Linewidth enhancement factor	-1.38
SRH recombination time constants τ_n, τ_p [ns]	0.79, 0.58
Spontaneous emission rec. coef. [10^{-10}cm^3/s]	1.85
Auger rec. coef. C$_n$, C$_p$ [10^{-29}cm^6/s]	6.70, 3.37

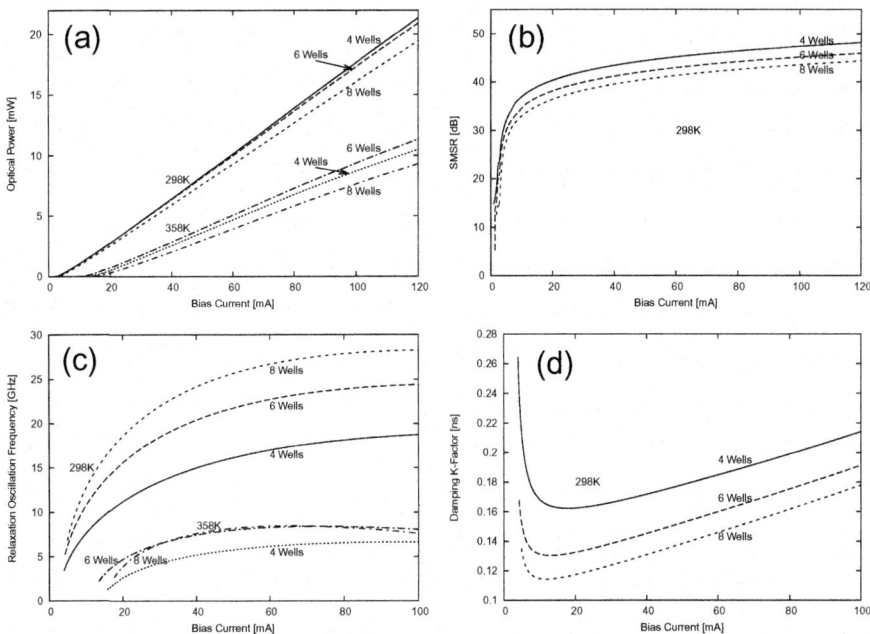

Fig. 4.6. (a) Simulated optical power and bias current dependence (*L-I* curve) under different heat sink temperatures for 1.55-μm InGaAsP/InP RW SL-MQW DFB laser diodes with different number of quantum wells. (b) Side-mode suppression ratio (SMSR) and bias current dependence for laser structures with different number of quantum wells at room temperature (25°C). (c) Relaxation oscillation frequency and bias current dependence (*L-I* curve) under different heat sink temperatures for laser structures with different number of quantum wells. (d) Damping K-factor and bias current dependence for laser structures with different number of quantum wells at room temperature (25°C).

ber of quantum wells. This explains the poor slope efficiency and the low-damping K-factor (defined in [46]) of the eight-well structure as shown in Fig. 4.6(a) and (d), respectively, as higher optical modal loss leads to lower-damping K-factors [43]. However, low damping K-factor leads to broad 3 dB modulation bandwidth [47]. Therefore, a trade-off must be made between the laser steady state and dynamic performance. Figure 4.7 shows the simulated IM and FM responses for structures with a different number of quantum wells under the same bias current. It is found that the parasitic FM response is insensitive to the number of quantum wells. As Fig. 4.7 also demonstrates that the structure with less quantum wells has higher IM response (due to the higher slope efficiency caused by the lower optical modal loss), the ratio between FM and IM responses, an indicator of the frequency chirp, is therefore lower for the structure with less quantum wells. This is in accordance with the measurement reported in [42]. Finally, TD optical signal waveform

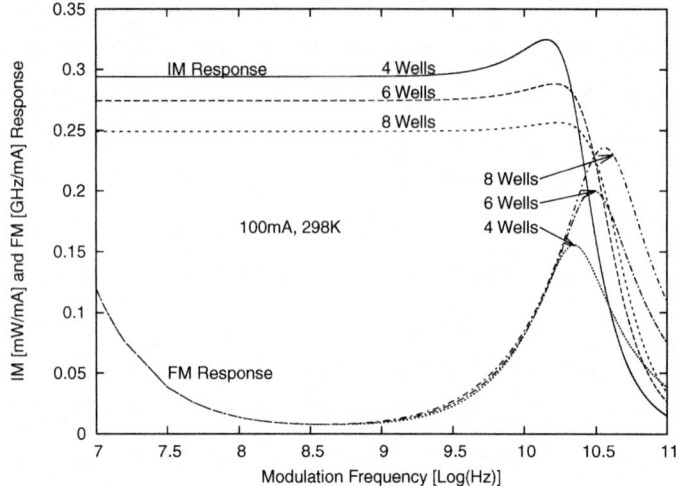

Fig. 4.7. Small-signal intensity modulation (IM) and frequency modulation (FM) responses for laser structures with different number of quantum wells under a bias current of 100 mA and at room temperature (25°C).

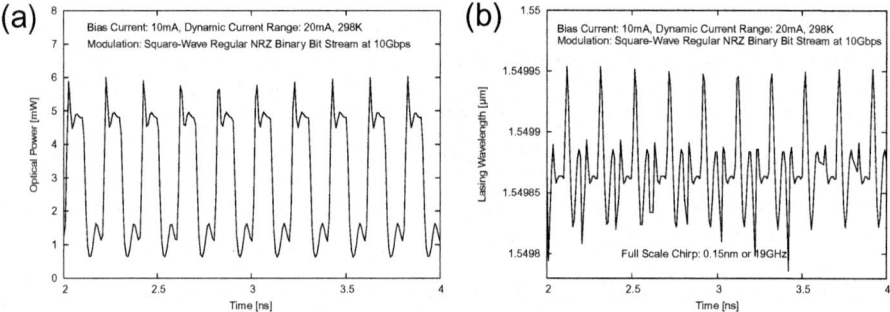

Fig. 4.8. (a) Simulated time-domain large-signal optical waveform under direct 10-Gbps modulation and at room temperature (25°C) (bias current = 10 mA; dynamic current range = 20 mA; square-wave regular NRZ binary bit stream with pulse width = 100 ps). (b) Simulated time-domain large-signal lasing wavelength chirp under direct 10 Gbps modulation and at room temperature (25°C) (bias current = 10 mA; dynamic current range = 20 mA; square-wave regular NRZ binary bit stream with pulse width = 100 ps).

and lasing wavelength chirp are simulated for the eight-well structure through the LS-AC analysis and the results are shown in Fig. 4.8(a) and (b). We find that reasonable optical waveform can be obtained when the laser is directly modulated under 10-Gbps binary stream and the full-scale frequency chirp is within 19 GHz. Therefore, a such designed device can be used as the directly modulated light source in 10-Gbps digital transmitters [48].

4.6 Summary

A comprehensive physics-based quasi-3D model for SL-MQW DFB laser diodes is developed. The model considers the optical field confinement, the carrier transport, and the heat transfer over the cross section of the laser. It also incorporates the LSHB effect, which leads to nonuniform distributions of the optical field, carrier concentration, current density, and temperature along the laser cavity. A rigorous model for the material gain and refractive index change is also adopted. It should be pointed out that although the conventional 1D model for DFB laser diodes can also produce accurate results given proper choice of modal parameters, the quasi-3D model presented here is capable of providing more detailed information such as the cross-sectional distributions of the optical field, carrier concentration, current density, potential, and temperature. Moreover, critical dependence of the laser performance on those design parameters such as the material compositions, the doping levels, and the cross-sectional geometrical structure, which are not considered in the simple 1D models, is accounted for in this model. On the other hand, this model can be viewed as an extension of the conventional 2D model for FP laser diodes as the longitudinal dependence is further taken into account. It is therefore capable of handling laser diodes with arbitrary complexity along the cavity, such as multiple-section DFBs/DBRs [49, 50], super grating DFBs/DBRs [51], sampled grating DBRs [52], complex-coupled DFBs [53, 54], and higher order grating DFBs [55]. A number of novel techniques are also proposed and demonstrated in the implementation of this model for efficient simulation of the device steady state and dynamic characteristics. By adopting these numerical techniques, we have made the physics-based quasi-3D simulation possible on PCs for SL-MQW distributed-feedback laser diodes.

References

1. M. Sargent III, M. O. Scully, and W. E. Lamb: *Laser Physics* (Addison-Wesley, Reading, MA 1974)
2. J. E. Bowers, C. A. Burrus, and R. J. McCoy: Electron. Lett. **21**, 812 (1985)
3. G. L. Tan, N. Bewtra, K. Lee, and J. M. Xu: IEEE J. Quantum Electron. **29**, 822 (1993)

4. R. Stratton: IEEE Trans. Electron Devices **19**, 1288 (1972)
5. N. Tessler and G. Eisenstein: IEEE J. Quantum Electron. **29**, 1586 (1993)
6. R. Nagarajan, M. Ishikawa, T. Fukushima, R. S. Geels, and J.E. Bowers: IEEE J. Quantum Electron. **28**, 1990 (1992)
7. S. L. Chuang: *Physics of Optoelectronic Devices* (Wiley, New York 1995)
8. D. Ahn, S. L. Chuang, and Y. C. Chang: J. Appl. Phys. **64**, 4056 (1988)
9. S. L. Chuang: Phys. Rev. B **43**, 9649 (1991)
10. C. H. Henry, R. A. Logan, and K. A. Bertness: J. Appl. Phys. **52**, 4457 (1981)
11. D. J. Channin: J. Appl. Phys. **50**, 3858 (1979)
12. H. S. Carslaw, and J. C. Jaeger: *Conduction of Heat in Solids* (Oxford Univ. Press, Oxford, U.K. 1959)
13. J. Piprek: *Semiconductor Optoelectronic Devices* (Academic Press, New York 2003)
14. M. Yamada, K. Sakuda: Appl. Optics **26**, 3474 (1987)
15. A. D. Sadovnikov, X. Li, and W. P. Huang: IEEE J. Quantum Electron. **31**, 1856 (1995)
16. X. Li, A. D. Sadovnikov, W. P. Huang, and T. Makino: IEEE J. Quantum Electron. **34**, 1545 (1998)
17. X. Li and W. P. Huang: IEEE J. Quantum Electron. **31**, 1848 (1995)
18. G. P. Li, T. Makino, R. Moore, and N. Puetz, K.W. Leong, H. Lu: IEEE J. Quantum Electron. **29**, 1736 (1993)
19. H. Lu, T. Makino, and G. P. Li: IEEE J. Quantum Electron. **31**, 1443 (1995)
20. J. Buus: IEE Proc. Part J **132**, 42 (1985)
21. T. Ohtoshi, K. Yamaguchi, C. Nagaoka, T. Uda, Y. Murayama, and N. Chinone: Solid-State Electron. **30**, 627 (1987)
22. K. B. Kahen: IEEE J. Quantum Electron. **24**, 641 (1989)
23. M. Ueno, S. Asada, and S. Kumashiro: IEEE J. Quantum Electron. **26**, 972 (1990)
24. Z. M. Li, K. M. Dzurko, A. Delage, and S. P. McAlister: IEEE J. Quantum Electron. **28**, 792 (1992)
25. P. Vankwikelberge, G. Morthier, and R. Baets: IEEE J. Quantum Electron. **26**, 1728 (1990)
26. L. M. Zhang and J. E. Carroll: IEEE J. Quantum Electron. **28**, 604 (1992)
27. A. Lowery, A. Keating, and C. N. Murtonen: IEEE J. Quantum Electron. **28**, 1874 (1992)
28. J. Hong, W. P. Huang, and T. Makino: IEEE J. Quantum Electron. **31**, 49 (1995)
29. W. Li, W. P. Huang, X. Li, and J. Hong: IEEE J. Quantum Electron. **36**, 1110 (2000)
30. M. Kito, H. Sato, N. Otsuka, N. Takenaka, and M. Ishino, Y. Matsui: IEEE Photonics Technol. Lett. **7**, 144 (1995)
31. J. E. A. Whiteaway, B. Garrett, G. H. B. Thompson, A. J. Collar, C. J. Armistead, and M. J. Fice: IEEE J. Quantum Electron. **28**, 1277 (1992)
32. J. E. A. Whiteaway, G. H. B. Thompson, A. J. Collar, and C. J. Armistead: IEEE J. Quantum Electron. **25**, 1261 (1989)
33. J. Kinoshita and K. Matsumoto: IEEE J. Quantum Electron. **24**, 2160 (1988)
34. C. E. Zah, R. Bhat, B. N. Pathak, F. Favire, W. Lin, M. C. Wang, N. C. Andreadakis, D. M. Hwang, M. A. Koza, T. P. Lee, Z. Wang, D. Darby, D. Flanders, and J. J. Hsieh: IEEE J. Quantum Electron. **30**, 511 (1994)

35. J. S. Osinski, P. Grodzinski, Y. Zhou, and P. D. Dapkus: Electron. Lett. **27**, 469 (1991)
36. E. Yablonovitch and E. O. Kane: OSA/IEEE J. Lightwave Technol. **4**, 504 (1986)
37. S. R. Selmic, T. M. Chou, J. P. Sih, J. B. Kirk, A. Mantie, J. K. Butler, D. Bour, and G. A. Evans: IEEE J. Selected Topics in Quantum Electron. **7**, 340 (2001)
38. T. Okuda, H. Yamada, T. Torikai, and T. Uji: IEEE Photonics Technol. Lett. **6**, 27 (1994)
39. R. Olshansky, V. A. Lanzisera, and ZP. M. Hill: IEEE J. Lightwave Technol. **7**, 1329 (1989)
40. M. Shibutani, T. Kanai, K. Emura, and J. Namiki: Proc. ICC'91, pp. 1176–1181, (1991)
41. R. Nagarajan, T. Fukushima, M. Ishikawa, J. E. Bowers, R. S. Geels, and L. Coldren: IEEE Photonics Technol. Lett. **4**, 121 (1992)
42. K. Uomi, M. Aoki, T. Tsuchiya, and A. Takai: IEEE J. Quantum Electron. **29**, 355 (1993)
43. K. Uomi, M. Aoki, T. Tsuchiya, and A. Takai: Fiber and Integrated Optics **12**, 17 (1993)
44. S. Seki, T. Yamanaka, W. Lui, Y. Yoshikuni, and K. Yokoyama: IEEE J. Quantum Electron. **30**, 500 (1994)
45. W. P. Huang, X. Li, and T. Makino: IEEE J. Quantum Electron. **31**, 842 (1995)
46. R. Olshansky, P. Hill, V. A. Lanzisera, and W. Powazinik: IEEE J. Quantum Electron. **23**, 1410 (1987)
47. J. E. Bowers, B. Hemenway, A. H. Gnauck, and P. Wilt: IEEE J. Quantum Electron. **22**, 833 (1986)
48. K. Uomi, T. Tsuchiya, H. Nakano, M. Aoki, M. Suzuki, and N. Chinone: IEEE J. Quantum Electron. **27**, 1705 (1991)
49. M. G. Davis and R. F. O'Dowd: IEEE J. Quantum Electron. **30**, 2458 (1994)
50. A. Tsigopoulos, T. Sphicopoulos, I. Orfanos, and S. Pantelis: IEEE J. Quantum Electron. **28**, 415 (1992)
51. N. Chen, Y. Nakano, K. Okamoto, K. Tada, G. I. Morthier, and R. G. Baets: IEEE J. Selected Topics in Quantum Electron. **3**, 541 (1997)
52. B. S. Kim, J. K. Kim, Y. Chung, and S. H. Kim: IEEE Photonics Technol. Lett. **10**, 39 (1998)
53. F. Randone, I. Montrosset: IEEE J. Quantum Electron. **31**, 1964 (1995)
54. A. Champagne, R. Maciejko, D. M. Adams, G. Pakulski, B. Takasaki, and T. Makino: IEEE J. Quantum Electron. **35**, 1390 (1999)
55. A. M. Shams, J. Hong, X. Li, and W. P. Huang: IEEE J. Quantum Electron. **36**, 1421 (2000)

5 Multisection Lasers: Longitudinal Modes and their Dynamics

M. Radziunas[1] and H.-J. Wünsche[2]

[1] Weierstraß-Institut für Angewandte Analysis und Stochastik, Mohrenstr. 39, 10117 Berlin, Germany, radziunas@wias-berlin.de
[2] Humboldt-Universität zu Berlin, Institut für Physik, Newtonstr. 15, 12489 Berlin, Germany, wuensche@physik.hu-berlin.de

5.1 Introduction

Multisection semiconductor lasers are useful for different purposes. Examples are wavelength tuning [1], chirp reduction [2], enhanced modulation bandwidths [3, 4], mode-locking of short pulses [5], and frequency-tunable self-pulsations [6, 7, 8, 9].

A deep understanding of these devices is required when designing them for specific functionalities. For these purposes, we have developed a hierarchy of models and a corresponding software *LDSL-tool* (abbreviation for (L)ongitudinal (D)ynamics in multisection (S)emiconductor (L)asers). This software is based on the system of hyperbolic partial differential traveling wave equations [10] for the spatio-temporal distribution of the optical field in the laser, nonlinearly coupled to ordinary differential equations for carrier densities and material polarization in the active parts of the devices [11]. Direct numerical integration of the model equations and a comprehensive data post-processing make our software a powerful tool well suited for simulation and analysis of different dynamical effects experimentally observed in semiconductor lasers [7, 8, 9, 12, 13, 14, 15]. In this chapter, we put special emphasis on the possibilities of *LDSL-tool* to analyze the dynamics in terms of longitudinal modes [8, 12, 13, 15, 16].

The concept of optical modes plays an outstanding role for understanding lasers in general. They represent the natural oscillations of the electromagnetic field and determine the optical frequency and the lifetime of the photons contained in a given laser cavity. In multisection lasers, the modes are exceptionally sensitive to variations of the carrier distributions. Not only the overall number of photons is involved in the dynamics, but also the internal spatial distribution of the photons. We will show how computation and analysis of instantaneous modes and field decomposition into these changing modal components help to understand the complicated dynamics of multisection lasers. This analysis allows us to determine the dimensionality of the observed dynamics as well as to understand sometimes curious temporal profiles and properties of the optical field power in these dynamical regimes.

Supplementing simulations with a consequent mode analysis provides a deeper insight into the nature of different operating regimes and allows one to understand the mechanisms of transitions between different regimes. This concept has been successfully used to design lasers with specific dynamics [12, 13], to analyze limitations of fabricated lasers operating at the required dynamical regime, and to improve the design of these lasers [8, 14, 15]. In this chapter, we shall illustrate the usefulness of mode analysis by a few examples. In particular, we shall consider two different self-pulsating regimes of the three-section structure sketched in Fig. 5.1(a). Multisection lasers of this type can serve as an optical clock in all-optical signal processing [6, 17].

The chapter is organized as follows. The model and the basic ideas of mode analysis are introduced step by step and illustrated with example calculations in Sects. 5.2 through 5.7. Details are suppressed here as much as possible in favor of clarity. Using all ideas introduced before, we analyze the dynamics of another laser example in Sects. 5.8 and 5.9. After summarizing conclusions in Sect. 5.10, we finish with mathematical details and a description of the adopted numerical schemes in the Appendix.

5.2 Traveling Wave Model

The calculation of spatio-temporal carrier and photon distributions in a semiconductor laser is, in general, a complicated multidimensional problem. Fortunately, the corresponding transverse distributions are mostly fixed by the device design and confined to regions small compared with the laser length. Under these circumstances, the transverse problem can be treated independently [18, 19] and will not be considered here. We investigate laser devices of this type consisting of m successively located sections S_k with lengths L_k [see Fig. 5.1(b)]. Here and in the following, the subscript indices $_k$ attribute quantities to the section S_k. The considered two counterpropagating optical

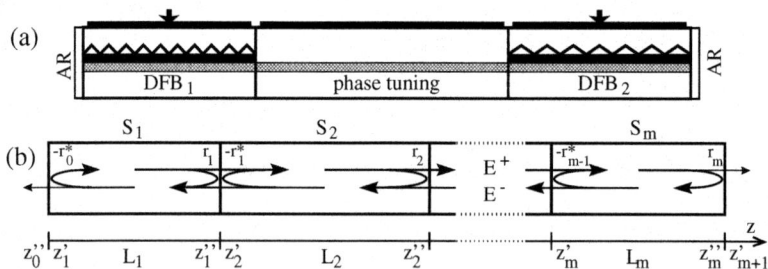

Fig. 5.1. Schemes of multisection semiconductor lasers. Panel a: scheme of the three-section laser treated as example in this chapter. Panel b: general scheme of a multisection laser composed of m sections with a sketch of counterpropagating optical waves. Details are described in the text.

waves are represented by their slowly varying amplitudes $E^+(z,t), E^-(z,t)$, a pair of complex valued functions of the longitudinal coordinate z, and the time t. The optical field drives the complex material polarization functions $p^+(z,t), p^-(z,t)$ and interacts with the real carrier density function $n(z,t)$. These quantities obey the traveling wave (TW) model equations (see, e.g., [10, 11]), which can be written in the compact form:

$$-i\frac{\partial}{\partial t}\Psi(z,t) = H(\beta, z)\Psi + F_{sp}, \tag{5.1}$$

$$\frac{\partial}{\partial t}n(z,t) = \mathcal{N}(n, |E|^2, \Re e(E^*p), z). \tag{5.2}$$

The wave function Ψ in (5.1) is the four-component column vector $\Psi = (E^+, E^-, p^+, p^-)^T$ of the optical and polarization amplitudes, where E satisfies boundary conditions at the interfaces of the sections and facettes. The superscripts ^{T} and * denote the transpose and complex conjugation, respectively. Accordingly, the evolution operator $H(\beta, z)$ is a 4×4 matrix containing first-order derivatives with respect to the space coordinate z within the sections. Details of this operator as well as of the boundary conditions are specified in the following Sect. 5.3. Here, we emphasize only that its dependence on the carrier densities and the optical fields is fully mediated by the distribution of the complex propagation parameter β defined in (5.3) along the optical waveguide. The inhomogeneity $F_{sp} = (F_{sp}^+, F_{sp}^-, 0, 0)^T$ represents the spontaneous emission noise added to the guided waves. It is a stochastic force in the sense of a Langevin calculus, which has been described elsewhere (see, e.g., [14, 20] for more details). The right-hand side of the carrier rate equation (5.2) comprises pumping by injection, spontaneous and stimulated recombination, and is also described in detail in Sect. 5.3.

More details of the model and of the used set of parameters will be specified in the next section. Readers more interested in examples and possible applications of our analysis can skip it and continue with Sect. 5.4.

5.3 Model Details and Parameters

5.3.1 Model Details

The linear operator $H(\beta, z)$ in (5.1) is defined as follows. If z belongs to the interior of a device section, it holds:

$$H(\beta, z) \stackrel{def}{=} \begin{pmatrix} H_0(\beta, z) + \frac{iv_{gr}\bar{g}}{2}\mathcal{I} & -\frac{iv_{gr}\bar{g}}{2}\mathcal{I} \\ -i\bar{\gamma}\mathcal{I} & (\bar{\omega} + i\bar{\gamma})\mathcal{I} \end{pmatrix}$$

$$H_0(\beta, z)E \stackrel{def}{=} v_{gr}\begin{pmatrix} i\partial_z - \beta & -\kappa^-(z) \\ -\kappa^+(z) & -i\partial_z - \beta \end{pmatrix}\begin{pmatrix} E^+ \\ E^- \end{pmatrix}$$

$$\beta(n, |E|^2) \stackrel{def}{=} \delta - \frac{i\alpha}{2} + \frac{ig'(n - n_{tr})}{2(1+\varepsilon_G|E|^2)} + \frac{\alpha_H g'(n - n_{tr})}{2(1+\varepsilon_I|E|^2)}. \tag{5.3}$$

Here, $\mathcal{I}$ is a 2×2 identity matrix, $\kappa^{\pm}$ are, in general, spatially dependent complex coupling factors between the counterpropagating fields in sections with Bragg grating, and v_{gr} is the group velocity. The parameters δ, α, g', n_{tr}, α_H, ε_G, and ε_I represent static detuning, internal optical losses, effective differential gain,[3] including the transverse confinement factor, transparency carrier density, Henry linewidth enhancement factor, gain, and index compression factors, respectively. The effective inclusion of the polarization equations is used to model the frequency dependence of the optical gain close to its maximum. It corresponds to a Lorentzian gain peak with amplitude $\bar{g} > 0$ and full-width at half-maximum $2\bar{\gamma}$ in frequency domain, which is centered at the optical frequency $\bar{\omega}$ [11]. At the same time, the polarization equations automatically cause a dispersion of the effective refractive index according to the Kramers–Kronig relation.

The boundary conditions for the function E at the interfaces and facettes are given by:

$$E^+(z'_{k+1},t) = -r_k^* E^-(z'_{k+1},t) + t_k E^+(z''_k,t)$$
$$E^-(z''_k,t) = r_k E^+(z''_k,t) + t_k E^-(z'_{k+1},t). \qquad (5.4)$$

Here, z'_k and z''_k indicate the left and right edges of the section S_k. Accordingly, z''_k and z'_{k+1} denote the joining edges of two neighboring sections S_k and S_{k+1} and correspond to the same spatial coordinate [see also the notations in Fig. 5.1(b)]. The complex coefficients r_k, $-r_k^*$ and the real number $t_k = \sqrt{1-|r_k|^2}$ denote forward-to-backward, backward-to-forward field reflectivities at the junction of sections S_k and S_{k+1}, and field transmission coefficients through this junction, respectively [see [20, 21] and Fig. 5.1(b)]. The quantities $E^-(z''_0,t)$ and $E^+(z'_{m+1},t)$ represent the amplitudes of the fields emitted at the facettes. Accordingly, $E^+(z''_0,t)$ and $E^-(z'_{m+1},t)$ are the amplitudes of possible incoming fields at the same facettes. In this chapter, we do not consider optical injections and set these values to zero.

By proper scaling of the wave function Ψ, $|E(z,t)|^2$ is the local photon density, i.e., the local power at z divided by $\hbar v_{gr}\sigma c_0/\lambda_0$, where $\hbar$ is the Planck constant, c_0 is the speed of light in vacuum, and λ_0 is the central wavelength. Now the function $\mathcal{N}$ on the right-hand side of the carrier rate equation (5.2) can be defined as follows:

$$\mathcal{N}(n,|E|^2,\Re e(E^*p),z) \stackrel{def}{=} \frac{I}{eL\sigma} + \frac{U'_F}{eL\sigma r_s}\left(\frac{1}{L}\int_S n\,dz - n\right) -$$
$$- (An + Bn^2 + Cn^3) - v_{gr}\left[\left(\frac{g'(n-n_{tr})}{1+\varepsilon_G|E|^2} - \bar{g}\right)|E|^2 + \bar{g}\Re e(E^*p)\right]. \quad (5.5)$$

Here, e is the electron charge and the parameters I, σ, U'_F, r_s, A, B, and C denote the injection current, the cross-section area of the active zone, the

---

[3] The linear gain $g'(n-n_{tr})$ in (5.3) and (5.5) can be replaced by the logarithmic model $g'n_{tr}\ln(n/n_{tr})$.

derivative of the Fermi level separation with respect to n, the series resistivity, and the recombination coefficients, respectively. As all parameters can have different values in different sections (see, e.g., Table 5.1), the operator H and the rate function $\mathcal{N}$ depend explicitly on the position z. The integration in (5.5) is also done within the corresponding section.

5.3.2 Parameters

In the following, we shall illustrate different methods of mode analysis by simulating and analyzing the three-section laser as represented in Fig. 5.1(a). This laser is completely anti-reflection coated and consists of two distributed feedback (DFB) sections and one phase tuning section integrated in between.

Table 5.1. Laser Parameters Used in the Simulations (φ is a Bifurcation Parameter).

	explanation	S_1	S_2	S_3	units
c_0/v_{gr}	group velocity factor	3.4	3.4	3.4	
L	length of section	250	400	250	μm
$\kappa^+ = \kappa^-$	coupling coefficient	130	0	130	cm^{-1}
α	internal absorption	25	20	25	cm^{-1}
g'	effective differential gain	9	0	9	10^{-17}cm^2
δ	static detuning (example I)	300	$-\pi/2L_2$	-170	cm^{-1}
	static detuning (example II)	300	$-\varphi/2L_2$	-80	cm^{-1}
α_H	Henry factor	-4		-4	
ε_G	gain compression factor	3		3	10^{-18}cm^3
ε_I	index compression factor	0		0	10^{-18}cm^3
$\bar{g}$	Lorentzian gain amplitude	200	0	200	cm^{-1}
$\bar{\omega}$	gain peak detuning	0		0	ps^{-1}
$2\bar{\gamma}$	FWHM of gain curve	50		50	ps^{-1}
I	current injection (example I)	70		9	mA
	current injection (example II)	70		70	mA
σ	cross-section area of AZ	0.45		0.45	μm^2
U_F'	differential Fermi voltage	5	5	5	10^{-20}Vcm^3
r_s	series resistivity	5	5	5	Ω
n_{tr}	transparency carrier density	1	1	1	10^{18}cm^{-3}
A	inverse carrier life time	0.3		0.3	ps^{-1}
B	bimolecular recombination	1	1	1	10^{-10}cm^3/s
C	Auger recombination	1	1	1	10^{-28}cm^6/s
λ_0	central wavelength		1.57		μm
r_0, r_3	facet reflectivity coefficient		0		
r_1, r_2	internal reflectivity coefficient		0		

It is supposed that the two DFB gratings are slightly detuned. The first section (DFB$_1$) is pumped well above threshold and provides gain for lasing.

In example I (Sects. 5.4 through 5.7), the second DFB section (DFB$_2$) is driven close to its gain transparency. That is, we have $I \approx e\sigma LR(n_{tr})$ and $n(z,t) \approx n_{tr}$ in this section, which, therefore, acts as a passive dispersive reflector. In example II (Sects. 5.8 and 5.9), both DFB sections are pumped above threshold, so that the device is operating as a phase controlled mode beating (PhaseCOMB) laser. The center section has no grating, its active layer does not couple to the laser emission, and, therefore, it serves as a passive phase tuning section. The contribution of this section to the field roundtrip phase is given by:

$$\varphi = -2\delta_2 L_2. \tag{5.6}$$

The parameters belonging to both modi of operation are collected in Table 5.1. A more detailed discussion concerning fabrication, applicability, and numerical simulation of such lasers can be found in [7, 8, 14] and in references therein.

A proper modeling of multisection lasers should take into account nonlinearities. Fortunately, these systems are dissipative; i.e., the solution always approaches an attractor of the system after a more or less extended transient time. Single-section lasers usually possess only one attractor: a stationary state, i.e., a continuous-wave (cw) emission with stationary carrier densities. In multisection devices, nonstationary attractors appear like periodic or even chaotic self-pulsations. Furthermore, multiple different attractors may coexist in certain parameter ranges. It depends on the initial conditions to determine which attractor is approached in these cases. *LDSL-tool* offers different types of initial conditions. Random initial values allow to discover possible multiple attractors in a given point of operation but require, in general, long transients toward the asymptotically stable solution. Another possibility is to use the final solution of a preceding simulation as initial values for the next one. When changing parameters in sufficiently small steps, the latter procedure yields, in general short transients and a smooth continuation of the solution. The nonsmooth transitions between the solutions indicate a bifurcation, where the given attractor changes qualitatively or loses stability. In the following examples, we use this type of initial conditions because it allows to study the dependence of attractors on parameters including their bifurcations.

5.4 Simulation of a Passive Dispersive Reflector Laser

To demonstrate the possibilities of a direct integration of the TW model, we first consider the laser structure of Fig. 5.1(a) with the parameters labeled "example I" in Table 5.1. Here, the third laser section is biased at gain transparency and operates as a passive distributed Bragg reflector. This modus of operation has been successfully used as optical clock in all-optical signal processing at 10 Gbit/s [6]. The behavior of the device depends sensitively on the contribution of the passive middle-section parameter φ (modulo 2π) to the optical round trip phase (see (5.6) and [7, 22]). Figure 5.2 demon-

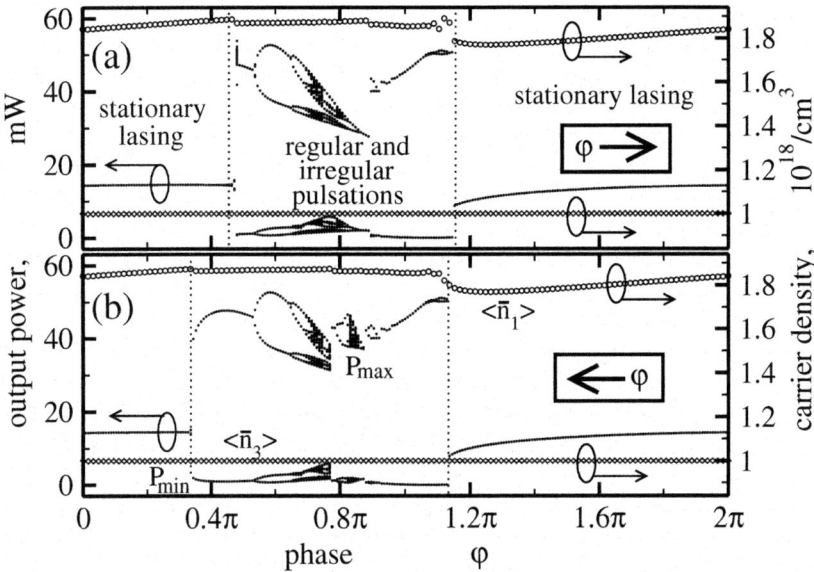

Fig. 5.2. Simulation example I. Panels a and b: increasing and decreasing, respectively, phase parameter φ. Bullets and diamonds: spatial-temporal averages $\langle \bar{n}_{1,3} \rangle$ of the carrier density in the DFB sections. Dots: relative maxima and minima of the power emitted from the left facet.

strates the complexity of phenomena obtained by simulation of such a laser when varying the phase parameter φ over one period. At certain values of φ, the optical output or the carrier density changes qualitatively in an abrupt manner. Such bifurcations are a genuine property of nonlinear systems (see, e.g., [16, 23, 24, 25] for a deeper discussion of possible bifurcation scenarios in multisection lasers).

In the given example, we tuned φ with small steps, using the final profiles of the functions Ψ and n as the initial values in the simulations at the new step. After finishing the transient regime, the output power maxima and minima and spatio-temporal averages of carrier densities in both DFB sections are recorded. Transitions between different operation regimes are observed at several values of φ. They are different for increasing φ (diagram a) and decreasing φ (diagram b). Obviously, this hysteresis appears because the system approaches different stable long-term solutions in dependence on the initial conditions.

Figure 5.2 represents the attractors at different φ by means of only few characteristics. A more detailed characterization of a self-pulsating solution is

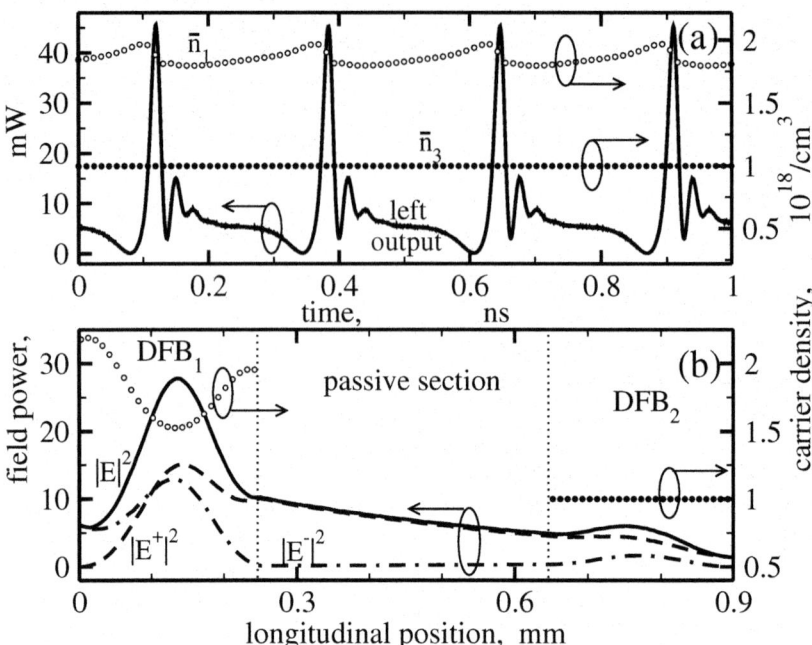

Fig. 5.3. Simulated behavior of optical fields and carrier densities in a laser at $\varphi = \pi$. Panel a: temporal variation. Solid: output from left facet. Empty and full bullets: sectional averages $\bar{n}_{1,3}$ of carrier density within the DFB sections. Panel b: axial distributions at the last instant of panel a. Dashed: $|E^+|^2$. Dash-dotted: $|E^-|^2$. Solid: total power. Bullets: carrier density.

given in Fig. 5.3 for the particular phase $\varphi = \pi$. Panel a of this figure visualizes the temporal evolution of the calculated optical output and of the sectionally averaged carrier densities $\bar{n}_1$ and $\bar{n}_3$. The carrier density of section S_3 does not couple to photons and remains nearly constant. The carriers in section S_1 and the optical output show periodic variations. Within each period, the output power exhibits a large first pulse followed by a damped oscillation. This substructure looks like a highly damped relaxation oscillation. However, the frequency of these damped oscillations is about 35 GHz, far beyond typical relaxation oscillation frequencies. Indeed, they are of a completely different nature as the mode analysis below will show.

The axial distribution of powers and carriers at the last instant of Fig. 5.3(a) is drawn in panel b of the figure. Note the spatial variations of the optical intensity and the carrier density in section S_1. High optical intensity inside this section and corresponding enhancement of the stimulated recombination imply a dip in the carriers. This effect is called spatial hole burning (for more details, see [14] and references therein).

The examples above show that our model is able to demonstrate a variety of bifurcations as well as nontrivial dynamical regimes possessing spatially nonuniform distributions of fields and carriers. In the following, we shall show how many of these effects can be deeper understood when analyzing the situation in terms of optical modes.

5.5 The Concept of Instantaneous Optical Modes

Although the concept of modes is basic in laser physics, there exist many different versions depending on the context. In general, the set of optical modes represents all possible natural oscillations of the optical field in a resonator. The standing waves in a usual Fabry–Perot cavity may serve as a well-known example. In the present context, we consider the longitudinal modes of hot multisection cavities. Mathematically speaking, they are the eigenfunctions and -values of the evolution operator $H(\beta, z)$ in (5.1), i.e., the sets of all complex valued objects $(\Theta(\beta, z), \Omega(\beta))$, which solve the spectral equation

$$H(\beta, z)\Theta(z) = \Omega\Theta(z). \tag{5.7}$$

We have explicitly stated that the operator H depends on the actual distribution of $\beta = \beta(z,t)$ along the cavity. Thus, Θ and Ω depend also on this distribution and vary, in general, with time. They represent the instantaneous mode spectrum of the hot compound cavity. A detailed description of methods used to solve this spectral problem for any fixed distribution of $\beta(z)$ is given in the Appendix.

For a moment, let us disregard the spontaneous emission noise and consider β not depending on time as in any stationary state of an ideal laser. In this case, $\Psi(z,t) = \exp(i\Omega t)\Theta(\beta, z)$ solves the full TW equation (5.1). Thus, real and imaginary parts of Ω represent frequency and decay rate of the optical mode, respectively. Figure 5.4 shows these quantities for different configurations. First, we have simulated a solitary single-section DFB laser with the parameters of section S_1, as given in Table 5.1. Its mode spectrum, in case of a stationary operation, is shown by the open bullets in Fig. 5.4. The single mode with $\Im m\Omega = 0$ keeps constant and carries the laser power. All other modes have positive $\Im m\Omega$ and decay if initially present. Their decay rates are very high so that they are damped out within a few picoseconds and play no role even under modulation of the laser in the GHz range.

Now we turn to the full multisection laser. In order to see the impact of the extended cavity, we keep the propagation function β in the DFB$_1$ section frozen at the solitary laser level, while assuming $n(z) = n_{tr}$ in the remaining sections. Panel a of Fig. 5.4 shows an increased number of side modes and a general reduction of their decay rates. When tuning the phase parameter φ, these modes are moving along the solid line, each one is replacing its next neighbor after one period.

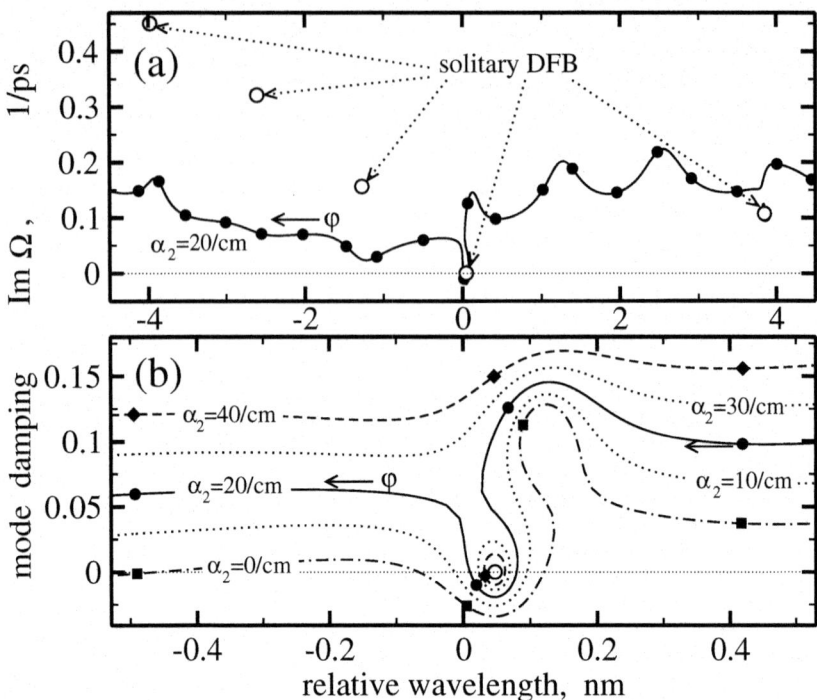

Fig. 5.4. Spectrum of optical modes. Abscissa: relative wavelength ($-\Re e\Omega \cdot \lambda_0^2/2\pi c_0$). Panel a: overview. Panel b: the modes close to resonance. Lines: location of eigenvalues for all possible φ and different optical losses α_2 in the passive section. Arrows: direction of increasing φ. Full symbols: modes at $\varphi = 0$. Empty bullets: modes of the solitary DFB laser.

The scenario close to the fundamental mode of the solitary laser is depicted in Fig. 5.4(b) for some fixed values of optical losses α_2 in the passive middle section. In the present example I, the combined second and third sections provide a dispersive delayed feedback of the field back into the active first section. The losses α_2 can be regarded as a measure of the feedback strength. At very low losses (large feedback), all modes are located on a single open curve, exhibiting a deep valley close to the resonance of the solitary laser. When increasing the phase φ, the modes move from right to left through this valley. With increasing losses (decreasing feedback), the valley becomes narrower and deeper. The edges of the valley touch each other at some α_2 between 20 and 30 cm^{-1} forming afterward an additional closed curve around the fundamental mode of the solitary DFB laser. This loop contains only one mode, which is moving around once per phase period. The diameter of the loop shrinks with further increasing losses. At the same time, the remaining part of the mode curve shifts upwards, continuously forming new shrinking loops when approaching higher modes of the solitary DFB laser. In the limit of infinite losses, the feedback is zero and all these closed loops collapse into the solitary laser modes.

The simulation example of the previous section corresponds to a middle loss level of 20 cm^{-1} (solid line in Fig. 5.4). A strong dependence of mode damping on φ appears here close to the fundamental mode of the solitary laser. Moreover, at a particular phase value, two neighbor modes located at the opposite valley edges come close to each other. Such a vicinity of two modes is a general indication of more complex dynamics [26, 27]. Therefore, we turn now to mode dynamics.

5.6 Mode Expansion of the Optical Field

In this section, we perform a mode analysis of the nonstationary state of the self-pulsating laser simulated in Sect. 5.4.

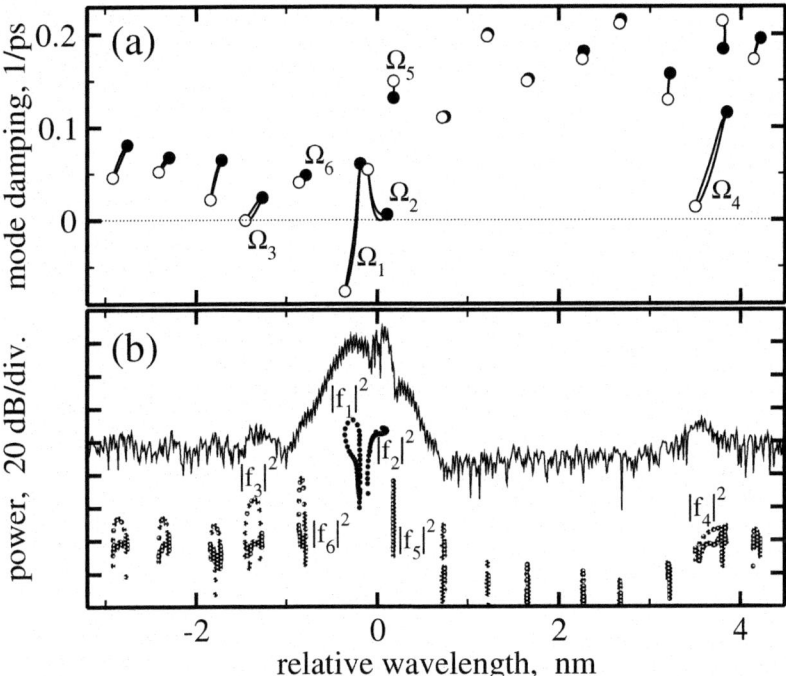

Fig. 5.5. Mode analysis of the pulsating state drawn in Fig. 5.3(a). Panel a: mode spectrum. Empty and full bullets: location of Ω at maximal and minimal sectional average $\bar{n}_1$, respectively. Lines: movement of Ω in between. Panel b: mode decomposition of the optical field. Symbols in lower part: modal contributions $|f_k|^2$ in a series of instants versus modal wavelength. Above: optical spectrum of the analyzed output field. Note the logarithmic scale.

Figures 5.5(a) and 5.6(a) present the results of mode computations with the distributions $\beta(z,t)$ calculated along one period of pulsation. Figure 5.5(a) shows the resulting variation of the mode spectrum in the wavelength-damping plane, whereas Fig. 5.6(a) draws the mode damping $\Im m \Omega$ versus time t. Both of these diagrams indicate that the damping of only three or four modes becomes negative or rather small. Accordingly, we expect that only few modes contribute to the dynamics.

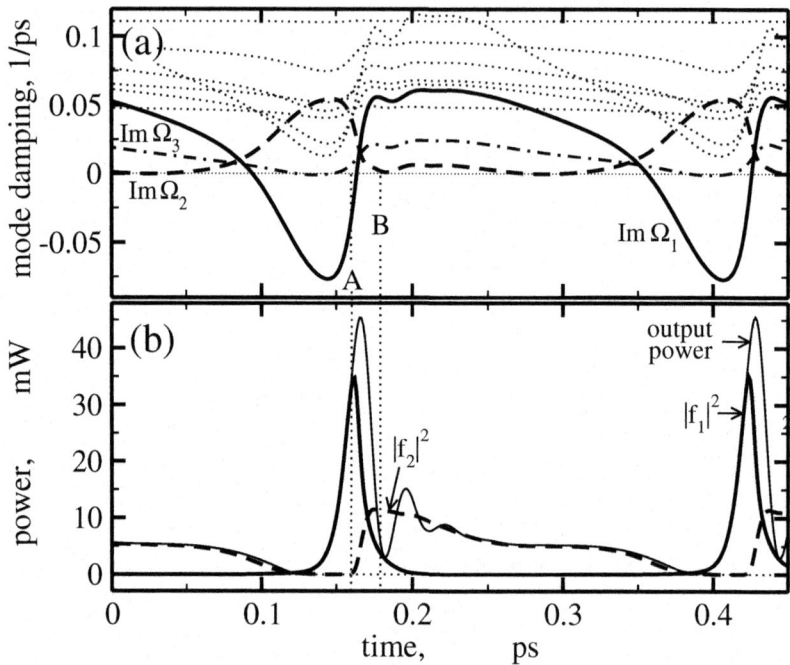

Fig. 5.6. Mode analysis of the pulsating state drawn in Fig. 5.3(a). Panel a: evolution of modal decay rates. Solid, dashed, and dash-dotted: modes that sometimes reach zero or negative damping. Thin dotted: modes with always positive damping. Labeling as in Fig. 5.5. Panel b: evolution of the output field power (thin solid) and modal amplitudes. Solid and dashed bold lines: the two most prominent modal contributions.

In order to quantify the different modal contributions, we express the calculated optical state vector $\Psi(z,t)$ at any instant t as a superposition of the instantaneous modes [12, 16]:

$$\Psi(z,t) = \sum_{k=1}^{\infty} f_k(t) \Theta^k \left(\beta(z,t), z \right), \tag{5.8}$$

where the complex modal amplitudes $f_k(t)$ can be calculated as described in the Appendix. After an appropriate scaling of eigenfunctions $\Theta^k(\beta, z)$, f_k represents the contribution of mode k to the outgoing field E^- at the left facet z_1':

$$E^-(z_1', t) = \sum_{k=1}^{\infty} f_k(t). \tag{5.9}$$

Figures 5.5(b) and 5.6(b) show that only the two modes with labels 1 and 2 are contributing to the analyzed dynamical regime, whereas all other modes remain suppressed. These two modes alternate during the pulsation. Mode 1 dominates the first part of the pulses, and mode 2 determines the long plateau between them. Beating between them causes damped oscillations of the total intensity, which were a puzzling detail of the simulation results. By more closely inspecting the behavior of the two dominating modes, we can discover further peculiarities. First, they exhibit a remarkably big variation of the damping rates. Second, the damping of mode 2 varies opposite to most other modes. Third, the intensity of mode 2 is steeply rising within a short time interval $[A, B]$, as shown in Fig. 5.6, although its damping is positive. At the same time, mode 1 starts to decay although still being undamped. These anomalies will be understood in the next section when we will discuss the driving forces for the mode dynamics.

5.7 Driving Forces of Mode Dynamics

In order to get insight into the different forces causing changes of the modal amplitudes f_k, we substitute the expression (5.8) into the TW equation (5.1) and multiply scalarly the resulting equation by the adjoint modes $\Theta^\dagger$ [see the discussion in [21] and (5.21) for an explicit expression of $\Theta^\dagger$]. Due to the orthogonality of the modes Θ^l and the adjoint modes $\Theta^{k\dagger}$, this procedure yields the following equations of motion for the modal amplitudes:

$$\frac{d}{dt} f_k = i\Omega_k f_k + \sum_{l=1}^{\infty} K_{kl} f_l + F_k$$

$$\text{where} \quad K_{kl} \stackrel{def}{=} -\frac{[\Theta^{k\dagger}, \partial_t \Theta^l]}{[\Theta^{k\dagger}, \Theta^k]}, \quad F_k \stackrel{def}{=} i\frac{[\Theta^{k\dagger}, F_{sp}]}{[\Theta^{k\dagger}, \Theta^k]}, \tag{5.10}$$

and $[\xi, \zeta]$ denotes the usual scalar product of four-component complex vector functions. Obviously, the evolution of the amplitude f_k is driven not only by the term with the complex eigenvalues Ω_k. In addition, the second term at the right-hand side of (5.10) contains the contributions proportional to the temporal changes $\partial_t \Theta^l$ of the mode functions. The third term in the same formula represents the part of spontaneous emission emitted into mode k. Whereas the first and third terms appear in all types of lasers, the second term is responsible for the peculiarities of the multisection lasers. It describes the redistribution of modal amplitudes due to the changes of the expansion basis. It disappears completely in an ideal single-section device with a spatially constant β. In this case, only the eigenvalues Ω_k change with β but not the eigenfunctions Θ^k. It can be concluded that the second term in (5.10) is generally small in well-designed single-section devices. As a consequence, the amplitudes of different modes evolve independently, each one fed by spontaneous emission.

The situation in multisection devices is different. Variations of β within one section, even if spatially constant there, generally change the β-profile of the whole device. These changes cause variations of the compound cavity modes, which, in turn, imply nonvanishing matrix elements K and their non-negligible contribution to (5.10). Applying the temporal derivative to (5.7) and using again the orthogonality of the adjoint modes one can express the corresponding matrix K_{kl} as follows:

$$K_{kl} = \frac{1}{\Omega_k - \Omega_l} \frac{[\Theta^{k\dagger}, (\partial_t H)\Theta^l]}{[\Theta^{k\dagger}, \Theta^k]} \quad (k \neq l) \tag{5.11}$$

$$K_{kk} = -\frac{\partial}{\partial t} \ln \sqrt{[\Theta^{k\dagger}, \Theta^k]}.$$

Hence, the coupling term K_{kl} can dominate over the damping term $\Im m \Omega_k$, when the separation $\Omega_k - \Omega_l$ between a mode pair becomes small. This effect is responsible for the growth of mode 2 within the time interval $[A, B]$ in Fig. 5.6, where its damping term $\Im m \Omega_2$ is still positive. It also explains the early decay of mode 1 within the same interval.

So far we have used (5.10) for a deeper understanding of the evolution of the mode amplitudes f_k, which have been determined by the optical state vector $\Psi(z, t)$ and the propagation factor $\beta(z, t)$ obtained in a preceding simulation. Alternatively, it is also possible to combine these equations directly with the carrier equations (5.2). This yields a set of ordinary differential equations equivalent to the original TW model. By truncating this system to the essential modes and to the essential moments of carrier distributions, one gets an approximate description of the dynamics in a finite dimensional phase space, which also opens the way for a path-following bifurcation analysis. A more detailed description of these mode-approximation techniques together with corresponding examples can be found in, e.g., [15, 16, 22, 23, 24, 26]. For a deeper discussion of the impact of nearly degenerate modes on the dynamics of multisection lasers as well as how to treat the case of degeneracy, we refer to [23, 24, 26, 27].

5.8 Mode-Beating Pulsations in a PhaseCOMB Laser

Let us turn now to example II, the Phase-Controlled Mode-Beating (Phase-COMB) regime of operation. Here both DFB sections are pumped above the threshold. Furthermore, the Bragg gratings are detuned by about the stop band width [cf. Fig. 5.8(a)] and the two adjacent inner DFB modes are selected for lasing via the phase tuning in the middle section. Beating of the two coupled modes leads to self-pulsations with a frequency determined by the spectral separation of the lasing modes [8, 26]. Practical realizations of this regime up to 80 GHz and its application in optical signal processing have been reported, e.g., in [17].

5.8.1 Simulation

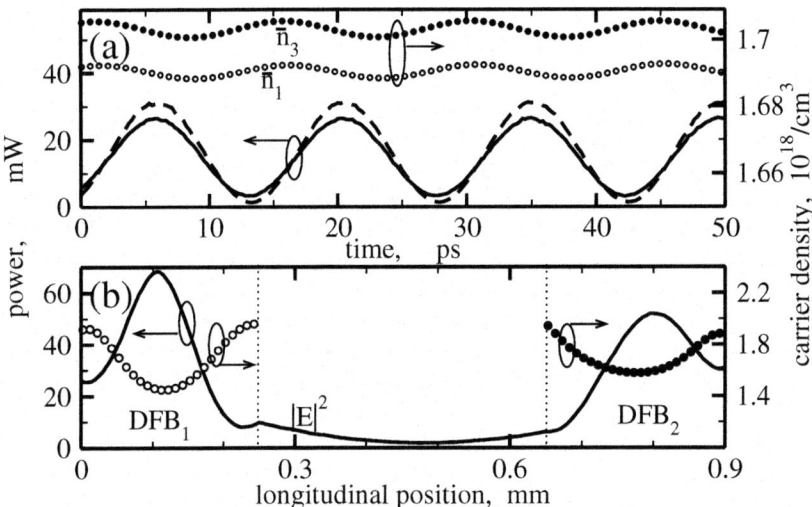

Fig. 5.7. Simulated behavior of optical fields and carriers in the PhaseCOMB laser at $\varphi = \pi$. Diagrams and notations as in Fig. 5.3. The dashed line in panel a is the output power at the right facet.

Figure 5.7 characterizes a typical self-pulsating state calculated for the phase parameter $\varphi = \pi$. Both DFB sections contain comparably high carrier densities and optical powers (panel b), they operate above the laser threshold. The emitted power (panel a) exhibits well modulated nearly sinusoidal oscillations with about 70 GHz frequency. The carrier densities are extremely weakly modulated with the same frequency.

5.8.2 Mode Decomposition

The motion of the eigenvalues during one pulsation period is shown in Fig. 5.8(c). It is smaller than the symbol size in accordance with the weak carrier modulation and in contrast to Fig. 5.3(a) of example I. The two lowest

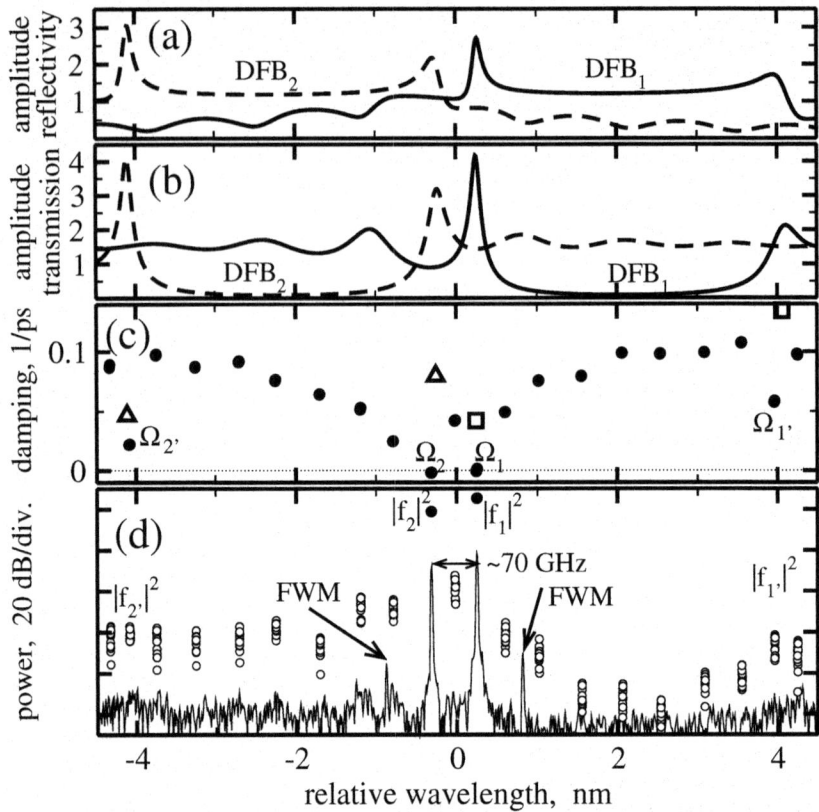

Fig. 5.8. Mode analysis of the pulsating state drawn in Fig. 5.7. Panels a and b: reflectivity and transmission spectra of DFB_1 (solid) and DFB_2 (dashed) sections. Panel c: mode spectrum of the PhaseCOMB laser (bullets). The mode spectra of noncoupled solitary DFB_1 (squares) and DFB_2 (triangles) sections are given for comparison. Panel d: optical spectra of the emission from the left facet (thin line) and modal contributions $|f_k|^2$ versus modal wavelength (bullets).

modes (labeled 1 and 2) have zero damping level, i.e., both are at threshold. Their expansion coefficients (panel d) are comparable, they exceed those of all other modes by orders of magnitude and represent well the two major peaks in the optical spectrum. We conclude that the self-pulsating state is carried by two modes lasing at the same time.

The spectral position of the two major modes agrees fairly well with the two adjacent inner resonance peaks of the reflectivity spectra of the DFB sections calculated in the given state of operation (panel a). Thus, each mode of this pair can be attributed to one of the two DFB sections. This conclusion is supported by the coinciding spectral positions of the modes of isolated DFB sections with the same carrier densities (open symbols in panel c).

The higher damping of these solitary DFB laser modes, on the other hand, indicates an important role of the interaction between both DFB sections in the self-pulsating state. To a certain extent, the strength of this interaction can be represented by the reflectivity of the corresponding opposite DFB section. These reflectivities are rather high over a wide spectral range (panel a). Thus, this useful interaction is accompanied by the formation of a Fabry–Perot-like cavity, which gives rise to the comb of only moderately damped side modes in panels c and d.

The calculated optical spectrum in panel d exhibits a few additional side peaks, which are small but well above the noise background. Similar features have been detected in experimental spectra. It seems to be obvious to attribute them to weakly damped side modes. However, the mode decomposition yields: There is no side mode at the position of the most prominent side peak. The real cause of these peaks is the four-wave mixing (FWM) of the two main modes. The presence of FWM peaks is generally an indication of a good overlap of the two waves and of their nonlinear interaction.

5.8.3 Spatio-temporal Properties of Mode-beating Self-pulsations

The results of the mode analysis presented above allow a deeper insight into the spatio-temporal behavior of the fields within the self-pulsating laser. Two modes $k = 1, 2$ dominate the expansion (5.8). The corresponding eigenvalues $\Omega_k(t)$ and eigenfunctions $\Theta^k(z,t)$ vary extremely weakly around their means $\bar{\Omega}_k$ and $\bar{\Theta}^k(z)$. Thus, it holds:

$$f_k(t) \approx f_k(0)\exp(i\omega_k t), \qquad k = 1,2,$$

where $\omega_k = \Re e \bar{\Omega}_k$ is the mean modal frequency. The intensities of both the forward and the backward traveling wave within the laser are well approximated by:

$$P(z,t) \approx |E_1(z)|^2 + |E_2(z)|^2 + 2|E_1(z)||E_2(z)|\cos\left(\Delta\omega\, t - \phi(z)\right). \quad (5.12)$$

Here:

$$E_k(z) \stackrel{def}{=} f_k(0)\bar{\Theta}_E^k(z), \quad \Delta\omega \stackrel{def}{=} \omega_2 - \omega_1, \quad \phi(z) \stackrel{def}{=} \arg(E_2(z)) - \arg(E_1(z))$$

are the initial optical mode amplitudes, the mean frequency difference between the two dominant modes, and the initial mode phase difference, respectively. Superscripts $\pm$ distinguishing between forward and backward are omitted for simplicity. All expressions hold for both the propagation directions.

The axial distributions of the amplitudes $|E_k(z)|$ and the phase difference $\phi(z)$ are drawn in Fig. 5.9 for the particular pulsation under consideration. Although the two modes originate from different DFB sections, their amplitudes are comparable over all the compound cavity. Thus, the beating term in

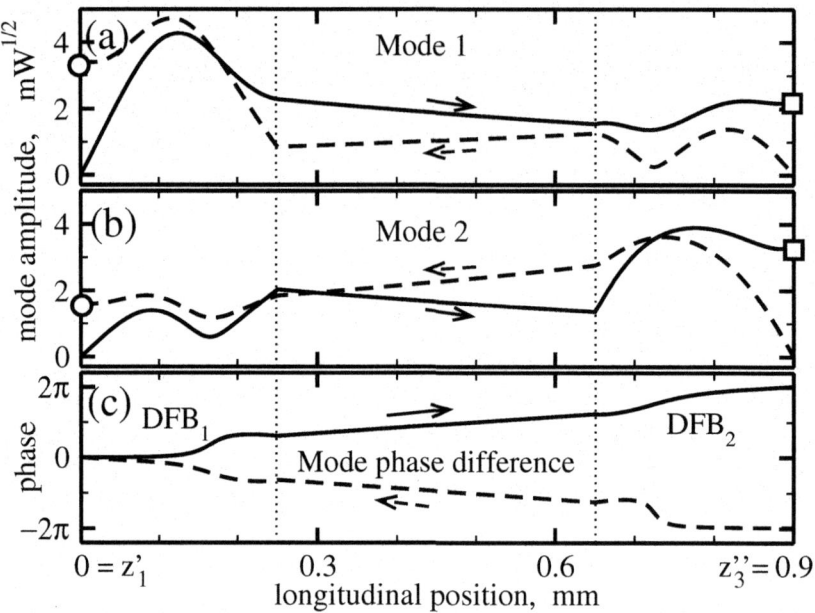

Fig. 5.9. Axial mode structure. Panels a and b: forward and backward traveling amplitudes $|E_k(z)|$ of the two main modes of Fig. 5.8. Panel c: phase differences $\phi(z)$ between them. Forward and backward propagating parts are given by solid and dashed lines, respectively. The open symbols indicate the amplitudes of mode waves emitted from the facets.

(5.12) causes everywhere a considerable modulation of the optical intensity, which, in turn, drives the carrier oscillations in the active sections.

Practical applications of self-pulsating lasers usually require a good extinction of the pulses; i.e., the outgoing amplitudes of both modes at the given facet should be comparable. In our example, this condition is best achieved at the right facet (open squares in Fig. 5.9). In this context, it is important that the section DFB_2 supports not only its own mode 2 but also amplifies the wave of mode 1. The corresponding transmission depends strongly on the wavelength [Fig. 5.8(b)]. It becomes very small if the two stop bands overlap too much. Thus, one should avoid this situation by properly adjusting the detuning of the DFB gratings in order to get self-pulsations with a high extinction.

The axial variation of the phase difference $\phi(z)$ between the two modes plays also an important role. It is a general rule that the corresponding phase difference between any two neighboring modes k and l ($\omega_l > \omega_k$) increases exactly by 2π over one round trip in propagation direction. In the present example, it increases by 4π [see line separation at the right facet z_3'' in Fig. 5.9(c)] because we have one weak mode between the two dominant ones.

Due to the increasing phase difference, the peak of the beating pulse travels cyclic around the cavity. A similar behavior is known from mode locked Fabry–Perot lasers. However, pulse speed and pulse length are very different in both cases. In a mode-locked laser, the speed equals the constant group velocity v_{gr}, the multimode pulses are very short and the length of the laser determines the repetition rate $v_{gr}/2L$. In our devices, the pulse peaks travel with v_{gr} only in the passive middle section, the two-mode pulses are nearly sinusoidal, and the frequency difference $\Delta\omega$ determining the repetition rate is impressed by the grating detuning but not by the device length.

Due to the high symmetry of our device, the single pass phase shift is nearly half the round-trip phase shift, i.e., 2π in our case. This explains the in-phase oscillations of both emitted powers shown in Fig. 5.7(a). At variance, anti-phase will appear if the dominant modes are neighboring or have an even number of modes in between.

5.9 Phase Control of Mode-beating Pulsations

After the detailed study of the case $\varphi = \pi$ in the previous section, we shall briefly sketch the use of mode analysis for understanding switching between qualitatively different types of mode-beating pulsations. In particular, we control these regimes by tuning the phase parameter φ.

5.9.1 Simulation of Phase Tuning

First, we report on simulation for both increasing and decreasing φ. The method of calculations is the same as already explicated in connection with Fig. 5.2. Selected results are summarized in panels a and b of Fig. 5.10. Obvious bifurcations appear at the positions B, D, and F when tuning φ in forward direction and at E, C, and A when tuning φ backward. Accordingly, at least two attractors coexist in the regions $[A, B]$, $[C, D]$, and $[E, F]$. Unique pulsating states are observed within the rest of the phase period.

5.9.2 Mode Analysis

It is not possible to give here a complete description of all phenomena observed in the simulation. Therefore, we confine the mode analysis to the case of increasing phase φ. Wavelengths and dampings of a few modes calculated in arbitrarily chosen instants t are presented in panels c and d, respectively. We did not determine mean values because the extremely weak oscillations of $\beta(z)$ usually imply accordingly small variations of the modes.

The mode system as a whole is 2π periodic in φ. However, the wavelength of each individual mode shifts down with φ and every mode arrives at the position of its former next neighbor after one period. As an example, the dominant modes 1_n, 2_n, and $2'_n$ at $\varphi = 2\pi$ are identical with modes 1, 2, or 2′ dominating at $\varphi = 0$. Consequently, at least one discontinuous jump of each operating mode must appear within every phase-tuning period.

5.9.3 Regimes of Operation

The pulsations with ~ 70 GHz frequency within the phase interval $[D, F]$ are determined by the two second-neighbor modes 1_n (gray bullet) and 2_n (black square). The damping of both these modes is practically zero. This case has already been extensively described in the previous section.

In the range $[B, D]$, the situation is a bit different. In connection with highly asymmetric carrier densities, the two closest neighbor modes 1 (open diamonds) and 2_n (black squares) are dominating here. They drive pulsations of only ~ 30 GHz frequency. Surprisingly, the dampings of the two dominant modes differ now considerably from zero and exhibit a large scatter. The reasons are as follows. The smaller mode separation implies an increase of the mode-coupling factor K, which allows a larger deviation of the main mode damping factors $\Im m\Omega$ from zero (see (5.10,5.11) and the discussion in Sect. 5.7). This increased coupling also gives rise to higher FWM peaks (panel c). Moreover, the lower frequency causes stronger modulations of the β-distribution and, in turn, a more significant shift of some modes during pulsations. These deeper modulations are also responsible for the scatter of mode dampings (interval $[B, D]$ in panel d) calculated at arbitrary instants during the pulsations.

Within the phase interval $[F, 2\pi]$ continued in $[0, B]$, three modes are dominating simultaneously. Their damping is close to zero at the same time (panel d). This cooperation of 3 modes with comparable intensities causes three large peaks in the optical spectrum (panel c) and two peaks at about 70 and 450 GHz in the rf power spectrum (panel a). In contrast to the previous cases, now the DFB$_2$ section is simultaneously supporting different stop band side modes 2 and 2′. It seems that this robust operation of three modes is possible only due to the nonuniform distribution of carriers within DFB sections, whereas spatial hole burning within DFB$_2$ tends to select mode 2′, the slightly stronger coupling between modes 1 and 2 due to the larger factor K in (5.10) and (5.11) allows it to support mode 2 as well. A deeper analysis of this interesting case is, however, beyond the scope of this presentation.

5.9.4 Bifurcations

Let us now discuss the mode analysis nearby the bifurcations. For brevity, we consider only bifurcation positions B and F. Panel b shows a general increase of the mean carrier densities in both DFB sections when approaching these bifurcations from the left. The corresponding higher gain is required to keep the respective dominant modes at threshold, although the phase condition becomes increasingly bad for these modes. In both cases, the increasing

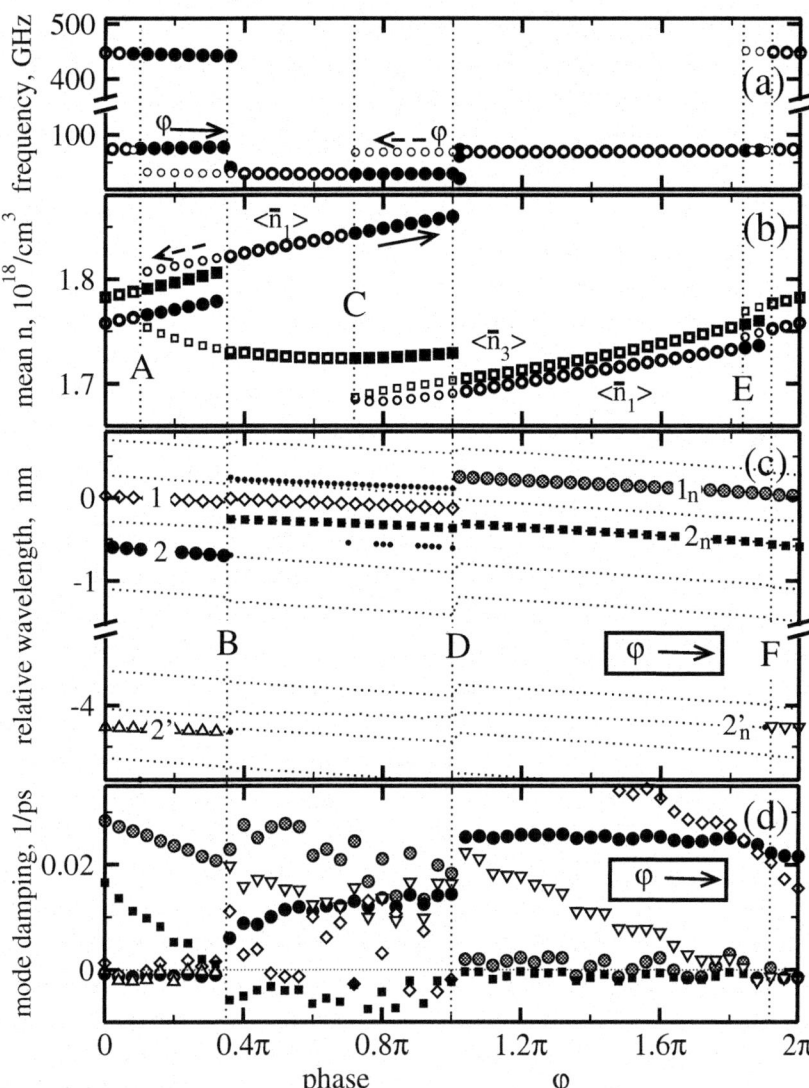

Fig. 5.10. Phase tuning characteristics of a PhaseCOMB laser. Panels a and b: simulations for decreasing (empty symbols) and increasing (full symbols) phase φ. Panel a: frequencies of the dominant peaks in the rf power spectrum (left facet); only side peaks suppressed by ≤ 10 dB are considered. Panel b: spatio-temporal averages of carrier densities in DFB$_1$ (bullets) and DFB$_2$ (squares). Panel c: wavelengths of all modes in the given spectral range (dotted lines). Big symbols indicate positions of dominant peaks (differing by ≤ 10 dB). The medium symbols in interval $[B, D]$ represent four-wave mixing side peaks, which are suppressed by ≤ 40 dB here. Panel d: damping $\Im m\Omega$ of important modes. The same symbols as in panel c are used but side modes are also shown.

gain also causes a gradual decrease of the damping of one side mode. The bifurcation appears when this side mode becomes undamped.

The bifurcation at the position B is accompanied by a sudden redistribution of the carriers. As a result, two formerly dominant modes become damped and a new dominant mode takes over. On the contrary, the change of carrier distributions is small at the position F. The system of dominant modes is only supplemented by an additional member, the mode whose damping approached to zero.

The scenarios of the other transitions between pulsating states are more complicated. Their analysis requires additional considerations that are beyond the scope of this chapter.

5.10 Conclusion

In this work, we discuss and demonstrate the possibilities of the software *LDSL-tool* studying the longitudinal mode dynamics. Computation of modes, decomposition of nonstationary fields into modal components, study of the mode spectra, of the relations and different transitions between modes, and of the mode dependencies on different parameters allow for a deeper understanding of the processes observed in multisection semiconductor lasers.

A Numerical Methods

In this appendix, we give a derivation of some formulas and briefly introduce the numerical methods used to solve the problems discussed above.

A.1 Numerical Integration of Model Equations

There is a variety of numerical methods to integrate different forms of the TW model (5.1) and (5.2). Examples are the transfer matrix method [20, 28], the power matrix method [29], finite difference schemes [10], the split step method [30], and the transmission line method [31]. These approaches do not use polarization equations but introduce dispersion effects by means of digital filter techniques. We prefer the polarization model because it yields a well defined analytic description of the field evolution operator $H(\beta, z)$, which we analyzed in this chapter.

The numerical integration of the partial differential TW equations is mostly based on the field propagation in time and space along characteristic directions ($v_{gr} t \pm z = \text{const}$) determined by the group velocity v_{gr}. This approach is also used in our work, where we treat the equations for the fields $E^{\pm}$ by means of second-order accuracy central finite difference schemes. The ordinary differential equations for the polarizations $p^{\pm}$, as well as for the carrier densities n, are integrated by means of central finite difference schemes.

A.1.1 Discretization of the Domain

The continuous functions $E(z,t)$, $p(z,t)$, and $n(z,t)$ are approximated by their discrete analogues. The values of such grid functions are determined on section-wise uniform temporal-spatial meshes represented schematically in Fig. 5.11.

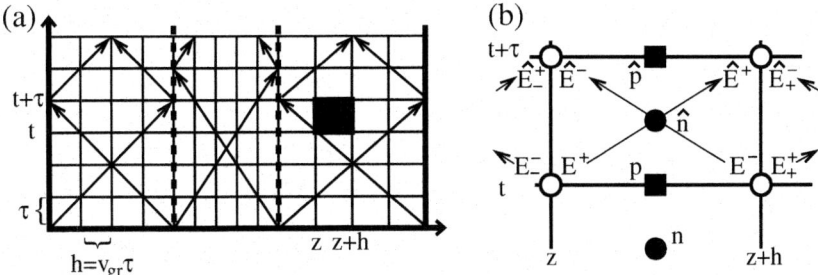

Fig. 5.11. Discretization of the spatial-temporal domain. Panel a: full domain overview. Panel b: positions (solid symbols) where grid functions are defined and their notations in the neighborhood of one cell (black box in panel a) of the mesh.

To generate the mesh for grid function E of the optical field, we discretize the time axis with a uniform step τ (horizontal thin lines). This time step should be sufficiently small in order to cover a required optical frequency range $\mathcal{F}$: $\tau \leq 1/\mathcal{F}$. The spatial axis is discretized with uniform steps $h = v_{gr}\tau$ (vertical thin lines). Junctions between sections should contain mesh points. Therefore, it is supposed that the length of each section is an integer multiple of $v_{gr}\tau$. The diagonal points of this mesh are located along the characteristic directions of the TW equations (slant arrows) in accordance with the physical field propagation. From the point of view of numerical analysis, such discretization is able to support a stable and convergent numerical scheme. Finally, the grid polarization and grid carrier density functions are considered on the meshes shifted by $h/2$ and $(h/2, \tau/2)$, respectively.

A.1.2 Numerical Schemes

Let E, $E_{\pm}$, $\hat{E}$, and $\hat{E}_{\pm}$ denote the grid field function values at the neighboring mesh points indicated by empty bullets in panel b of Fig. 5.11. Similarly, by p, $\hat{p}$, n, and $\hat{n}$, we denote grid polarization and carrier density function values at the positions denoted by full bullets and squares in the same figure.

Assume we know the values of grid function n (at the time layer $t-\tau/2$) and $E^{\pm}$, $p^{\pm}$ (at the time layer t). The grid function $\hat{n}$ is a solution of the finite difference scheme approximating (5.2) at the positions $(z+h/2, t)$:

$$\frac{\hat{n}-n}{\tau} = \mathcal{N}\left(\frac{\hat{n}+n}{2}, P, P_p, z+\frac{h}{2}\right)$$
$$P \stackrel{def}{=} \sum_{\nu=\pm} \frac{|E^\nu|^2+|E^\nu_\nu|^2}{2}, \qquad P_p \stackrel{def}{=} \Re e \sum_{\nu=\pm} \frac{(E^\nu+E^\nu_\nu)^* p^\nu}{2}. \tag{5.13}$$

The integral term used in the definition of the function $\mathcal{N}$ should be replaced by the sectional average of the corresponding grid carrier density function.

To find the grid functions $\hat{E}^\pm$ and $\hat{p}^\pm$, we are solving the difference schemes approximating (5.1) at the positions $(z+h/2, t+\tau/2)$:

$$\frac{\hat{E}^\pm - E^\pm}{h} = -\left(i\beta(\hat{n},\tilde{P}) + \frac{\bar{g}}{2}\right)\frac{\hat{E}^\pm + E^\pm}{2} - i\kappa^\mp \frac{\hat{E}^\mp + E^\mp}{2} + \frac{\bar{g}}{2}\frac{\hat{p}^\pm + p^\pm}{2}$$
$$\frac{\hat{p}^\pm - p^\pm}{\tau} = \bar{\gamma}\frac{\hat{E}^\pm + E^\pm}{2} + (i\bar{\omega}-\bar{\gamma})\frac{\hat{p}^\pm + p^\pm}{2}, \quad \tilde{P} \stackrel{def}{=} \sum_{\nu=\pm} \frac{|\hat{E}^\nu|^2 + |E^\nu|^2}{2}. \tag{5.14}$$

In addition, at the edges of sections, the grid optical field functions $\hat{E}^\pm$ are satisfying the boundary-junction conditions given in (5.4). In the case of nonzero spontaneous emission (i.e., $F_{sp}^\pm \neq 0$), we add properly normalized uncorrelated complex random numbers to the newly found optical fields $\hat{E}^\pm$ at each spatial grid point.

The scheme (5.13) is, in general, nonlinear with respect to $\hat{n}$ and is resolved by an iterative procedure. The already computed grid function n is used as an initial approximation for $\hat{n}$ in the nonlinear terms and in sectional averages. As the carrier density varies slowly, a single iteration step is sufficient here to get a good approximation of $\hat{n}$. A similar procedure is used to resolve the schemes (5.4) and (5.14) for the optical fields in cases $\varepsilon_{G,I} \neq 0$; i.e., when β is effectively dependent on the photon density $\tilde{P}$.

A.2 Computation of Modes

The vector-function $\Theta(\beta, z)$ can be split into field $\Theta_E = (\Theta_E^+, \Theta_E^-)^T$ and polarization $\Theta_p = (\Theta_p^+, \Theta_p^-)^T$ parts. The spectral problem (5.7) together with the boundary conditions (5.4) can be written as:

$$\begin{cases} H_0(D,z)\Theta_E(\beta,z) = 0 & D(\beta,\Omega) \stackrel{def}{=} \beta + \frac{\Omega}{v_{gr}} - \frac{i\bar{g}}{2}\frac{i(\Omega-\bar{\omega})}{\bar{\gamma}+i(\Omega-\bar{\omega})} \\ \Theta_E(\beta,z) \text{ satisfy (5.4)} & \text{scaling: } \Theta_E^-(\beta, z_1') = 1 \\ \Theta_p(\beta,z) = \frac{\bar{\gamma}}{\bar{\gamma}+i(\Omega-\bar{\omega})} \Theta_E(\beta,z). \end{cases} \tag{5.15}$$

This system should be considered separately in each laser section.

A.2.1 Transfer Matrices and Characteristic Function

Assume that β and Ω are fixed and we know the value of function Θ_E at the left edge z_k' of the section S_k. To find the function $\Theta_E(\beta,z)$ at any other z

5 Multisection Lasers: Longitudinal Modes and their Dynamics

of the same section, we need to solve a system of ODE's [given by the first line of (5.15)] in this section. For β and κ constant within S_k, the solution of this linear problem is given by transfer matrices M_k^S:[4]

$$\Theta_E(\beta, z) = \begin{pmatrix} \cos\eta - \frac{iD_k(z-z'_k)}{\eta}\sin\eta & -\frac{i\kappa_k^-(z-z'_k)}{\eta}\sin\eta \\ \frac{i\kappa_k^+(z-z'_k)}{\eta}\sin\eta & \cos\eta + \frac{iD_k(z-z'_k)}{\eta}\sin\eta \end{pmatrix} \Theta_E(\beta, z'_k)$$

$$\stackrel{def}{=} M_k^S(z,\beta,\Omega)\Theta_E(\beta, z'_k)$$

$$\eta \stackrel{def}{=} (z-z'_k)\sqrt{D_k^2 - \kappa_k^+\kappa_k^-} \in \mathbf{C}. \tag{5.16}$$

Boundary-junction conditions (5.4) and the scaling of Θ imply the following conditions on the eigenfunction $\Theta_E(\beta, z)$ at the edges of laser sections:

$$\Theta_E(z'_{k+1}) = \frac{1}{t_k}\begin{pmatrix} 1 & -r_k^* \\ -r_k & 1 \end{pmatrix}\Theta_E(z''_k) \stackrel{def}{=} M_k^J \Theta_E(z''_k) \quad 0 < k < m$$

$$\Theta_E(z'_1) = \begin{pmatrix} -r_0^* \\ 1 \end{pmatrix} \quad \Theta_E(z''_m) = c\begin{pmatrix} 1 \\ r_m \end{pmatrix} \quad c \in \mathbf{C}. \tag{5.17}$$

Here, the transfer matrices M_k^J show how the functions $\Theta_E(\beta, z)$ are propagated through the junction of the sections S_k and S_{k+1}.

We know now how the function Θ_E is transfered through sections and junctions of the sections. Thus, the consequent multiplication of the matrices M^S and M^J allows us to construct an overall transfer matrix $M(z, \beta, \Omega)$ determining the propagation of Θ_E from the left facet of the laser z'_1 to an arbitrary position z. Moreover, the assumed scaling of Θ [known value of $\Theta_E(\beta, z'_1)$] together with the easy relation between Θ_p and Θ_E [as given in (5.15)] allows one to determine completely the eigenfunction $\Theta(\beta, z)$ for arbitrary $z \in S_k$:

$$\Theta_E(\beta, z) = M_k^S(z)\left[M_{k-1}^J M_{k-1}^S(z''_{k-1}) \cdots M_1^J M_1^S(z''_1)\right]\begin{pmatrix} -r_0^* \\ 1 \end{pmatrix}$$

$$\stackrel{def}{=} M(z, \beta, \Omega)\begin{pmatrix} -r_0^* \\ 1 \end{pmatrix}$$

$$\Theta_p(\beta, z) = \frac{\bar{\gamma}}{\bar{\gamma}+i(\Omega-\bar{\omega})}\Theta_E(\beta, z). \tag{5.18}$$

The value of the vector function Θ_E at the right facet z''_m of the laser can be defined by means of the transfer matrix $M(z''_m, \beta, \Omega)$ or immediately taken from (5.17). The equality of these complex two-component vectors implies a complex algebraic characteristic equation:

$$\mathcal{M}(\Omega;\beta) \stackrel{def}{=} (-r_m, 1)M(z''_m, \beta, \Omega)\begin{pmatrix} -r_0^* \\ 1 \end{pmatrix} = 0, \tag{5.19}$$

---

[4] For nonuniform β or κ within S_k, we find the transfer matrices in small subintervals between z'_k and z, where β or κ are approximated by constants. These subintervals can correspond to the spatial steps of the numerical scheme. The product of these matrices of all subintervals gives an approximation of required transfer matrix.

where $\mathcal{M}(\Omega;\beta)$ is a nonlinear complex algebraic characteristic function of the complex variable Ω. Here β is fixed and is treated as a parameter. The set of complex roots Ω of this function coincides with the set of eigenvalues of the original spectral problem. With any Ω satisfying (5.19), one can easily reconstruct the corresponding eigenfunction Θ by means of (5.18). Thus, in order to solve the spectral problem, we are looking for the roots of the function $\mathcal{M}(\Omega;\beta)$, which is analytic in all complex domain with exception of some singular points.

A.2.2 Finding Roots of the Characteristic Equation

The complex equation (5.19) has an infinite number of roots, which, in general, need not be isolated [23, 27]. Among these roots, we are looking for a finite number of those with low damping (negative or small positive $\Im m\Omega$) and $\Re e\Omega$ from some certain frequency range. For simplicity, we assume that the derivative $\partial_\Omega\mathcal{M}$ computed at these roots remains nonzero, i.e., these roots remain isolated from each other.[5] To find the roots Ω_r, $r=1,\ldots,q$, we apply Newton's iteration scheme:

$$\Omega_{r(j+1)} = \Omega_{r(j)} - \mathcal{M}(\Omega_{r(j)};\beta)\Big/\partial_\Omega\mathcal{M}(\Omega_{r(j)};\beta) \qquad j\to\infty. \qquad (5.20)$$

Here, $\{\Omega_{r(j)}\}_{j=0}^\infty$ is a sequence of approximations converging rapidly to Ω_r, if only $|\partial_\Omega\mathcal{M}(\Omega_{r(j)};\beta)|$ is not too small.

The initial approximation $\Omega_{r(0)}$ is determined by parameter continuation (homotopy method) from known previous solutions for a similar laser with the same length and the same number of sections. The parameters are changed in small steps toward the parameter set of the actual laser, and Newton's iteration scheme is applied in each step. The step size is chosen inversely proportional to the maximal distance between corresponding roots of the previous two steps. This step adjustment allows one to guarantee a good initial approximation for Newton's scheme at the next parameter step.

If suitable previous solutions are not available, we start the parameter continuation from a Fabry–Perot-type laser ($\kappa^\pm = \bar{g} = 0$, $r_j = 0$ for $0 < j < m$ and $r_0 = r_m = 1$). In this case, the related transfer matrix M has a simple diagonal form and all roots of the characteristic function can be written explicitly:[6]

$$\Omega_r = \left(\pi r + \frac{\pi}{2} - \sum_{k=1}^m \beta_k L_k\right)\Big/\left(\sum_{k=1}^m L_k/v_{gr,k}\right) \qquad r\in\mathbf{N}.$$

---

[5] Mode degeneracy appears only at singular parameter constellations. The very proximity to it, however, modifies the laser dynamics [24, 26].

[6] Aside from these roots, the characteristic function with $\bar{g} \neq 0$ still has an infinite number of roots concentrated nearby to $\bar{\omega}_k + i\bar{\gamma}_k$, $k=1,\ldots,m$ (see [23]). Until $\bar{\gamma}$ are large enough, the impact of these "polarization modes" can be neglected.

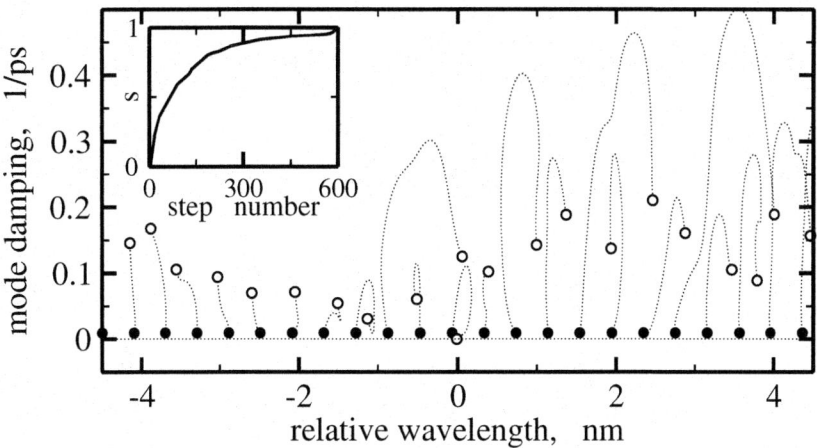

Fig. 5.12. Evolution of modes during parameter continuation from a simple Fabry–Perot-type laser (full bullets, $s=0$) toward the three-section DFB laser of Example I (empty bullets, $s=1$). Insert: relative distance s from the initial parameter set versus step number.

Evolution of modes from these values during parameter continuation is illustrated in Fig. 5.12.

A.3 Mode Decomposition

Assume we have computed the functions $E(z,t)$ and $p(z,t)$ in a given instant t. Using also computed distribution $\beta(z,t)$ and following (5.18), we find q eigenfunctions $\Theta(\beta(z,t),z)$. The corresponding eigenfunctions $\Theta^\dagger$ of the adjoint problem read as:

$$\Theta^\dagger(\beta,z) = \begin{pmatrix} \Theta_E^\dagger(\beta,z) \\ \Theta_p^\dagger(\beta,z) \end{pmatrix} = \left(\Theta_E^{-*}, \Theta_E^{+*}, \frac{v_{gr}\bar{g}}{2\bar{\gamma}}\Theta_p^{-*}, \frac{v_{gr}\bar{g}}{2\bar{\gamma}}\Theta_p^{+*}\right)^T. \quad (5.21)$$

Exploiting the orthogonality of Θ and $\Theta^\dagger$, we find the modal amplitudes $f_k(t)$ appearing in (5.8):

$$f_k(t) = \left[\Theta^{k\dagger}(\beta(z,t),z), \begin{pmatrix} E(z,t) \\ p(z,t) \end{pmatrix}\right] \Big/ \left[\Theta^{k\dagger}(\beta(z,t),z), \Theta^k(\beta(z,t),z)\right].$$

To estimate the scalar product of the grid functions E and p with the analytically given eigenfunction $\Theta^{k\dagger}$, we use the trapezoidal rule for approximate integration. The scalar product in the denominator is given by the analytic formula[7]:

---

[7] We assume that β is constant in each section. If not, we split the sections into smaller subsections where β can be treated as constant and use this formula again.

$$\left[\Theta^{\dagger},\Theta\right] = \sum_{r=1}^{m}\left(1+\frac{v_{gr,r}\bar{g}_r\bar{\gamma}_r}{2(\bar{\gamma}_r+i(\Omega-\bar{\omega}_r))^2}\right)\frac{i\left[\kappa_r^-\Theta_E^{-2}-\kappa_r^+\Theta_E^{+2}\right]\big|_{z_r'}^{z_r''}+\chi_r}{2(D_r^2(\Omega)-\kappa_r^+\kappa_r^-)}$$

$$\chi_r \stackrel{def}{=} 2D_r(\Omega)L_r[2D_r(\Omega)\Theta_E^+\Theta_E^-+\kappa_r^-\Theta_E^{-2}+\kappa_r^+\Theta_E^{+2}]\big|_{z_r'}.$$

$D(\Omega)$ is defined in (5.16), $\Theta_E^{\pm}$ at the section edges z_r' and z_r'' is found from formula (5.18).

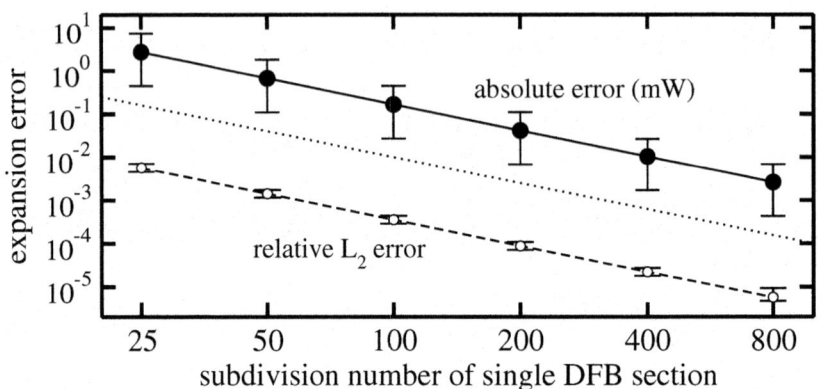

Fig. 5.13. Variation of expansion errors with increasing number of subdivisions. Full bullets: time averaged absolute L_2 space error. Empty bullets: corresponding relative error. Error bars indicate the fluctuations of these errors during the pulsations shown in Fig. 5.3. The dotted line is given by relation $100(\text{subdivision number})^{-2}$ and indicates a quadratic decay of errors.

To control the precision of our mode expansion (if spontaneous emission $F_{sp}^{\pm}$ is neglected), to be sure that none of the important modes has been lost and to check the precision of our numerical integration scheme, we compute the absolute and relative L_2 space errors:

$$\max_z\left\{\left|E(z,t)-\sum_{k=1}^{q}f_k(t)\Theta_E^k(\beta,z)\right|\right\}, \quad \left\|E(z,t)-\sum_{k=1}^{q}f_k(t)\Theta_E^k(\beta,z)\right\|\Big/\|E(t)\|.$$

Possible origins of errors are the insufficient finite number of modes used in the field decomposition, errors when estimating integrals of grid functions, and the limited precision of the numerical scheme used for the integration of the TW model. The quadratic relation between errors and discretization step (increased number of section subdivisions) indicates the dominance of integration errors. At the same time, this dependence shows an expected correction of computed fields and modes implied by the second-order precision numerical scheme. This quadratic decay of errors also indicates the presence of all important modes in the truncated field expansion (5.8).

Acknowledgment

The work of M.R. was supported by DFG Research Center "Mathematics for key technologies: Modeling, simulation, and optimization of real-world processes".

References

1. N. Chen, Y. Nakano, K. Okamoto, K. Tada, G. I. Morthier, and R. G. Baets: IEEE Journ. Sel. Topics Quant. Electron. **3**, 541 (1997)
2. M. C. Nowell, L. M. Zhang, J. E. Carroll, and M. J. Fice: IEEE Photon. Technol. Lett. **5**, 1368 (1993)
3. O. Kjebon, R. Schatz, S. Lourdudoss, S. Nilsson, B. Stalnacke, and L. Backbom: Electron. Lett. **33**, 488 (1997)
4. U. Feiste: IEEE J. Quantum Electron. **34**, 2371 (1998)
5. D. A. Yanson, M. W. Street, S. D. McDougall, I. G. Thayne, and J. H. Marsh: Appl. Phys. Lett. **78**, 3571 (2001)
6. B. Sartorius, C. Bornholdt, O. Brox, H. J. Ehrke, D. Hoffmann, R. Ludwig, and M. Möhrle: Electron. Lett. **34**, 1664 (1998)
7. M. Radziunas, H.-J. Wünsche, B. Sartorius, O. Brox, D. Hoffmann, K. Schneider, and D. Marcenac: IEEE J. Quantum Electron. **36**, 1026 (2000)
8. M. Möhrle, B. Sartorius, C. Bornholdt, S. Bauer, O. Brox, A. Sigmund, R. Steingrüber, M. Radziunas, and H.-J. Wünsche: IEEE J. Sel. Top. Quantum Electron. **7**, 217 (2001)
9. O. Brox, S. Bauer, M. Radziunas, M. Wolfrum, J. Sieber, J. Kreissl, B. Sartorius, and H.-J. Wünsche: IEEE J. Quantum Electron. **39(11)**, 1381 (2003)
10. J. E. Carroll, J. E. A. Whiteaway, and R. G. S. Plumb: *Distributed Feedback Semiconductor Lasers* (IEE, U.K. 1998)
11. U. Bandelow, M. Radziunas, J. Sieber, and M. Wolfrum: IEEE J. Quantum Electron. **37**, 183 (2001)
12. M. Radziunas, H.-J. Wünsche, O. Brox, and F. Henneberger: SPIE Proc. Ser. **4646**, 420 (2002)
13. H.-J. Wünsche, O. Brox, M. Radziunas, and F. Henneberger: Phys. Rev. Lett. **88(2)**, 23901 (2002)
14. H.-J. Wünsche, M. Radziunas, S. Bauer, O. Brox, and B. Sartorius: IEEE J. Sel. Top. Quantum Electron. **9(3)**, 857 (2003)
15. S. Bauer, O. Brox, J. Kreissl, B. Sartorius, M. Radziunas, J. Sieber, H.-J. Wünsche, and F. Henneberger: Phys. Rev. E **69**, 016206 (2004)
16. M. Radziunas and H.-J. Wünsche: SPIE Proc. Ser. **4646**, 27 (2002)
17. C. Bornholdt, J. Slovak, M. Möhrle, and B. Sartorius: Proceedings OFC **TuN6**, 87 (2002)
18. G. P. Agrawal and N. K. Dutta: *Long-Wavelength Semiconductor Lasers* (Van Nostrand Reinhold, New York 1986)
19. O. Hess and T. Kuhn: Prog. Quant. Electr. **20(2)**, 85 (1996)
20. D. D. Marcenac: Fundamentals of laser modelling. PhD Thesis, St. Catharine's College, University of Cambridge (1993)
21. B. Tromborg, H. E. Lassen, and H. Olesen: IEEE J. Quantum Electron. **30**, 939 (1994)

22. U. Bandelow, H.-J. Wünsche, B. Sartorius, and M. Möhrle: IEEE J. Sel. Top. Quantum Electron. **3**, 270 (1997)
23. J. Sieber: Longitudinal Dynamics of Semiconductor Lasers. PhD Thesis, Faculty of Mathematics and Natural Sciences II, Humboldt-University of Berlin (2001)
24. J. Sieber: SIAM J. on Appl. Dyn. Sys. **1(2)**, 248 (2002)
25. N. Korneyev, M. Radziunas, H.-J. Wünsche, and F. Henneberger: SPIE Proc. Ser. **4986**, 480 (2003)
26. H. Wenzel, U. Bandelow, H.-J. Wünsche, and J. Rehberg: IEEE J. Quantum Electron. **32**, 69 (1996)
27. J. Rehberg, H.-J. Wünsche, U. Bandelow, and H. Wenzel: Z. Angew. Math. Mech. **77(1)**, 75 (1997)
28. M. G. Davis and R. F. O'Dowd: IEEE J. Quantum Electron. **30**, 2458 (1994)
29. L. M. Zhang and J. E. Carroll: IEEE J. Quantum Electron. **28**, 604 (1992)
30. B.-S. Kim, Y. Chung, and J.-S. Lee: IEEE J. Quantum Electron. **36**, 787 (2000)
31. A. J. Lowery: Proc. Inst. Elect. Eng. J **134(5)**, 281 (1987)

6 Wavelength Tunable Lasers: Time-Domain Model for SG-DBR Lasers

D. F. G. Gallagher

Photon Design, 34 Leopold St., Oxford OX4 1TW, U.K., dfgg@photond.com

6.1 The Time-Domain Traveling Wave Model

Early models of laser diode behavior [1, 2, 3] were based on a rate equation analysis that considered only the intensity of the optical fields, and so could not deliver any information on the spectral behavior of the laser and, indeed, for more complex structures such as distributed feedback (DFB) lasers and distributed Bragg reflector (DBR) lasers, could not deliver a realistic response even of the intensity.

Second-generation laser models [4, 5, 6] attempted to model the detailed optical fields of the resonating laser modes — in other words, to simultaneously obey both phase matching conditions and photon/electron rate equations. Those models were fundamentally frequency-domain models, although extensions to deliver some time-domain results were developed. Such models were very successful in advancing the understanding of laser diode behavior, providing good light-current characteristics, spectral information such as side-mode suppression ratio and linewidth, relative intensity noise (RIN) spectra, amplitude modulation (AM), and frequency modulation (FM) characteristics. However, they could not simulate the fastest phenomena in a laser, for example, mode locking when time scales are so short that cavity modes do not have a chance to build up.

This study is founded on a time-domain traveling wave (TDTW) model that is based on the so-called advection equations, as put forward by Carroll et al. [7, 8]. The essence of this model is that the fields are propagated in the time domain as photons with an amplitude and phase and moving forward each timestep δt by a distance determined by the group velocity: $\delta z = v_g \delta t$. This is a very elegant technique because it delivers most of the spectral characteristics of the device, but avoids all the complications of finding laser cavity modes in the frequency domain. The basic TDTW method is based on the advection equations:

$$\left(\frac{1}{v_g}\frac{\partial}{\partial t} + \frac{\partial}{\partial z}\right) A(z,t) = -(i\delta - g)A(z,t) + i\kappa_{BA} B(z,t) + F_P(z,t) \quad (6.1)$$

$$\left(\frac{1}{v_g}\frac{\partial}{\partial t} - \frac{\partial}{\partial z}\right) B(z,t) = -(i\delta - g)B(z,t) + i\kappa_{AB} A(z,t) + F_P(z,t) \, . (6.2)$$

- A and B are the amplitudes of the forward and backward traveling waveguide modes, respectively.
- A fast oscillation $\exp(i\omega_0 t)$ has been removed from the fields, where ω_0 is some central optical frequency near the lasing operation.
- g is the optical gain seen by the waveguide modes, net of losses due to material absorption and scattering.
- δ is the detuning of the mode from some central wavelength, typically the Bragg wavelength. Changes of refractive index with carrier density and temperature are included here.
- κ_{BA} and κ_{AB} are grating coupling coefficients, modeling distributed coupling of the forward and backward traveling modes.
- $F_p(z,t)$ is a stochastic noise term, representing spontaneous emission into the lasing mode.

The stochastic noise term is crucial to the functioning of the TDTW model — just as in a real laser, it is the spontaneous emission that starts the laser lasing. The TDTW model will not produce any optical signal unless this term is included or photons are introduced from some other source.

To simulate a device of finite length, a structure is broken into one or more z elements (see Fig. 6.1). At each timestep, (6.1) is applied to each z element to determine the fields exiting the z element at the end of the timestep as a function of the fields entering the z element at the start of the timestep and the spontaneous emission events occurring in the z element during the timestep.

Procedurally, it is more convenient to write (6.1) in matrix form:

$$\begin{bmatrix} A(z+\delta z, t+\delta t) \\ B(z, t+\delta t) \end{bmatrix} \quad (6.3)$$
$$= \exp[(g-i\delta)\delta z] \begin{bmatrix} m_{11} & m_{12} \\ m_{21} & m_{22} \end{bmatrix} \begin{bmatrix} A(z,t) \\ B(z+\delta z, t) \end{bmatrix} + \begin{bmatrix} F_{P,a} \\ F_{P,b} \end{bmatrix},$$

with the matrix coefficients:

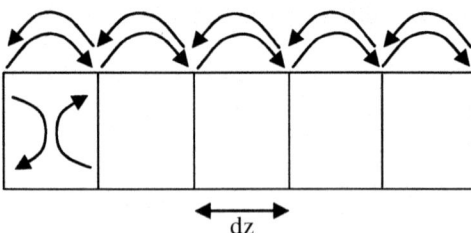

Fig. 6.1. A device is divided into many z elements — photons (energy) propagate a distance $dz = v_\sigma \, dt$ at each timestep.

$$m_{12} = \kappa_{BA}\delta z$$
$$m_{21} = \kappa_{AB}\delta z$$
$$m_{11} = \sqrt{1 - m_{12}m_{12}^*}$$
$$m_{22} = \sqrt{1 - m_{21}m_{21}^*} \,. \tag{6.4}$$

The grating coefficients κ_{AB} and κ_{BA} may represent real-index gratings, loss gratings, and gain gratings or any combination:

$$\kappa_{AB} = \kappa_{\mathrm{r}} + \kappa_{\mathrm{l}} - \kappa_{\mathrm{g}}^* \,,$$
$$\kappa_{BA} = \kappa_{\mathrm{r}}^* + \kappa_{\mathrm{l}}^* + \kappa_{\mathrm{g}} \,. \tag{6.5}$$

The forcing term F_P in (6.1) results from the spontaneous emission terms, because these emission events have a phase that is random with respect to the lasing fields and is given by:

$$F_P = \sqrt{h\nu \frac{N_{\mathrm{e}}}{\tau_{\mathrm{r}}} \frac{\beta_{\mathrm{NA}}}{2} \delta t} \, , i_n(t, z) \tag{6.6}$$

where $i_n(t, z)$ is a random number with inverse-normal probability distribution, i.e., we assume that it is uncorrelated in time and space. N_{e} is the carrier density, τ_{r} is the radiative spontaneous recombination lifetime of the carriers, and $h\nu$ is the photon energy. The zero time-correlation for i_n implies a white noise source. Apart from an unrealistic spontaneous spectrum well below threshold, this assumption has little effect on the simulation, as above threshold, the optical spectrum is dominated by the frequency response of the cavity and the gain curve of the active medium. The zero space-correlation implies that the z-sampling interval is significantly longer than the diffusion length, which is usually the case, but is not important close to or above threshold. β_{NA} is the coupling of the spontaneous emission into the numerical aperture of the waveguide mode, and the 2 in the denominator accounts for the fraction coupling in one direction. Note that, whereas the spontaneous coupling coefficient conventionally includes a part that takes account of the fraction of the spectrum that falls within the linewidth of the laser mode, here we must include only the spatial part because the spectral part is taken into account implicitly in the TDTW model.

An alternative expression used by some authors is:

$$F_P = \sqrt{h\nu g(N_{\mathrm{e}}) n_{\mathrm{sp}}(N_{\mathrm{e}}) \delta t} \, i_n(t, z) \,, \tag{6.7}$$

where n_{sp} is the so-called spontaneous inversion factor. This avoids the need to determine β_{NA} and n_{sp} is generally around 1 to 5. However, n_{sp} has a pole at transparency ($g(N) = 0$), which is troublesome numerically.

6.1.1 Gain Spectrum

As it stands, the model has no way of defining the gain spectrum. An infinite impulse response (IIR) digital filter may be used to implement a Lorentzian gain response [9, 10]:

$$g(\omega) = g(\omega_0)/\sqrt{1 + \tau^2(\omega - \omega_0)}. \tag{6.8}$$

The IIR filter of this is given by:

$$A(iz, t) = m_a c_1(t - \delta t) - m_b c_2(t - 2\delta t) + m_c A(iz, t - \delta t)$$
$$B(iz - 1, t) = m_a d_1(t - \delta t) - m_b d_2(t - 2\delta t) + m_c B(iz - 1, t - \delta t)$$
$$\underline{c}(t) = \exp[(g - i\delta)\delta z][M_t]\underline{F}_P(t)$$
$$\underline{F}_P(t) = \{A(iz - 1, t), B(iz, t)\}$$
$$\underline{d}(t) = \exp[(g - i\delta)\delta z][M_t]\underline{F}_P(t)$$
$$\underline{c} = (c_1, c_2) \qquad \underline{d} = (d_1, d_2). \tag{6.9}$$

The matrix M is given by (6.3) and it is the same for $c(t - \delta t)$ and $c(t - 2\delta t)$ etc. The m_a, m_b, m_c terms define the filter coefficients and, thus, the gain function. These are given by:

$$m_a = (K + 1 + g_{\mathrm{pk}}\delta z/2)/(K + 1 - g_{\mathrm{pk}}\delta z/2)$$
$$m_b = \mathrm{e}^{\mathrm{i}\Theta}(K - 1 - g_{\mathrm{pk}}\delta z/2)/(K + 1 - g_{\mathrm{pk}}\delta z/2)$$
$$m_c = \mathrm{e}^{\mathrm{i}\Theta}(K - 1 + g_{\mathrm{pk}}\delta z/2)/(K + 1 - g_{\mathrm{pk}}\delta z/2)$$
$$\Theta = -2\pi v_c \delta t (\lambda_{\mathrm{pk}} - \lambda_0)/\lambda_0^2 \qquad K = 2\tau v_{\mathrm{g}}/\delta z, \tag{6.10}$$

where g_{pk} is the amplitude gain peak (1/2 of power gain) seen by the laser's waveguide mode and λ_{pk} is the position of the gain peak relative to the reference wavelength $\lambda_0 = 2\pi v_c/\omega_0$.

This digital filter will match the position of the gain peak and the gain curvature at the peak. It also conveniently falls to unity at the Nyquist frequencies $\pm 1/(2\delta t)$. However, it is not ideal and especially does not match the high absorption at photon energies above the band gap, as shown in Fig. 6.2; thus, it is not possible with this filter to obtain a function that is good both well above and well below the band gap simultaneously.

Gain saturation may be readily included in this model by assuming that the saturation is slow relative to δt. Thus, we can write the gain g_{pk} in (6.10) as:

$$g_{\mathrm{pk}} = \frac{g_{\mathrm{pk,lin}}(N_{\mathrm{e}})}{(1 + \varepsilon_g(|A|^2 + |B|^2))}. \tag{6.11}$$

The gain g_{pk} is evaluated for each z element and the intensities A^2, B^2 should be averaged over the z element. A more sophisticated gain saturation model can also include carrier heating effects by modifying λ_{pk}, assuming the carrier heating occurs slowly relative to δt:

$$\lambda_{\mathrm{pk}} = \lambda_{\mathrm{pk},0} + \Delta N_{\mathrm{e}}(a_{\mathrm{pk}} + \Delta N_{\mathrm{e}} b_{\mathrm{pk}}) \qquad \Delta N_{\mathrm{e}} = N_{\mathrm{e}} - N_{\mathrm{o}}, \tag{6.12}$$

where N_{o} is the transparency carrier density and a_{pk}, b_{pk} are constants defining the gain peak shift.

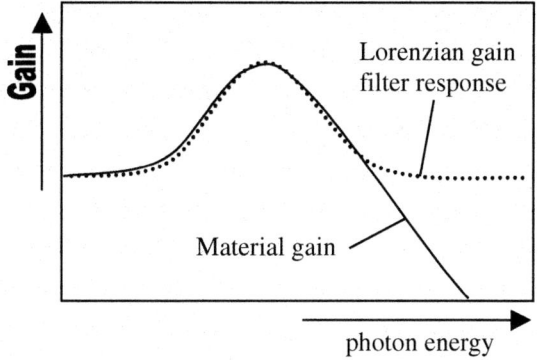

Fig. 6.2. Digital gain filter response compared with real material gain profile.

6.1.2 Noise Spectrum

A more advanced model for the noise spectrum is simply achieved by adding a correlation in time, which is implemented using a digital filter of the form:

$$i_n(t + \delta t) = i_n(0) + a/\delta t e^{i\omega \delta t} i_n(t) , \qquad (6.13)$$

where ω defines the peak wavelength of the spontaneous spectrum, a defines the bandwidth, and $i_n(0)$ is the inverse normally distributed random number. However, this will only have significant impact well below threshold and can be ignored in most cases.

6.1.3 Carrier Equation

Considering initially only a 1-dimensional (z direction) model, the carrier density N in a *z-element* is given by:

$$\frac{dN_e}{dt} = -\frac{\Gamma G(N_e, \lambda)}{V_N} \frac{P_\#}{(1 + \varepsilon P_\# / V_P)} - \frac{N_e}{\tau_s(N_e)} + \frac{J}{q_e d} + F_N \qquad (6.14)$$

where $P_\#$ is the photon number for the z element, V_P is some characteristic waveguide mode volume for the z element, Γ is the confinement factor of the mode to the active layer, $G(N_e, \lambda)$ is the material (power) gain, τ_s is the carrier lifetime including both radiative and nonradiative processes, J is the injection current density, and d is the active layer thickness. V_N is the active layer volume so that carrier number is given by $N V_N$. τ_s is the spontaneous lifetime of the carriers, including both radiative and nonradiative terms, well approximated as:

$$1/\tau_s = 1/\tau_r + 1/\tau_{nr} \quad 1/\tau_r = BN_e \quad 1/\tau_{nr} = 1/\tau_N + CN_e^2 ,$$

where B quantifies the stimulated emission process, which is in practice close to bimolecular. C quantifies the Auger processes, which is predominantly a 3-particle process, i.e., with a recombination rate proportional to N^3. Other nonradiative recombination processes, which are generally unimolecular, for example, recombination via defect sites, are quantified by τ_N.

In the time domain, we do not know the wavelength needed to calculate the $G(N_e, \lambda)$ term. It is possible to implement a digital filter to compute this term in a similar manner to the photon gain. However, it is more simple and more efficient to apply quantum conservation — each photon generated by stimulated emission is matched by a carrier lost, taking care to account also for the number of photons lost by absorption or scattering:

$$\delta P_\# = P_{\#,\text{out}}(t+\delta t) - P_{\#,\text{in}}(t)$$
$$= \delta P_{\#,\text{stim}} - 0.5[P_{\#,\text{in}}(t) + P_{\#,\text{out}}(t+\delta t)]\alpha_{\text{mode}}\delta z$$
$$\delta N_e V_N = -\delta P_{\#,\text{stim}}, \qquad (6.15)$$

where $P_\#$ and $N_e V_N$ are the photon and carrier numbers, respectively, $P_{\#,\text{in}}$ and $P_{\#,\text{out}}$ are the photon numbers entering and exiting the z element during the time step, and α_{mode} represents the mode loss due to material absorption and scattering.

6.1.4 Carrier Acceleration

One disadvantage of any time-domain method is that one has to compute for perhaps a considerable time for the device to settle before measuring some steady-state property, such as optical power at a given drive current. This is especially a problem in generating a light-current curve where many steady-state measurements must be made. This problem may be partially diminished, however. For a simple Fabry–Perot laser, the resonance frequency is proportional to $1/\sqrt{\tau_N \tau_P}$. The resonance frequency is a good guide to convergence rate, so we can accelerate convergence by decreasing either τ_P or τ_N. As τ_P is already close to δt, only a decrease of carrier lifetime is practical. Although the resonance frequency suggests we can get acceleration proportional merely to $\sqrt{\tau_N}$, where carriers are used merely for tuning and do not experience stimulated recombination, an acceleration linear with $1/\tau_N$ may be achieved. Reducing the carrier lifetime while maintaining the steady-state conditions is readily achieved by a modification of (6.14). The carrier density change at time step, with acceleration, is given by:

$$\delta N_e = \delta ts \left(\frac{\Gamma G(N_e, \lambda)}{V_N} \frac{P_\#}{(1+\varepsilon P_\#/V_P)} - \frac{N_e}{\tau_s(N_e)} + \frac{J}{q_e d} \right) + \delta t F_N, \qquad (6.16)$$

where s is an acceleration factor. Notice that we do not accelerate the noise term F_N. Experience suggests that s can be as much as 20 or 40 without introducing stability problems.

6.1.5 Extension to Two and Three Dimensions

The TDTW method may be readily and efficiently extended to 2D and even 3D. Figure 6.3 shows a schematic cross-section of a laser diode.

It is assumed to have three independent active layers, for example three quantum wells. Each active layer is divided into many "computation cells" laterally, and each cell is given an independent carrier density and temperature. Taking diffusion between neighboring cells into account, the carrier equation (6.14) is readily extended and becomes:

$$\frac{dN_i}{dt} = \frac{\Gamma_i G(N_i, \lambda)}{V_N} \frac{P_\#}{[1 + \varepsilon P_\#(\Gamma_i/\Gamma_{\text{ave}})/V_P]} - \frac{N_i}{\tau_s(N_i)} + \frac{J_i}{q_e d}$$
$$+ (N_{i+1} - N_i) D\left[(N_i + N_{i+1})/2\right]$$
$$+ (N_{i-1} - N_i) D\left[(N_i + N_{i-1})/2\right] + F_N \,, \qquad (6.17)$$

where Γ_i is the confinement factor of the mode to the ith cell, Γ_{ave} is the average of Γ_i, $D(N)$ is the carrier-density-dependent diffusion coefficient, and J_i is the current density injected into the cell.

The photon equations require a little more care — the method can apply only one set of Lorentzian gain filter parameters to the waveguide mode; but now the carrier density varies from cell to cell and, therefore, so too will g_{pk}, λ_{pk}, and the gain curvature K. A little algebra gives us an averaged set of filter parameters:

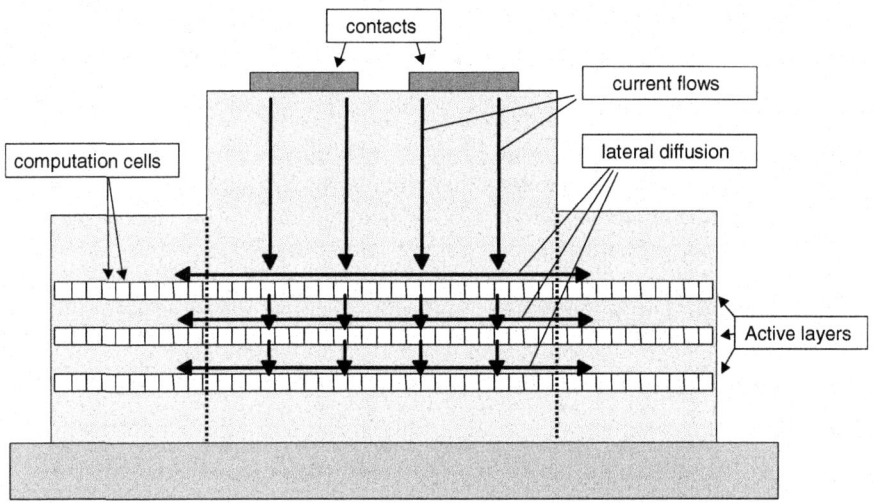

Fig. 6.3. Cross-section schematic of ridge laser with three active layers.

$$g_{\text{pk}} = \sum_i \Gamma_i g_{\text{pk},i}$$

$$g_2 = \sum_i \Gamma_i g_{2,i}$$

$$K = \frac{\lambda^2}{\pi v_c \delta t}\sqrt{g_2/g_{\text{pk}}}$$

$$\lambda_{\text{pk}} = \left(\sum_i \Gamma_i g_{2,i}\lambda_{\text{pk},i}\right)/g_2 , \tag{6.18}$$

where g_2 is the gain curvature defined by

$$g(\lambda) = g_{\text{pk}} + g_2(\lambda - \lambda_{\text{pk}})^2 . \tag{6.19}$$

6.1.6 Advantages of the TDTW Method

The method has many advantages for the simulation of laser diodes and traveling wave amplifiers. The TDTW model has formed the basis of a powerful commercial simulator CLADISS-2D [11].

- Frequency-domain methods require one to locate and follow longitudinal modes individually, thus if 20 modes are lasing simultaneously, then one must solve a self-consistent set of equations with all 20 modes included. TDTW does not locate or track individual longitudinal modes, if there are longitudinal resonances, then these become apparent when the time-domain fields are Fourier-transformed. This leads to several benefits:
  - Avoids convergence problems attempting to locate longitudinal modes.
  - If the laser itself is unstable, then this is readily seen in an unstable time evolution, rather than a failed simulation. If the laser is close to unstable, then a frequency-domain method will experience convergence problems.
  - No need to determine which longitudinal modes are required for the simulation.
- The method lends itself readily to complex structures, with multiple sections; even Y-branched geometries are straightforward.
- The method can efficiently take transverse carrier profiles into account, so extensible to 2D and 3D simulations (but with certain limitations — see below).
- The algorithm is very fast, for example, CLADISS-2D can simulate a DBR laser divided into 16 z elements at a rate of 5900 timesteps per second on a 2.4-GHz Pentium-4 personal computer (PC) in 1D and 1850 steps per second in 2D (15 discrete carrier densities in transverse direction).
- Is stable well below threshold — very low power densities are awkward for a frequency-domain method. Indeed, well below threshold, the optical power is not concentrated in longitudinal modes at all, requiring a different approach and more complexity for the frequency-domain method.

- Frequency-domain methods can be readily extended to give time-evolution information [4] but always assume steady-state along the length of the optical cavity, and so cannot model the fastest time effects such as mode-locking — TDTW has no such limitation.
- TDTW is a stochastic model — all noise effects are realistically reproduced, and so the method can deliver RIN spectra and even linewidths in a natural way.
- Multiple transverse modes may be readily modeled (but see limitations below).

6.1.7 Limitations of the TDTW Method

Nothing is perfect in life, even TDTW! Here are some of the main limitations of the method:

- The stochastic basis is sometimes inconvenient, for example, if one wants to generate a light-current curve, one wants the average or expected optical power, and the superimposed noise must be reduced by long simulation times.
- The Lorentzian gain filter is not realistic at energies substantially above the band gap. In contrast, the gain spectrum $g(\lambda)$ can be chosen freely in the frequency-domain methods.
- The method imposes a relationship between the time and space steps $\delta z = v_g \delta t$. A uniform section is generally broken up into an integer number of z elements of length δz, so that δt may be chosen appropriately. However, as the same δt applies to all sections of the device, it is not, in general, possible to obey this equation for more than one section, introducing an error in the group velocity (note, however, that the phase velocity can be readily matched everywhere).
- The timestep δt implies a free spectral range limit of $\Delta f = 1/\delta t$. To increase Δf, one must decrease δt, however, this implies a decrease in δz too — so simulation time is proportional to Δf^2.
- Calculating a linewidth in the time domain is, in reality, impractical for linewidths typically seen $\sim$ 10 MHz; to measure a 10-MHz linewidth, a simulation must run for many times 0.1 ms to average out the noise.
- All transverse modes must have the same group velocity. (More complex algorithms can tackle this restriction, see e.g., [12]).

6.2 The Sampled-Grating DBR Laser

6.2.1 Principles

Wavelength-controlled laser diodes, such as the DFB laser, have traditionally had very little tuning capability. Today's wavelength division multiplexed

(WDM) datacom transmission systems carry 40 or more wavelength channels at 100-GHz channel spacing — 0.8 nm at $\lambda = 1.55\,\mu$m. To construct such a system with untunable lasers requires fabrication of many pretuned devices. A laser that can tune over the whole, or at least a large part, of the WDM band substantially reduces the number of unique parts required to build it and the number of spares that a telecoms carrier must stock to keep it in service. Furthermore, tunability increases reliability as one spare laser can take over the job of any one of the other 40 lasers. In addition, the tunability permits flexible λ-routing — the route of the signal through the network being controlled by its wavelength. Thus, there are strong incentives to develop optical sources that can be tuned over the whole WDM band.

A number of designs have been proposed as tunable lasers. The simple DFB laser where a grating extends over the whole length of the device, can be tuned by up to 7 nm or so if fitted with multiple electrodes and designed with a sublinear gain medium [13]. Another design is the tunable DBR laser [14] with a gain section, a phase-tuning section and a grating section, which can achieve 12 nm or more. These devices are restricted to tuning by a factor $\Delta\lambda/\lambda$, given by $\Delta\mu/\mu$, where μ is the effective index of the waveguide and $\Delta\mu$ is the change in effective index achieved by , e.g., carrier injection.

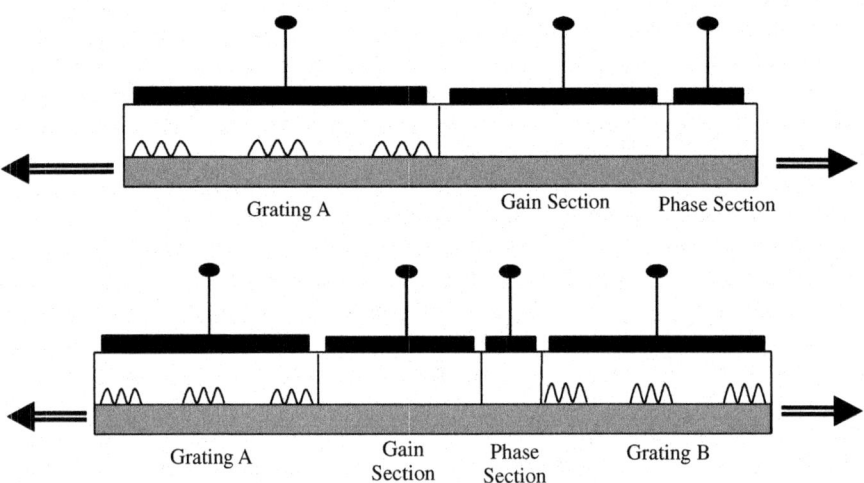

Fig. 6.4. (*top*) A three-section SG-DBR design with one sampled grating. (*bottom*) A four-section SG-DBR design with two sampled gratings, each with a slightly different sampling pitch.

The sampled-grating DBR laser (SG-DBR) is emerging as one of the main contenders for a tunable optical source. This is because it can achieve a tuning factor much greater than $\Delta\mu/\mu$. There are two principle configurations as shown in Fig. 6.4.

The more basic configuration consists of three sections:

- A DBR grating that is periodically suppressed so that only short bursts of grating remain.
- A gain section that provides amplification of the laser mode.
- A phase-tuning section that provides fine control of the phase matching condition for the laser cavity — to experience significant amplification, a photon must undergo exactly $2\pi n$ ($n = 0, 1, 2, 3...$) phase change as it travels one loop of the cavity.

The response of the sampled grating can be readily understood by applying the convolution theorem:

$$\text{FT}[a(z)b(z)] = \frac{1}{2\pi}\text{FT}[a(z)] * \text{FT}[b(z)],$$

where $*$ denotes convolution and FT denotes a Fourier transform.

An infinitely long grating has a sinusoidal first-order refractive index variation:

$$\delta\mu(z) = A\sin(\Lambda z) = A(e^{i2\pi z/\Lambda} - e^{-i2\pi z/\Lambda})/2.$$

The Fourier transform of this is trivial — two delta functions at frequencies $\pm 2\pi/\Lambda$. A short grating may be written as the product of an infinite grating and a square function as shown in Fig. 6.5.

The Fourier transform of the square function of width w, is the sinc-function $2\sin(wk/2)/k$, where k is the wavenumber. Thus, the response of a finite grating of periodicity $\Lambda = 2\pi/k_1$ is given by the convolution of the FT of $\sin(k_1 x)$, a delta function at $\pm k_1$, i.e.:

$$F(k) = \frac{\sin[w(k \pm k_1)/2]}{(k \pm k_1)/2}.$$

The sampling imposes one additional feature on the spectrum. The sampling function denoted $\text{SF}(z/L)$ may be considered as a convolution of the hat

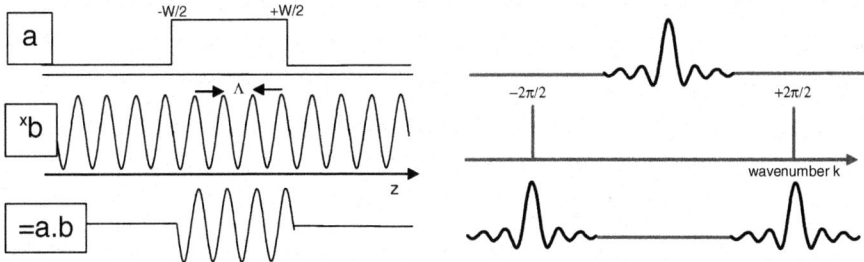

Fig. 6.5. (*left*) A short section of grating decomposed into its fundamental spatial frequency and a "hat function." (*right*) The Fourier spectra of the spatial functions.

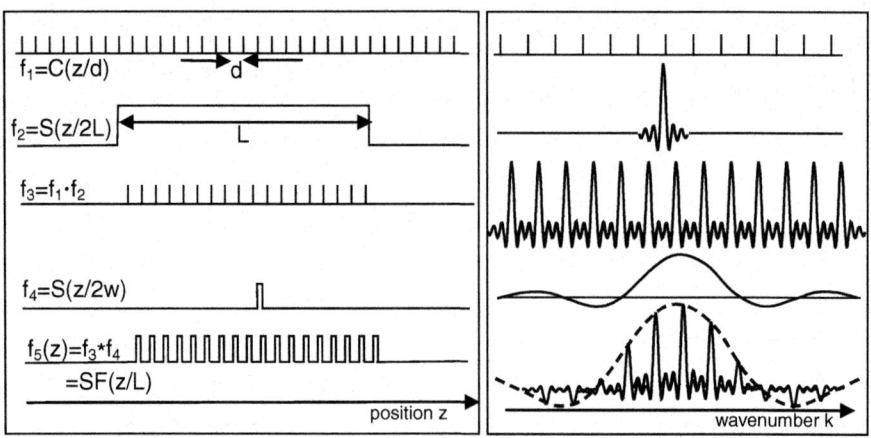

Fig. 6.6. (*left*) Decomposition of a finite number of "samples"; (*right*) associated spectral response.

function (square pulse, width w) denoted $S(z/2w)$ and a "comb", of pulse spacing d, denoted $C(z/d)$, multiplied by a function defining the total length of the sampled grating L – $S(z/L)$:

$$SF(z/L) = S(z/2w) * (C(z/d)S(z/2L)) .$$

This is illustrated in Fig. 6.6.

In summary then, the sampled grating of burst length w repeated with a period d and overall length L gives us a response spectrum that is a comb of sinc functions $\sin(Lk/2)/(k/2)$ spaced by $\Delta k = 2\pi/d$, and the whole comb having an envelope $F_4(k) = \sin(wk/2)/(k/2)$.

The FWHM of the sinc function is approximately 1.88 radians. Observing that a wavenumber change of δk is associated with a wavelength change $\delta\lambda = -\delta k \lambda^2/(2\pi)$, giving:

Full width of each SG resonance : $\quad \delta\lambda_{\mathrm{fw}} = 1.88\,\lambda^2/(\pi\mu L)$ (6.20)

Full width of envelope : $\quad \Delta\lambda_{\mathrm{env}} = 1.88\,\lambda^2/(\pi\mu w)$ (6.21)

SG resonance spacing : $\quad \Delta\lambda_{\mathrm{SG}} = \lambda^2/(2\mu d) .$ (6.22)

Putting some numbers in here, a typical sampled grating with $w = 5\,\mu\mathrm{m}$, $d = 45\,\mu\mathrm{m}$, and $L = 400\,\mu\mathrm{m}$ in a waveguide of effective index 3.5 operated around $\lambda = 1.55\,\mu\mathrm{m}$ gives us a reflection spacing of 8.5 nm and an envelope FWHM of approximately 80 nm. In practice, the gain spectrum of the SG-DBR laser's gain section may be narrower than this envelope, and so, must also be taken into account.

Figure 6.7 illustrates the computed reflection spectrum from a typical sampled grating (dimensions in caption). The sinc envelope is clearly visible.

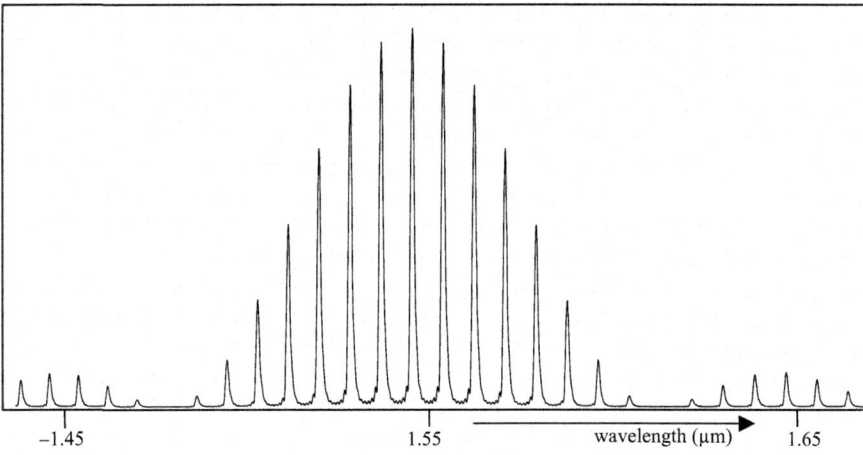

Fig. 6.7. Reflection spectrum (optical power) from a sampled grating. 5-µm grating samples separated by 35 µm with an overall length of 320 µm.

The spectrum was generated using the TDTW model. Note that it is important to divide each sample into several z elements to obtain the envelope; here we have divided each 5-µm burst into four elements giving a z step of 1.25 µm. If the envelope is not important, for example, if the gain bandwidth is narrower, then one can use just one element per grating sample. The spectrum was generated by injecting an optical signal into one end of the device, while sweeping its wavelength slowly over time and monitoring the reflected light, thus the x axis is actually the time coordinate.

6.2.2 Reflection Coefficient

The discussion in the previous section gives an idea of the principles of the sampled grating. Next, we will determine more precisely the reflection coefficient of such a grating. The coupled wave equations are:

$$\frac{\partial}{\partial z}A(z) = -(i\delta - g)A(z) + i\kappa_{BA}B(z) \qquad (6.23)$$

$$\frac{\partial}{\partial z}B(z) = -(i\delta - g)B(z) + i\kappa_{AB}A(z) \, . \qquad (6.24)$$

The solution of this is well known:

$$A(z) = a_1 e^{ikz} + a_2 e^{-ikz}$$

$$B(z) = \frac{(ik+\gamma)a_1 e^{ikz} - (ik-\gamma)a_2 e^{-ikz}}{\kappa}$$

$$a_1 = \frac{1}{\left[1 + \frac{(ik+\gamma)}{(ik-\gamma)} \exp(2ikL)\right]}$$

$$a_2 = 1 - a_1 \,, \tag{6.25}$$

with the solutions for k given by:

$$k^2 = \kappa\kappa^* - \gamma^2$$

$$\gamma = -i\delta + \alpha/2$$

$$\delta = \beta_\mathrm{m} - \pi/\Lambda \,,$$

where β_m is the propagation constant of the mode, α is the mode (power) loss, Λ is the grating periodicity, and assuming $\kappa_{FB} = \kappa_{BF} = \kappa$.

This gives a (amplitude) reflection coefficient:

$$r(\lambda) = \frac{2ik/\kappa}{\left[1 + \frac{(ik+\gamma)}{(ik-\gamma)}\right] \exp(2ikL)} \,. \tag{6.26}$$

Using the convolution theorem, we can modify this, assuming resonances of the sampled grating are sufficiently far apart to not interfere, giving:

$$r(\lambda) = \sum_n \frac{2ik/\kappa}{\left[1 + \frac{(ik+\gamma_n)}{(ik-\gamma_n)} \exp(2ikL)\right]} \,, \tag{6.27}$$

summing over all resonances of the sampled grating of overall length L, and where the resonance term δ is now determined by the sampling period d according to:

$$\delta_n = \beta_\mathrm{m} - n\pi/d \,.$$

6.2.3 The Three-section SG-DBR Laser

The SG-DBR laser with just one sampled grating achieves SG-DBR resonance selection by the interaction of the sampled grating, the Fabry–Perot resonances of the laser cavity, and the gain envelope [15]. The spacing of the Fabry–Perot resonances is given by:

$$\Delta\lambda_\mathrm{FP} = \lambda^2/2\mu L_\mathrm{cav} \,,$$

where μ is the effective index of the waveguide mode and L_cav is the cavity length. For a typical device length of 700 μm this gives a Fabry–Perot spacing of 0.5 nm. This is less than one-tenth the spacing of the sampled-grating

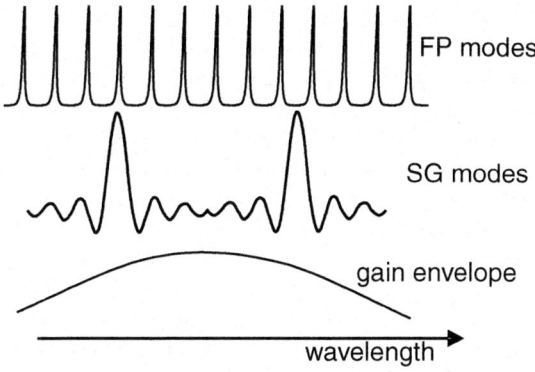

Fig. 6.8. Response of three-section SG-DBR.

reflections and, in general, a Fabry–Perot mode will match one SG resonance better than the others, providing a mechanism for selecting just one SG resonance. This is illustrated in Fig. 6.8.

A phase-tuning section [Fig. 6.4 (top)] permits one to shift the comb of Fabry–Perot modes along and so select different sampled-grating resonances. With ten FP modes per SG mode, a shift of just 0.5 nm in the FP modes will sweep through all SG modes, providing substantial "tuning gain". The design is, however, somewhat troublesome in practice due to poor mode discrimination.

We will simulate a three-section SG-DBR laser constructed of 5×10-μm grating samples spaced by 80 μm, with an InGaAsP gain section of 250 μm and a phase section of 50 μm. Figure 6.9 shows the calculated optical intensity along the length of the device. Here, we have run a simulation until the device was lasing and then switched off the optical noise sources and run the simulation a short while longer to allow the device to stabilize before capturing the profiles. Leaving the noise in would have significantly obscured the structure of the profiles.

Figure 6.10 shows a TDTW simulation of the device with the tuning section current swept to shift the Fabry–Perot modes. The tuning current is swept by 30 mA in the 8-ns time window shown. The central wavelength has been set to 1.55 μm. Note that the spontaneous lifetime of the tuning sections is rather low at 1 ns to speed up the simulation time. A more realistic lifetime of 5 ns to 10 ns would give a corresponding decrease in tuning current. The device initially jumps to the next mode as desired, but then jumps back because of poor mode discrimination. The spectral width of each SG resonance in this case is approximately 1 nm and the FP mode spacing is only 0.8 nm, so it is likely that more than one SG resonance will be excited. Making the grating longer helps narrow the SG resonance, but, unfortunately, also reduces the FP mode spacing.

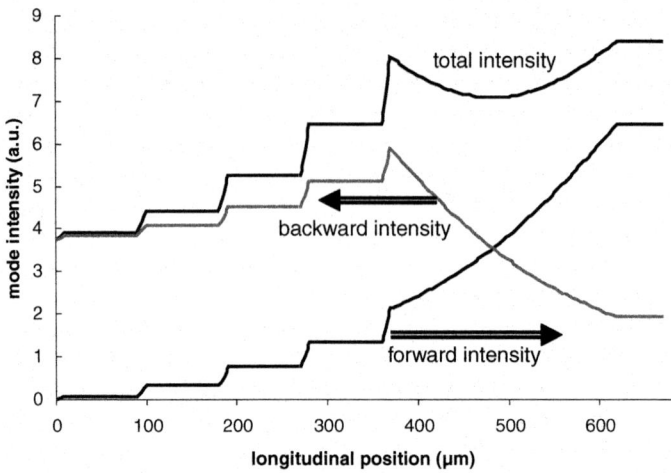

Fig. 6.9. Optical intensity along laser cavity.

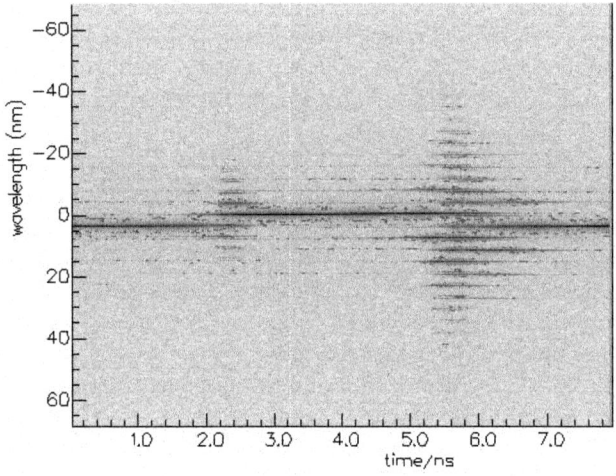

Fig. 6.10. Time evolving spectrum of a three-section SG-DBR laser as the tuning section current is swept.

6.2.4 The Four-section SG-DBR Laser

To get better mode discrimination, a second sampled grating can be used at the other end of the laser [16], as depicted in Fig. 6.4 (bottom). The grating mode spacing can then be controlled in each grating. Typically, the second grating is designed with a slightly different resonance spacing, so that only at widely spaced wavelengths do the resonances of each grating match up. This

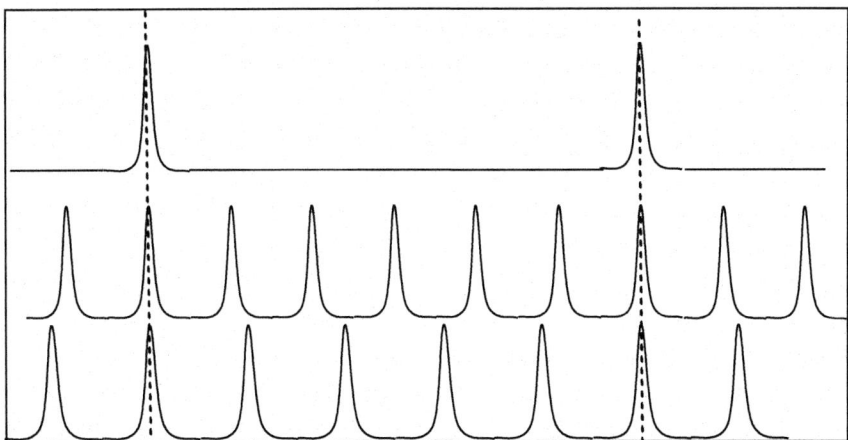

Fig. 6.11. Illustrating the interaction of two sampled gratings in a four-section SG-DBR laser. The bottom and middle curves represent the reflection from the left and right sampled gratings;, respectively. The top plot represents the resulting cavity gain. Lasing action occurs when strong reflections from both gratings coincide, indicated by dashed vertical lines.

creates a "Vernier scale" effect, where a small shift in one grating (by tuning) leads to a large change in lasing frequency. This is illustrated in Fig. 6.11.

To obtain lasing, the loop gain of the device, given by the product of each grating reflection and of the gain:

$$g_{\text{Loop}} = R_1 R_2 \exp(2gL_g) \, ,$$

must equal unity (see Fig. 6.12).

If the spacing of the left and right gratings are d_l and d_r, respectively, then we have resonance spacings $\Delta\lambda_l$ and $\Delta\lambda_r$, given by:

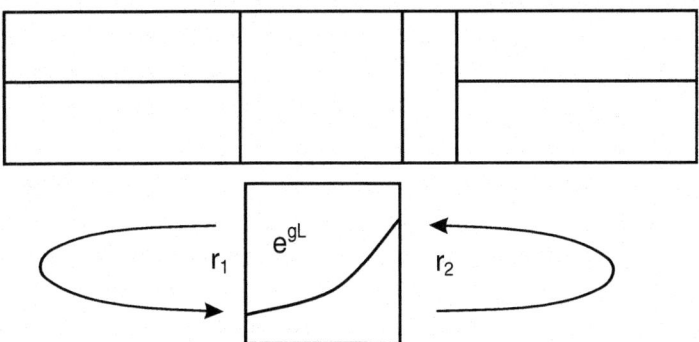

Fig. 6.12. Threshold condition.

$$\Delta\lambda_1 = \lambda^2/(2\mu d_1)$$
$$\Delta\lambda_2 = \lambda^2/(2\mu d_2) .$$

The spacing between coinciding resonances, in terms of the number of SG resonances, is given by:
$$N_{\text{res}} = \overline{d}/(d_1 - d_2) . \tag{6.28}$$

Thus detuning the second grating by one-tenth gives us a "free spectral range" of ten resonances. Clearly, the smaller the detuning, the larger the free spectral range. However, there is a tradeoff as the mode discrimination becomes poorer.

This factor $d/(d_1 - d_2)$ also gives us the tuning amplification of the four-section SG-DBR laser. Injecting carriers into one of the sampled gratings causes a shift $\delta\lambda$ in the resonances (it also changes the resonance spacing, but not by a significant amount). But, because of the Vernier effect, the resonance of the laser shifts by $\delta\lambda.d/(d_1 - d_2)$.

6.2.4.1 Side Mode Suppression

There are three principle side-mode phenomena to consider. First, there are the Fabry–Perot cavity-like modes at a spacing:

$$\Delta\lambda_{\text{FP}} = \lambda^2/2\mu L_{\text{cav}} . \tag{6.29}$$

The second is suppression of adjacent sampled-grating resonances by the Vernier effect. The third is suppression of the coincident resonance in the next free-spectral range given by N_{res} above.

The side-mode suppression can be conveniently quantified using the "normalized loss margin," M [17], defined as the difference in cavity loss between the two modes divided by the threshold gain of the lower threshold mode:

$$M = \Delta\alpha/\Gamma g_{\text{th}} , \tag{6.30}$$

where Γ is the gain confinement factor.

Taking the cavity-mode suppression first, this suppression ratio is defined solely by the ratio of the effective cavity length L_{cav} to the length of the sampled gratings and the κL product. Thus, to obtain good suppression of cavity side modes, we require the grating length is of the order of the cavity length.

The suppression of neighboring "free-spectral range" windows is achieved by the gain envelope and the response of the finite length grating samples, as given in (6.21) above. In general, these side modes are not a problem because they are so widely spaced that both the gain curve and the sample response provide more than adequate discrimination.

The main challenge in the four-section SG-DBR laser is the suppression of adjacent modes of the sampled gratings. The amplitude gain of the side mode relative to main mode is given approximately by:

$$g_{\text{sm}} = \frac{\sin^2(L\delta\beta/4)}{(L\delta\beta/4)^2}, \tag{6.31}$$

where the tuning mismatch $\delta\beta = 2\pi\mu\delta\lambda/\lambda^2$ is given by:

$$\delta\beta = \frac{2\pi\mu\Delta\lambda_{\text{SG}}}{\lambda^2 N_{\text{res}}} \cdot = \frac{\pi(d_1 - d_2)}{d^2}$$

6.2.5 Results

Using the TDTW model outlined in Sect. 6.1 and implemented in CLADISS-2D, we simulate the four-section SG-DBR laser here. The device represents a buried heterostructure InGaAsP laser oscillating around 1.55-μm wavelength. The simulations use the 2D/3D algorithm described in Sect. 6.1 to also take lateral diffusion/hole burning effects into account; although it is not expected that significant lateral effects occur in the buried heterostructure. For the tuning sections, a linewidth enhancement factor was not used since this relates the real index change to the imaginary index change (gain) and we have very low gain here. Instead, the change in index with carrier density $d\mu/dN_e$ was defined directly.

6.2.5.1 Varying One Grating Current

Figure 6.13 shows the instantaneous wavelength (derived from $\omega = d\phi/dt$) of the optical output as the tuning current of grating A is varied over the simulation time of 20 ns. Note that the spontaneous lifetime of the tuning sections is rather low at 1 ns to speed up the simulation time. A more realistic lifetime of 5 ns to 10 ns would give a corresponding decrease in tuning current. The tuning current was varied from 1 mA to 30 mA, which, with a 5-ns lifetime, would correspond to 0.2 mA and 6 mA. We will write tuning currents from now on in the notation $I/5$ mA, so here $I_1 = 1/5$ mA to 30/5 mA to remind the reader that the simulated currents are not actual currents.

The figure also shows the optical power exiting the right-hand side. The simulation clearly shows large jumps between sample-grating orders plus much smaller cavity mode jumps. Where the cavity modes jump, a strong beating can be seen between the two modes in the power plot. Note also the substantial tailing off of the intensity as the device is detuned.

Figure 6.14 shows the time-evolving spectrum of the same simulation. These time-evolving spectra are one of the most powerful features of the TDTW method and provide a highly informative picture of the device operation. The plot is generated by taking slices of typically 256 to 512 timesteps and applying fast Fourier transformation (FFT) to get a coarse spectrum, which is plotted as a vertical line of varying intensity. The next slice is then extracted, Fourier-transformed, and then another vertical line plotted, etc. until the whole time-domain simulation has been displayed. Controlling the

Table 6.1. Parameters Used in the Simulations.

All Sections

Parameter	Symbol	Value
Active layer width	w_a	$2.5\,\mu\text{m}$
Active layer thickness	d_a	$50\,\text{nm}$
Mode effective index (phase)	μ_m	3.2569
Mode effective index (group)	$\mu_{m,gp}$	3.486
Mode confinement factor	Γ	0.0609
Facet reflectivities (device)	R_1, R_2	0, 0

Gain Section

Parameter	Symbol	Value
Section length	L_s	$450\,\mu\text{m}$
Transparency carrier density	N_o	$1.5\text{e}18\,\text{cm}^{-3}$
Differential gain	G	$7.0\text{e-}6\,\text{cm}^3/s$
Gain compression	ε_g	$1\text{e-}15\,\text{cm}^3$
Auger recombination rate	C	$1\text{e-}28\,\text{cm}^6/s$
Gain curvature	g_2	$500\,\mu\text{m}^{-2}$
Radiative recombination rate (from $dN_e/dt = -B.N_e^2$)	B	$1\text{e-}20\,\text{cm}^3/s$
Linewidth enhancement factor	α_H	3

Phase Tuning Section

Parameter	Symbol	Value
Section length	L_s	$165\,\mu\text{m}$
Non radiative lifetime	τ_s	$1\,\text{ns}$
Change of refractive index with carrier density	$d\mu/dN_e$	$-5\text{e-}20\,\text{cm}^3$

Grating Sections

Parameter	Symbol	Value
Section length	L_g	$450\,\mu\text{m}, 360\,\mu\text{m}$
Non radiative lifetime	τ_s	$1\,\text{ns}$
Change of refractive index with carrier density	$d\mu/dN_e$	$-5\text{e-}20\,\text{cm}^3$
No. of grating samples		10, 9
Sample length	w	$5\,\mu\text{m}$
Sample period	d_1, d_2	$45\,\mu\text{m}, 40\,\mu\text{m}$
Coupling coefficient	κ	$0.015\,\mu\text{m}^{-1}$

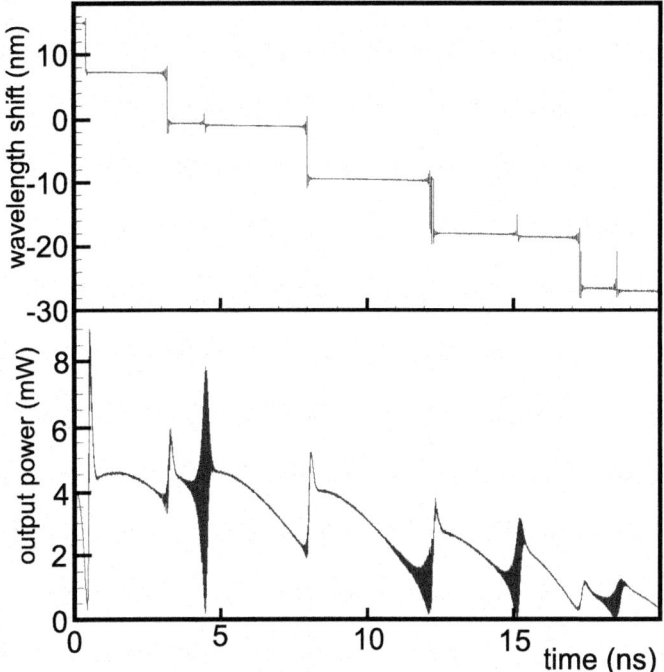

Fig. 6.13. (*top*) Instantaneous wavelength of SG-DFB as tuning current of grating A is varied. (*lower*) The corresponding optical power.

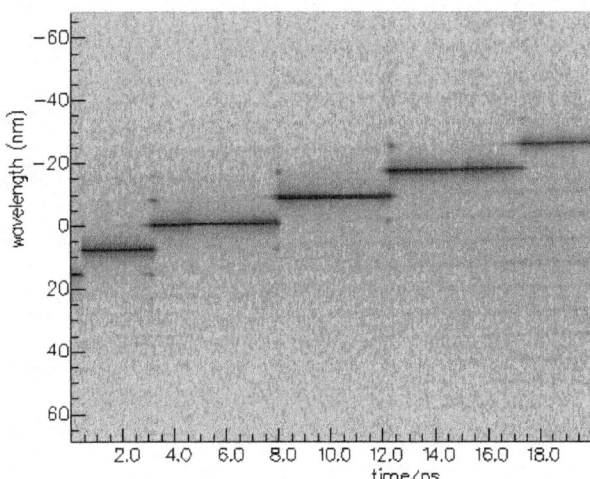

Fig. 6.14. Time-evolving spectrum of SG-DBR lasers as the current to grating A is varied.

FFT size provides a convenient tradeoff between spectral resolution and time resolution.

The figure clearly shows the grating order hops, although the cavity hops are below the spectral resolution of the plot. Faint lines can also be seen from other grating orders.

6.2.5.2 Varying Both Grating Currents Together

To fine-tune the SG-DBR laser, the currents in the sampled gratings may be altered together, keeping the current density in each grating equal, i.e., so

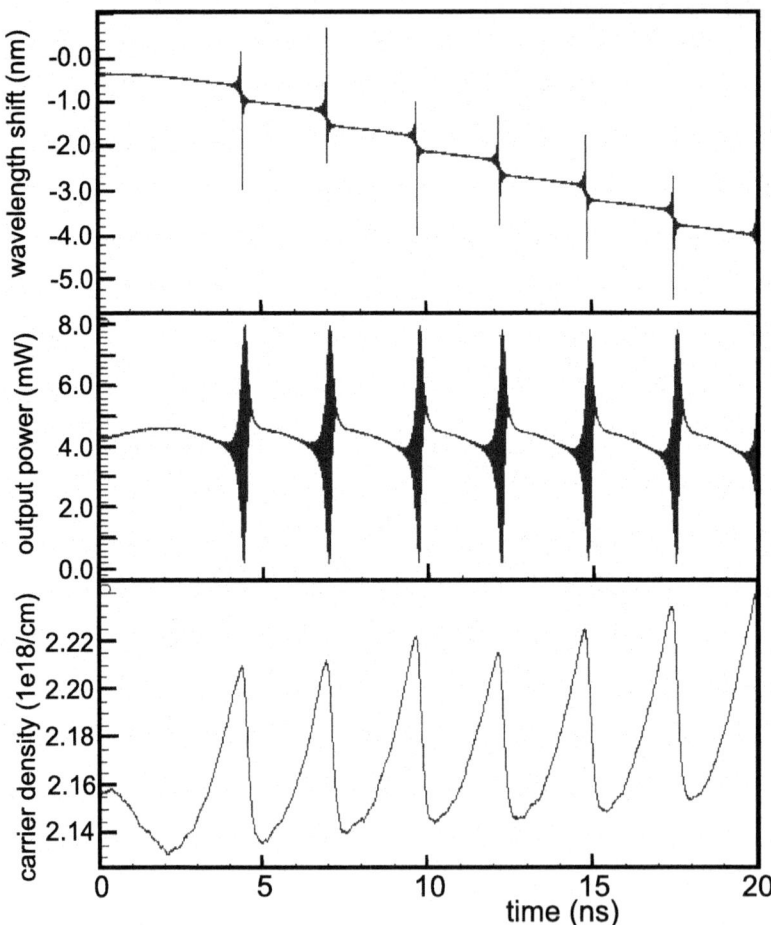

Fig. 6.15. (*top*) Instantaneous wavelength of the SG-DBR laser as tuning current of gratings A and B are varied together. (*middle*) The corresponding optical power. (*bottom*) The carrier density in the gain section.

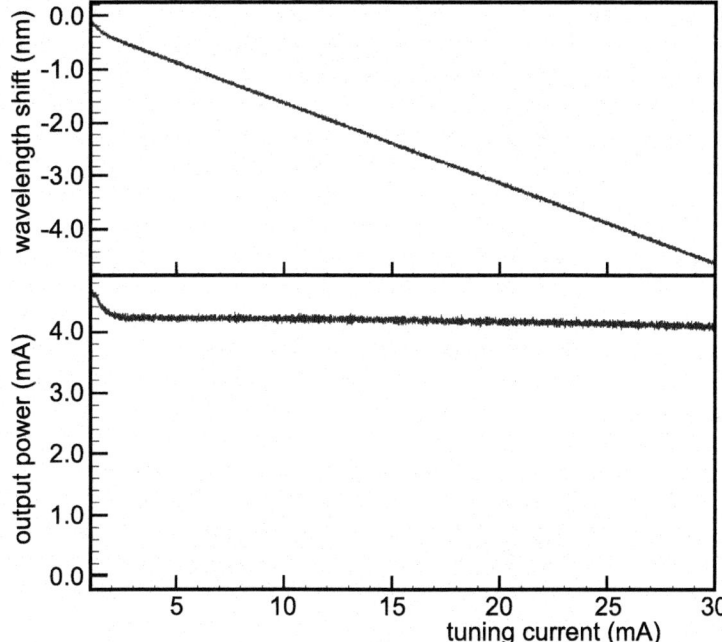

Fig. 6.16. (*top*) Instantaneous wavelength of the SG-DBR laser as tuning current of both gratings and phase section are varied in a consistent manner. (*bottom*) The corresponding optical power.

that $I_{g1}/L_{g1} = I_{g2}/L_{g2}$, where g1, g2 signify the two gratings. The device response is shown in Fig. 6.15, where $I1$ was ramped from $1/5$ mA to $30/5$ mA and $I4$ from $0.8/5$ mA to $24/5$ mA. Clearly the device now does not exhibit any grating mode hops but it is still hopping from one cavity mode to another. To prevent this we need to also vary the phase-tuning current. Ideally, the effective index of the waveguide mode along the whole laser would increase and decrease uniformly, but we do not have free control of the effective index in the gain section, so the phase section must do the job of both gain and phase sections. Thus, we have the current density relation:

$$J_{\mathrm{ph}} = J_{\mathrm{gr}}(L_{\mathrm{ph}} + L_{\mathrm{ga}})/L_{\mathrm{ph}} ,$$

where ph, gr, and ga, denotes the phase tuning, grating, and gain sections, respectively. Figure 6.15 shows a simulation of the resulting response — $I1$, $I3$, and $I4$ varied from $1/5$ mA, $4/5$ mA, and $1/5$ mA to $30/5$ mA, $43/6$ mA, and $24/5$ mA giving a continuous tuning range of 4 nm.

6.2.5.3 Tuning Matrix

To understand how the tuning currents in grating 1 and grating 2 interact, we must run simulations with a matrix of different I_1, I_4 values [I_1 = current

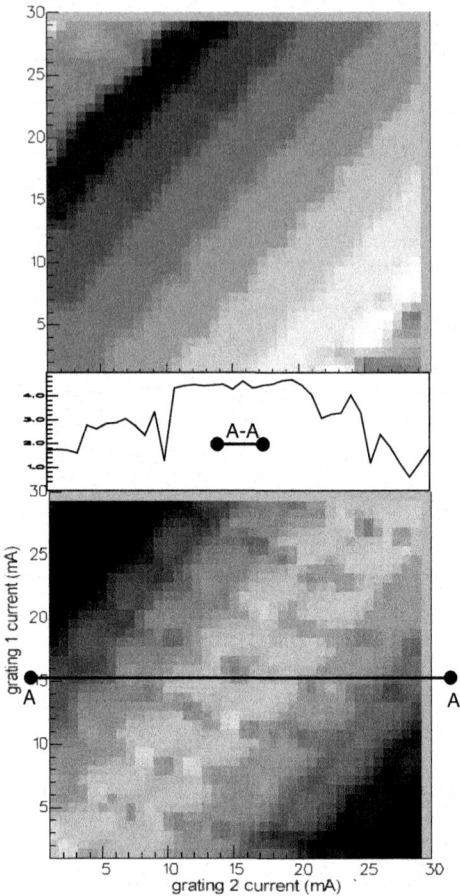

Fig. 6.17. (*top*) Variation of wavelength with $I1$, $I4$. Intensity of plot is proportional to wavelength shift. (*bottom*) Intensity variation with $I1$, $I4$. (*center*) Intensity at $I1 = 15$ mA.

in section 1 (grating 1), I_4 = ditto section 4, (grating 2)]. Figure 6.17 shows the results of such a simulation. Each simulation was run for 0.25 ns to allow settling time, and then the wavelength and output power were measured. To achieve convergence in such a short settling time, the carrier acceleration technique described above was used with a carrier acceleration factor of four. Also, each simulation was started from the state of the previous simulation. Thus, a whole matrix of 1600 simulations could be performed in just a few hours on a 2-GHz PC.

The top plot in the figure clearly shows that regions of stable grating order occur in diagonal stripes as expected from the theory. The lowest plot shows the intensity variation decaying away from the diagonal as the lasing wavelength shifts away from the gain peak. The center plot in the figure shows the intensity along the cross-section AA. The intensity plot is somewhat uneven. This is probably due to the fact that the phase-tuning current was not varied so that the laser was at a different distance from cavity resonance in each simulation.

6.2.5.4 Linewidth

A tunable laser is no use unless the tuned output has good wavelength stability — in other words, a narrow linewidth. A typical DFB laser with a grating strength κL of 1.5 has a linewidth of a few megahertz at 10 mW. A tunable laser needs to at least match this sort of linewidth. Fortunately, the long grating lengths required in a SG-DBR laser also narrows the grating bandwidth (within one sampled-grating resonance) and so also the linewidth.

Calculating the linewidth of a SG-DBR laser using the TDTW method directly is very inefficient. A linewidth of 1MHz implies variations in timescales of $1\,\mu s$. So, a simulation must be run for many times $1\,\mu s$. The problem is compounded by the fact that the linewidth is related to the noise sources of the laser and because the TDTW is a stochastically driven model; to estimate the expectation value of the linewidth, the simulation must be run many times until the noise is averaged out. A superior way is to use a small signal approach. A steady-state solution is found using the TDTW method and then a small signal (perturbation) analysis is performed around this steady-state solution. The theory for obtaining the linewidth via a small signal approach is given by Morthier et al. [4] and implemented in their original CLADISS laser model. Morthier presents the theory for a 1D laser model, and the extension to a 2D carrier density is relatively simple.

A laser has several sources of noise:

1. Phase noise — the spontaneously emitted photons will have a random phase relative to the laser mode. The resulting jumps in the phase of the output have an associated frequency spectrum. The linewidth induced by the phase noise is dependent on the power in the laser; in fact it is known to be proportional to $1/P$. Phase noise is often the dominant contribution to linewidth, particularly at low power.
2. Intensity noise — the spontaneously emitted photons will induce fluctuations in the output power. These power fluctuations have a spectral content, and so induce linewidth broadening.
3. Carrier fluctuations — the random nature of spontaneous recombination induces fluctuations in the carrier density, and this, in turn, induces fluctuations in the refractive index, changing the resonant frequency of the laser.
4. Thermal fluctuations — random thermal fluctuations induce additional refractive index changes, which again, in turn, alter the resonant frequency of the laser.

Figure 6.18 shows the calculated linewidth of the SG-DBR laser as a function of reciprocal power. The simulation takes only phase and intensity noise into account. Clearly very narrow linewidths can be obtained by the SG-DBR laser. This is thanks to the very long grating lengths and large cavity size — in this case, it is the length of the whole sampled grating rather than the length of each burst that is important. The plot clearly shows the $1/P$ dependence.

6.2.5.5 Modulation Response

A time-domain model is a natural approach to obtaining the modulation response of a laser. However, care must be taken in doing so. First, the modulation signal must be decided. The most direct approach is to use a sinusoidal modulation of the currents with gradually increasing modulation frequency. This approach is illustrated in Fig. 6.19. Here, we have modulated the SG-DBR laser with a ±5 mA amplitude sine wave on top of the 25 mA bias current (into the gain section).

A better way to get the modulation is via the impulse response function. In this case, we send a short impulse into the drive contact of the laser (in this case, the gain section), follow the resulting oscillations of the device in the time domain and then take the Fourier transform of the optical power response — the power spectrum. As the spectral content of an impulse is flat over all frequencies, then the power spectrum of the impulse response gives use immediately the modulation response of the laser. Figure 6.20 shows

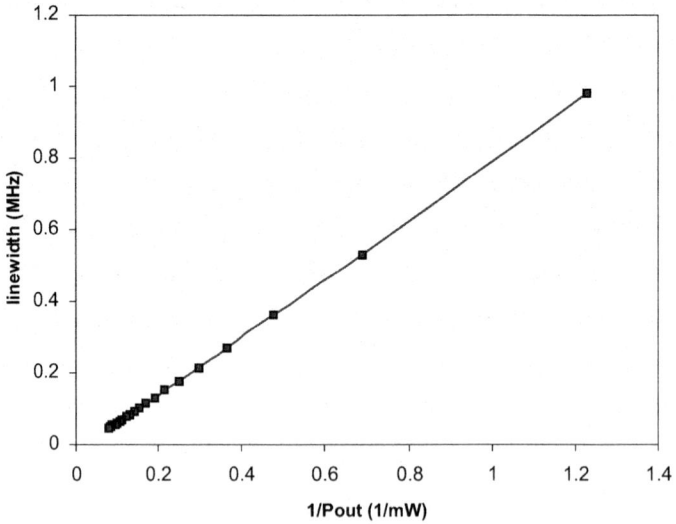

Fig. 6.18. Linewidth of the SG-DBR laser as a function of output power. The power is varied by altering the current to the gain section.

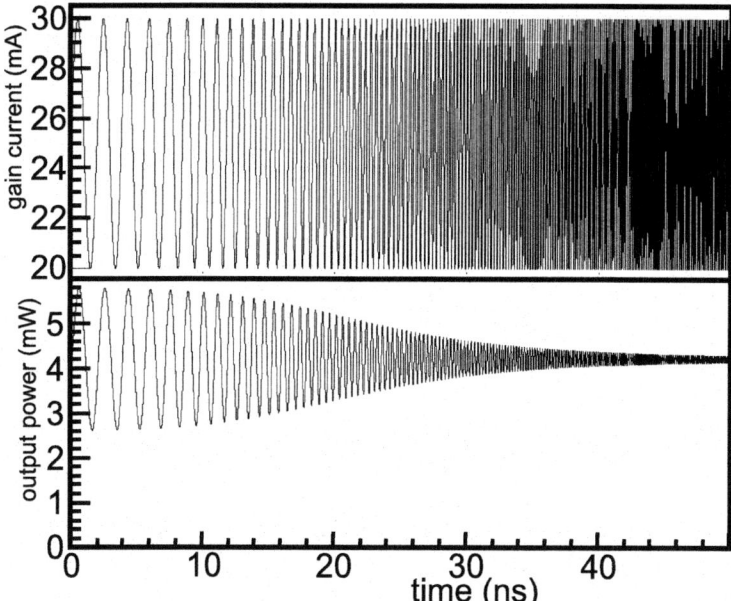

Fig. 6.19. Modulation response of SG-DBR using a chirped frequency drive current. *Top* is the drive current, and *lower* is the resulting output power.

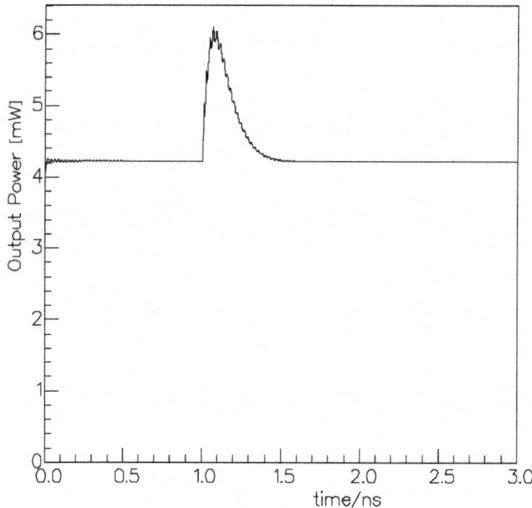

Fig. 6.20. Impulse response of the SG-DBR laser.

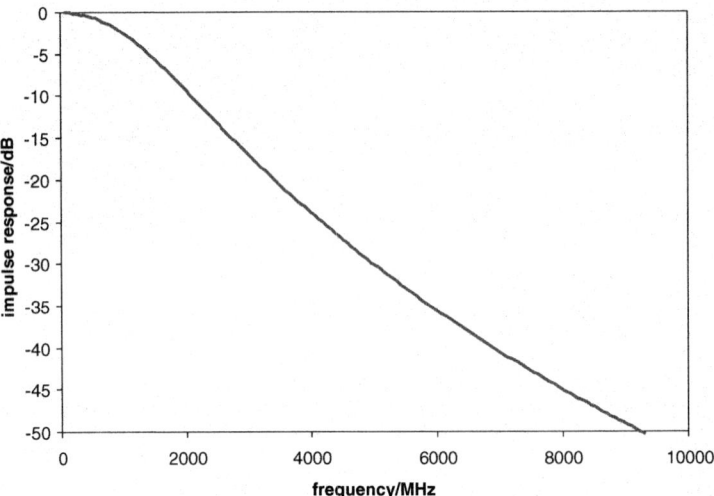

Fig. 6.21. Impulse response of the SG-DBR laser.

the impulse response of the SG-DBR laser. The device has been biased at 25 mA and then an impulse of 1 mA.ns superimposed at $t = 1$ ns. In an impulse response, the noise does not play any significant role and, indeed, only disrupts the response with unwanted noise. Thus, we start from a stable lasing state with photons in the cavity and turn off the noise sources before beginning the simulation.

Figure 6.21 shows the resulting power spectrum of this response, calculated using FFTs as:

$$R(\Omega) = \int_{t=0}^{\infty} P(t)e^{i\Omega t} dt ,$$

where $P(t)$ is the optical power exiting the device, in this case, from the right-hand facet.

Clearly, the SG-DBR laser is very slow, with a 3-dB bandwidth of barely 1 GHz. This compares with bandwidths of short Fabry–Perot lasers of 20 GHz or more.

6.3 The Digital-Supermode DBR Laser

6.3.1 Principle of Operation

An alternative design to the Vernier-based principle of the SG-DBR laser has been a proposed by Reid et al. at Marconi [18]. The structure of this so-called digital-supermode DBR (DS-DBR) laser is shown in Fig. 6.22 below. On the right side, it has a sampled grating as before or any other grating that has a comb reflection function, such as the phase grating [19]. On the

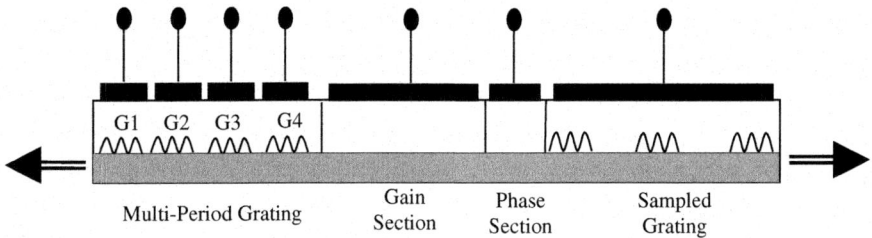

Fig. 6.22. Schematic of the DS-DBR laser — using a collection of gratings on the left, each tuned to a different resonance of the sampled grating on the right.

left, it has a set of five or more gratings, each tuned to one of the resonances of the sampled grating. In between, it has a gain section and a phase-tuning section.

Each of the sections of grating on the left have a contact that allows current to be injected into the gratings. As in the SG-DBR laser, the gratings are operated at wavelengths much longer than the band gap wavelength so that they are, to all intents and purposes, transparent, with the injected current altering only the effective index of the mode propagating along the waveguide. The reflection response of each grating section is designed to just overlap with its neighbor.

Initially, none of the gratings are pumped. Then, one grating is pumped with sufficient current to shift its Bragg wavelength to coincide with that of one of its neighbors. At this new wavelength, there are now two gratings reflecting, giving double the feedback. This, then, preferentially selects one resonance of the sampled grating.

Alternatively, one can bias all gratings and then decrease the current in, say, G2 (see figure) to move it toward G3 resonance, and increase the current in G4 to move it also toward G3. The selected wavelength then has a reflection three times the background — nine times in power terms.

6.3.2 Simulations

We will simulate a device with similar construction to that given in Sect. 6.2. In the table below, we list only the parameters different to Sect. 6.2.

To illustrate the operation of the multiperiod grating, we run a simulation to study the reflection of the grating in isolation. A time-domain simulation using an injected optical source, chirping from -40 nm to $+40$ nm (about $1.55\,\mu$m), gives us a reflection spectrum. The results are shown in Fig. 6.23. At the top, we can see the combined response from the five gratings — somewhat "chaotic" as each grating interferes with the next. Switching the leftmost grating on moves its reflection spectrum 8 nm to shorter wavelength. This can be seen in the "G1 ON" line — subgrating 1 on. The G1 grating peak simply moves away from the others and still no reflection dominates.

Table 6.2. Parameters of the DS-DBR Laser.

Parameter	Symbol	Value
Gain Section length	L_s	$300\,\mu m$
Gain Section current	I_{ga}	$25\,mA$
Phase Section Length	L_{ph}	$300\,\mu m$
Phase Section current	I_{ph}	$4\,mA$
M-P Grating Kappa	κ_{mp}	$0.01/\mu m$
M-P Grating Length (each) (5 gratings)	L_{gmp}	$60\,\mu m$
M-P Grating Period	Λ_{gmp}	$0.2353\,\mu m$
		$0.2366\,\mu m$
		$0.2379\,\mu m$
		$0.2393\,\mu m$
		$0.2406\,\mu m$

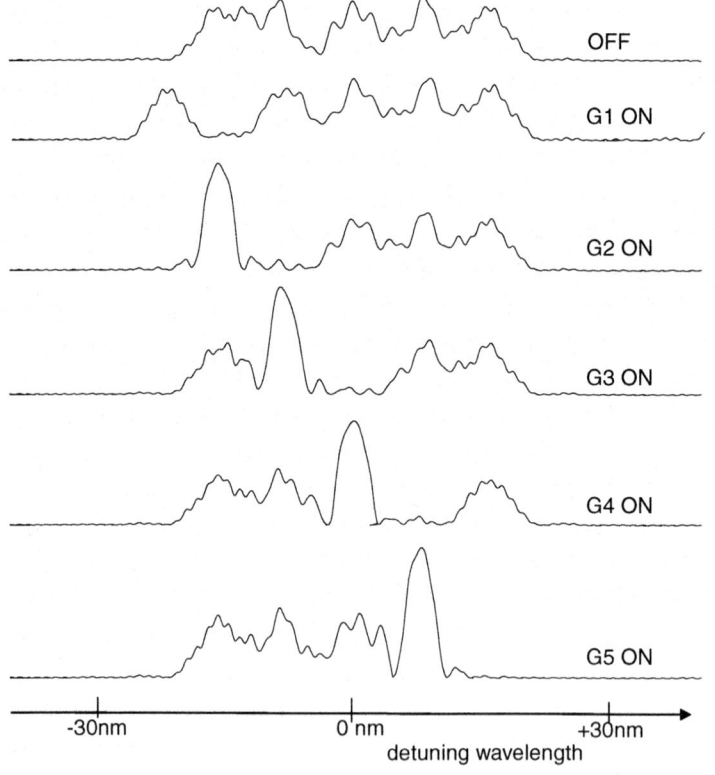

Fig. 6.23. Reflections off a multi-period grating as each grating section is turned on in turn. The *top* plot shows the broad reflection generated by the five subgratings.

However when we turn on G2, its peak moves to coincide with G1, creating a peak double the height of the others. If this is made to coincide also with a reflection from the sampled grating at the other end of the laser, then the laser will oscillate at this wavelength. The figure shows how turning on the other gratings in turn can readily select the position of the reflection peak. This is a "quasi-digital" tuning because the precise tuning current is not critical, giving a more tolerant operation, but at the expense of more electrical contacts than the Vernier-based SG-DBR laser.

We can study the tuning effect of the grating current by generating a time-evolving reflection spectrum as the tuning current of one of the gratings is varied. First we require a broadband source so that we can measure the reflection simultaneously over the whole bandwidth of the grating. This is done by using an optical impulse just 0.1 ps in duration and repeated every 20 ps. This creates a fine comb of wavelengths. In Fig. 6.24, we see a time-evolving reflection spectrum as the current in G3 is gradually ramped from 0 mA to 20/5 mA, i.e., right through the expected resonance match at 8/5 mA.

The simulations show that the grating is working as planned, and so all that remains is to put the multi-period grating into a laser device. Because the multi-period grating has a peak reflection only two times that of its background reflection, we shall need a somewhat broader gain spectrum than before. We can achieve this simply by shortening the gain section, causing

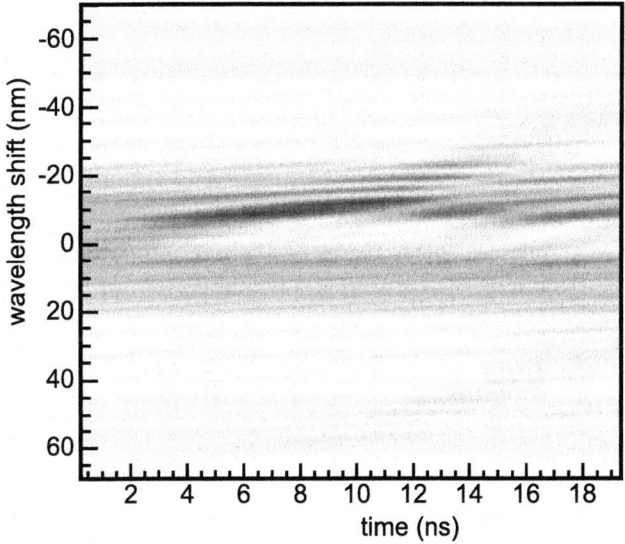

Fig. 6.24. Showing time-evolving reflection spectrum for a multi-period grating. The grating was illuminated by short pulses having a broad spectral content. Over the simulation time, the current G3 is ramped from zero to 20/5 mA. A strong reflection can be clearly seen where grating 3 crosses over grating 2.

the magnitude of the gain peak to increase and noting that in this model, the gain curvature is a user-defined parameter that does not vary with carrier density. (The curvature dependence with carrier density is more complex in reality, but the net effect here is to obtain a broader gain curve as desired.)

Figure 6.25 shows the resulting output of the laser, as each grating is switched on in turn. To speed up the simulation, a four-times carrier acceleration factor has been used. Notice the significant turn-on delay, suggesting that there is marginal stability in this device. Notice also that at 8 ns the device should switch back to -17-nm wavelength offset but has still not switched 2 ns later. The problem is even better illustrated by the time-evolving spectrum shown in Fig. 6.26. This uncovers two problems that would not be apparent from the instantaneous wavelength plot: first, a side mode rising early in the simulation and, second, multiple sampled-grating resonances appearing toward the end of the simulation — these will disappear when the device eventually switches to the G2-on state. However, these do show the potential problems with this design. The design would be substantially helped by using a broader gain bandwidth.

6.4 Conclusions

This chapter has presented an introduction to the time-domain traveling wave (TDTW) model for the simulation of laser diodes, illustrating its application to the design of tunable lasers. The method is simple in concept and straightforward to implement in software. It is numerically stable and is readily extended to 2D and even 3D simulations. It rapidly provides light-current curves, stable even far below threshold; it provides a good estimation of the side-mode suppression ratio (SMSR) and RIN spectra. And of course it provides time evolution of all of these things. It is readily adapted to devices with complex longitudinal structure — it can be used as easily with a traveling-wave amplifier as with a laser oscillator configuration, even a ring laser. Even where a quantity is not inherently time evolving, such as a spectrum or light-current curve, the method is fast and efficient. It is even able to simulate mode locking — see Chap. 7. The main limitations of the method are that (1) it assumes all modes are traveling at the same group velocity, (2) its stochastic nature makes some simulations inefficient, (3) it is not appropriate for any phenomenon with a long time-scale such as a linewidth. For this latter case, we can apply well-known small signal approaches around a steady state found using TDTW.

The method has been illustrated with application to sampled-grating tunable lasers, where it readily captures all of the tuning and spectral characteristics of the laser.

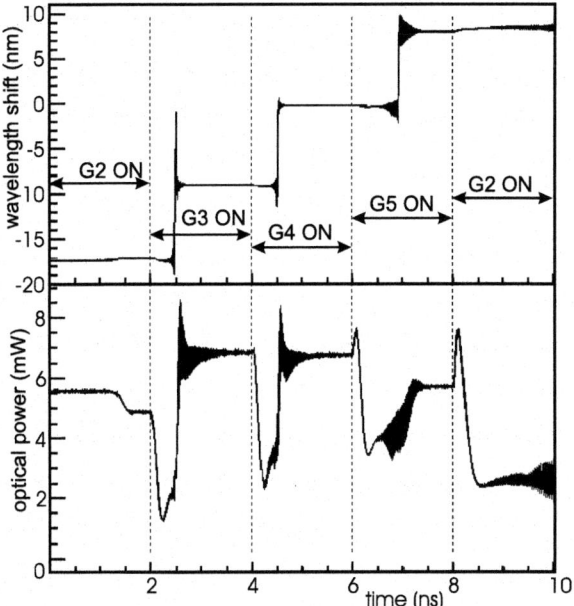

Fig. 6.25. Digital tuning of a multi-period tunable laser. Gratings G2, G3, G4, and G5 are each tuned on in turn for 2 ns at the times indicated by the arrows. The *upper* plot shows the resulting output, and the *lower* plot shows the output power from the sampled grating end. Note the significant switching time lags.

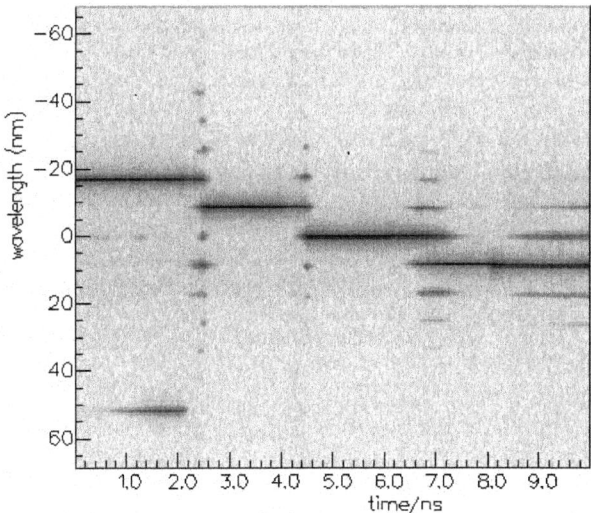

Fig. 6.26. Time evolving spectrum, as the multi-period grating is tuned (operation currents are given in Fig. 6.25).

References

1. G.H.B. Thompson: *Physics of Semiconductor Laser Devices* (Wiley, Chichester 1980)
2. G. P. Agrawal and N. K. Dutta: *Semiconductor Lasers* (Van Nostrand Reinhold, New York 1993)
3. J. E. Carroll: *Rate Equations in Semiconductor Electronics* (Cambridge University Press, Cambridge 1985)
4. P. Vankwikelberge, G. Morthier, and R. Baets: J. Quantum Electron. **26**, 1728–1741 (1990)
5. U. Bandelow, R. Schatz, and H.-J. Wunsche: Photon Technol. Lett **8**, 1–3 (1996)
6. G. Morthier and P. Vankwikelberge: *Handbook of Distributed Feedback Laser Diodes* (Artech, London 1997)
7. L. M. Zhang, S. F. Yu, M. C. Nowell, D. D. Marcenac, J. E. Carroll, and R. G. S. Plumb: J. Quantum Electron. **30**, 1389–1395 (1994)
8. C. F. Tsang, D. D. Marcenac, J. E. Carroll, and L. M. Zhang: IEE Proc. Optoelectron. **141**, 89–96 (1994)
9. J. Carrol, J. Whiteaway, and R. Plumb: *Distributed Feedback Semiconductor Lasers* (IEE, U.K. 1998) Chapter 7.9
10. E. C. Ifeachor and B. W. Jervis: *Digital Signal Processing*, 2nd ed. (Prentice Hall, New York 2002) Chapter 8
11. See www.photond.com
12. M. Giovannini and I. Montrosset: Opt. Quantum Electron **35**, 119–131 (2003)
13. P. I. Kuindersma, W. Scheepers, J .M. H. Cnoops et. al.: Tunable three-section strained MQW PA-DFB's with large single mode tuning range and narrow linewidth. *Proc. 12th IEEE International Semiconductor Laser Conference*, Davos, Switzerland (1990), pp. 248–249
14. F. Delorme, G. Terol, H. de Bailliencourt, et al.: IEEE J. Sel. Top. Quantum Electron. **5**, 480–480, 1999
15. C. K. Gardiner, R. G. S. Plumb, P. J. Williams, and T. J. Reid: IEE J. Proc. Optoelectron. **143**, 24–30 (1996)
16. S. A. Wood, R. G. S. Plumb, D. J. Robbins, N. D. Whitbread, and P. J. Williams: IEE Proc. Optoelectron. **147**, 43–48 (2000)
17. V. Jayaraman, Z.-M. Chuang, and L. A. Coldren: J. Quant. Electron. **29**, 1824–1827 (1993)
18. D. C. J. Reid, D. J. Robbins, A. J. Ward, et al.: ThV5, Proc. OFC 2002
19. G. Sarlet, G. Morthier, R. Baets, D. J. Robbins, and D. C. J. Reid: IEEE Photon. Tech. Lett. **11**, 21–23 (1999)

7 Monolithic Mode-Locked Semiconductor Lasers

E. A. Avrutin[1], V. Nikolaev[1,2], and D. Gallagher[3]

[1] Department of Electronics, University of York, York YO10 5DD, U.K., eaa2@ohm.york.ac.uk
[2] A. F. Ioffe Physico-Technical Institute, St.Petersburg 194021, Russia
[3] Photon Design Ltd., 34 Leopold St., Oxford OX4 1TW, U.K.

We discuss current and potential problems in theory and numerical modeling of monolithic mode-locked semiconductor lasers, present some recent advances in the field, compare the advantages and shortcomings of time- and time-frequency domain modeling approaches, and show some examples of applications of simulation software to modeling specialized lasers for high-speed communications.

7.1 Background and General Considerations

Taken most generally, **mode locking** (ML) is a regime of laser generation whereby the laser emits light in several modes with a constant phase relation, i.e., with constant and precisely equidistant frequencies. Usually, the term is used more specifically, referring to what is, rigorously speaking, *amplitude-modulation* (AM) mode locking, meaning that the phases of the (longitudinal) modes are not only constant but also may be considered approximately equal. In such a regime, the laser emits a train of ultrashort (shorter than the roundtrip) optical pulses with a repetition frequency F near the cavity roundtrip frequency or its harmonic, $F \approx M v_g/2L$ (v_g being the group velocity of light in the laser waveguide, L the cavity length and M the harmonic number, or the number of pulses coexisting in the cavity; in the simplest and most usual case, $M = 1$).

ML is usually achieved either by modulation of the laser net gain at a frequency F or its (sub)harmonic (*active* ML) or by exploiting nonlinear properties of the medium, usually by introducing a *saturable absorber* (SA) into the laser cavity (*passive* ML). The combination of these methods is known as *hybrid* ML; if the external modulation is in the form of short pulses, the corresponding regime is referred to as *synchronous* ML.

ML in monolithic semiconductor laser diodes is attracting considerable interest, first, for an increasing number of practical applications in microwave optoelectronics (e.g., microwave over fiber), high-speed optical time-division multiplexed (OTDM) and wavelength-division-multiplexed (WDM) communications; a reasonably up-to-date overview is presented in [1], and a summary and some updates are given below). Second, from the scientific point of view, it is attracting interest as an important prototype system in nonlinear dynamics and as a manifestation of high-speed nonlinearities in active semiconductor media. Theoretical models, ideally with predictive capabilities, are therefore useful for detailed understanding of the physics of these devices, analysis of their behavior and ultimately, optimization of laser design and operation regimes.

Monolithic mode-locked semiconductor lasers are, essentially, edge-emitting semiconductor lasers and so their modeling can benefit significantly from the progress in the modeling of semiconductor lasers in general. However, there are several significant specific features that are to be borne in mind when approaching the task of modeling these devices. Namely:

1. The design of a mode-locked laser is almost by definition relatively complex, with at least two separately biased sections, one of which is the forward-biased gain section, and the other, the modulator section in actively mode-locked lasers or the saturable absorber section in passively mode-locked ones, is reverse-biased. Depending on application, other elements can be included as well, as discussed below, which also has to be taken into account in the model. On the other hand, mode-locking essentially involves *longitudinal* phenomena — locking of longitudinal modes or circulation of pulses along the cavity. This means that many questions may be answered using numerical models operating in just one spatial dimension plus time, unlike the case in, say, high-power broad-area structures or vertical cavity lasers.

2. Dynamics of ML semiconductor lasers can be rich. These lasers are known to be prone to dynamic instabilities of various kinds, including self-pulsing envelopes on top of the ML pulse train and slow, chaotic oscillations of the pulse parameters. Hybrid ML may give rise to additional dynamic complexities, particularly outside the range of stable hybrid ML operation.

These complex dynamics of ML lasers include, generally speaking, several very different characteristic time scales: the pulse scale (subpicosecond to a few picoseconds); the cavity roundtrip scale (usually units to tens of picoseconds in realistic lasers of practical interest), the electron–photon resonance oscillations (or self-pulsing) scale (tens to hundreds of picoseconds), and, finally, the slow transient scale, including supermode competition and, in some regimes, chaotic oscillations (typically hundreds of picoseconds to tens or even hundreds of nanoseconds).

The device behavior may be sensitive to a large number of parameters, some of which are not very well known (for example, fast nonlinearities may play an important role, as is often the case with short pulses in semiconductor media). Relatively modest changes in parameters of both the forward- and reverse-biased sections (e.g., current, or saturable absorber recovery time in hybrid or passive mode-locking) can change the device

behavior qualitatively, moving it from one operation regime to another (some examples are shown below).
3. Comparison with experiments, which is the ultimate test for any model, is not always straightforward; for example, some instabilities may be difficult to register experimentally and may just show as increased noise [2].

It is therefore important that a model applied to these lasers is sufficiently:

1. Physically accurate, if the results of modeling are to be used in device design. In particular, to compare the modeling to experiment, it is desirable to have a physical model, not just for the forward-biased (gain) sections as is the case in most other semiconductor lasers, but for the reverse-biased absorbers or modulators, too; the physical processes governing the behavior of this section are very different from those responsible for the behavior of the gain section.
2. Generic, to accommodate the different laser designs and operation regimes.
3. Numerically efficient, to deal with the differing timescales without excessive loss of accuracy and allow The researcher to scan through a broad range of parameter values in manageable time.
4. Transparent and instructive, to help understand the physical origin of the complex dynamics and show the effect of parameters on the laser behavior.

The structure of this chapter reflects the features above. In Sect. 7.2, we review the main applications of ML semiconductor lasers and the way modeling can be used to help design lasers for these applications. Section 7.3 describes the main approaches to ML laser modeling, and Sect. 7.4 discusses a particularly important laser design, application, and dynamic modeling approach in more detail. Section 7.5 presents the recent results in microscopic modeling of the laser parameters with the emphasis on saturable absorber recovery time. Section 7.6 contains the brief summary and some assessment of possible future developments.

7.2 Modeling Requirements for Specific Laser Designs and Applications

In addition to the functionally essential components — the amplifier and saturable absorber or modulator — mode-locked laser cavities, depending on the intended applications, can include additional elements [1] . First, additional passive sections can be added for decreasing threshold or improving tuning properties. It has been shown that lasers with passive sections in the cavity can exhibit shorter pulses and less chirp than all-active lasers [3]; numerically simulated spectral and threshold properties were in good agreement with the experimental observations. The specific length of the passive section to optimize the improvement of pulse parameters, and the extent of this improvement, can be the object of further modeling.

In addition, stabilizing the operating wavelength of mode-locked lasers may become an important issue, particularly in a WDM environment. For this purpose, distributed Bragg reflector (DBR) lasers are natural candidates; the properties of the DBR are to be included in the model for any design optimization. Besides stabilizing the wavelength, the DBR may be expected to help achieve at least two other important purposes. First, DBR lasers are known to produce pulses with a weaker and better controlled chirp than Fabry–Perot ones, which is an issue for many possible applications. Second, the restricted spectral width of the pulse can be seen as a manifestation of a lower quality factor of mode-locking oscillations; we can thus expect the introduction of the DBR (and any subsequent increase in its spectral selectivity, by means of a smaller coupling factor) to increase the locking range, in terms of modulation frequencies and amplitudes, in case of a hybridly mode-locked operation. At the same time, the introduction of the DBR restricts the spectral width of the mode-locked emission and as such poses a lower limit on the achievable pulse duration. This tradeoff between the pulse duration, on the one hand, and the locking range and the chirp, on the other, may be a good subject for modeling, as are other issues such as pulse chirp, jitter, and noise. Some aspects of modeling Fabry–Perot and DBR pulse sources for optical communications will be considered in more detail in Sect. 7.4. We note that modeling locking range may be of particular importance for applications such as all-optical clock recovery.

A somewhat different set of requirements is posed by application of the same type of laser in analogue rather than digital applications, as a source of microwave signals on an optical carrier [1]. For these microwave-over-fiber applications, short pulse generation may be unnecessary, indeed undesirable, as it would only waste power on producing unwanted harmonics of the microwave signal, and so the use of a highly selective DBR cavity is preferable. The chirp of pulses is not an issue in this case, but stability is as any dynamic instability would dramatically increase the microwave linewidth — a detailed (in)stability analysis could then be an issue for modeling.

The idea of using a pulsating *optical* source, such as a mode-locked laser, for generation of a very high-frequency *electrical* signal becomes particularly attractive at (sub)terahertz frequencies [4], where the range of alternative purely electrical methods is limited. An ideal mode-locked source operating at these very high frequencies should, after detection, provide a stable microwave spectrum, ideally without unwanted "nonharmonic" satellite lines. Two main routes of achieving terahertz-repetition-rate mode-locking have been investigated. The first of these is harmonic ML involving the use of a spectrally selective cavity, which can incorporate Bragg gratings [5], potentially including specially designed sampled ones [6], or single or multiple internal reflectors [4]. The other route is known as multiple- or asymmetric colliding pulse mode-locking (CPM) and uses saturable absorbers positioned in one (asymmetric CPM [7]) or several positions (multiple CPM [8], integer fractions of the cavity length). In development of both harmonic and multiple CPM, modeling has been extensively used to both understand the laser behavior and optimize the laser design for pure harmonic mode-locking oscillations; it led the research that resulted in the fabrication of the world's fastest mode-locked laser with the pulse repetition frequency of 2.1 THz using harmonic ML design with intracavity reflectors [4].

A completely different approach to applications of mode-locked lasers treats them as sources, not of short pulses, but of a comb of precisely equidistant (phase-locked) frequency lines [9, 10]. By wavelength-stabilizing such a laser by external optical injection and then wavelength-domain-demultiplexing its output, a number of wavelengths at the International Telecommunication Union (ITU)-specified channels have been simultaneously generated [9, 10]. The requirements of laser radiation are different in this class of applications: A broad spectrum maximizing the number of channels is required, but chirp is not a problem (on the contrary, it can be advantageous if it helps broaden the spectrum). Modeling can be useful here in determining the requirements to the external optical injection, as well as optimizing the laser for broad-spectrum generation.

A more detailed overview of applications of mode-locked lasers is given in [1].

7.3 Overview of Dynamic Modeling Approaches

Any model of mode-locked laser dynamics should account for pulse shortening by modulation (active/hybrid ML) and/or saturable absorption (passive/hybrid ML) and for pulse broadening by saturable gain and cavity dispersion (including gain/loss dispersion and group velocity/phase dispersion). In addition, if spectral properties are to be accounted for accurately, self-phase modulation needs to be included in the model. Several approaches have been used so far.

7.3.1 Time-Domain Lumped Models

Conceptually the simplest, and historically the oldest, models of mode-locked lasers are **time-domain lumped models**, based on the approximation that the pulse width is much smaller than the repetition period, and the gain experienced by a ML pulse over one roundtrip in the cavity is small. The amplification and gain/group velocity dispersion, which in reality are experienced by the pulse simultaneously, may then be approximately treated in two independent stages. This allows the substitution of the distributed amplifier in the model by a lumped *gain element* performing the functions of amplification and self-phase-modulation. Mathematically, this element can be described by a nonlinear integral operator acting on the complex pulse shape function $A(t)$, t being the local time. Neglecting dispersion and fast nonlinearities and using the fact that the gain recovery time $\tau_G \gg \tau_p$ (pulse duration), one may obtain an approximate explicit expression for this operator (see, e.g., [1] and references therein):

$$\hat{G} A(t) = \exp\left(\frac{1}{2}G(t)\right) A(t) \approx (1 + G(t)) A(t) \qquad (7.1)$$

$$G(t) = G_i(1 + i\alpha_G)\exp(-U(t)/U_G), \qquad U(t) = \int_{-\infty}^{t} |A(t)|^2 dt,$$

where $G_i = \Gamma \int g(z, t \to -\infty)dz$ is the total amplification in the gain element at the time before the arrival of the pulse, α_G is the linewidth enhancement factor in the amplifier, and $U_G = 1/(v_g dg/dN)$ is the saturation energy of the amplifier. We note that (7.1) is, strictly speaking, valid only in a hypothetical unidirectional ring cavity. A more accurate expression, taking into account a realistic tandem laser geometry with an end reflector, is given by Khalfin et al. [11].

The saturable absorber, if any, is also considered as a lumped element, described by an operator similar to (7.1):

$$\hat{Q} A(t) = \exp\left(-\frac{1}{2}Q(t)\right) A(t) \approx \left(1 - \frac{1}{2}Q(t)\right) A(t) \qquad (7.2)$$

$$Q(t) = Q_i(1 + i\alpha_A)\exp(-U(t)/U_A),$$

with the total initial absorption $Q_i = \Gamma \int a(z, t \to -\infty)dz$, linewidth enhancement factor α_A, and absorber saturation energy $U_G = 1/(v_g A_a)$. Equations (7.1) and (7.2) take into account only "slow" saturable gain/absorption and SPM characterized by a relaxation time τ_{SA}; the expression (7.2) holds if τ_{SA} is much longer than the pulse duration τ_p. Some lumped time-domain models [12] do include "fast" effects in the SA (SA nonlinearities) as an equivalent fast absorber characterized by an operator $\hat{Q}_F$ functionally similar to (7.2) and with an equivalent absorption:

$$Q_F(t) = Q_{iF}\left(1 - \epsilon_A |A(t)|^2\right). \qquad (7.3)$$

Gain nonlinearities may in principle be included in the same way.

Finally, in a lumped model, the dispersion of material gain and refractive index, together with any artificial dispersive elements present in the cavity, such as a DBR grating, are combined in a lumped *dispersive element*. In the frequency domain, its effect on the pulse is written as:

$$\hat{D} A^T(\omega) = \exp\left[\frac{1}{1 - i(\omega - \omega_p)/\omega_L} + i\left(\varphi_0 + \frac{D(\omega - \omega_0)^2}{\omega_L^2}\right)\right] A^T, \qquad (7.4)$$

where A^T is the Fourier transform of the complex pulse shape $A(t)$, and ω_p and $\omega_L \ll \omega_p$ the peak frequency and the bandwidth of the dispersive element (defined by the gain curve of the amplifier and the frequency selectivity of a grating element, if it is present in the cavity). The value of ω_p may change during the pulse (due to gain curve changes with carrier density), which modifies the dispersive operator [12], although in the majority of papers on the subject, the effect is neglected. ϕ_0 denotes the phase shift introduced by the element, and D denotes the equivalent dispersion (including the group velocity dispersion of the waveguide and the effective dispersion of the external grating element, if any). To rewrite the operator (7.4) in the time domain, one may expand the first term around the reference frequency ω_0, noting that $|\omega_p - \omega_0| \ll \omega_0$. Then, after a standard transformation $(\omega - \omega_0)A^T \div id/dt\, A$, (7.4) becomes a differential operator; if the exponential is expanded keeping the first two terms, the operator is reduced to second order.

The main mode-locking equation is obtained by writing out the condition that the shape of the pulse is conserved over the repetition period. In the operator notation introduced above:

$$\hat{A}\hat{D}\hat{Q}\hat{Q}_F\, A(t) = A(t + \delta T), \tag{7.5}$$

where δT is the shift of the pulse or detuning between the repetition period and the roundtrip of the "cold" cavity (or its fraction in case of locking at harmonics of the fundamental frequency). In between the pulses, gain and SA are allowed to recover with their characteristic relaxation times. This allows one to calculate the values of gain and saturable absorption at the onset of the pulse, given the pulse energy and repetition period (the only point at which this latter parameter enters a lumped time-domain model). A further significant simplification in the time-domain model is made if the pulse energy is much smaller than $U_{G,A}$ and the pulse duration is much smaller than τ_{SA} so that both gain and absorber recovery can be neglected during the pulse. Then, the exponentials in the operators (7.1)-(7.3), as well as in (7.4), may be expanded, keeping terms up to the second order in (7.1) and to the first order in (7.2,7.3). Then, following the route pioneered by H. Haus in the first papers on the subject [13] and later adapted to diode lasers [14, 12], the mode-locking equation (7.5) is rewritten as a complex second-order integro-differential equation known as the master equation of mode locking, which permits an analytical solution of the form:

$$A(t) = A_0 \left(\cosh \frac{t}{\tau_p}\right)^{-1+i\beta} \tag{7.6}$$

known as the *self-consistent profile* (SCP). Assembling the terms proportional to the zeroth, first, and second power of $\tanh(t/\tau_p)$ in the mode-locking equation, one obtains three complex, or six real, equations [14, 12] for six real variables: pulse amplitude A_0, duration measure τ_p, chirp parameter β, optical frequency shift $\Delta\omega = \omega_p - \omega_0$, repetition period detuning δT, and phase shift ϕ_0. These equations, generally speaking, cannot be solved analytically but still allow some insight into the interrelation of pulse parameters. For example, it can be deduced [12] that the pulse duration may be considerably shortened by the presence of a fast SA, and the achievable pulse durations are estimated about ten times the inverse gain bandwidth, decreasing with increased pulse energy.

By requiring the net small-signal gain before and after the pulse to be negative, so that noise oscillations are not amplified, the self-consistent profile approach also allows the parameter range of the stable ML regime to be estimated. The conclusions are that:

1. Increasing the dispersion parameter D increases the parameter range for ML
2. CPM configurations, linear or ring, increase the pulse stability and lead to shorter pulses by increasing the parameter s.

The SCP model has also been successfully used to explain, at least qualitatively, the shape of the pumping current dependence of the repetition frequency detuning from the "cold-cavity" roundtrip frequency in passively mode-locked DBR laser diodes [15]. Indeed, experimentally the detuning as measured in [15] shows a minimum value in its dependence on the gain section current. As the pulse energy is approximately proportional to the excess current over threshold, this experimental result is in at least qualitative agreement with the parabolic dependence on the pulse energy predicted by the SCP model:

$$\frac{\Delta F}{F} \approx \frac{1}{4}\left(-(Q^{(i)} - sG^{(i)})\frac{E}{U_a} + \frac{1}{4}Q^{(i)}\left(\frac{E}{U_a}\right)^2\right). \tag{7.7}$$

When applied more quantitatively, however, the self-consistent pulse profile model may not be too accurate and cannot adequately describe details of pulse shape and spectral features — indeed, the pulse shape given by the expression (7.6) and the corresponding spectrum (both of which have the shape of a hyperbolic secant) are always symmetric, which, in general, need not, and often is not, the case in practice.

Still, due to their relative simplicity and efficiency, it may be argued that lumped models may be useful as **blocks** in models of complex optoelectronic **systems** where a mode-locked laser is used as a source and where the approximate picture the model gives is sufficient to predict the system behavior. The instructive nature of these models may make them particularly suitable wherever the modeling is used primarily for tutorial purposes.

7.3.2 Distributed Time-Domain Models

A more modern approach is offered by **time-domain traveling wave** (TDTW) models, also called **distributed time-domain** models, which treat the propagation of an optical pulse through a waveguide medium. A model then starts with decomposing the optical field in the laser cavity into components propagating right and left in the longitudinal direction (say, z):

$$E(r,t) = \Phi(x,y)\left(E_R \exp(-i\beta_0 z) + E_L \exp(i\beta_0 z)\right) \exp(i\omega_0 t), \quad (7.8)$$

with Φ being the transverse/lateral waveguide mode profile and ω_0 and $\beta_0 = n(\omega_0)\omega_0/c$ being the reference optical frequency and the corresponding wavevector, respectively. Then, for slowly varying amplitudes $E_{R,L}$, one obtains the reduced equations:

$$\pm\frac{1}{v_g}\frac{\partial E_{R,L}}{\partial z} + \frac{\partial E_{R,L}}{\partial t} = \left(\hat{g} + i\hat{\Delta\beta} - a_i\right)E_{R,L} + iK_{LR,RL}E_{L,R} + F_n(z,t). \quad (7.9)$$

Here, the operators $\hat{g}$ and $\hat{\Delta\beta}$ describe the modal gain and the variable part of the (real) propagation constant, respectively, and include both the saturation nonlinearity and frequency dependence (dispersion; hence the operator nature of these parameters); a_i is the unsaturable loss. At z values within SA sections, the gain value g is naturally changed to $-a$, a being the saturable absorption coefficient. The terms containing the coupling constants $K_{LR,RL}$ need to be included only if a grating is present at the given value of z (and time t), i.e., either in DBR sections if any, or in SA sections if a standing-wave induced grating is essential. In the former case, for index gratings without chirp or phase shift, it is sufficient to use a single real constant $K_{LR} = K_{RL} = K$. The final term is the noise source that leads to the self-starting of the model and is essential for modeling of noise and pulse jitter. $g(N,S)$ is the peak gain value, determined by the local carrier density $N(z,t)$ and photon density $S(z,t) = E_R^2(z,t) + E_L^2(z,t)$. Usually, the simplest linear relation is sufficient for bulk active layers, and a logarithmic approximation for those using single or very few quantum wells:

$$g_{bulk}(N,S) = \frac{A(N-N_0)}{1+\varepsilon S}, \quad g_{QW}(N,S) = \frac{AN_0}{1+\varepsilon S}\ln\frac{N}{N_0}, \quad (7.10)$$

where N_0 and A are the carrier density and gain cross section (at transparency if the logarithmic expression is used), and ε is the gain compression factor. For multiple quantum well lasers, use of both approximations has been reported, although the logarithmic one is probably more accurate. In saturable absorber sections, similar equations are used but with different parameters.

The real propagation constant $\hat{\Delta\beta}$ includes the time-dependent (nonlinear) phase shifts usually related to gain by means of Henry's linewidth enhancement factor α_H, which can take different values α_G, α_A in the two sections The exact forms of implementing it vary and deserve some discussion. Most straightforwardly, the phase shifts can be written as $\hat{\Delta\beta} = \alpha_H(\hat{g} - g_{ref})$, the latter parameter being the reference value of peak gain, which can be taken as either zero or the threshold gain value. This expression implicitly assumes, however, that the same linewidth enhancement factor describes the carrier-dependent "quasilinear" gain and the ultrafast nonlinearities described by the gain compression parameter ε. As the physics of the gain (or saturable absorption) compression is different from that of carrier density dependence of gain, this assumption is not necessarily physically justified. It may therefore be more accurate to rewrite the phase shift terms as $\hat{\Delta\beta} = \alpha_H(\hat{g}(1+\varepsilon S) - g_{ref})$, so that the alpha-factor applies only to the carrier dependence for which it is, strictly speaking, introduced, and not the nonlinearity. The parameter ε can then, if necessary, be made complex to account for ultrafast refractive index nonlinearity, if any, and the relation between its real and imaginary parts need not be the same as the alpha-factor. The difference between these two expressions is not important for laser designs and regimes not intended for ultrafast pulse generation and so can be easily overlooked. However, in the mode-locking case, where intense short pulses are formed and so the correct treatment of ultrafast nonlinearities becomes important, the precise way of dealing with ultrafast nonlinearities of gain and index can have a very significant effect on the results. The approximations involved in the model thus need to be considered very carefully. In addition to the above, $\hat{\Delta\beta}$, like gain, can be dispersive (frequency-dependent) as described below, hence, the operator sign; see below for the discussion of the implementation of dispersion.

When modeling very short (subpicosecond) pulses, additional modifications to modeling the optical properties of the material may be needed. First, the operator $\hat{\Delta\beta}$ may include an additional term describing group velocity dispersion of the passive structure [16], and second, the nonlinearities represented by the gain compression factor need not be considered instantaneous but can be given their own relaxation time (~ 0.1 ps) as in [8]. When mode-locking pulses generated are a few picoseconds in duration, however, neither is particularly important.

For detailed quantitative optimization, tabulated dependences taken from experiments or detailed many-body calculations will be required to represent $\hat{g}$ and $\hat{\Delta\beta}$ accurately.

At the laser facets, standard reflection/transmission boundary conditions are imposed on $E_{R,L}$.

The field propagation equations are coupled with coordinate-dependent rate equations for the carrier density:

$$\frac{d}{dt}N(z,t) = \frac{J(z,t)}{ed} - N\left(BN + \frac{1}{\tau_{nr}} + CN^2\right) - v_g \text{Re}\left(E_L^* \hat{g} E_L + E_R^* \hat{g} E_R\right) \quad (7.11)$$

(J/ed — pumping term, J — current density, e — elementary charge, d — active layer thickness, B — bimolecular recombination constant, τ_{nr} — nonradiative recombination rate, C — Auger recombination rate). Carrier capture dynamics can be taken into account by adding an extra equation for carrier densities in the contact layers, but its significance for mode-locked lasers can be expected to be very modest in all cases except where direct current modulation is involved.

Lasing spectra are, as is typical in time-domain modeling, calculated by fast Fourier transforming the calculated temporal profiles, discarding the initial turn-on transient to describe steady-state mode locking.

An important point in any numerical implementation of (7.9) is the implementation of *gain dispersion* in the operator $\hat{g}$. In the mode-locked laser constructions realized so far, the spectrum of mode-locked lasers, although broad, is still usually significantly narrower than that of gain/absorption, meaning that only the top of the gain curve needs to be represented accurately. Therefore, in all the studies reported so far, the dispersion has usually been approximated by either a parabola or a Lorentzian curve, which can be implemented numerically in time domain using several different approaches. The oldest one, to the best of our knowledge, involves a convolution integral formula:

$$\hat{g} E = g\omega_L \cdot \int_0^\infty E(t-\tau) \exp(-\omega_L \tau) d\tau, \qquad (7.12)$$

ω_L being the gain linewidth parameter (which can be made complex, the imaginary part representing the static or dynamic shift of the gain maximum with respect to the reference frequency; the pre-integral factor then contains only the real part of ω_L). The other approach involves rewriting the integral relation between the input and output of the spectral filter representing gain in a small segment of a laser in a differential form [17]. Finally, some versions of TDTW models [18] use the transmission-line approach, which is fundamentally equivalent to (7.9) and models the gain spectrum using effective circuit elements. The most modern methods of simulating gain spectra, those using specialized spatiotemporal digital filters for ultra-broadband simulation [19], have not, to the best of our knowledge, been necessary for modeling mode-locking lasers so far; however, applying those to mode-locking can be an interesting future development, particularly with subpicosecond pulse applications in mind.

The approach of [17] also involves the use of the notion of *instantaneous complex frequency*, assuming piecewise-exponential variation of fields, as opposed to straightforward finite-difference integration of (7.9). This improves the speed and the accuracy of the calculations (compared with finite difference modeling) considerably for those laser constructions that admit relatively long (>0.1 ps) simulation timesteps. From the mode-locking perspective, this is most useful for DBR mode-locked lasers, which emit relatively long pulses; with Fabry–Perot mode-locked lasers (which can emit subpicosecond pulses), the numerical approach of [17] does allow for longer timesteps and thus very efficient simulation, but to ensure numerical accuracy, shorter steps may still be preferable.

Distributed models are very powerful and general, and their use is not restricted to ML lasers, although the essentially distributed nature of these devices, with an emphasis on timescales shorter than the roundtrip, makes ML lasers an ideal showcase for the strength of TDTW models. TDTW models are much more accurate and generic than lumped models in analyzing various aspects of laser behavior and have been successfully used by several research groups to analyze mode-locked lasers. For example, the in-house package *LasTiDom* developed at the University of Glasgow by the first author of this chapter [20] was used alongside experiments to reproduce the experimentally observed wealth of dynamic phenomena (pure mode-locking, combined mode-locking + Q-switching behavior, pure Q-switching, ML with chaotic envelope), predict the spectral behavior of high-power long-cavity lasers and the advantages of integrated active-passive cavity cavities over all-active lasers, study the dynamics of external locking and clock recovery by passively mode-locked lasers, and design lasers for harmonic ML at sub-THz and THz frequencies using both multiple CPM and compound-cavity approaches [1, 8].

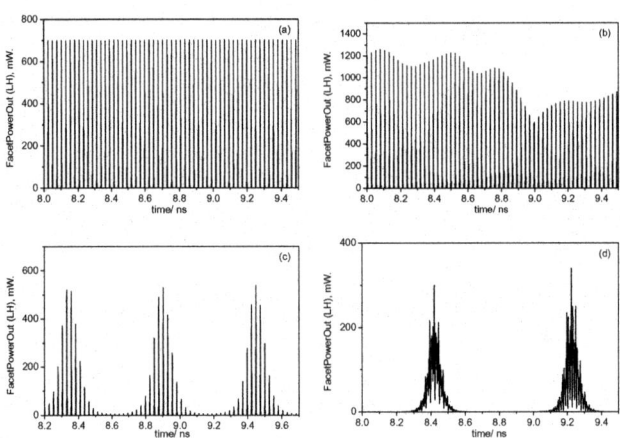

Fig. 7.1. Illustration of the advantages of the TDTW models (here, the commercial package CLADISS-2D): modeling different dynamic regimes of a DBR laser passively mode locked at 40 GHz. Active layer width/thickness 9 and 0.4 μm, respectively, other parameters as in Table 1 except $\alpha_G = 1$, $\alpha_A = 0$. Different dynamic regimes are shown: (a)–(c): absorber recovery time τ_{SA}=10 ps, current I =300 mA - good quality mode locking (a); 440 mA — unstable ML with a chaotic envelope (b); 150 mA — combined ML/Q-switching regime (c). (d) τ_{SA}=100 ps, current 150 mA — Q-switching.

TDTW simulations have been taken to a new level by the advent of commercial laser simulators that use modifications of TDTW models in their solvers [21, 22]. Although they are by necessity slightly less flexible than in-house software, these can offer considerable advantages as regards both a user-friendly front end and a carefully optimized efficient solver.

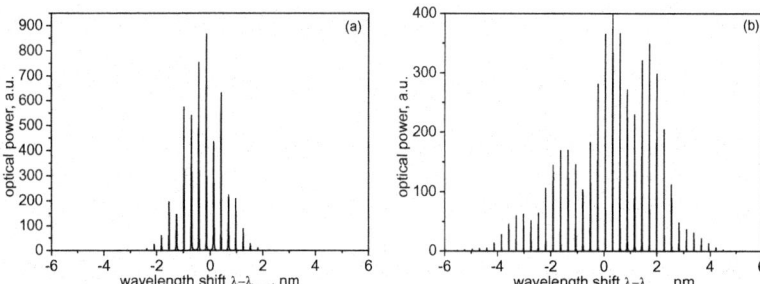

Fig. 7.2. Spectra of steady-state mode locking calculated using CLADISS-2D. Parameters as in Fig. 7.1, current 300 mA. $\alpha_A = 0$. (a) $\alpha_G = 0$, (b) $\alpha_G = 1$. Notice the considerable spectral shift and broadening caused by the self-phase modulation in the latter case.

The software we use here is CLADISS-2D by Photon Design [21]. It offers (in addition to all of the versatility that can be expected of a TDTW model in dealing with different ML laser regimes — Fig. 7.1 — and constructions) also several important new possibilities. For example, the option for implementing several active layers with different properties makes it possible to analyze the possibilities of a mode-locked laser with an active layer consisting of several quantum wells of different thickness in order to broaden the gain spectrum. The built-in spectral analysis instrument allows both a spectrum over a given time interval (Fig. 7.2) and a time-dependent spectrum to be calculated. It also includes a mode tracker facility that makes it possible for the researcher to track the dynamics of individual modes (Fig.7.3), which helps analyze the spectral properties of the device and can be potentially important for WDM applications.

An attractive advantage of the *solver* (as opposed to front end) used in CLADISS-2D is the sophisticated algorithm for treating gain dispersion, which allows for the calculation precision to be maintained with relatively long integration steps, as illustrated by Fig. 7.4 . In contrast, with the simple finite difference algorithm, the integration steps have to be kept to $\Delta z = 1 - 2\mu$m to predict the pulse durations accurately enough. As the speed of a TDTW model scales approximately with the inverse square of the simulation step, this can make an important difference.

The main limitation of distributed time-domain models is that even with an optimized solver, they still tend to pose considerable requirements on

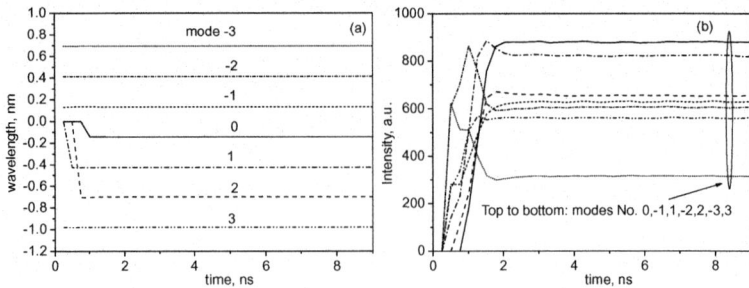

Fig. 7.3. Example of the peak ("mode") tracker facility output in CLADISS-2D. Parameters as in Fig. 7.4, time window for spectral calculation 1.021 ns. The switch-on transient in mode intensities is seen, followed by steady-state ML operation.

the CPU time. This is mainly because time and space steps are essentially related in these models ($\Delta z = v_g \Delta t$) and need to be sufficiently short (typical spatial step being $\Delta z = 1-5$ μm, although 10 μm may be acceptable for some applications as shown above) to reproduce the pulse characteristics faithfully.

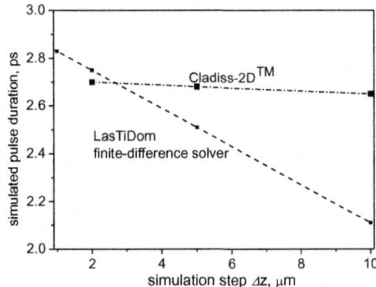

Fig. 7.4. Dependence of the simulated pulse width in a DBR mode-locked laser on the simulation step in different models. Parameters as in Sect. 4, Table 1, except $\alpha_H = 0$ in both gain and SA sections. Absorber relaxation time 10 ps, current $I/I_{th} = 3$ (with the active area thickness and width of 0.04 and 9 μm respectively, this corresponds to 350 mA).

7.3.3 Static or Dynamic Modal Analysis

The approach of following the dynamics of modes is logically developed in the technique of **static or dynamic modal analysis** [1, 23]. In this approach, instead of (7.8), one uses an expansion of the optical field in the form:

$$E(r,t) = \Phi(x,y) \sum_k E_k(t) u_k(z) \exp\left[i\left(\int \omega_k dt + \varphi_k(t)\right)\right], \qquad (7.13)$$

where E_k, u_k, ω_k, and φ_k are, respectively, the amplitude, mode profile, frequency, and phase of the longitudinal mode k. The fast laser dynamics (of the order of, and shorter than, roundtrip, that is, on the pulse scale) is then contained in the instantaneous values of amplitudes and phases, whereas the slow dynamics (on the scale longer than roundtrip) is determined by the dynamics of complex mode amplitudes $\mathbf{E}_k = E_k(t)\exp\left(i\varphi_k(t)\right)$, which satisfy complex rate equations:

$$\frac{\partial}{\partial t}\mathbf{E}_k = \left[\frac{1}{2}g_k^{net} + i(\Omega_k - \omega_k)\right]\mathbf{E}_k + \sum_m G_m, \mathbf{E}_{k+m} + F_k^{(L)}(t) \ . \qquad (7.14)$$

g_k^{net} and Ω_k are the net gain and cold-cavity eigenfrequency of the k-th mode, and G_m are the mode coupling parameters that include contributions due to four-wave mixing in gain and SA (passive/hybrid ML) sections, and to modulation (active/hybrid ML). The expressions for these parameters are, in general, cumbersome and are given elsewhere [1, 23].

This approach is less powerful and universal than the TDTW one due to (1) the inherent assumptions of weak-to-modest nonlinearity, modulation, and dispersion, and (2) the explicit introduction of slow and fast timescales obstructing the analysis of regimes such as clock recovery. Its advantage is that the timesteps can be much longer (and the number of variables can be smaller) than in the TDTW model. This makes the modal analysis method particularly efficient in analyzing, say, long-scale dynamics of external locking of DBR hybridly mode-locked lasers [23] (Fig. 7.5). It also has the logical advantage of describing "normal" steady-state ML as a steady-state solution and is particularly well suitable for lasers with spectrally selective elements such as compound cavity or DBR designs.

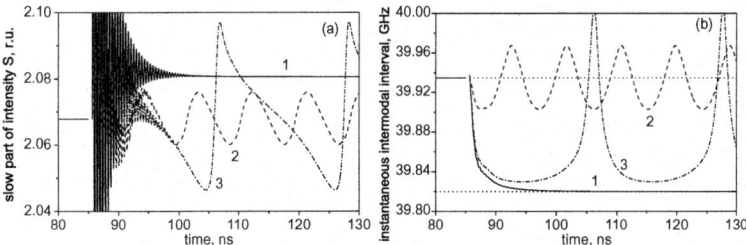

Fig. 7.5. Illustration of advantages of dynamic modal analysis [23]: fast and efficient modeling of long-term transients in locking a passively mode-locked laser to an external signal. 1 — successful locking (strong modulation), 2—wave mixing (very weak modulation), 3—quasi-locking instability (weak modulation). Parameters as in [23].

7.4 Example: Mode-Locked Lasers for WDM and OTDM Applications

7.4.1 Background

As optical sources capable of producing sub-picosecond optical pulses at multigigahertz rates, mode-locked lasers are natural candidates for high-speed communications. A single channel in a future system is likely to have a bit rate of either 10 or 40 GBit/s, with further optical time-domain (OTDM) and/or wavelength domain (WDM) multiplexing of channels. One possible scheme [24] includes four channels with a bit rate of 40 GBit/s each, time-division multiplexed to achieve an aggregate bit rate of 160 GBit/s. Wavelength division multiplexing is a possibility as well. The specifications of the laser performance for this group of applications typically include:

1. Sufficiently high output power (>0 dBm average).
2. A well-stabilized bit rate near 40 GBit/s.
3. A wavelength stabilized at the ITU grid.
4. A minimum pulse duration, particularly for OTDM applications; typically 2–3 ps [24]. If OTDM is not required, a longer pulse duration of 5–6 ps may be acceptable.
5. Near transform-limited operation.
6. Low pulse jitter (e.g., ¡0.3 ps [24]).
7. A sufficiently low relative intensity noise (RIN) (e.g., $<3\%$ [24]).

7.4.2 Choice of Modeling Approach

The modeling in this case can be expected to be applicable to both actively and hybridly mode-locked lasers (the hybrid regime with external modulation being necessary for precise stabilization/tuning of the repetition rate as well as for jitter and noise reduction), and it can give quantitative results on pulse parameters, including stability and noise/jitter directly comparable with experiments. Besides, in a realistic, relatively long device, the gain and loss per pass may not necessarily be small, so a large-signal model is desirable. Between them, these constraints make the distributed time-domain approach the most suitable one.

7.4.3 Parameter Ranges of Dynamic Regimes: The Background

The last two specifications of Sect. 7.4.1 mean that, at the very least, any trace of dynamic instability in the laser operation needs to be eliminated — indeed, experimentally, such instabilities, although not essentially stochastic by nature, manifest themselves as increased noise and jitter. A typical map of dynamic instabilities is presented in Fig. 7.6. The coordinates have been chosen to represent the two independent tuning mechanisms present in a mode-locked laser: the gain section current and the saturable absorber voltage — an important effect of increasing the latter is known to be the reduction in the saturable absorber relaxation time, as explained in more detail in Sect. 7.5.

As we are mainly interested in a qualitative picture in this generic background study, we use the results from [25] obtained with a different parameter set from the rest of this section, e.g., $N_{0G} = N_{0A} = 1.2 \times 10^{18} \mathrm{cm}^{-3}$, $\epsilon_G = 3 \times 10^{-17} \mathrm{cm}^3$, $\epsilon_A = 1 \times 10^{-16} \mathrm{cm}^3$. We analyzed the simplest conceptually possible laser design: a fully active Fabry–Perot cavity with a saturable absorber (80 µm long in this case), to allow for the most straightforward comparison with the simplest instability analysis method possible: the lumped rate equations [26]. Those cannot, of course, be used to analyze mode locking, as they only apply to "slow" dynamics on a timescale longer than the roundtrip. They can, however, predict the range of the self-pulsating (SP) instability at low currents and moderately low absorber recovery times, which can be expected to be one of the limitations for achieving high-quality mode locking. The range of parameters for this instability is inside the thin curve in Fig. 7.6. The distributed time domain modeling does indeed show SP or mode-locking with SP envelope in a similar range of parameters, although the presence of ML does influence the exact range in which SP is observed. In addition, TDTW modeling predicts a range of unstable mode locking at high currents associated with the incomplete depletion of gain by the mode-locking pulse, leading to the emergence of multiple competing pulse trains.

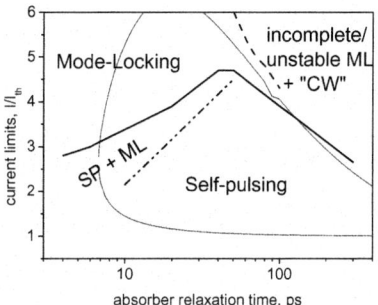

Fig. 7.6. Parameter ranges in a Fabry–Perot tandem-cavity laser with a saturable absorber.

The results here have been received mainly using the in-house package, *LasTiDom*, developed by the first author, as it is specially tailored for analysis of pulsating output. The results are in agreement with the predictions of the commercial package CLADISS-2D if we set the linewidth enhancement factors in both the gain and the saturable absorber sections to zero. With nonzero alpha-factors, the predictions of the two models may differ (particularly regarding the onset of chaotic envelope modulation), because of the different conventions taken concerning the ultrafast refractive index nonlinearities, as discussed above.

7.4.4 Choice of Cavity Design: All-Active and Active/Passive, Fabry–Perot and DBR Lasers

At first glance, most of the specifications detailed in Sect. 7.4.1 could be achieved with a Fabry–Perot laser. This concerns, not just the pulse duration (which can be as small or 1–2 ps or even shorter) but also the operating wavelength, which in such a laser can be coarsely tuned by temperature and finely tuned by the gain section current (Fig. 7.7) and thus can be adjusted to fit the ITU specification. However, this restricts the laser operating point,

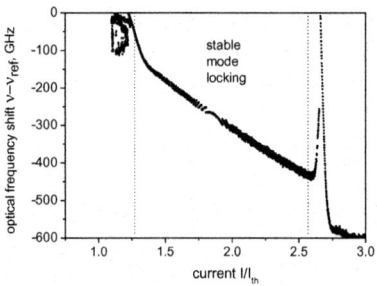

Fig. 7.7. Shift of optical frequency at the peak of the ML pulse from the reference value with current in a Fabry–Perot laser. Parameters as in Table 1, except the passive section is terminated by a cleaved facet. Absorber relaxation time: 15 ps.

and besides, the shift in the wavelength is accompanied by an increase in chirp, so that achieving the nearly transform limited operation is very problematic (the laser analyzed in Fig. 7.7 contains a passive section (Fig. 7.8) in the cavity that is known to reduce threshold and chirp [3], but even in this case, the chirp is significant). Therefore, a DBR construction, with better chirp properties and a better fixed wavelength, is more appropriate. Fig. 7.9 shows simulated mode-locking pulse width for a given excess current above threshold in DBR constructions with different values of the coupling coefficient. Other parameters are as in Table 1, except that the lengths of passive and Bragg reflector sections of the laser L_{pas}, L_{DBR} are adjusted to keep both the product $KL_{DBR} = 0.6$ (and thus the threshold) and the effective cavity length (and thus the repetition rate) approximately constant for all K values. As can be expected, the smaller the coupling coefficient (and thus the passband of the DBR), the broader the pulses emitted — in our simulations, the minimum pulse duration achievable in a laser is well approximated by $\tau_p \approx \tau_{p(Fabry-Perot)} + X/K, X \approx 200$ ps/cm . Thus, coupling coefficients of

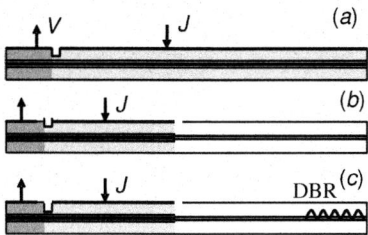

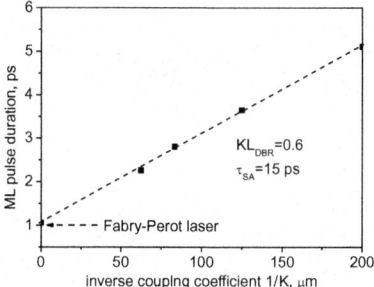

Fig. 7.8. Schematic of mode-locking laser constructions referred to in this chapter: (a) all-active laser used for simple background analysis of Fig. 7.6; (b) active-passive (extended cavity) mode-locking construction used in Figs. 7.7 and 7.9; (c) extended-cavity DBR construction used in the rest of Sect. 7.4 and for calculating Fig. 7.4.

Fig. 7.9. The minimum pulse width over the current range of stable mode-locking in a DBR laser vs. the inverse coupling constant $1/K$ of the DBR section ($1/K = 0$ corresponds to the Fabry–Perot laser of Fig. 7.7). Absorber relaxation time 15 ps, current $I/I_{th} = 3$, other parameters as in Table 1.

~ 50 cm^{-1} are sensible (and indeed are often used) for straightforward single-channel transmission, but higher values in excess of 100 cm^{-1} are desirable where OTDM is envisaged. We note also that large coupling, with realistic linewidth enhancement factors, may also increase the shift of the operating wavelength from the Bragg value (related to chirp) and decreases the locking range under hybrid mode-locked operation (discussed in more detail below). In the simulations below, we mainly use the value of 120 cm^{-1} to allow a compromise between the pulse duration, on the one hand, and chirp and locking range, on the other; the same value has recently been used by other researchers [27].

7.4.5 Passive Mode Locking

To illustrate the possibilities of TDTW modeling in the simulations of passive mode locking, we use a typical set of parameters for a laser with three to five

Table 7.1. The Main Parameters Used in the Calculations, Unless Specified Otherwise in the Text.

notation	meaning	SA	Gain section	passive	DBR	units
L	length	40	440	540	50	μm
Γ	confinement factor	0.07	0.07	0.07	0.07	
a_i	internal (dissipative) loss	10	10	0	0	cm^{-1}
n_g	group refractive index	3.6	3.6	3.6	3.6	
N_0	transparency carrier density	5×10^{18}	1.2×10^{18}			cm^{-3}
A	cross-section	9×10^{-16}	4×10^{-16}	0	0	cm^2
α_H	linewidth enhancement factor	1	3			
ϵ	nonlinear compression coefficient	3×10^{-17}	2×10^{-17}			cm^3
τ_{nr}	nonradiative recombination time	10-30 ps	10 ns			
B	bimolecular recombination coefficient	2.5×10^{-10}	2.5×10^{-10}			cm^3/s
C	Auger recombination coefficient	5×10^{-29}	5×10^{-29}			cm^6/s
r	amplitude reflectance, left facet	0.565				
	right facet				0	
$\Delta\Omega$	gain spectrum width parameter	3×10^{13}	3×10^{13}			s^{-1}
K	Bragg coupling coefficient				120	cm^{-1}

quantum wells in the active layer (Table 1); we use a linear gain-current relation, although similar results can be obtained with a logarithmic one. Again, most simulations in this section were performed using the specialized in-house package *LasTiDom*. With zero alpha-factors, the results agree well with those from the the commercial package, CLADISS-2D (as demonstrated by Fig. 7.4 above). With nonzero alpha-factors, the predictions of the two models are qualitatively similar, but quantitatively can differ considerably, presumably due to the difference in handling the ultrafast nonlinearities (in *LasTiDom*, the alpha-factor acts only on the carrier-dependent gain, with the ultrafast gain compression factor treated as purely real).

Figure 7.10 shows a calculated laser transient, clearly showing the various time scales mentioned in Sect. 1: relaxation oscillations at switch-on, followed by a slower transient during which the steady-state mode-locking is achieved, resulting in a stream of pulses 2–3 ps long (Fig. 7.11). Here and elsewhere, "relative units" of light intensity mean measuring this parameter in units of saturation photon density of the amplifier section $S_g = 1/(v_g A_G \tau(N_{th}))$, where the effective recombination time at threshold includes all recombination channels: $1/\tau(N_{th}) = 1/\tau_{nr,G} + BN_{th} + CN_{th}^2$. As in the first papers on

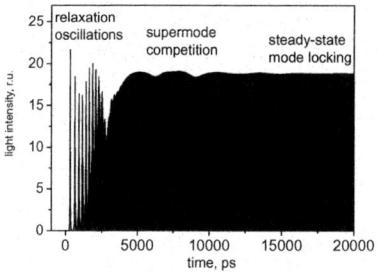

Fig. 7.10. Transient behavior of mode-locking operation. Note the relaxation oscillations followed by the slower supermode competition transient. Absorber relaxation time 15 ps, current $I/I_{th} = 3.7$, other parameters as in Table 1.

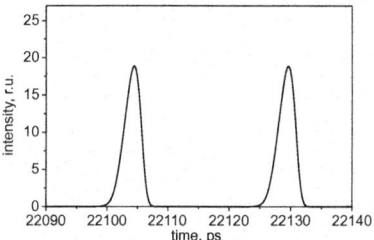

Fig. 7.11. Fragment of the steady-state mode-locking pulse train resulting from the transient shown in Fig. 7.10.

the subject [28], the two main parameters we vary are the applied current and the absorber relaxation time τ_{SA}, the latter representing one of the effects of the reverse bias on the absorber. In the DBR lasers, as in the Fabry–Perot ones, we find that the range of currents for stable mode locking decreases with the increase in τ_{SA}, and it virtually disappears at $\tau_{SA} = 30$ ps. Figure 7.12 contains the dependence of some of the most important radiation parameters on current for fixed values of τ_{SA}. The way these parameters (except the repetition period) have been calculated was by means of applying a current in the shape of a long (200 ns) ramp corresponding to a quasi-stationary increase of current. Parameters of each individual pulse in the laser output were registered by the software as simulations progressed and are plotted as a small dot in Fig. 7.12(a), (b), and (d). Because the timing separation (period) between the pulses (25 ps) is very small compared with the duration of the current ramp (200 ns), the dots are visually joined to form lines in the figure, at least for stable ML. The onset of the two kinds of instabilities is clearly seen: the self-pulsing at low currents and the chaotic envelope at high currents. However, we wish to point out that the use of current ramps for studying operation regimes, although very efficient numerically (and particularly easy to implement and use in a commercial package such as CLADISS-2D), may not always be accurate. Indeed, to represent a truly quasi-stationary change in the operation regime, a current ramp must be slower than any transient time of the laser operation. As these transients, particularly at the onset of chaotic modulation, can take tens of nanoseconds, this requirement can be difficult to meet. We therefore complemented our results by calculating a set of runs for fixed current values and by discarding the initial transient for each current value to analyze the limit cycle and calculate time-averaged pulse parameters. The results were used to check the borders of regimes (shown as vertical lines), to calculate the repetition rates plotted in Fig. 7.12(c) (the repetition rate was determined as the inverse of the time-averaged repetition period; the lines in Fig. 7.12(c) are just to guide the eye), and to check the other parameters of stable ML. The latter are represented as small rectangles in Fig. 7.12(a), (b), and (d); as expected, they are, to a good accuracy, the same as those given by the ramp simulation throughout the parameter range for stable ML. The current limits for self-pulsation instability at low currents are also validated; however, in predicting the onset of chaotic modulation at high currents, the ramp method is seen to be less accurate.

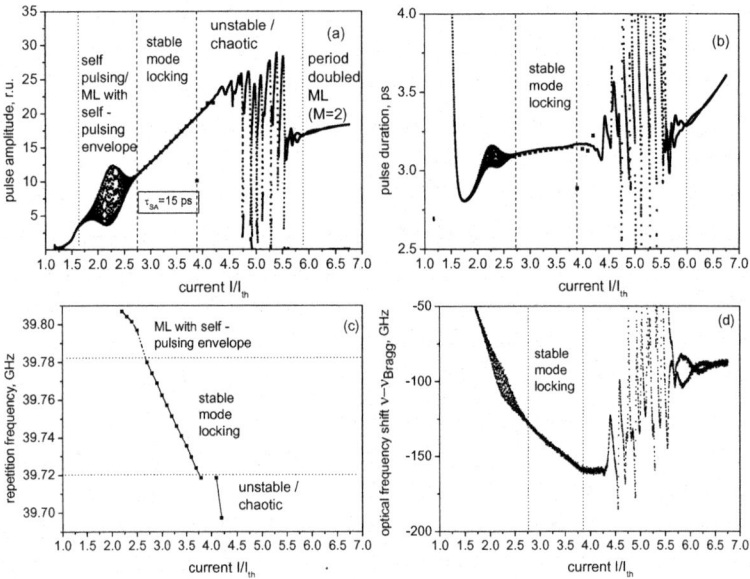

Fig. 7.12. Simulated current dependences of mode-locking parameters: (a) pulse amplitude, (b) pulse duration, (c) repetition rate, and (d) optical frequency shift at the peak of the pulse. Dots/scatter: quasi-stationary simulations using a current ramp 200 ns long; small rectangles: time-averaged results for fixed current values. Parameters as in Table 1.

In agreement with experimental observations and early estimates using a frequency-domain model [29], the shortest pulses are observed at the onset of self-pulsation. The dependence of both the pulse repetition rate and the pulse duration on current is qualitatively the same as that typically observed experimentally (for example, in [5, 24]) in lasers of a design similar to that simulated here. The pulse duration sees a minimum at the onset of self-pulsations and increases with current during steady mode locking; the repetition rate has a maximum at low currents and decreases in a fairly linear fashion over much of the useful range of currents. The latter fact is in agreement with most experiments and with the linear part of (7.7) (the minimum at higher currents and subsequent growth described by the quadratic part of (7.7) is presumably not reached in the simulations, nor is it always observed experimentally). Quantitatively, the rate of decrease in the repetition frequency with current shown by the simulations is slightly slower than that observed experimentally in [15, 5], and noticeably slower than that observed in [24] (where a frequency range of 500 MHz was covered with current varied over a range comparable with that simulated here — to the best of our knowledge, no numerical simulations have so far reproduced tuning as broad as this in a laser of this design).

7.4.6 Hybrid Mode Locking

At present, neither computational model at our disposal includes a detailed model for the reverse-biased saturable absorber. Therefore, voltage modulation, inherent in hybrid mode locking, can be either represented as a sinusoidal modulation applied to the unsaturated absorption (in the in-house package) or to an artificially large (to account for the small carrier lifetime) *ac* bias current applied to the absorber section (as in [27]), which is also possible with a commercial solver such as CLADISS-2D. Here, we select the former option, which we believe is closer to the experimental situation. The schematic dynamics of successful and unsuccessful locking are shown in Fig. 7.13. The parameter plotted in Fig. 7.13 — the time lag between the optical pulse and the nearest modulation peak (unsaturated absorption minimum) — has been chosen as the clearest illustration of the locking process and the difference between locked and unlocked states. Under locked operation (Curve 1 in Fig. 7.13), the laser pulse position stabilizes, after an initial transient, at a given distance from the minimum of the modulation sinusoid (ahead of it if the modulation period is smaller than the free-running period as in Fig. 7.13, lagging behind it in the opposite case as in Fig. 7.14). The predicted time to lock (the characteristic transient time of Curve 1) is of the order of nanoseconds to tens of nanoseconds, depending on the detuning between the modulation and free-running (passive) mode-locking frequencies and the modulation amplitude. Under unlocked operation, the timing of the pulse relative to the absorption minimum changes quasi-periodically (Curve 2). As in Fig. 7.12, each ML pulse is represented by a dot in Fig. 7.13. The scatter of these dots in the case of successful locking also illustrates the timing jitter of the locked operation. The jitter is about 0.1 ps, if determined as the standard deviation of pulse-to-pulse period. It is reasonably constant across the locking range (except possibly the very edges of the range), as in the experiments.

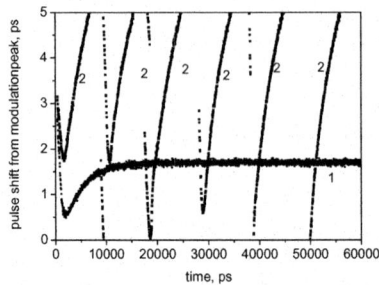

Fig. 7.13. An illustration of locking dynamics under hybrid mode-locking operation: dynamics of the timing of the ML pulse relative to the nearest modulation peak. Relative modulation amplitude $\delta a = \Delta a/a_0 = 0.1$, absorber relaxation time 15 ps, current $I/I_{th} = 3.0$, other parameters as in Table 1. Curve 1: successful locking (modulation period 25.13 ps). Curve 2: unsuccessful attempt at locking (frequency mixing regime, modulation period 25.11 ps).

Figure 7.14 represents the dependence of different pulse parameters on the modulation amplitude at a given frequency detuning (calculated using slow quasi-static decrease of the amplitude over 100–200 ns of simulations). As the amplitude is decreased from the initial high value, the pulses broaden and become less intense; simultaneously, the time difference between the pulse and the minimum of the modulation signal is progressively increased, until finally, the locking is lost. To calculate the pulse parameters in the locked regime, a "ramp" (gradual change) of modulation amplitude was used, but to calculate the locking range accurately, long simulation runs for fixed parameter values (as in Fig. 7.13) have been found to be essential, for the same reason as discussed above in connection with determining the onset of chaotic instability in passive mode locking.

By determining the minimum modulation amplitude at which the repetition frequency of the lasing pulses coincides with the modulation period, or the average time lag between the pulse peak and the nearest absorption minimum (modulation peak) remains constant, we can find the locking range in terms of the modulation amplitude and frequency detuning. We note that this definition is, strictly speaking, not the same as the one used in experiments, where the locking range is determined as the parameter interval for operation with noise and/or jitter below an agreed value. However, the increase in the noise and jitter at the edges of the experimental locking region (presumably associated with, or acting as a precursor for, the loss of synchronization) is so steep that the two definitions are not expected to differ significantly. In agreement with earlier simulations using dynamic modal analysis [23] and

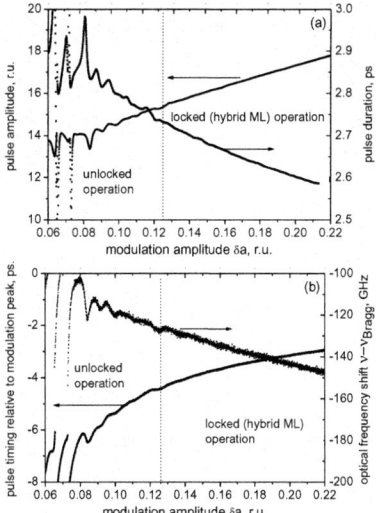

Fig. 7.14. The simulated dependences of the mode-locking pulse parameters on the relative modulation amplitude $\delta a = \Delta a/a_0$ under hybrid ML. (a) pulse amplitude and duration, (b) the timing of the ML pulse relative to the nearest modulation peak and the optical frequency at pulse peak. Modulation period $T_{mod}=25.18$ ps, the rest of parameters as in Fig. 7.13.

similarly to predictions for other pulsating lasers (e.g., Q-switched lasers), we find that the width of the locking range scales approximately linearly with the modulation amplitude in a broad range of (small to moderate) amplitude values. At very large modulation amplitudes ($\delta a \sim 0.4$-0.5), simulated pulse trains can acquire a chaotic envelope, which may account for jitter increase at high modulation seen in some experiments. Quantitatively, the locking range determined in our calculations (Fig. 7.15) is of the same order as (slightly smaller than) that typically measured experimentally in DBR mode-locked lasers, for example, by [24] (tens to hundreds of megahertz depending on the laser parameters and modulation amplitude). However, not only is the simulated locking range narrower than the one typically measured in lasers giving a similar pulse width, but also one important feature of the experimentally observed locking picture is not well reproduced by our (and, to the best of our knowledge, any other reported) simulations. Indeed, experimental studies tend to yield a very asymmetric locking range, with slowing the laser down being much easier than speeding it up. This is a very general feature, reported by several experimental teams [24, 30, 31] in Fabry–Perot and DBR, monolithic and external cavity, and actively and hybridly mode-locked lasers. Simulations either do not reproduce this asymmetry at all, or predict an asymmetry weaker than (and not even necessarily of the same sign as) that measured experimentally. This, as well as the modest (compared with experiments) current tuning of the repetition rate under passive mode locking, remain important questions for further investigations. It may be the case that both are caused by using simple phenomenological approximations for gain and saturable absorption as opposed to a more accurate representation of the full carrier-density-dependent gain and absorption spectra.

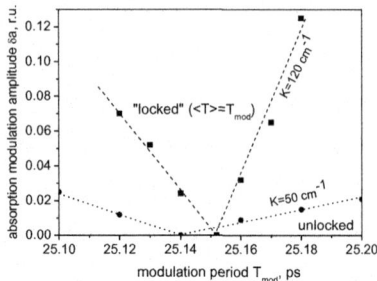

Fig. 7.15. The locking cone (relative modulation amplitude $\delta a = \Delta a / a_0$ vs. modulation period) of hybrid mode locking. Parameters as in Fig. (7.13).

7.5 Modeling Semiconductor Parameters: The Absorber Relaxation Time

From the examples above, it can be concluded that one-dimensional time-domain models are well suited for the description of the dynamics of light in a mode-locked laser; however, a correct set of parameters representing the semiconductor medium is needed for the model to be used for quantitative prediction and optimization. In most cases, it should be sufficient to describe the properties of both the gain and the saturable absorber section using only one true dynamic variable: the coordinate-dependent carrier density as in (7.10). Its dynamics are described and related to optical properties using phenomenological parameters, the most important of which are the nonradiative and spontaneous recombination rates, peak gain, gain bandwidth, linewidth enhancement factor (intraband contribution only if the shift of gain peak with carrier density is included), and the gain nonlinearity factor. Most of the above can be represented as relatively simple analytical functions of carrier density (and, if needed, carrier temperature, which is the second dynamic parameter that can be introduced for modeling gain dynamics in subpicosecond lasers as in [2, 31]), and they can be fitted to detailed microscopic calculations or to experiments.

The parameters of the gain sections of ML lasers are largely the same as those of other laser types and as such widely analyzed and relatively well known (with the possible exception of the nonlinearity parameters). The SA parameters are less common and therefore less intensely studied, so that there is some scope for further research even in regard to relatively straightforward parameters such as the effective recovery time, rather important for laser dynamics (see Fig. 7.6). The physics of the SA recovery in the most commonly used absorbers, those produced by reverse biasing a section of a (usually quantum well) laser, is the sweepout of carriers by the electric field. Recently, we have developed a microscopic model for the sweepout process that treats both thermal and tunneling carrier escape on the same footing [32].

In this model, the carrier escape current is found using the formalism of propagating states and reflection/transmission coefficients. To find the current generated by carriers leaving the biased quantum well, it is sufficient to know the associated outside probability flux. Considering the escape probability flux as a result of interference of all different processes, which account for penetration of the external probability flux (associated with the wide optical-confinement layer) into the quantum well and back, we derived the the following formula for the escape current:

$$J = \frac{e}{2\hbar\pi} \int \frac{|t(E)|^4 P(E)}{|1 - r_+ r_-(E)|^2} dE . \qquad (7.15)$$

Here, t is the transmission coefficient that describes the penetration of an electron or hole from outside to some point Z inside the quantum well, and r^+ and r^- are the reflection coefficients for a carrier propagating from Z in two different directions. The integration is over the appropriate region of the energy E associated with the carrier motion along the growth axis. The function P gives the particle density at all states with fixed E, and in the quasi-equilibrium is:

$$P(E) = \frac{m_\parallel}{\beta\hbar^2\pi} \ln\left(1 + \exp(\beta(\mu - E))\right), \qquad (7.16)$$

where $m_\parallel$ is the carrier in-plane effective mass, μ is the quasi-Fermi level, and β is the inverse temperature measured in energy units. The quantity $P(E_{QW})$ gives the particle density at the quantized 2D level E_{QW} inside the quantum well. The escape time is then given as $\tau = en/J = \sum_s P(E_{QW}^s)$, where n is the density inside the quantum well and the sum is taken over all well-localized (quasi-)bound levels.

The calculated times show a more complex dependence on the electric field than the often postulated exponential one, but they can be down to several picoseconds at high enough fields and particularly in optimized quantum well structures (Fig. 7.16).

In GaAlAs materials, the escape times for carriers of both signs are of the same magnitude, with the smaller effective mass facilitating the electron tunneling being to some extent offset by the potential barrier seen by the tunneling electron being higher than that facing the hole. Repeating the calculations for InGaAsP materials, which have a deeper well for holes than

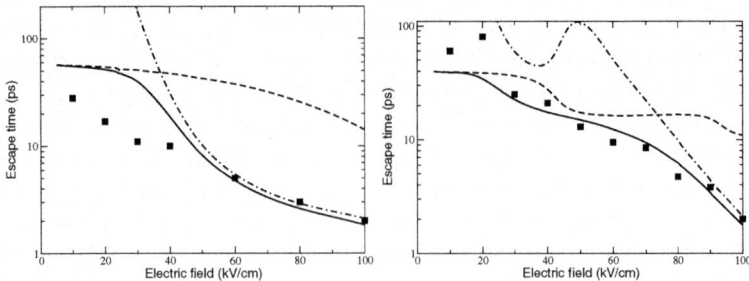

Fig. 7.16. Electron (left) and hole (right) sweepout time in GaAs/Al$_{0.2}$Ga$_{0.8}$As quantum well saturable absorbers vs. applied electric field: solid-total, dashed — thermal, dash-dotted — tunneling, dots — experiments from literature.

electrons, we find, as could be expected, that the hole tunneling is considerably slower and thus likely to be the main limiting factor for the device response (Fig. 7.17).

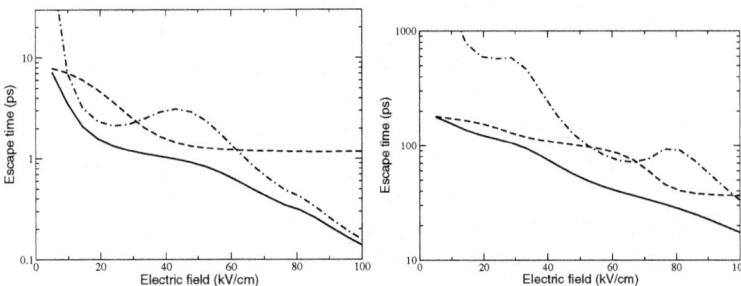

Fig. 7.17. Electron (left) and hole (right) sweepout time from an InGaAs quantum well with barriers of an InGaAsP compound lattice matched to InP.

In a practical device, the field is reduced by the photocarrier screening so the sweepout time is dependent on the carrier densities in the waveguide layer surrounding the quantum well active region. The dynamics of carrier densities are governed by two characteristic times: the time of the carrier escape from the waveguide layer and the RC time of the external circuit supplying carriers to the depleted regions bordering this layer. The former of these must, strictly speaking, be obtained by treating the kinetics of the carriers in the waveguide layer self-consistently. By way of a preliminary study, we have implemented a simplified model, modifying the analysis proposed for bulk saturable absorber by Uskov et al. [33] to the case of a quantum-well absorber embedded in a waveguide layer and approximating the waveguide carrier escape time by a constant. The results show that to maintain absorber relaxation times of 10–20 ps at currents several times above threshold, the waveguide escape time should be as low as single picoseconds — with longer barrier escape, any increase in current (and thus in optical power and absorber carrier density) simultaneously increases the absorber recovery time. In terms of the regime diagram of Fig. 7.6, this means that with an increase in current, the laser operating point moves, not vertically upward, but upward and to the right, which significantly decreases the range of stable mode locking. We believe this effect is important in any quantitative comparison with experiments; furthermore, accurate studies of the absorber dynamics will be necessary for this.

7.6 Directions for Future Work

As follows from the previous two sections, the most immediate route of progress in quantitative modeling of mode-locking devices appears to be improving the microscopic foundations of dynamic modeling. Calculation of the absorber recovery time as discussed in the previous section can be seen as the first step in this direction, but much more work is required, as it is well known that varying the absorber bias voltage changes, not just the recovery time, but also the unsaturated absorption (through Quantum-Confined Stark and Franz–Keldysh effects) and possibly the saturation cross section. The use of accurate models of optical as well as kinetic (recombination parameters) properties of both gain and saturable absorber sections (including effects of nonequilibrium carrier distribution so that ultrafast nonlinearities of both gain/saturable absorption and the refractive index in both sections can be calculated accurately) will also be needed if true predictive capability is to be expected from the model. In most cases of practical interest, it should be enough to tabulate the most important parameters (as discussed at the start of the previous section) as functions of quasi-equilibrium distribution parameters: carrier density and, if needed, carrier temperature.

Several cases can be conceived, however, where the reduction of the population dynamics to those of carrier density and temperature could become inadequate. These include, first, an as yet hypothetic situation where a laser emits very short pulses (<0.1 ps). In this case, the adiabatic exclusion of polarization (implicit in most models) is inaccurate, and the introduction of parameters such as the linewidth enhancement factor becomes tenuous. Instead, description in terms of density matrix or semiconductor Bloch equations will become more appropriate. So far, pulses as short as this have not been generated by monolithic ML laser diodes, but there appears to be no fundamental reason why they should not be produced in some advanced future construction, for example, with an artificially flattened gain spectrum; modeling tools will need to be updated to analyze such a regime correctly.

Another case that needs special care is that of advanced active media such as quantum dots (QDs). In this case, the carrier density is not a very meaningful parameter because the distribution function that describes the dot level occupation (in itself an approximation) can deviate significantly from equilibrium. A recent preliminary study [34] has combined the dynamic modal analysis approach with rate equations for population of individual dot energy levels to analyze the potential of these lasers for active ML. The results are promising, with subpicosecond pulses predicted for some parameter values, but the experimental implementation will rely on the successful realization of a construction monolithically integrating an efficient modulator with the QD gain region. The nonequilibrium distribution effects, including strong hole-burning, that are associated with QD lasers, are shown to affect the characteristics of the ML regime.

7.7 Summary

We have presented an overview of the current state of the art in modeling mode-locked semiconductor lasers. The distributed time-domain approach has been identified as the most versatile and reliable way of modeling these devices, although both the traditional semi-analytical mode-locking theory and the time-frequency "dynamic modal analysis" method can be of use in some specific problems. By analyzing DBR lasers for high-speed operation in more depth, we have shown that the existing dynamic models with phenomenological material parameters can reproduce the general features of the experimentally observed laser behavior and can be used to compare different options. However, more work on the microscopic theory of the active media, particularly saturable absorbers and modulators, is needed before reliable quantitative predictive capability is achieved.

Acknowledgment

One of the authors (EA) is grateful to the authors of [24, 27] for the use of printouts of these talks, and his colleagues at A.F. Ioffe Institute, St. Petersburg, Russia, and Universities of Glasgow and York (particularly E. L. Portnoi, J. M. Arnold, and J. H. Marsh) for numerous discussions on various aspects of mode locking.

References

1. E. A. Avrutin, J. H. Marsh, and E. L. Portnoi: IEE Proc.-Optoelectron. **147**, 251 (2000)
2. M. Schell, M. Tsuchiya, and T. Kamiya: IEEE J. Quantum Electron. **32**, 1180 (1996)
3. F. Camacho, E. A. Avruitin, P. Cusumano, A. S. Helmy, A. C. Bryce, and J. Marsh: IEEE Photon. Technol. Lett. **9**, 1208 (1997)
4. D. A. Yanson, M. W. Street, S. D. McDougall, H. Marsh, and E. A. Avrutin: IEEE J. Quantum Electron. **38**, 1 (2002)

5. H. F. Liu, S. Arahira, T. Kunii, and Y. Ogawa: IEEE J. Quantum Electron. **32**, 1965 (1996)
6. B. S. Kim, Y. Chung, and S. H. Kim: IEEE J Quantum Electron. **35**, 1623 (1999)
7. T. Shimizu, I. Ogura, and H. Yokoyama: Electron. Lett. **33**, 1868 (1997)
8. J. F. Martins-Filho, E. A. Avrutin, C. N. Ironside, and J. S. Roberts: IEEE J. Select. Topics in Quantum Electron **1**, 539 (1995)
9. H. Sanjoh, H. Yasaka, Y. Sakai, K. Sato, H. Ishii, and Y. Yoshikuni: IEEE Photon. Technol. Lett. **9**, 818 (1997)
10. M. Teshima , K. Sato, and M. Koga: IEEE J. Quantum Electron. **34**, 1588 (1998)
11. V. B. Khalfin, J. M. Arnold, and J. H. Marsh: IEEE J. Select. Topics Quantum Electron. **1**, 523 (1995)
12. J. A. Leegwater: IEEE J.Quantum Electron. **32**, 1782 (1996)
13. H. A. Haus: IEEE J.Quantum Electron. **11**, 736 (1975)
14. R. G. M. P. Koumans and R. Van Roijen: IEEE J.Quantum Electron. **32**, 1782 (1996)
15. S. Arahira and Y. Ogawa: IEEE J. Quantum Electron. **33**, 255 (1997)
16. E. A. Avrutin, J. M. Arnold, and J. H. Marsh: IEEE Proc. Pt. J. **143**, 81 (1996)
17. J. Caroll, J. Whiteaway, and D. Plumb: *Distributed Feedback Semiconductor lasers*, Chapter 7 (IEE, London 1998)
18. L. Zhai, A. J. Lowery, and Z. Ahmed: IEEE J. Quantum Electron. **31**, 1998 (1995)
19. M. Kolesik and J. V. Moloney: IEEE J. Quantum Electron. **37**, 936 (2001)
20. http://www-users.york.ac.uk/eaa2/research/lastidom
21. http://www.photond.com
22. http://www.vpiphotonics.com/CMActivePhotonics.php
23. E. A. Avrutin, J. M. Arnold, and J. H. Marsh: IEEE J. Select. Topics in Quantum Electron. **9**, 844 (2003)
24. B. Huettl, R. Kaiser, W. Rehbein, H. Stolpe, M. Kroh, S. Ritter, R. Stenzel, S. Fidorra, R. Sahin, G. Jakumeit, and R. Heidrich: *3rd International Workshop on Semiconductor Laser Dynamics* (Berlin, 15–17 September 2004)
25. E. A. Avrutin and V. V. Nikolaev: *3rd International Workshop on Semiconductor Laser Dynamics* (Berlin, 15–17 September 2004)
26. E. A. Avrutin: IEE Proc.-J. **140**, 16 (1993)
27. U. Bandelow: *3rd International Workshop on Semiconductor Laser Dynamics* (Berlin, 15–17 September 2004)
28. L. M. Zhang and J. E. Carroll: IEEE J. Quantum Electron. **31**, 240 (1995)
29. J. Paslaski and K. Y. Lau: Appl. Phys. Lett. **59**, 7 (1991)
30. R. Ludwig, S. Diez, A. Ehrhardt, L. Kuller, W. Pieper, and H. G. Weber: IEICE Transact. Electron. **E81C**, 140 (1998)
31. S. Bischoff, J. Mork, T. Franck, S. D. Brorson, M. Hofmann, K. Frojdh, L. Prip, and M. P. Sorensen: Quant. and Semiclassical Optics **9**, 655 (1997)
32. V. V. Nikolaev and E.A. Avrutin: IEEE J. Quantum Electron. **39**, 1653 (2003)
33. A. V. Uskov, J. R. Karin, R. Nagarajan, and J. E. Bowers: IEEE J. Select. Topics in Quantum Electron. **1**, 552 (1995)
34. C. Xing and E. A. Avrutin: *Conference on Semiconductor and Integrated Optoelectronics (SIOE'2003)* (Cardiff, U.K., April 2003)

8 Vertical-Cavity Surface-Emitting Lasers: Single-Mode Control and Self-Heating Effects

M. Streiff[1], W. Fichtner[1], and A. Witzig[2]

[1] Integrated Systems Laboratory, ETH Zurich
 ETH Zentrum, Gloriastrasse 35, CH-8092 Zürich, Switzerland,
 fichtner@iis.ee.ethz.ch
[2] ISE Integrated Systems Engineering AG,
 Balgriststrasse 102, CH-8008 Zürich, Switzerland

A comprehensive device simulator for vertical-cavity surface-emitting lasers (VCSEL) is presented. In order to demonstrate its practical use as a design tool, the task of finding a device structure with maximum single-mode power is performed. Single-mode control is an important aspect in VCSEL design for a wide range of applications. For instance, in order to minimize the pulse dispersion in a fiber optic link, it is desirable to keep the spectrum of the laser emission as narrow as possible, and therefore restrict the emission ideally to a single mode at a given optical power level. Sensing applications constitute an additional field where single-mode control is essential. Pure single-mode operation is needed in this case to achieve ultimate spectral detection resolution.

Section 8.1 gives a detailed description of the VCSEL structure and the design parameters in question. Two design concepts that are expected to enhance the single-mode behavior of the VCSEL are described: One of them uses a metallic absorber and the other employs an anti-resonant structure. From the device description and the "design plan", the modeling requirements are formulated. The effectiveness of both metallic absorber and anti-resonant structure is significantly compromised by an intricate interplay of electronic, thermal, and optical effects. As a conclusion, the proposed design task can only be handled by a 2D model that takes these effects into account in a self-consistent manner. Such a model is implemented in the DESSIS VCSEL device simulator, which will be presented in Sect. 8.2. The two device concepts are investigated and compared in Sect. 8.3.

8.1 VCSEL Device Structure

The device structure that will be investigated is based on [1]. As [1] presents a purely optical benchmark, it had to be augmented with additional electronic specifications.

The structure is detailed in Fig. 8.1 and Table 8.1. The VCSEL is designed in the AlGaAs/GaAs material system with an InGaAs active region that emits at 980 nm. It is an etched mesa structure with an Al_2O_3 aperture. The

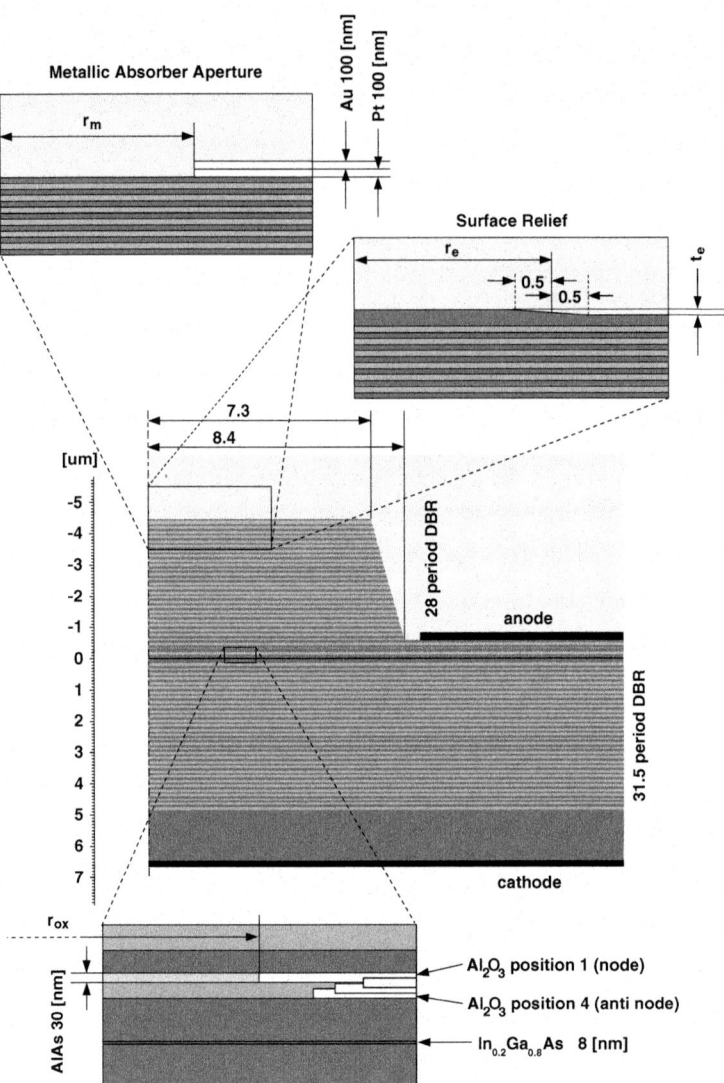

Fig. 8.1. VCSEL device structure. Insets show Al_2O_3 aperture geometry, metallic absorber, and surface relief option for optical resonator.

Table 8.1. Material Composition and Thicknesses of Epitaxial Layers in Basic VCSEL Device Structure.

	thickness [nm]	material	
air		air	
28 period DBR	69.49	GaAs	
	79.63	$Al_{0.8}Ga_{0.2}As$	
	69.49	GaAs	
	49.63-x	$Al_{0.8}Ga_{0.2}As$	
oxide aperture	30.00	AlAs	$r < r_{ox}$
	30.00	Al_2O_3	$r > r_{ox}$
	x	$Al_{0.8}Ga_{0.2}As$	
	136.49	GaAs	
λ cavity	8.00	$In_{0.20}Ga_{0.80}As$	
	136.49	GaAs	
31.5 period DBR	69.49	$Al_{0.8}Ga_{0.2}As$	
	79.63	GaAs	
substrate		GaAs	

position	x
1	49.63
2	33.09
3	16.54
4	0.00

one-wavelength-thick λ-cavity at the center of the device is enclosed with a bottom $\lambda/4$ distributed Bragg reflector (DBR) with 31.5 mirror pairs and a top DBR with 28 mirror pairs. Note that λ is equivalent to the wavelength in vacuum. A 30-nm-thick AlAs layer can be placed in four different positions in the vertical direction inside the $\lambda/4$ layer in the top DBR closest to the cavity (see bottom inset in Fig. 8.1). This AlAs layer can be oxidized to an arbitrary lateral depth. In this way, a low refractive index, insulating aperture for both the optical field and the electrical current is formed. The active region is given by a single strained $In_{0.20}Ga_{0.80}As$ 8-nm-thick quantum well embedded in GaAs barriers at the center of the λ cavity. The anode side is connected with an intracavity contact. The top DBR can therefore be left undoped and the free-carrier absorption kept to a minimum. The intrinsic region is assumed p-doped at 10^{17} cm^{-3}. The current spreading layer next to the anode is p-doped with $2 \cdot 10^{18}$ cm^{-3} and the cathode side n-doped with $2 \cdot 10^{18}$ cm^{-3}.

Two variants of the basic VCSEL structure described above will be investigated in Sect. 8.3. One variant has a metallic absorber composed of 100-nm Pt and 100-nm Au placed on top of the anode–side DBR. The metallic absorber has an aperture with a given radius. In the other variant, the top layer of the anode DBR stack is $3\lambda/4$ thick instead of the original $\lambda/4$. A surface relief is etched into the top layer. This forms an off-center anti-resonant structure outside a given radius. The thicker top layer is used to accommodate the surface relief without exposing any aluminium-containing layers to air. This would lead to the formation of an unwanted oxide layer and deteriorate the optical properties of the anti-resonant structure.

The following design parameters will be varied in the course of finding a device structure with maximum single-mode power:

- Vertical position and radius r_{ox} of Al_2O_3 aperture.
- Radius of aperture r_m in metallic absorber.
- Depth t_e and radius r_e of surface relief.

A useful VCSEL model has to cover several physical aspects. Bulk carrier transport and carrier transport at heterojunctions and quantum wells have to be modeled in the AlGaAs/InGaAs material system taking into account lattice self-heating and cooling effects. Furthermore, generation of carriers by absorption of light and carrier recombination due to stimulated and spontaneous emission have to be considered as well as nonradiative carrier recombination caused by Auger and Shockley–Read–Hall (SRH) processes. Moreover, the optical modes have to be determined for a dielectric function:

$$\varepsilon_r(\mathbf{r},\omega) = \varepsilon'_r(\mathbf{r},\omega) + i\varepsilon''_r(\mathbf{r},\omega) \qquad (8.1)$$

with an arbitrary spatial profile. The dielectric function is complex valued in the frequency domain due to the presence of optical gain and loss that have to be calculated for the InGaAs active region. Here, it is assumed that the dielectric function is sensitive to temperature, doping, and extrinsic carrier densities, respectively, via the influence that these factors have on the bandstructure of the material.

In order to predict the behavior of the VCSEL device structure subject to the design parameter space given in Fig. 8.1, a useful VCSEL model has to describe spatially resolved quantities self-consistently; namely, electrical potential, electron and hole densities, local temperature, and optical intensity. This is accomplished by expanding all model equations by finite elements (FE). The input parameters to the model equations in this discretized form are the local physical material parameters. The topology of the device is then given by the mesh employed. This approach is in contrast to compact, computationally less costly VCSEL models with variables that describe global quantities. The input parameters to such models are "effective" parameters such as confinement factors and quantum efficiencies [2]. As the FE model is by its nature sufficiently general, it can actually be used to compute the parameters for such a compact model for the VCSEL structures outlined in Fig. 8.1. The DESSIS VCSEL device simulator, which will be presented in Sect. 8.2, follows an FE approach. In practice, the limitation of the FE approach lies in the maximum size of the discretized problem that can be accommodated on a given computer and the maximum permissible time to numerically solve the associated system of matrix equations. It will be shown in Sect. 8.3 that it is feasible to use FE to handle design tasks involving realistic VCSEL device structures.

8.2 Device Simulator

The simulator model is based on the photon rate equation approach. That is, the optical field is decomposed into a given number of modes, and the temporal evolution of the mean electromagnetic energy for each mode is described by a separate rate equation. This approach assumes that the shapes of the optical modes depend on the instantaneous value of the dielectric function. The time dependence of the dielectric function usually derives from the current modulation of the VCSEL device. Therefore, the model will break down if the inverse of the laser modulation frequency approaches the photon roundtrip time in the optical resonator. This time is very short in a VCSEL cavity (< 1 ps) due to its small size, as long as distant reflections external to the cavity can be neglected. The model will, thus, remain valid up to high modulation frequencies (10 GHz corresponds to a modulation period of 100 ps) [3]. The detailed description of the model and its implementation can be found in [4] and [5]. Related work is covered in [6, 7, 8].

8.2.1 Optical Model

The derivation of the equations for the optical model starts with the vectorial wave equation in the time domain:

$$\nabla \times \frac{1}{\mu_r(\mathbf{r},t)} \left(\nabla \times \mathbf{E}(\mathbf{r},t) \right) + \frac{1}{c^2} \frac{\partial^2}{\partial t^2} \left[\varepsilon_r(\mathbf{r},t,\tau) * \mathbf{E}(\mathbf{r},t-\tau) \right] = \mathbf{F}(\mathbf{r},t) , \quad (8.2)$$

with the permeability $\mu_r(\mathbf{r},t)$ [3], the permittivity $\varepsilon_r(\mathbf{r})$, the speed of light c, and the electric field $\mathbf{E}(\mathbf{r},t)$. The star denotes convolution in τ. In the active region, spontaneous emission enters (8.2) by the source term $\mathbf{F}(\mathbf{r},t)$ and optical gain and loss by the dielectric function $\varepsilon_r(\mathbf{r},t)$. The mode expansion:

$$\mathbf{E}(\mathbf{r},t) = \sum_{\mu=1}^{N} a_\mu(t) e^{+i \int_0^t \omega'_\mu(\tau) d\tau} \boldsymbol{\Psi}_\mu(\mathbf{r},t) + c.c. , \quad (8.3)$$

with the complex-valued scalar coefficient $a_\mu(t)$, and the real-valued scalar function $\omega'_\mu(t)$, transforms (8.2) into a set of N decoupled scalar differential photon rate equations:

---
[3] Normally, in optical problems, $\mu_r(\mathbf{r})$ can be set to one. However, in this treatment, $\mu_r(\mathbf{r})$ will be used to formulate "open" boundary conditions and is thus included in the derivations.

$$\frac{d}{dt}S_\mu(t) = -2\omega_\mu''(t) \cdot S_\mu(t) + R_\mu^{sp}(t) \tag{8.4}$$

and N eigenproblems:

$$\nabla \times (\nabla \times \boldsymbol{\Psi}_\mu(\mathbf{r},t)) = \frac{(\omega_\mu'(t) + i\omega_\mu''(t))^2}{c^2} \cdot \varepsilon_r(\mathbf{r},t,\omega_{ref}') \cdot \boldsymbol{\Psi}_\mu(\mathbf{r},t), \tag{8.5}$$

with $\mu = 1, \ldots, N$ at each point in time t. There is one photon rate equation for each one of the N optical modes. The photon rate equations describe the time evolution of the mean electromagnetic energy $S_\mu(t)$ in each mode. The source term $R_\mu^{sp}(t)$ is the electromagnetic power radiated into mode μ due to spontaneous emission in the active region. As optical gain and loss are present in the dielectric function, (8.5) is a generalized, non-Hermitian, complex symmetric eigenproblem. The eigenvector $\boldsymbol{\Psi}_\mu(\mathbf{r},t)$ of this problem is the optical mode shape of mode μ. The real part of the square root of the eigenvalue is the angular velocity $\omega_\mu'(t)$, and the imaginary part is the modal rate of change $\omega_\mu''(t)$ of the electromagnetic field of mode μ. As mentioned earlier, it is assumed that the terms $\varepsilon_r(\mathbf{r},t)$, $S_\mu(t)$, $\omega_\mu'(t)$, $\omega_\mu''(t)$, and $\boldsymbol{\Psi}_\mu(\mathbf{r},t)$ are all functions of time that vary with a period longer than the photon roundtrip time in the optical resonator.

8.2.2 Electrothermal Model

For the bulk electrothermal simulation, a thermodynamic model is employed [9, 10] that accounts for self-heating in the device. It is assumed that the charge carriers are in local thermal equilibrium with the lattice. The basic equations of the thermodynamic model comprise the Poisson equation:

$$\nabla \cdot (\varepsilon \cdot \nabla \phi) = -q\left(p - n + N_D^+ - N_A^-\right) \tag{8.6}$$

for the electrostatic potential, and the continuity equations for the electrons, holes and the local heat:

$$\nabla \cdot \boldsymbol{j}_n = q\left(R + \frac{\partial}{\partial t}n\right) \tag{8.7}$$

$$-\nabla \cdot \boldsymbol{j}_p = q\left(R + \frac{\partial}{\partial t}p\right) \tag{8.8}$$

$$-\nabla \cdot \boldsymbol{S} = H + c_{th}\frac{\partial}{\partial t}T. \tag{8.9}$$

The electron and hole densities n and p, the lattice and carrier temperature T, and the electrostatic potential ϕ are the unknowns of this system of nonlinear equations. The electron and hole current densities and the conductive heat flow are denoted by $\boldsymbol{j}_n$, $\boldsymbol{j}_p$ and $\boldsymbol{S}$, respectively. All of these quantities are, in fact, functions of $\mathbf{r}$ and t; but the arguments will be omitted in the following to keep notation simple. Furthermore, N_D^+ and N_A^- are the ionized donor and acceptor concentrations, q is the elementary charge, ε the DC permittivity, and c_{th} the total heat capacity of the semiconductor.

The net recombination rate R in the nonactive bulk region is assumed equal to the radiative recombination rate due to spontaneous emission R^{sp} plus the nonradiative recombination rate composed of the Auger R^a and SRH R^{srh} rates [11]:

$$R^{sp} = C_{sp} \cdot (np - n_i^2) \quad (8.10)$$

$$R^a = (C_n n + C_p p) \cdot (np - n_i^2) \quad (8.11)$$

$$R^{srh} = \frac{np - n_i^2}{\tau_p \cdot (n + n_i) + \tau_n \cdot (p + n_i)}. \quad (8.12)$$

The coefficients C_{sp}, τ_n, and τ_p are the spontaneous emission coefficient, and the minority carrier lifetimes for electrons and holes, respectively. The parameter n_i is the intrinsic carrier density. As R^{sp} and R^{srh} contribute only little to the total current when the VCSEL is in the lasing state, the corresponding coefficients are set constant. The Auger coefficients assume a temperature dependence:

$$C_n(T) = A_n + B_n^{(1)} \cdot \left(\frac{T}{T_0}\right) + B_n^{(2)} \cdot \left(\frac{T}{T_0}\right)^2 \quad (8.13)$$

$$C_p(T) = A_p + B_p^{(1)} \cdot \left(\frac{T}{T_0}\right) + B_p^{(2)} \cdot \left(\frac{T}{T_0}\right)^2, \quad (8.14)$$

with $T_0 = 300$ K. The coefficients in the above equations can vary strongly and often depend on device processing conditions. They are therefore regarded as fitting parameters in the following.

The net heat generation rate H is given by:

$$H = \frac{\mathbf{j}_n \cdot \mathbf{j}_n}{q n \mu_n} + \frac{\mathbf{j}_p \cdot \mathbf{j}_p}{q p \mu_p} + qR\left(\phi_p - \phi_n + T(P_p - P_n)\right) - \mathbf{j}_n \cdot T \nabla P_n - \mathbf{j}_p \cdot T \nabla P_p. \quad (8.15)$$

The first two terms represent the electron and hole Joule heating rates where μ_n and μ_p are the electron and hole mobilities. The second term is the recombination heat with quasi-Fermi potentials ϕ_n and ϕ_p for electrons and holes. Due to the high carrier densities, Fermi–Dirac distribution functions are employed. The last term is the Thomson/Peltier heating rate. The absolute thermoelectric powers P_n and P_p are approximated by the analytical formulas for a nondegenerate semiconductor and parabolic energy bands:

$$P_n = \frac{k_B}{q}\left(\log \frac{n}{N_c} - \frac{5}{2}\right) \quad (8.16)$$

$$P_p = \frac{k_B}{q}\left(-\log \frac{p}{N_v} + \frac{5}{2}\right), \quad (8.17)$$

with Boltzmann's constant k_B and the effective density of states N_c, N_v for the conduction and valence band.

In the bulk region, (8.6–8.9) are completed by the flux equations for the charge carriers and the conductive heat:

$$\mathbf{j}_n = -qn\mu_n \left(\nabla \phi_n + P_n \nabla T\right) \tag{8.18}$$

$$\mathbf{j}_p = -qp\mu_p \left(\nabla \phi_p + P_p \nabla T\right) \tag{8.19}$$

$$\mathbf{S} = -\kappa_{th} \nabla T \,, \tag{8.20}$$

with the total thermal conductivity κ_{th} of the semiconductor.

Hetero-interfaces can be modeled using a thermionic emission model [12]. However, the carrier transport in the bottom–side DBR mirror of the VCSEL, shown in Fig. 8.1, is rendered by transport through a homogeneous region, except for the DBR layers closest to the λ cavity. The homogeneous region is attributed an effective conductivity for heat, electrons, and holes, and an effective heat capacity. This is justified by the fact that the net carrier recombination rate in the DBR mirror away from the λ cavity is comparatively low. Figure 8.2 shows that the minority carrier density reaches thermal equilibrium very quickly moving away from the λ cavity into the DBR mirror. Hence, also the net Peltier/Thomson effect will be low in a region where one wide and one narrow band gap material are alternated in a periodic fashion, as is the case in an AlGaAs DBR mirror stack. Additionally, the forward-biased n-doped hetero-interfaces graded over 20 nm merely introduce an additional series resistance due to a small potential barrier formed by a space charge region caused by carrier diffusion from the wide band gap to the lower band gap material, as shown in Fig. 8.2. However, to appropriately account for the carrier injection from the DBR mirror region into the λ cavity, the first one or two periods of the DBR periods have to be included in the electrothermal model. The key advantage to substitute the DBR mirror stack with a homogeneous material is the reduced computational effort to solve the electrothermal system of equations in a 2D FE formulation that is envisaged here. The effective parameters of the homogeneous material can easily be determined in a 1D simulation taking all hetero-interface into account, or be derived from measurements.

The remaining hetero-interfaces in the λ cavity are assumed graded except the quantum well, which is treated as a scattering center for carriers. In the quantum well, the carrier density bound to a quantum well n^{2D}, p^{2D} is connected to the mobile carrier density n, p above the quantum well via a net carrier capture rate [8, 13, 14] given by:

$$C^{cap} = \left(1 - e^{\frac{q \cdot \left(\phi_n^{2D} - \phi_n\right)}{k_B T}}\right) \left(1 - \frac{n^{2D}}{N^{2D}}\right) \frac{n^{3D}}{\tau_e} \tag{8.21}$$

for the electrons and similarly for the holes with the electron quasi-Fermi potentials ϕ_n and ϕ_n^{2D} for the mobile and bound electrons, respectively. The

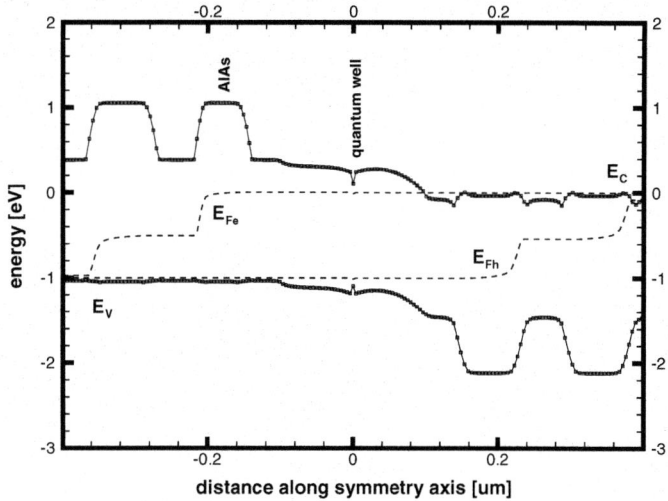

Fig. 8.2. Cut of energy band diagram along symmetry axis of VCSEL device structure for a forward bias of 1.0 V. Conduction and valence bands, quasi-Fermi energies for electrons and holes are shown. Peaking in conduction band on n-doped side and valence band on p-doped side is caused by carrier diffusion from wide band gap to low band gap material.

parameter N^{2D} is the sum of all quantum well states per unit volume and τ_e an effective electron scattering time. The quasi-Fermi potentials ϕ_n^{2D} and ϕ_p^{2D} govern the occupation of the combined states of all conduction subbands or valence subbands, respectively. The assumption of Fermi distributions is not true in general. It breaks down, for instance, in the case of strong spectral hole burning in which carriers are distributed in energy space according to a nonequilibrium distribution. However, this effect can be included via a gain saturation term [15].

In the quantum well region the continuity equation (8.7) is replaced by the following set of equations:

$$\nabla \cdot \mathbf{j}_n = q \left(R + C^{cap} + \frac{\partial}{\partial t} n \right) \quad (8.22)$$

$$\nabla \cdot \mathbf{j}_n^{2D} = q \left(R^{nr} + R^{sp} + R^{st} - C^{cap} + \frac{\partial}{\partial t} n^{2D} \right) \quad (8.23)$$

for the electrons and equally for the holes. The quantum well regions correspond to the active region of the device. Therefore, the net recombination rate consists of stimulated radiative recombination and optical generation R^{st}, spontaneous radiative recombination R^{sp}, and nonradiative recombination (Auger and SRH) R^{nr}. The heat generated by the recombination processes is taken into account as part of H in (8.9). Equation (8.23), together with current equations similar to (8.18) and (8.19), using the appropriate parameters for lateral current flow, describe carrier transport in the quantum well plane. For the Poisson equation (8.6), the electron and hole densities n and p are equal to the sum of free and bound electrons n^{2D} and holes p^{2D}.

8.2.3 Optical Gain and Loss

The computation of optical gain and loss in the dielectric function $\varepsilon_r(\mathbf{r}, \omega'_{ref})$ [4] for the active region is based on solving the stationary Schrödinger equation in 1D to obtain the subband structure of the quantum well. A parabolic band approximation is used for the bandstructure of electrons, light and heavy holes and flat bands are assumed in real space. The latter is only valid if the diode is turned on in the forward-conducting state, which is true for a laser diode above threshold. That is, the model presented here does not adequately render subthreshold operation.

Using Fermi's Golden Rule R^{st} and R^{sp} in (8.23), the local optical material gain:

$$r^{st}(\mathbf{r}, E) = C_0 \sum_{i,j} \int |M_{i,j}|^2 \rho^{el}(\mathbf{r}, E')(f_i^C(\mathbf{r}, E') + f_j^V(\mathbf{r}, E') - 1) \cdot L(E, E') dE' \qquad (8.24)$$

and the local spontaneous emission:

$$r^{sp}(\mathbf{r}, E) = C_0 \frac{c}{n(\mathbf{r})} \rho^{opt}(\mathbf{r}, E) \sum_{i,j} \int_0^\infty |M_{i,j}|^2 \rho^{el}(\mathbf{r}, E') f_i^C(\mathbf{r}, E') f_j^V(\mathbf{r}, E')$$
$$\cdot L(E, E') dE' \qquad (8.25)$$

are determined [16] with:

$$C_0 = \frac{\pi q^2}{n(\mathbf{r}) c \varepsilon_0 m_0^2 \omega'_\mu}. \qquad (8.26)$$

The term $|M_{i,j}|^2$ is the optical matrix element, $\rho^{el}(\mathbf{r}, E')$ the reduced density of states, $\rho^{opt}(\mathbf{r}, E)$ the optical mode density function, $f_i^C(\mathbf{r}, E')$ and $f_j^V(\mathbf{r}, E')$ the local Fermi–Dirac distributions of electrons and holes, $L(E, E')$ is the linewidth broadening function, $n(\mathbf{r})$ is the refractive index, m_0 the free electron mass, and ε_0 the permittivity of the vacuum. Integration is over energy space and summation is over the electron and hole subbands of the quantum well. The contribution of the local material gain (absorption) in the active region is introduced in the dielectric function as follows:

$$\varepsilon^{st}(\mathbf{r}, \omega') = \left(n(\mathbf{r}, \omega')^2 - \frac{c^2}{4\omega'^2} r^{st}(\mathbf{r}, \hbar\omega')^2 \right) + i \frac{n(\mathbf{r}, \omega')c}{\omega'} r^{st}(\mathbf{r}, \hbar\omega'). \qquad (8.27)$$

---

[4] The dependence of the dielectric function on t as given in (8.5) is implicitly assumed but will be omitted for clarity in this section

Outside the active region, the dielectric function uses static parameters. That is, constant complex refractive indices are assumed for optical absorption induced by carriers or metallic regions. The refractive index is approximated by a value constant with respect to ω' and a linear dependence on temperature:

$$n(\mathbf{r}) = n_0(\mathbf{r}) \cdot (1 + \alpha_n \cdot (T - T_0)). \qquad (8.28)$$

8.2.4 Simulator Implementation

The simulator assumes a rotationally symmetric VCSEL device structure that is discretized with FE in the cross-section perpendicular to the wafer surface, as shown in Fig. 8.3.

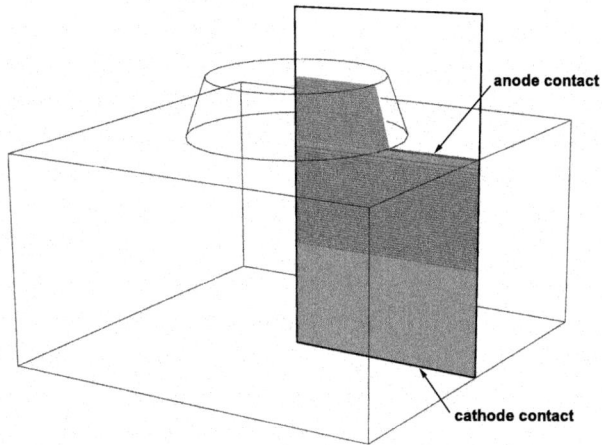

Fig. 8.3. VCSEL device structure is discretized with FE in the cross-section perpendicular to the wafer surface.

The electrothermal equations (8.6–8.9) and (8.18–8.19) are expanded with rotationally symmetric 2D FE by the box method. Furthermore, one scalar photon rate equation (8.4) for each optical mode is added. Only the mean electromagnetic energy $S_\mu(t)$ and the modal rate of change $\omega''_\mu(t)$ for each mode μ in the photon rate equation are strongly coupled to the electrothermal system of equations by the optical material gain (absorption) given in (8.24) and (8.27). The optical modes $\boldsymbol{\Psi}_\mu(\mathbf{r}, t)$, in contrast, are only weakly coupled. Therefore, the electrothermal and photon rate equations are solved inside a single Newton–Raphson scheme [17] to ensure reliable convergence. A direct LU (lower/upper triangular matrix) factorization algorithm [18] is used to solve the linear equations.

The optical problem in (8.5) is discretized with 2D body-of-revolution (BOR) FE. The BOR expansion yields the two lowest-order linearly polarized fundamental LP01 (HE11) and first-order LP11 (TE01 + TM01 + HE21) modes immediately (Fig. 8.4). The VCSEL mode designation is explained in Fig. 8.5. As the FE basis functions, first- and second-order hybrid edge-node elements are used. Radiating losses from the VCSEL cavity are taken into account by lining the boundary of the computational domain with a lossy perfectly matched layer (PML) [4]. The generalized variational functional of (8.5), expanded in the mentioned FE, is solved using the Jacobi–Davidson subspace iteration method. The solution of the Jacobi correction equation is determined by the iterative biconjugate stabilized method [19, 20].

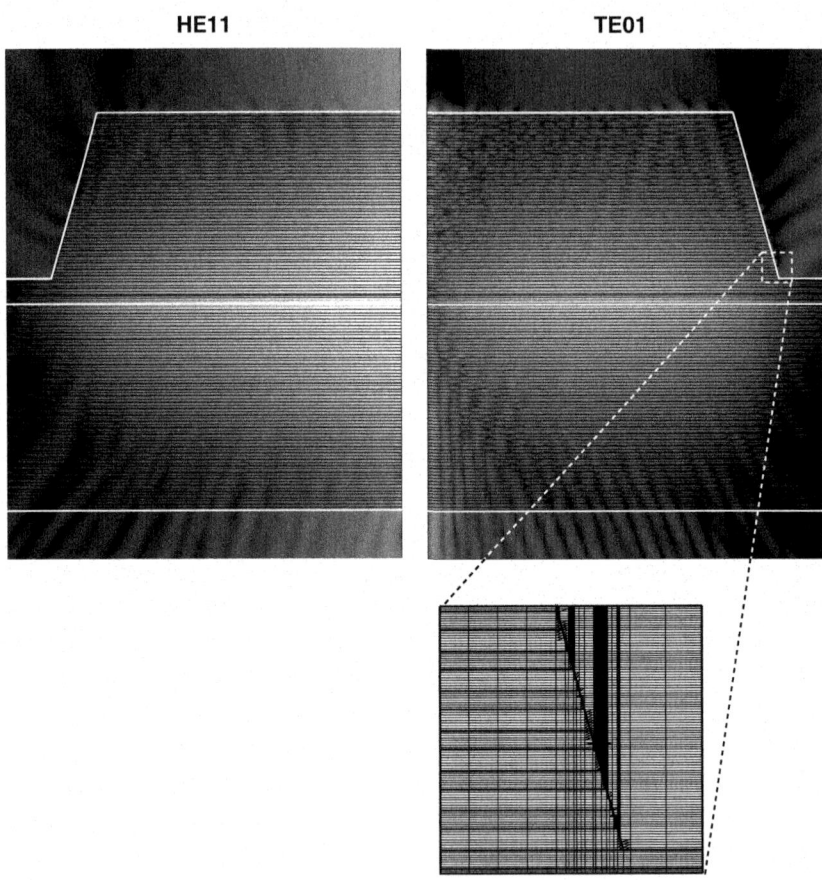

Fig. 8.4. Normalized optical intensity of fundamental HE11 and first-order TE01 modes on logarithmic gray scale. Inset shows portion of mesh used to discretize the optical problem. The optical cavity is of the type shown in Fig. 8.1, with an oxide aperture in position 1 (node) and $r_{ox} = 1$ μm.

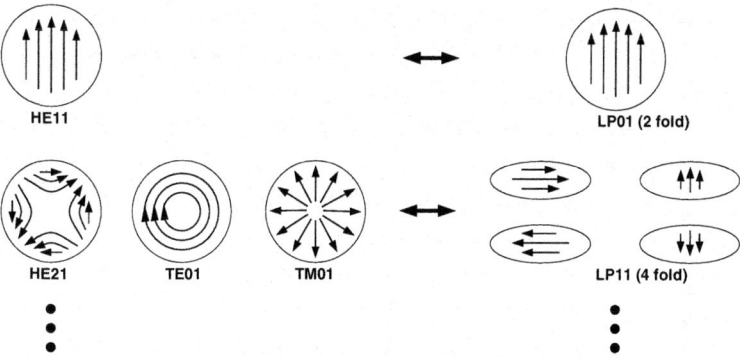

Fig. 8.5. Schematic of correspondence between hybrid and LP modes in weakly guiding approximation. Arrows indicate direction of electric field.

The strong coupling between the electrothermal and the optical problem is represented in the photon rate equations. They are integrated in the Newton–Raphson scheme, whereas the solutions of the eigenproblems (8.5) are merely updated in an exterior loop. In this way, self-consistent solutions of the electronic, thermal, and optical device equations are obtained. Good convergence of the Newton–Raphson process is maintained even at threshold where the electromagnetic energy stored in the optical modes changes by several orders of magnitude. Figure 8.6 shows a summary of the simulation flow.

As mentioned earlier, the complexity of the electrothermal device equations does not permit a 2D FE discretization of the individual DBR mirror hetero-interfaces. Instead, the DBR mirrors are represented as homogeneous regions with effective material parameters. In contrast, due to the long-ranging wave nature of the electromagnetic radiation in the VCSEL resonator, the DBR hetero-interfaces have to be resolved for the optical problem. Consequently, two separate meshes, a coarser one for the electrothermal problem and a finer one for the optical problem, are used. Linear interpolation translates variables between the two meshes. The insets in Figs. 8.4 and 8.7 show representative examples of meshes used for the optical and electrothermal problem, respectively.

In order to adapt the size of the computational problems to the individual physical problems, the three simulation domains shown in Figs. 8.4 and 8.7 are defined. The largest area is covered by the thermal domain, shown on the right of Fig. 8.7 where only the thermal equations are solved. This is to ensure that the contacts to the thermal bath held at the ambient temperature are sufficiently remote from the core VCSEL device. In this way, the influence of the thermal contacts on the temperature distribution in the VCSEL device can be reduced. The area shown on the left of Fig. 8.7 defines the extent of the electrothermal domain, where, in addition to the thermal equations, the

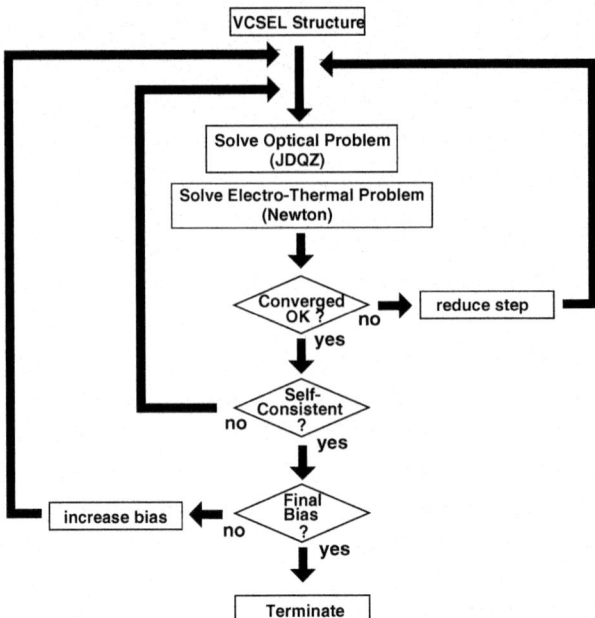

Fig. 8.6. Flow chart describing the procedure for self-consistent coupled electrothermo-optical simulation as implemented in DESSIS.

electronic equations are solved. This domain is deliberately chosen smaller to reduce the size and run-time of the computational problem. Figure 8.4 shows the extent of the optical domain where the optical equations are solved.

8.3 Design Tutorial

Simulations of the device structure shown in Fig. 8.1 are performed using the DESSIS simulator and the results discussed. The tutorial is arranged step-by-step. In Sect. 8.3.2, the properties of the cold cavity optical resonator are investigated in detail. The effect of the position of the oxide aperture in the cavity is analyzed. Building onto this basic structure, the effect of two different design concepts aiming at enhancing the single-mode behavior of the original device are assessed. The first concept uses a metallic absorber and the second one an integrated anti-resonant structure. Once some familiarity has been gained with the cold cavity problem, the performance of the different approaches are compared in a self-consistently coupled electrothermo-optical simulation in Sects. 8.3.3 and 8.3.4.

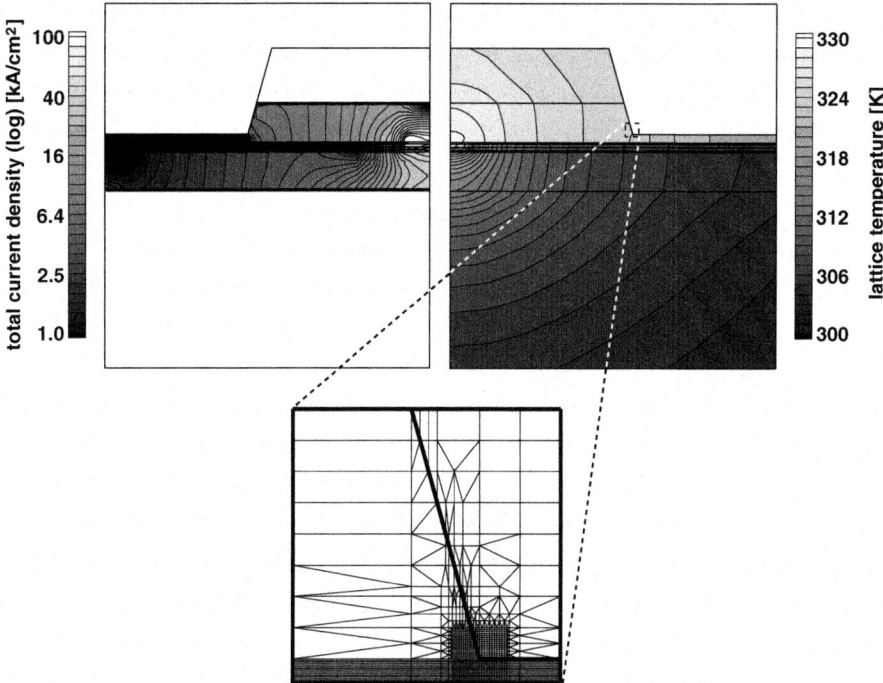

Fig. 8.7. Total current density distribution on logarithmic gray scale and lattice temperature profile on linear gray scale. Terminal current is 5.3 mA and output power 1.3 mW. Inset shows portion of mesh used to discretize the electrothermal problem. The VCSEL structure is of the type shown in Fig. 8.1, with an oxide aperture in position 1 (node) and $r_{ox} = 1$ μm.

8.3.1 Single-Mode Control in VCSEL Devices

Two basic categories of single-mode control strategies can be distinguished. The first category comprises methods that employ some sort of lateral optical guide to cut off higher-order modes. Methods of the second category introduce mode-selective loss or gain by integrating filter structures in the resonant cavity.

Methods belonging to the first category are based on the fact that the existence of higher-order modes depends critically on the lateral geometrical dimensions of the optical guide. For example, VCSEL devices with oxide apertures below a critical radius will operate in the fundamental HE11 mode and will no longer support higher-order modes. Unfortunately, rather narrow oxide apertures are required for this, which increase the electrical resistance of the device. As a consequence, the maximum single-mode output power is often limited by the increased heating. Additionally, narrow oxide apertures are a device reliability hazard and can cause significant optical scattering losses with respect to the total optical losses of the cavity.

The second category introduces some kind of integrated mode-selective filtering. Several approaches have been described, for instance, in [21], and more recently in [22, 23, 24]. This relaxes the requirements with respect to the lateral mode confinement. That is, larger oxide apertures can be used, and therefore higher laser currents, without causing excessive heating. One idea is to tailor the current injection such that the radial optical gain profile is restricted to a small region at the center of the device where the fundamental mode sits. A method to direct current injection is proton implantation. The other approach is based on deliberately introducing losses to higher-order modes. This can easily be done by using ring-shaped metallic absorber apertures. As the fundamental mode is the only one to have its optical intensity confined to the center of the device, it will experience less absorption in the metal than all other higher-order modes. First, this leads to increased losses for the unwanted modes, and second, to some screening of radiation coming from higher-order modes once they turn on. An even more efficient technique introduces a relief by etching the top surface of the optical resonator. By choosing the depth of this relief appropriately higher-order modes can be made to reflect out-of-phase from the etched semiconductor air interface. In this way, losses for higher-order modes can be increased substantially.

8.3.2 VCSEL Optical Modes

The properties of the cold cavity optical resonator are investigated in detail. Numerical experiments comparing results for different mesh densities, using second-order FE [25], and verification with other methods [5] suggest that the error in the results shown in Figs. 8.8–8.11 lies within ± 0.2 nm for the wavelength and ± 1% for the quality factor. The quality factor is obtained solving (8.5) and is defined as [26]:

$$Q_\mu = \frac{\omega'_\mu}{2 \cdot \omega''_\mu} , \qquad (8.29)$$

where ω'_μ is the angular velocity and ω''_μ is the modal rate of change of the optical mode μ.

Figures 8.8 and 8.9 show an analysis of the effect of the vertical and lateral oxide aperture position on the wavelength and the quality factor of the fundamental HE11 and the first-order TE01 optical mode. The optical simulation parameters as defined in (8.28) assuming $T_0 = 300$ K are summarized in Table 8.2, and will be used throughout the entire tutorial section. Also, refer to Fig. 8.1 for the definition of the geometry parameters used. The intracavity contact design allows the anode–side mirror to be left undoped. Negligible absorption is therefore assumed in the optical resonator material, except for the active and the metal absorber regions. Optical gain and absorption in the active region will be calculated and taken into account in the coupled

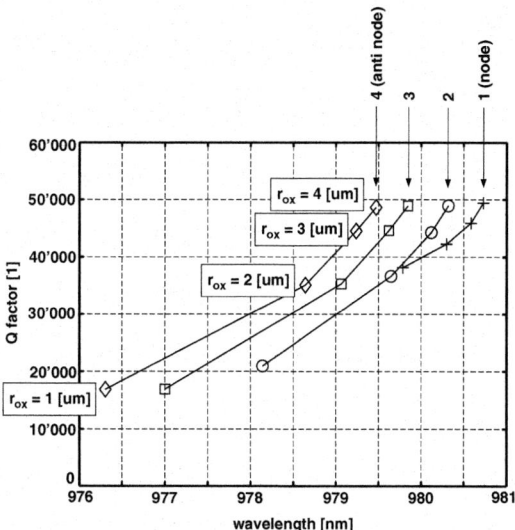

Fig. 8.8. Wavelengths and quality factors for HE11 mode. Values are plotted for different oxide aperture radii r_{ox} and different oxide aperture positions: position 1 +, position 2 ○, position 3 □, position 4 ◊.

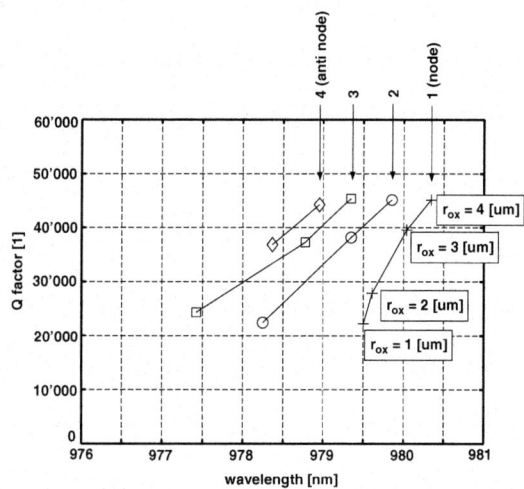

Fig. 8.9. Wavelengths and quality factors for TE01 mode. Values are plotted for different oxide aperture radii r_{ox} and different oxide aperture positions: position 1 +, position 2 ○, position 3 □, position 4 ◊. Please note that TE01 modes are cut off for oxide apertures with $r_{ox} = 1$ μm at positions 2 and 3, and furthermore, for $r_{ox} = 1$ μm, 2 μm at position 4.

Table 8.2. Complex Refractive Indices and Temperature Coefficients.

material	$n+ik$ [1]	α_n $\left[\frac{1}{K}\right]$
air	1.0	0.0
Au	$0.20 + i5.50$	0.0
Pt	$2.99 + i5.17$	0.0
GaAs	3.53	$2.0 \cdot 10^{-4}$
$Al_{0.8}Ga_{0.2}As$	3.08	$2.0 \cdot 10^{-4}$
AlAs	2.95	$1.2 \cdot 10^{-4}$
Al_2O_3	1.60	0.0
$In_{0.20}Ga_{0.80}As$	3.54	$3.7 \cdot 10^{-4}$

electrothermo-optical simulation discussed in Sects. 8.3.3 and 8.3.4 and are not considered in this section.

If the oxide aperture is moved from the node position, where the electromagnetic field has an intensity minimum, to the anti-node position, where it has a maximum, the quality factor decreases sharply for apertures with a small radius. For apertures with a large radius, the quality factor remains almost the same and only the wavelength shifts to lower values. The strong decrease in the quality factor for apertures at or close to the anti-node position is caused by the rapid increase of scattering losses that the optical field experiences for narrow aperture radii.

The strong confinement exerted by such an aperture even leads to cutting off the TE01 mode in some cases, as shown in Fig. 8.9. This effect can, in fact, be used to obtain single-mode operation of a VCSEL device, as will be shown in Sect. 8.3.4. However, this approach is usually compromised by excessive heating that current flow causes through a narrow aperture. Additionally, narrow anti-node apertures cause excessive optical losses and, hence, results in VCSEL devices with high threshold currents and poor efficiency. The remainder of this tutorial will, therefore, focus on oxide apertures in the node position, which is the technologically relevant case.

Similarly, the properties of the enhancements to the original design are investigated. The geometry of the two variants, metallic absorber and integrated anti-resonant structure, are described in Fig. 8.1. Figures 8.10–8.13 show the effect of varying the radius of the oxide aperture versus either the radius of the metal aperture or the radius of the edge of the surface relief. For reference, data of the original structure with an oxide aperture at position 1 (node) is also shown.

The key feature is that the quality factor not only decreases for narrow oxide apertures, as was the case in the original structure, but also for larger apertures from a certain radius onwards. This is due to the additional loss caused by either increased absorption in the metal or out-of-phase reflection from the etched semiconductor air interface for part of the optical mode. The affected portion of the optical mode is determined by the lateral extent of the structure causing that loss. Therefore, as the radii r_m (metallic

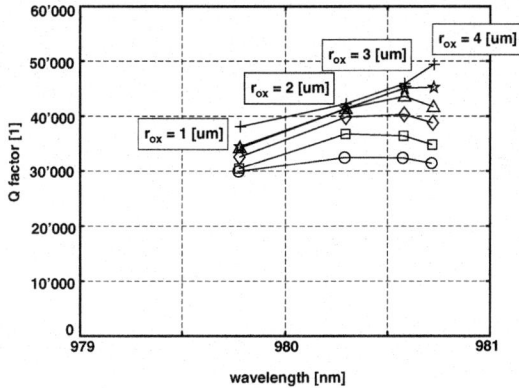

Fig. 8.10. Wavelengths and quality factors for HE11 mode for an oxide aperture at position 1 (node). Values are plotted for different oxide aperture radii r_{ox} versus radii r_m of the metallic absorber: original structure + (no metallic absorber), $r_m = 1.5$ μm ○, $r_m = 2.0$ μm □, $r_m = 2.5$ μm ◊, $r_m = 3.0$ μm △, $r_m = 3.5$ μm ⋆.

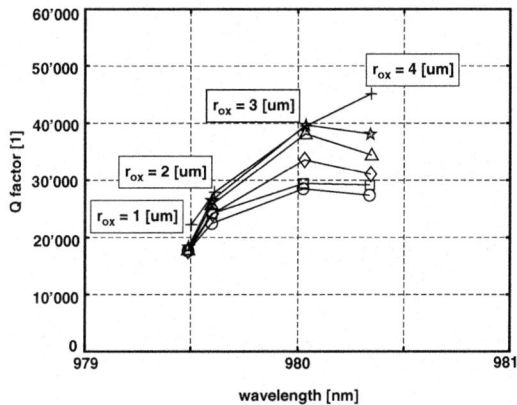

Fig. 8.11. Wavelengths and quality factors for TE01 mode. Oxide aperture at position 1 (node). Values are plotted for different oxide aperture radii r_{ox} versus radii r_m of the metallic absorber: original structure + (no metallic absorber), $r_m = 1.5$ μm ○, $r_m = 2.0$ μm □, $r_m = 2.5$ μm ◊, $r_m = 3.0$ μm △, $r_m = 3.5$ μm ⋆.

absorber) and r_e (anti-resonant structure) are decreased, the effect becomes more pronounced. However, decreasing r_m and r_e causes the quality factor to decrease in general. Furthermore, the anti-resonant structure has a stronger effect than the metallic absorber.

For the cold cavity, an interesting figure of merit is the relative difference between the quality factors of the HE11 and TE01 optical modes. Maximizing this figure will yield a device structure that exhibits maximal discrimination between HE11 and TE01 optical modes in the "cold cavity sense." That is, neglecting that the active region will expose the optical mode to optical gain

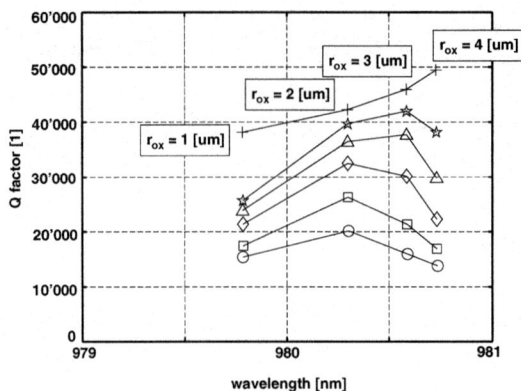

Fig. 8.12. Wavelengths and quality factors for HE11 mode. Oxide aperture at position 1 (node). Values are plotted for different oxide aperture radii r_{ox} versus radii r_e of the surface relief: original structure + (no metallic absorber), $r_e = 1.5$ μm ○, $r_e = 2.0$ μm □, $r_e = 2.5$ μm ◊, $r_e = 3.0$ μm △, $r_e = 3.5$ μm ⋆.

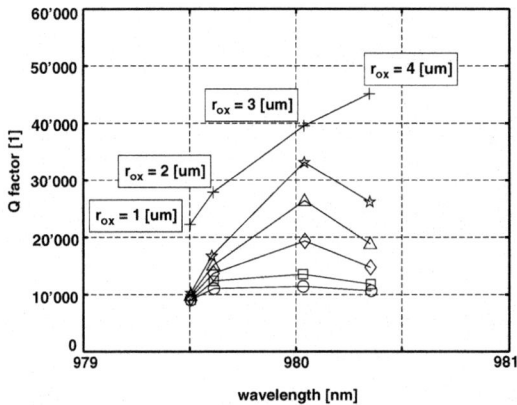

Fig. 8.13. Wavelengths and quality factors for TE01 mode. Oxide aperture at position 1 (node). Values are plotted for different oxide aperture radii r_{ox} versus radii r_e of the surface relief: original structure + (no metallic absorber), $r_e = 1.5$ μm ○, $r_e = 2.0$ μm □, $r_e = 2.5$ μm ◊, $r_e = 3.0$ μm △, $r_e = 3.5$ μm ⋆.

with a certain spatial profile, and that the optical problem will change over the operation range of the VCSEL device.

Although investigating the cold cavity is a necessary initial design task to be carried out, it is by no means sufficient to optimize a VCSEL structure for maximum single-mode power emission because it neglects the interplay between electronic, thermal, and optical effects. In principle, the entire parameter space would have to be scanned by coupled electrothermo-optical simulations, as will be used in Sects. 8.3.3 and 8.3.4. However, in order to keep the problem manageable within the framework of this tutorial, the design task will focus on finding an optimum radius r_{ox} of the oxide aperture for a given geometry of the anti-resonant structure and metallic absorber. This is an interesting exercise because, by changing the radius of the aperture, the temperature and optical gain profiles changes. Thereby, the interaction between electronic, thermal, and optical effects can be adjusted. The nominal radius of the oxide aperture is set at $r_{ox} = 4.0$ µm for additional cold cavity simulations that were performed to determine r_e and t_e for the surface relief and r_m for the metallic absorber. The relative difference between the quality factors of the HE11 and TE01 optical modes was maximized for r_m and r_e between 2.5 µm and 3.5 µm and a depth t_e of 72 nm. In the following, r_m of the metallic absorber will be set to 2.5 µm and $r_e = 2.5$ µm, $t_e = 72$ nm will be chosen for the surface relief. The values for r_e and r_m are deliberately set to values at the lower end of the range obtained from the cold cavity simulations because this is expected to approximately compensate the shift of the optical mode profiles toward the symmetry axis of the VCSEL device due to thermal lensing.

Figure 8.14 compares the discrimination between HE11 and TE01 optical modes for the basic VCSEL structure and the enhanced variants versus the radius r_{ox} of the oxide aperture. In Fig. 8.15, the HE11 output coupling efficiency is shown for the same parameter variation. The output coupling efficiency describes the optical power radiated through the top of the VCSEL structure with respect to the total optical loss of the resonator comprising material and radiation loss. An output coupling efficiency of 1.0 would correspond to the ideal case in which all optical power generated would be coupled through the top of the VCSEL device.

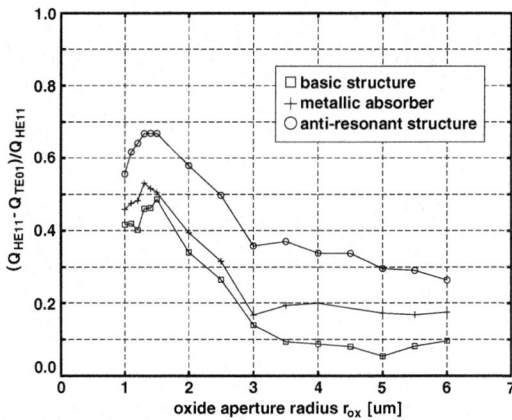

Fig. 8.14. Relative difference between the quality factors of the HE11 and TE01 optical modes for the basic VCSEL structure and the enhanced structures with surface relief and metallic absorber.

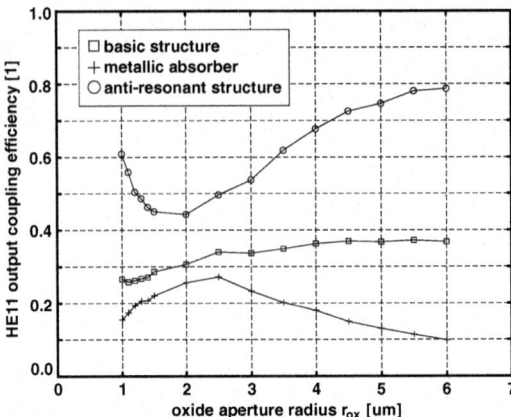

Fig. 8.15. Output coupling efficiency for the basic VCSEL structure and the enhanced structures with surface relief and metallic absorber.

Figure 8.14 shows that the anti-resonant structure provides a stronger enhancement of the mode discrimination than the metallic absorber. The latter is only slightly superior to the enhancement the basic structure provides already. Moreover, in all cases, a maximum mode discrimination is achieved for an r_{ox} between 1.0 μm and 2.0 μm. In Fig. 8.15, it can be seen that for the basic structure a larger radius r_{ox} leads to less scattering losses and, therefore, to a higher output coupling efficiency. Figure 8.15 also reveals the main disadvantage of the metallic absorber: The output coupling efficiency is lower than in the other two variants due to the screening effect that the metal layer exhibits. The HE11 mode is increasingly affected for $r_{ox} > 2.5$ μm. Compared with the basic structure for narrow r_{ox}, some scattered light that is normally radiated through the top is also absorbed by the metal layer. The anti-resonant structure shows the opposite behavior. The larger r_{ox}, the higher the output coupling efficiency. Furthermore, for narrow r_{ox}, an increase in efficiency is observed.

In the subsequent coupled electrothermo-optical simulations, it is expected that device structures that offer the best compromise between mode discrimination and output coupling efficiency will be candidates for maximum single-mode power emission.

8.3.3 Coupled Electrothermo-Optical Simulation

The parameters used for the coupled electrothermo-optical simulations are summarized in Tables 8.2, 8.3, 8.4 and 8.5.

Figure 8.16 shows the simulation results for a VCSEL structure with surface relief and $r_{ox} = 3$ μm. The rollover at high currents is mainly caused by increased Auger recombination as a consequence of self-heating. The curves

Table 8.3. Parameters for $Al_xGa_{1-x}As$. Last Column Shows Reference. Note that E_g is Given for 300 K.

$Al_xGa_{1-x}As$			
bandgap:			
E_g $x < 0.45$ (dir.)	$1.424 + 1.247 \cdot x$	$[eV]$	[27]
E_g $x \geq 0.45$ (ind.)	$1.900 + 1.250 \cdot x + 0.143 \cdot x^2$	$[eV]$	[27]
shrinkage ΔE_g	$-5.5 \cdot 10^{-4} \frac{1}{K} \frac{T^2}{204K+T}$	$[eV]$	[27]
eff. masses:			
electron	$\frac{m_{el}}{m_0} = 0.067 + 0.083 \cdot x$	$[1]$	[16]
heavy hole	$\frac{m_{hh}}{m_0} = 0.500 + 0.290 \cdot x$	$[1]$	[16]
light hole	$\frac{m_{lh}}{m_0} = 0.087 + 0.063 \cdot x$	$[1]$	[16]
mobility:			
electron (bulk)	$8500 - 6000 \cdot x$	$[\frac{cm^2}{Vs}]$	[27]
hole (bulk)	$400 - 250 \cdot x$	$[\frac{cm^2}{Vs}]$	[27]
electron (DBR)	3000	$[\frac{cm^2}{Vs}]$	-
hole (DBR)	140	$[\frac{cm^2}{Vs}]$	-
carrier scattering:			
SRH (p,n region)	$\tau_{srh,e} = \tau_{srh,h} = 0.1$	$[ns]$	-
SRH (i region)	$\tau_{srh,e} = \tau_{srh,h} = 1$	$[ns]$	-
Auger (electrons)	$A_n = 1.0 \cdot 10^{-30}$,	$[\frac{cm^6}{s}]$	-
	$B_n^{(1)} = 0$, $B_n^{(2)} = 0$	$[\frac{cm^6}{s}]$	-
Auger (holes)	$A_p = 1.0 \cdot 10^{-30}$,	$[\frac{cm^6}{s}]$	-
	$B_p^{(1)} = 0$, $B_p^{(2)} = 0$	$[\frac{cm^6}{s}]$	-
spontaneous emission	$C_{sp} = 0$	$[\frac{cm^3}{s}]$	-
thermal :			
conductivity (bulk)	$0.44 - 1.79 \cdot x + 2.26 \cdot x^2$	$[\frac{W}{Kcm}]$	[27]
conductivity (DBR)	0.11	$[\frac{W}{Kcm}]$	[2]

marked with circles are from a self-consistent electrothermo-optical simulation. The emission is single-mode up to a current of 11.5 mA when the TE01 mode starts to lase. If the optical problem is not solved self-consistently, but is only computed once at the beginning of a simulation run, the curves marked with crosses are obtained. In this case, the VCSEL device remains single-mode over the entire operation range.

The reason for the difference in the two results can be understood from Fig. 8.17. The modal optical gain contributed by the active region and the loss (radiation and material loss) contributed by the remaining optical resonator are plotted for the HE11 and the TE01 modes. The self-consistent result is shown at the top of Fig. 8.17 and the non-self-consistent one at the bottom.

For each mode, the modal gain generated by the active region rises up to the point where it compensates the loss of the remaining optical resonator. At this point the corresponding mode starts to lase. Clearly, heating effects lead to a perturbation of the optical problem such that significant deviations

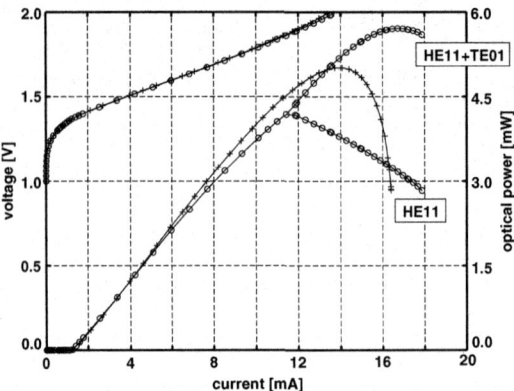

Fig. 8.16. Current voltage and current optical power characteristics for VCSEL structure with surface relief and $r_{ox} = 3$ μm. Curves with + show result if optical problem is solved only once at the beginning of the simulation. Curves with ○ show results for self-consistent electrothermo-optical solution.

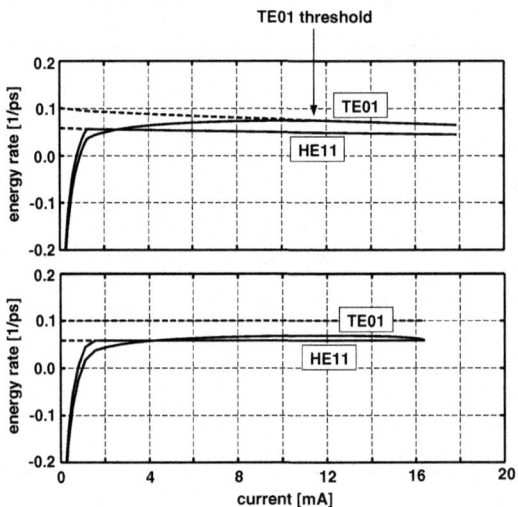

Fig. 8.17. Modal optical gain given by the active region (solid lines) and optical loss given by the remaining device structure (dashed lines) for HE11 and TE01 optical modes versus terminal current. Top: Self-consistent electrothermo-optical solution. Bottom: Optical problem is solved only once.

Table 8.4. Parameters for $In_{0.2}Ga_{0.8}As$ Quantum Well Active Region. Last Column Shows Reference.

$In_{0.2}Ga_{0.8}As$ (QW active region)			
bandgap:			
E_g	1.197	$[eV]$	[28]
shrinkage ΔE_g	$-5.0 \cdot 10^{-4} \frac{1}{K} \frac{T^2}{53K+T}$	$[eV]$	[28]
eff. masses:			
electron	$\frac{m_{el}}{m_0} = 0.063$	[1]	[29]
heavy hole	$\frac{m_{hh}}{m_0} = 0.500$	[1]	[29]
light hole	$\frac{m_{lh}}{m_0} = 0.074$	[1]	[29]
linewidth broadening	$\Gamma = 0.020$	$[eV]$	-
mobility:			
electron (QW)	7000	$[\frac{cm^2}{Vs}]$	-
hole (QW)	325	$[\frac{cm^2}{Vs}]$	-
carrier scattering:			
SRH	$\tau_{srh,e} = \tau_{srh,h} = 100$	$[ns]$	-
Auger (electrons)	$A_n = 3.0 \cdot 10^{-28}$,	$[\frac{cm^6}{s}]$	-
	$B_n^{(1)} = 0, B_n^{(2)} = 3.0 \cdot 10^{-28}$	$[\frac{cm^6}{s}]$	-
Auger (holes)	$A_p = 3.0 \cdot 10^{-28}$,	$[\frac{cm^6}{s}]$	-
	$B_p^{(1)} = 0, B_p^{(2)} = 3.0 \cdot 10^{-28}$	$[\frac{cm^6}{s}]$	-
QW capture	$\tau_{qw,e} = \tau_{qw,h} = 1$	$[ps]$	[13]
thermal conductivity	0.4	$[\frac{W}{Kcm}]$	[27]

Table 8.5. Thermal and Electrical Terminal Parameters.

ambient temperature	300	$[K]$	-
thermal contact surface resistances:			
top contact	0.1	$[\frac{Kcm^2}{W}]$	-
bottom contact	0.01	$[\frac{Kcm^2}{W}]$	-
electrical contact resistances:			
top contact	1	$[\Omega]$	-
bottom contact	1	$[\Omega]$	-

as shown in Figs. 8.16 and 8.17 develop. These deviations can be as severe as causing the complete absence of the threshold of a second-order mode. It is therefore essential that a self-consistent electrothermo-optical model is employed for the optimization task that is envisaged in this tutorial. Most heating in the device can be attributed to Joule heating. Further contributions come from Thomson/Peltier heat and recombination heat in the λ cavity. Heating affects the refractive index according to (8.28) and reduces scattering losses. Via the effect of thermal lensing, depending on the design of the VCSEL structure, material absorption and modal gain are altered.

Further to the results in Fig. 8.16, hole current, total heat power, temperature, refractive index close to the active region, normalized intensity, hole density, optical material gain in the active region, wavelength tuning of the HE11 and TE01 modes, and the maximum of the optical gain are given in Figs. 8.18 and 8.19, respectively. Current crowding at the oxide aperture $r_{ox} = 3$ μm is visible in Fig. 8.18. Strong thermal lensing and spatial hole burning can be observed in Fig. 8.19, respectively.

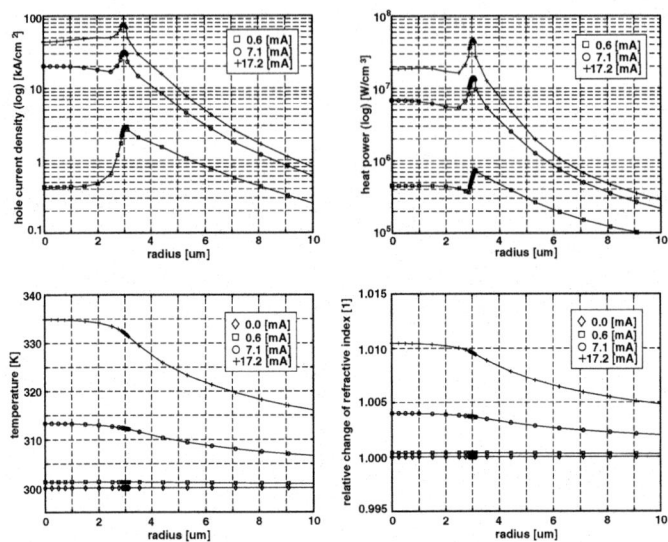

Fig. 8.18. Hole current density, total heat power density, temperature, and relative refractive index distributions in a plane at 50 nm from the quantum well on the anode side. Distributions are shown at different terminal currents for a VCSEL structure with surface relief and $r_{ox} = 3$ μm.

8.3.4 Single-Mode Optimization Using Metallic Absorbers and Anti-Resonant Structures

In this section, an optimum radius r_{ox} of the oxide aperture is determined using self-consistent simulations of the type described in Sect. 8.3.3. The results are shown in Fig. 8.20 for the original structure, in Fig. 8.21 for the metallic absorber, and in Fig. 8.22[5] for the surface relief enhancement of the optical resonator.

It is obvious that the VCSEL device structure with the anti-resonant structure beats the other candidates by far: 4.4-mW single-mode power is achieved for an oxide aperture of 4 μm compared with 2.1 mW for the best original VCSEL structure with an aperture of 2 μm. The best VCSEL device

---

[5] Please note that the scale changes in Fig. 8.22.

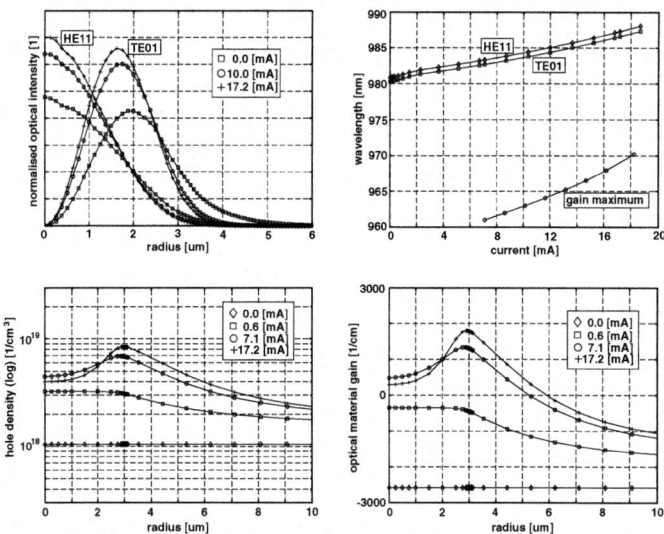

Fig. 8.19. Thermal lensing and wavelength tuning due to heating in optical resonator, and spatial hole burning in active region for a VCSEL structure with surface relief and $r_{ox} = 3$ μm. Normalized optical intensity distributions of HE11 and TE01 optical modes. Tuning of HE11 and TE01 mode resonances, and optical gain maximum versus current. Distribution of quantum well bound hole density and optical material gain.

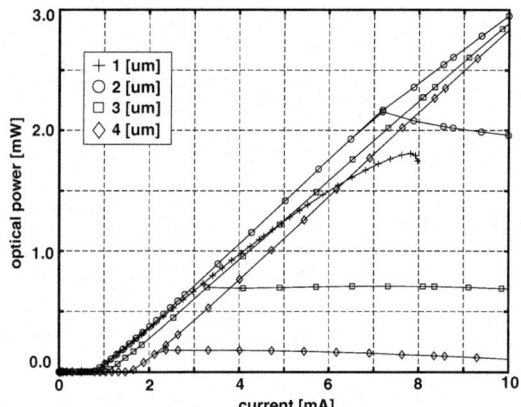

Fig. 8.20. Optical power versus current for HE11 and TE01 modes. Values are plotted for different oxide aperture radii r_{ox}. Correspondence between branches and optical modes is analogous to Fig. 8.16.

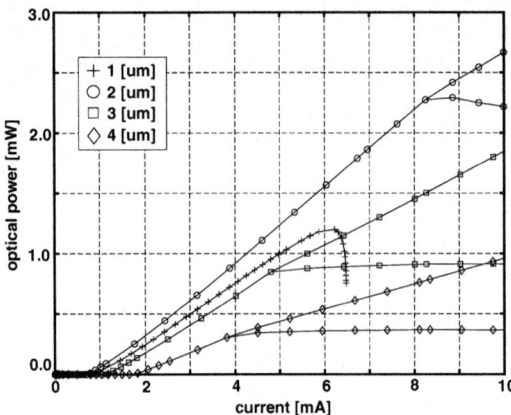

Fig. 8.21. Optical power versus current for HE11 and TE01 modes; r_m of the metallic absorber is set to 2.5 μm. Values are plotted for different oxide aperture radii r_{ox}. Correspondence between branches and optical modes is analogous to Fig. 8.16.

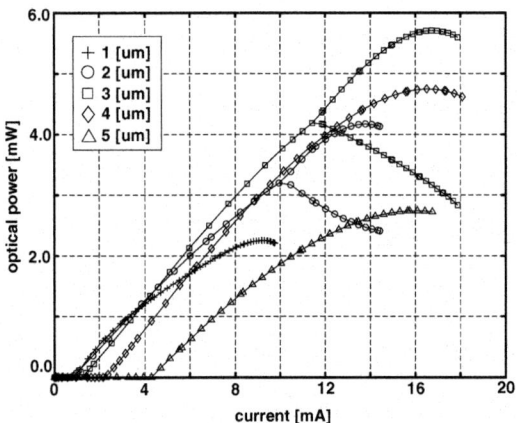

Fig. 8.22. Optical power versus current for HE11 and TE01 modes; r_e of the surface relief is set to 2.5 μm and $t_e = 72$ nm. Values are plotted for different oxide aperture radii r_{ox}. Correspondence between branches and optical modes is analogous to Fig. 8.16.

structure with the metallic absorber reaches 2.3 mW with an aperture of 2 μm and performs only slightly better than the best original structure.

For $r_{ox} = 1$ μm, all devices show pure single-mode operation and maximum single-mode power is limited by self-heating. By increasing the aperture radius self-heating can be lowered and higher power levels can be reached before rollover occurs. At the same time, higher-order modes are no longer cut off and consequently limit the maximum single-mode power. The larger the aperture, the lower the discrimination between the fundamental and higher-order modes and the lower the maximum single-mode power becomes.

The enhanced structures manage to push the TE01 threshold to higher values compared with the original structure due to the higher discrimination between HE11 and TE01 modes. In the case of the metallic absorber with r_{ox} = 4 μm in Fig. 8.21, 4 mA are reached compared with 2.1 mA in Fig. 8.20. Unfortunately, this effect is not exploited effectively by the metallic absorber. Although TE01 thresholds are now higher, the absorber increasingly screens radiation as the oxide aperture is made larger. This has already been observed in Fig. 8.15 at the end of Sect. 8.3.2 and is seen here to gradually lower the slope of the current versus optical power curve in Fig. 8.21 and compromise the maximum single-mode power that can be reached.

The anti-resonant structure (Fig. 8.22) pushes the TE01 threshold to even higher values because the optical resonator offers highest discrimination between HE11 and TE01 modes of the three design variants. For oxide apertures with 4 μm and 5 μm, TE01 thresholds are not present at all. Here the rollover has an additional contribution from lateral carrier leakage [30] that originates from the misalignment of the HE11 mode with the optical gain profile. Figure 8.19 shows the effect of thermal lensing on the optical modes and reveals strong spatial hole burning in the optical gain profile at higher currents. The overlap of the optical gain profile with the HE11 mode determines the spatial profile of the current sunk by stimulated emission in the HE11 mode. If that profile does not match the one given by the current injection, the optical mode will start to deplete carriers in the active region. Moreover, the mismatch is increased by the effect of thermal lensing due to self-heating of the device. In order to maintain the HE11 modal gain, additional current has to be provided to counterbalance the carrier depletion. This additional current cannot be turned into photons and is therefore lost as lateral leakage current. This, together with increased Auger recombination at higher temperatures, leads to the rollover observed in Fig. 8.22 for larger oxide apertures.

8.4 Conclusions

A comprehensive device simulator for VCSEL devices was presented. Further to this, its practical significance as a design tool was demonstrated. The task was to optimize a device structure to achieve maximum single-mode emission. The exercise, although not carried out exhaustively, gives essential insights into designing a VCSEL device for the envisaged goal. Only marginal improvements are possible with the proposed metallic absorber structure. It only slightly improves mode discrimination over the original structure and is limited by its inherently low HE11 output coupling efficiency. In contrast, high mode discrimination and high HE11 output coupling efficiencies are possible using an anti-resonant structure. The necessity of an additional tightly controlled masking and etch step to fabricate a surface relief make this the more expensive solution. A further increase in single-mode power would be expected if the current injection could be concentrated further to the symmetry axis of the device without affecting the optical field.

Acknowledgment

The authors would like to thank Michael Pfeiffer, Lutz Schneider (Integrated Systems Laboratory, ETH Zurich), Oscar Chinellato, Peter Arbenz (Institute of Computational Science, ETH Zurich), Wei-Choon Ng (ISE Integrated Systems Engineering Inc., San Jose), and Paul Royo (Avalon Photonics Ltd., Zurich) for helpful discussions and support.

References

1. P. Bienstman, R. Baets, J. Vukusic, A. Larsson, M. J. Noble, M. Brunner, K. Gulden, P. Debernardi, L. Fratta, G. P. Bava, H. Wenzel, B. Klein, O. Conradi, R. Pregla, S. A. Riyopoulos, J. F. P. Seurin, and S. L. Chuang: IEEE J. Quantum Electron., 37(12), 1618–1631 (2001)
2. L. A. Coldren and S. W. Corzine: *Diode Lasers and Photonic Integrated Circuits* (Wiley, New York 1995)
3. G. A. Baraff and R. K. Smith: Phys. Rev. A 61(4), 043808 (2000)
4. M. Streiff, A. Witzig, and W. Fichtner: IEE Proc. Optoelectron. 149(4), 166–173 (2002)
5. M. Streiff, A. Witzig, M. Pfeiffer, P. Royo, and W. Fichtner: IEEE J. Select. Topics Quantum Electron. 9(3), 879–891 (2003)
6. B. Klein, L. F. Register, M. Grupen, and K. Hess: OSA Optics Express 2(4), 163–168 (1998)
7. Y. Liu, W.-C. Ng, F. Oyafuso, B. Klein, and K. Hess: IEE Proc. Optoelectron. 149(4), 182–188 (2002)
8. B. Witzigmann, A. Witzig, and W. Fichtner: IEEE Trans. Electron Devices 47(10), 1926–1934 (2000)
9. ISE Integrated Systems Engineering AG, Zurich, Switzerland, *ISE TCAD 9.0 Manuals* (1995-2003)
10. K. Kells: General Electrothermal Semiconductor Device Simulation, Ph.D. dissertation, Eidgenössische Technische Hochschule Zürich, Zurich, Switzerland (1994)
11. S. Selberherr: *Analysis and Simulation of Semiconductor Devices*, (Springer, Vienna 1984)
12. D. Schroeder: *Modelling of Interface Carrier Transport for Device Simulation* (Springer, Vienna 1994)
13. M. Grupen and K. Hess: IEEE J.Quantum Electron. 34(1), 120–140 (1998)
14. M. S. Hybertsen, M. A. Alam, G. E Shtengel, G. L. Belenky, C. L. Reynolds, D. V. Donetsky, R. K. Smith, R. F. Baraff, R. F. Kazarinov, J. D. Wynn, and L. E. Smith: Proc. SPIE 3625, 524–534 (1999)
15. B. Witzigmann, A. Witzig, and W. Fichtner: Proc. LEOS Ann. Meeting, Vol. 2, 659–660 (1999)

16. S. L. Chuang: *Physics of Optoelectronic Devices* (Wiley, New York 1995)
17. R. E. Bank, D. J. Rose, and W. Fichtner: IEEE Trans. Electron Devices ED-30(9), 1031–1041 (1983)
18. O. Schenk: Scalable Parallel Sparse LU Factorization Methods on Shared Memory Multiprocessors, Ph.D. dissertation, Eidgenössische Technische Hochschule Zürich, Zurich, Switzerland (2000)
19. P. Arbenz and R. Geus: Numer. Linear Algebra Applicat. 6(1), 3–16 (1999)
20. D. R. Fokkema, G. L. Sleijpen, and H. A. van der Vorst: SIAM Journal of Scientific Computing 20(1), 94–125 (1998)
21. R. A. Morgan, G. D. Guth, M. W. Focht, M. T. Asom, K. Kojima, L. E. Rogers, and S. E. Callis: IEEE Photon. Technol. Lett. 4(4), 374–376 (1993)
22. H. Martinsson, J. A. Vukusic, and A. Larsson: IEEE Photon. Technol. Lett. 12(9), 1129–1131 (2000)
23. N. Ueki, A. Sakamoto, T. Nakamura, H. Nakayama, J. Sakurai, H. Otoma, Y. Miyamoto, M. Yoshikawa, and M. Fuse: IEEE Photon. Technol. Lett. 11(12), 1539–1541 (1999)
24. H. J. Unold, S. W. Z. Mahmoud, M. Grabherr, R. Michalzik, and K. J. Ebeling: IEEE J. Select. Topics Quantum Electron. 7(2), 386–392 (2001)
25. O. Chinellato, P. Arbenz, M. Streiff, and A. Witzig: Future Generation Comput. Syst., in press
26. J. D. Jackson: *Classical Electrodynamics*(Wiley, New York 1975)
27. S. Adachi, ed: *Properties of Aluminium Gallium Arsenide*, (INSPEC, London 1993)
28. P. Bhattacharya, ed: *Properties of Lattice-Matched and Strained Indium Gallium Arsenide* (INSPEC, London 1993)
29. M. Levinshtein, S. Rumyantsev, and M. Shur, eds: *Handbook Series on Semiconductor Parameters*, volume 2 (World Scientific, Singapore 1999)
30. C. Wilmsen, H. Temkin, and L. A. Coldren, eds: *Vertical-Cavity Surface-Emitting Lasers* (Cambridge University Press, Cambridge 1999)

9 Vertical-Cavity Surface-Emitting Lasers: High-Speed Performance and Analysis

J. S. Gustavsson, J. Bengtsson, and A. Larsson

Photonics Laboratory, Department of Microtechnology and Nanoscience, Chalmers University of Technology, SE-412 96 Göteborg, Sweden, anders.larsson@mc2.chalmers.se

In this chapter, we present an advanced, yet efficient, time-domain model for vertical-cavity surface-emitting lasers (VCSELs). The spatially quasi-three-dimensional model accounts for important interrelated physical processes, which dynamically change the distributions of the modal optical fields, carrier density, and temperature. To exemplify the capabilities of the model, we simulate the high-speed performance of surface-modified fundamental-mode-stabilized VCSELs of wavelength 850 nm. Modulation bandwidth, relative intensity noise (RIN), bit-error rate (BER), laser linewidth, etc. ,are calculated and compared with the performance of conventional multimode VCSELs. Section 9.1 gives a brief introduction to the VCSEL, followed by a discussion of its important high-speed characteristics in Sect. 9.2. The dynamic model is presented in Sect. 9.3, the simulation example is shown in Sect. 9.4, and finally, a conclusion is given in Sect. 9.5.

9.1 Introduction to VCSELs

The VCSEL is known as a low-cost light source with many attractive performance characteristics for optical communication applications. Low-cost manufacturing is enabled by the surface emission, which allows for on-wafer testing and screening. Other attractive features are, in particular, high efficiency and excellent high-speed modulation characteristics under low current and power conditions, a low-divergence circular beam for efficient fiber coupling, and easy integration in one and two dimensional arrays for parallel transmission of data. The VCSEL is, today, an established light source for data transmission in short-distance links, interconnects, and local networks (LANs, SANs, etc.). In these applications, the VCSEL is on-off modulated for the transmission of digital signals. Recent work on analog modulation of VCSELs indicates that VCSELs are suitable light sources also for the transmission of RF and microwave signals in, e.g., radio-over-fiber (RoF) networks used in antenna remoting in cellular systems for mobile communication. Common to all these applications is that they put certain requirements on the high-speed modulation performance. With higher data rates and modulation frequencies, the requirements become more demanding and it becomes more difficult to identify VCSEL designs that fulfill the requirements.

The VCSEL is characterized by a complex interplay between optical, electrical, and thermal effects. All these effects and their interdependencies are also of importance for the dynamic behavior. Therefore, the design optimization for high-speed modulation is a difficult task. Numerical modeling provides a valuable tool for this purpose, provided that all relevant physical effects are accounted for with sufficient accuracy. In this chapter, we present such a model and apply it to the simulation of the digital and small-signal modulation response of single and multimode VCSELs. As an example of the capabilities of the model, we take a VCSEL of great practical interest, which also provides a challenging modeling task: the surface-modified fundamental-mode-stabilized VCSEL. This is a conventional VCSEL where the surface of the top DBR mirror has a shallow ($\sim$ 60 nm, at wavelength 850 nm) etched mesa, which can prevent the onset of lasing in higher-order modes. More details on the structure of this VCSEL can be found in Sect. 9.4.2. To give an idea of the complexity of accurately modeling the VCSEL, we now give a few examples of phenomena that must be taken into account. A more thorough description of how these phenomena are treated is given in Sect. 9.3.

The VCSEL is a high-speed laser because of its high photon density and small active volume. Oxide-confined VCSELs, with polyimide and proton implantation to improve the electrical properties, have achieved modulation bandwidths exceeding 20 GHz [1]. Of primary interest in fiber-optic communication applications is the amplitude (intensity) modulation of the VCSEL as the modulation current changes. However, also the phase (frequency) of the emitted light changes as the refractive index in the active region changes with carrier density (anti-guiding). At low modulation frequencies (<10 MHz), the variations in temperature also contribute to index fluctuations (thermal lensing). The strong spatial hole burning (SHB) of carriers in the active region, from intense local stimulated recombination, not only affects the static but also the dynamic behavior of VCSELs. The nonuniform intensity distribution implies a nonuniform stimulated recombination rate, which makes the laser respond differently at different locations in the active region. The dynamic effects of SHB have been studied and discussed in detail in [2, 3], examples of which are strong distortion at frequencies below 2 GHz and increased damping of the relaxation oscillation (RO) at low output powers, respectively. Moreover, single and multimode VCSELs differ in their dynamic behavior. The reason for this is the strong modal competition in the multimode laser through the common carrier reservoir. Particularly sensitive to mode competition are modes with a large overlap of their respective mode profiles and where a significant variation of the spatial distribution of carriers takes place during a cycle of the current modulation, i.e., <2 GHz. Therefore, an individual mode in a multimode VCSEL can have a modulation that differs drastically from the current modulation, in contrast to the total output power. Thus, if multimode VCSELs are to be used in short-distance fiber-optic communication, it is important that all modes are transmitted with equal efficiency. Unfortunately, as the higher-order modes have a larger divergence, it is difficult to avoid mode-selective losses induced by the spatial filtering in the coupling between the laser and the fiber. Single-mode VCSELs have the advantage of being less sensitive to spatial filtering, which is a major reason why single mode emission is attractive in VCSELs. Figure 9.1 shows a streak camera image of the modal dynamics of a VCSEL operating at 10 Gbit/s.

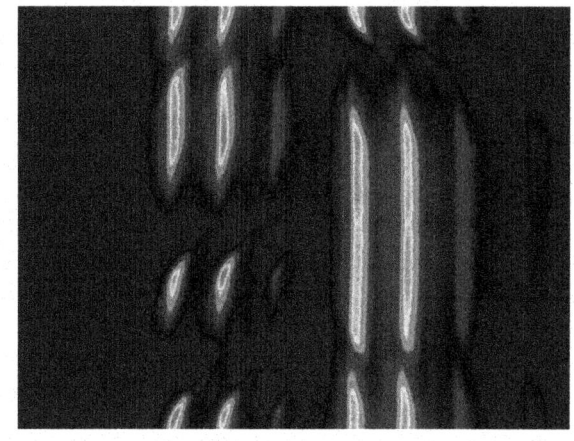

Fig. 9.1. Streak camera image of the modal dynamics of an oxide-confined VCSEL with a 4-μm oxide aperture, operating at 10 Gbit/s. This image shows that the VCSEL is multimode, with two dominant modes during most of the modulation cycle.

9.2 Important Characteristics of VCSELs

Desirable features of VCSELs in more demanding dynamic applications are high speed, low distortion, and a narrow linewidth. These properties are influenced by a number of physical processes that can be categorized as resonance and damping phenomena, nonlinear effects, and noise.

9.2.1 Resonance and Damping: Modulation Bandwidth

There are a number of important factors that influence the speed of a VCSEL. The maximum modulation bandwidth of the VCSEL is limited by either parasitic, thermal, or damping effects. The parasitic effects induce a low-frequency roll-off in the modulation response, which can originate from contact parasitics, distributed parasitics, and carrier transport effects in the device. The latter is normally caused by the finite time for the carriers to be captured by the quantum wells, while being diffusively transported across the separate confinement heterostructure (SCH), see Fig. 9.4. This accumulates carriers in the barrier regions of the SCH, which results in a capacitance. A number of design methods are used to minimize the parasitic effects. These include using materials with a low dielectric constant under the top bondpad and increasing the distance between the top bondpad and the conducting bottom mirror [4]. The thermal effects can limit the maximum obtainable RO frequency at elevated currents from strong self-heating in the device. This is then normally caused by the resistive heating in the p-doped mirror, which, for example, reduces the (differential) gain and increases the leakage of carriers from the quantum wells. From a small-signal analysis of the rate equations describing the interaction between the carriers in the active region and the photons in the cavity, which is outlined, for example, in [5], it turns out that both the square of the resonance frequency and the damping factor of the modulation transfer function are approximately proportional to the photon density or output power. Hence, when the output power is increased to increase the resonance frequency, the modulation response flattens by the increased damping. At some point, the damping becomes large enough such that the modulation response drops below the 3-dB limit already at frequencies below the resonance frequency. The maximum modulation bandwidth of the laser is then limited by damping effects. There are additional factors that significantly contribute to an increased damping in the VCSEL. For example, the gain in the quantum wells is suppressed at high levels of photon densities, which is attributed to spectral hole burning and hot carrier effects. The former occurs at intense stimulated recombination, whereby the carriers in the conduction and valence bands are depleted at the energy levels involved in the stimulated emission. The hot carrier effect is a consequence of the spectral hole burning. Carriers with higher energies try to fill the depleted energy levels, giving up their excess energy primarily as kinetic energy to the entire ensemble of carriers. We can interpret this increased energy of the carriers as an increase of their temperature above that of the rest of the semiconductor. As the carriers redistribute among the energy levels, in response to their higher temperature, they will populate more of the energy states far from the band edges. The energy states close to the band edges, which are the ones involved in the stimulated recombination, are, thus, further depleted and the gain is further suppressed.

The gain suppression effectively increases the damping. Similar effects are caused by SHB, particularly at lower output powers, and by the carrier transport in SCH. In summary, a high-bandwidth VCSEL is obtained from a small and efficient device, providing good thermal management and having low parasitics. It is important to realize, however, that there is a trade-off in the size and achievable output power of the device.

In a typical multimode VCSEL, the modes are essentially independent of each other at modulation frequencies >2 GHz. Therefore, the multimode laser can effectively be viewed as a single mode laser, having only one resonance frequency, where the intensity distribution is the incoherent superposition of the different modes. When comparing single mode with multimode VCSELs with the same oxide aperture diameter, the multimode VCSELs normally reach a higher modulation bandwidth at high output powers, cf. Fig. 4 in [4]. This is due to a lower damping from a lower photon density in the active region. The lower photon density is a result of the coexistence of spatially separated modes, sharing the available carriers.

9.2.2 Nonlinearity

The dynamic performance of the VCSEL also depends on the linearity of the relation between the output power and the applied current modulation. A deviation in the linearity results in distortion of the signal. A measure of the nonlinearities are the harmonic and intermodulation distortion values. The former is the output power at the harmonics of the modulation frequency, whereas the latter is the output power at the sum and difference frequencies when the current modulation contains more than one frequency. Analog (continuous) and digital modulation are the most common forms of modulation. The former is particularly sensitive to the nonlinearities exhibited by the laser. At very low modulation frequencies (<1 MHz), the output power will reflect the steady-state characteristics and its linearity. Primary sources of distortion are then the variation in temperature affecting the gain (and somewhat also the waveguiding properties), gain suppression at high photon densities, and SHB. The degree of carrier depletion resulting from SHB, which is particularly pronounced in single mode VCSELs, increases with photon density due to intense stimulated recombination. As the modal gain must be equal to the modal loss, the carrier density in regions of low intensity increases strongly to compensate for the reduced gain in the regions of high intensity. This results in an increasing number of carriers that are not used for stimulated recombination. Some VCSEL structures also suffer from significant nonlinear leakage currents, whereby a part of the applied current either does not reach the quantum wells or leaks out of the quantum wells. The leakage depends strongly on the barrier height of the quantum wells and it increases with temperature. When increasing the modulation frequency, it was found in [2] that the distortion in the VCSEL is dominated by the dynamic effects of SHB at frequencies <2 GHz. At higher frequencies, the distortion originates predominantly from the nonlinear effects of the RO. This provided that the VCSEL is not influenced by parasitics. The maximum distortion occurs at the RO frequency, and is strongly influenced by its damping. Consequently, as for a VCSEL with a high modulation bandwidth, low distortion is also obtained from a small and efficient device, providing good thermal management and having small parasitics. The effects of SHB can be reduced using, for example, an oxide aperture with a tapered tip [6].

When comparing single mode with multimode VCSELs with the same oxide aperture diameter and modulation, the dynamic effects of SHB are somewhat less pronounced in the multimode VCSELs from the spatially more uniform stimulated recombination rate. However, the nonlinear effects of the RO are enhanced by the lower damping.

9.2.3 Noise

The output from the VCSEL exhibits fluctuations in the intensity and frequency. The former provides a noise floor, which limits the smallest practical signal power of the output, and the latter induces a finite spectral broadening of the laser source. The noise originates mainly from the temporal uncertainty of the processes that supply or consume carriers and photons, such as carrier diffusion and photons escaping the cavity, and processes that create or annihilate carriers and photons, such as stimulated emission and absorption. Under normal working conditions, thermal noise from the mirror resistances can be neglected due to the high resistance in the p-doped mirror. The noise from the current supply also contributes to the noise; however, it has been shown that if a large series resistance is used, the contribution can be made arbitrarily small [7]. At certain working conditions, flicker $(1/f)$ noise that occurs in electronic devices (whose cause is still disputed) can significantly contribute to the noise [8].

A measure of the intensity noise is the relative intensity noise (RIN) and is defined by $\text{RIN} = 2S_P/\langle P\rangle^2$ in the frequency domain, where S_P is the double-sided spectral density of the output power fluctuations and $\langle P\rangle^2$ is the average output power. The frequency noise (FN) is defined by $\text{FN} = 2S_\nu$ in the frequency domain, where S_ν is the double-sided spectral density of the frequency fluctuations. The linewidth of the VCSEL can be estimated by $\Delta\nu = S_\nu(0)$ [8]. The RIN and FN both peak at the RO frequency and strongly depend on the damping. Note that the frequency noise is coupled to the intensity noise because fundamental physical relations dictate that the fluctuations in refractive index and gain of any material are interrelated. Once again, a good VCSEL with a low RIN and FN is obtained from a small and efficient device. In fact, it has been shown that such a VCSEL can exhibit sub-Poissonian noise in the output at low frequencies and high output powers [9]—[11], a phenomenon referred to as amplitude squeezing [7]. This means that the intensity noise level is lower than the lowest possible had the photon emission from the VCSEL been a pure Poisson process, as are the basic noise generating processes in the VCSEL. Further, we can also calculate the intensity spectrum of the laser, i.e., the double-sided spectral density of the complex electric field. The intensity spectrum is typically Lorentzian-shaped with a full-width half-maximum that correspond to the linewidth. It also exhibits satellite peaks that are separated from the main peak by the RO frequency, and multiples thereof if the damping is low, cf. Fig. 9.25. The linewidth of the laser was found to be approximately inversely proportional to the photon density, or output power. This relation is also obtained from a small-signal analysis of the rate equations, added with Langevin noise terms [5]. In reality, a linewidth floor or a rebroadening is often observed at high output powers. Possible causes are noise in the current supply, thermal effects, SHB, and gain suppression.

In a multimode VCSEL, the intensity noise of an individual mode is larger than that of the total output, especially at lower frequencies. This is a phenomenon referred to as mode-partition-noise (MPN), which results from having a common carrier reservoir for all modes. The feedback mechanism in the laser is, then, less effective in suppressing intensity fluctuations of individual modes, while being strong for the summed modes. When comparing single-mode with multimode VCSELs with the same oxide aperture diameter and at the same output power, the peak intensity noise is lower for the single-mode VCSELs because of their higher damping.

9.3 VCSEL Model

It is a challenging task to model the VCSEL. The complex layer structure with its high refractive index contrast makes it difficult to calculate the optical fields. Moreover, the small dimensions of the VCSEL forces the various physical processes to take place within a very limited volume where strongly varying currents, temperatures, and optical intensities can occur. Therefore, the nonlinear processes of current and heat transport, and their interaction with the optical fields, are strongly interdependent. Figure 9.2(a) shows the physical processes accounted for in the presented dynamic model. In the numerical implementation, the interdependencies are accounted for by iteratively finding a self-consistent solution for each point in time and space. For the algorithm to be efficient in terms of computational time, it is important that the computational window is the smallest possible that still enables reliable modeling of the physical process under study. Therefore, different windows are used for modeling the carrier and heat transport, and the optical fields, illustrated in Fig. 9.2(b). As shown, we assume a circular symmetric VCSEL structure with radial coordinate r and longitudinal coordinate z. The heat transport uses a very large computational window — the actual light generation takes place in a very small region in the upper left corner of the figure.

9.3.1 Current Transport

By properly designing the VCSEL, the effects of electrical parasitics can be made small, and they are, therefore, neglected in the model. In different regions of the VCSEL, different carrier transport mechanisms dominate. In the

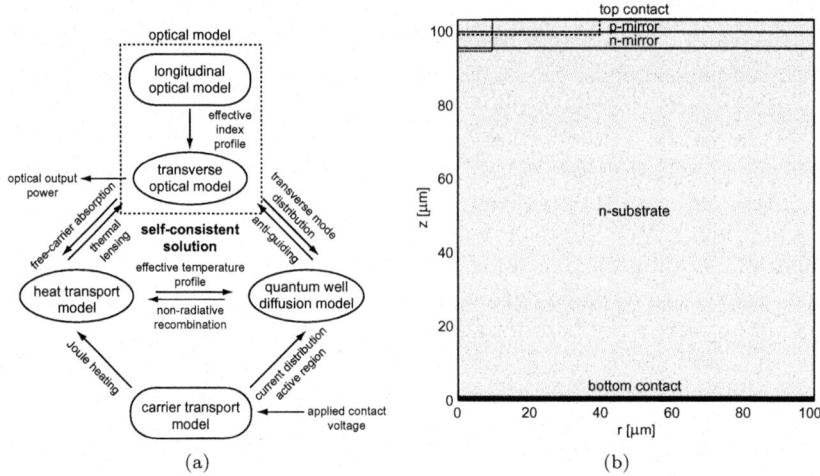

Fig. 9.2. (a) Flow chart of the dynamic model illustrating the interdependencies between the physical processes. (b) Computational windows used for the optical fields (dotted), current transport (dashed), and heat transport (entire region).

mirror regions, the transport is mainly caused by drift, whereas in the active region, the transport is mainly caused by diffusion. Of particular importance is to accurately determine the distribution of the current density injected into the active layer. This is primarily determined by the p-doped mirror region, due to the current crowding oxide aperture and the significantly lower electrical conductivity than in the n-doped region. Therefore, we use a computational window, like the one shown by the dashed line in Fig. 9.2(b), where the n-doped region is simply replaced with a ground plane. The p-doped mirror region is more carefully treated by replacing the complex conducting layer structure with an effective homogeneous medium with a highly anisotropic conductivity, to approximately model the large band discontinuities in the longitudinal direction. We calculate the electrostatic potential V in this region by solving the Laplace equation,

$$\nabla \cdot (\overline{\sigma} \nabla V) = 0, \tag{9.1}$$

where $\overline{\sigma}$ is the electrical conductivity tensor, describing the anisotropy of the material. The finite element method is applied using a Delaunay triangulation algorithm from which a discretization and adherent solution are shown in Figs. 9.3(a) and (b), respectively. The current density distribution $\vec{J}$ is then obtained from the relation $\vec{J} = -\overline{\sigma} \nabla V$, where the diffusion current has been neglected. Strictly, the obtained solution is a stationary solution; however, the small dimensions involved compared with the wavelength corresponding to the current modulation frequencies ($\lambda_{\mathrm{mod}} = c/\nu_{\mathrm{mod}} \sim 3\,\mathrm{cm}$ at $\nu_{\mathrm{mod}} = 10\,\mathrm{GHz}$) and the fast relaxation time for a current flow perturbation justify a quasi-stationary approximation. This implies that the current density distribution in the mirrors does not change its shape with time. This is definitely not the case in the active region where the slower process of diffusion dominates the carrier transport.

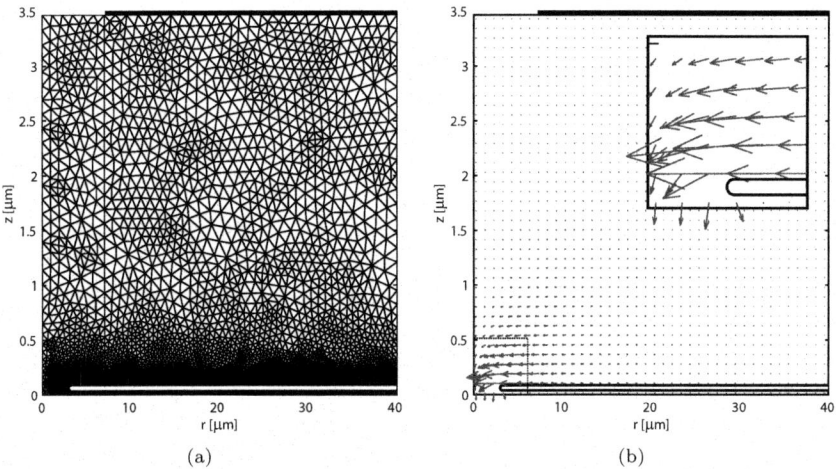

Fig. 9.3. Simulation of the current transport of a VCSEL with a 6-μm oxide aperture, (a) discretization mesh, (b) numerical solution for the current density distribution; the inset shows the vicinity of the oxide aperture in detail.

The carrier distribution in the active region is schematically illustrated in Fig. 9.4 along the longitudinal direction. Due to the capturing mechanism of the quantum wells, the electron and hole concentrations are higher in the quantum wells located closer to the n- and p-doped side, respectively. As a consequence, the gain provided from these quantum wells can be significantly lower and even absorbing if the structure contains more quantum wells than the carrier injection can handle. Normally, one to five quantum wells are used to optimize the performance. We assume that the quantum wells are few enough so that the conditions are roughly equal in all wells, and treat the quantum wells as one well with a thickness equal to the total thickness of the quantum wells. Macroscopically, the active region is assumed to be charge neutral, whereby the total number of electrons and holes are equal. An ambipolar carrier continuity equation can therefore be used to describe the effective carrier transport and recombination in the quantum wells. Further, to include carrier transport effects in the SCH, an additional carrier continuity equation is needed. The coupled rate equations for the effective carrier density in the SCH, N_b, and the quantum wells, N_w, take the form

$$\frac{\partial N_b}{\partial t} = D_b \nabla_\perp^2 N_b - \left[\frac{1}{\tau_b} + \frac{1}{\tau_{bw}}\right] N_b + \frac{d_w}{d_b} \frac{N_w}{\tau_{wb}} + \frac{J}{qd_b} \qquad (9.2)$$

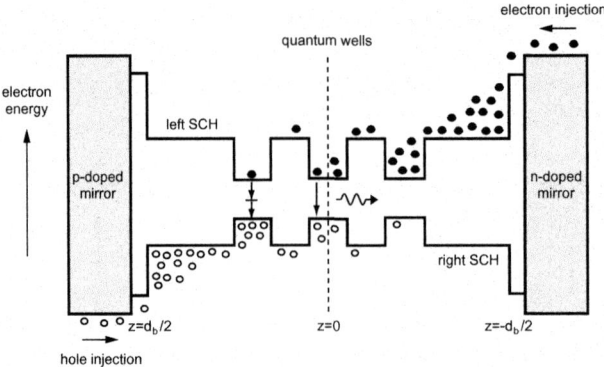

Fig. 9.4. Illustration of the electron and hole distributions in the quantum wells and energy barriers of the active region.

$$\frac{\partial N_w}{\partial t} = D_w \nabla_\perp^2 N_w - \left[\frac{1}{\tau_w} + \frac{1}{\tau_{wb}}\right] N_w + \frac{d_b}{d_w} \frac{N_b}{\tau_{bw}} -$$

$$- BN_w^2 - CN_w^3 - \frac{1}{d_w} \sum_{m,n=0,1}^{M,N} \frac{c_0 n_a \langle u \rangle}{\mathrm{Re}(\langle \epsilon_i \rangle)} g I_{mn} |\psi_{mn}|^2, \qquad (9.3)$$

where the carrier transport caused by drift has been neglected. J is the current density injected into the active region and q is the unit of electrical charge. D_b and D_w are the diffusion coefficients, d_b and d_w are the total thicknesses, and τ_b and τ_w are the nonradiative carrier lifetimes for Shockley–Read–Hall recombination of the barriers and quantum wells, respectively. Due to the small extension of the effective layers, the diffusion in the longitudinal direction has been dropped. τ_{bw} is the average carrier lifetime in the barriers before being captured by the quantum wells. Similarly, τ_{wb} is the average carrier lifetime in the quantum wells before escaping by thermionic emission. B is the spontaneous emission coefficient and C is the Auger recombination coefficient. The last term in (9.3) describes the net stimulated recombination, where a summation is performed over all excited modes and their individual overlap with the gain in the quantum wells [12]. Some device structures suffer from a large carrier leakage, i.e., carriers escaping the SCH, due to a small band gap discontinuity between the SCH and cladding layers. This is particularly the case for short-wavelength VCSELs (650 nm) and should then be considered. This thermionic emission can be modeled by replacing J with $J - J_{leak}$, where J_{leak} is a function that depends exponentially on temperature and carrier density [13]. The coupled rate equations (9.2) and (9.3) are solved by a finite discretization of time and space utilizing the Crank–Nicholson implicit method together with Newtons iterative method of solving a set of nonlinear equations.

A more accurate model of the 3D current transport in the VCSEL structure employs a self-consistent solution of the Poisson equation that relates the electrostatic potential to the charge density, continuity equations for the electrons and holes, and drift-diffusion equations for the electrons and holes [13]. In addition, at the hetero-interfaces, where a discontinuity in the valence or conduction band appears, thermionic emission and tunneling currents contribute to the transport. The detailed calculation of energy bands and Fermi levels in multilayered structures requires extensive computational effort. The simple current transport model that is used in this work reduces this need and can still predict the important effects of current crowding for different device geometries. For a VCSEL with a very low p-doped mirror resistivity, the n-doped mirror will start influencing the current distribution by its finite electrical conductivity. A better approach is then to solve the Laplace equation in both the p- and n-doped regions, and use the voltage drop across the active layer, corresponding to the carrier density, as a boundary condition [13]. This, then, requires a self-consistent solution of the Laplace equation and the rate equations for the carrier density in the active region, which increases the computational time.

9.3.2 Heat Transport

The heat generated in the VCSEL is normally very localized and can have steep gradients in the source distribution. However, the strongly diffusive nature of the heat transport is very forgiving in which the resulting temperature distribution is generally quite smooth. Due to the diffusive length scale of the heat transport, a relative large computational window is required. We, therefore, use a window like the one shown in Fig. 9.2(b), where the mirror regions are replaced with an effective homogeneous medium with a somewhat anisotropic thermal conductivity. The temporal evolution of the temperature distribution T is governed by the heat flux equation,

$$\rho C_p \frac{\partial T}{\partial t} = \nabla \cdot (\overline{\kappa} \nabla T) + Q, \qquad (9.4)$$

where ρ and C_p are the mass density and specific heat per unit mass, respectively. The thermal conductivity tensor $\overline{\kappa}$ describes the anisotropy of the material, and Q is the local heat generation rate. In the model, we include three heat sources. The first, and normally dominant, heat source is the resistive heating from the carrier transport in the p-doped mirror. The heat generated by Joule heating is calculated from $Q_{Joule} = \sigma_r \left[\partial V/\partial r\right]^2 + \sigma_z \left[\partial V/\partial z\right]^2$, where σ_r and σ_z are the electrical conductivities in the radial and longitudinal directions, respectively. The Joule heating becomes increasingly important at higher current levels because it scales as the square of the applied current. This is especially pronounced in VCSELs with small oxide apertures, where temperature increases of >100 K are not unusual. The second heat source is free-carrier absorption of light in the doped mirrors, which is computed from $Q_{fca} = \hbar \omega_0 \sum_{m,n=0,1}^{M,N} R_{abs}^{mn}$, where ω_0 is the nominal angular oscillation frequency and R_{abs}^{mn} is the local free-carrier absorption rate for mode (m,n). The

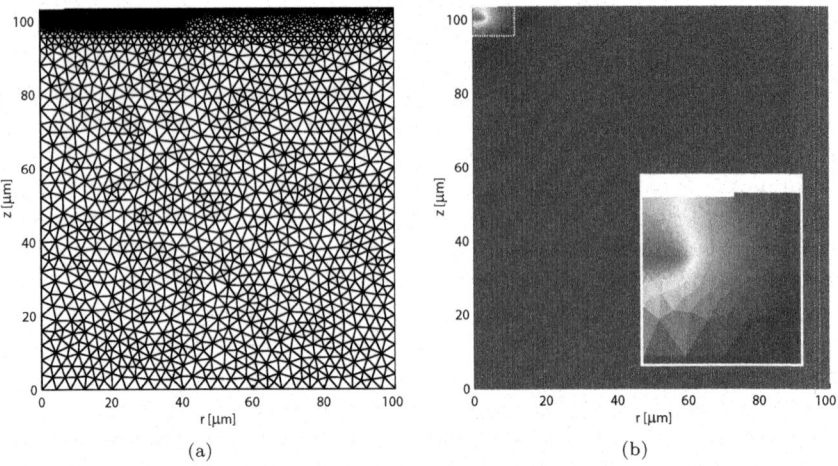

Fig. 9.5. Simulation of the heat transport of a VCSEL with a 6-μm oxide aperture, (a) discretization mesh, (b) numerical solution for the temperature distribution; the inset is a magnification of the dotted area.

latter is determined by the local free-carrier absorption coefficient and the modal intensity. The heating from free-carrier absorption, which is generally much higher in the p-doped mirror, scales approximately linearly with the applied current above threshold. The third heat source is the nonradiative Shockley–Read–Hall recombination in the active region, and it is approximately calculated from $Q_{nr} \approx qE_g [N/\tau_{nr}]$, where E_g is the band gap, N is either N_b or N_w, and τ_{nr} is either τ_b or τ_w. The heating from this nonradiative recombination saturates for current values above the threshold current because the carrier density is then clamped to its threshold value. Its relative importance is therefore mainly at low applied currents. Longer (1300 and 1550 nm) wavelength VCSELs also suffer from strong nonradiative Auger recombination, which contributes to the device heating, and should then be considered. Due to the diffusive time scale of the heat transport, the temporal evolution of the temperature distribution is a relatively slow process. When the VCSEL is current modulated with frequencies greater than 10 MHz, the thermal time constant, which is in the microsecond regime, becomes much longer than the time scale of the source fluctuations, and the temperature distribution will then be based on the time integrated value of the heat sources. Most coding schemes used in digital communication systems, for example, do not allow long sequences of "ones" or "zeros", and the time integrated value of the heat sources will then remain constant for a VCSEL emitting such bit sequences. In this case, the nonfluctuating temperature distribution can be found by solving for the steady-state solution of (9.4), where Q is then the time averaged heat generation rate. The temperature distribution is solved by again applying the finite element method using a Delaunay triangulation algorithm, from which a discretization and adherent solution are shown in Figs. 9.5(a) and (b), respectively. As a final note, longer wavelength VCSELs have shown a considerably stronger temperature dependence, therefore a model for these VCSELs should include temperature dependence in additional processes such as Auger recombination.

9.3.3 Optical Fields

The optical fields in the VCSEL are governed by the vectorial wave equation derived from Maxwells equations. To an excellent approximation, the propagation is paraxial, with a direction perpendicular to the epitaxial layers and a field polarization parallel to the plane of the layers. Under these circumstances, the scalar wave equation can be used to describe the single electrical field component, F,

$$\nabla^2 F - \frac{\epsilon}{c_0^2}\frac{\partial^2 F}{\partial t^2} = 0, \tag{9.5}$$

where ϵ is the relative permittivity. Due to the strong intensity variation in the DBR mirrors even within a single layer, the detailed layer structure must be taken into account in the calculations. A huge number of discretization points is therefore necessary had the finite difference or finite element method been applied. In order to reduce the computational time and effort, especially for vectorial solutions with geometries lacking in symmetry, several methods have been developed. Examples are the effective index method [14] and the weighted index method [15], which are scalar methods and decouple a 2D calculation for a circular symmetric case into two 1D calculations. They rely on a well behaved geometry and disregard diffraction losses. Beam propagation methods have also been applied [16, 17]. A very successful vectorial method, which can handle noncircular-symmetric geometries, utilizes eigenmode expansions of the optical field [18, 19]. In this work, we apply an effective index technique based on [20], which enables us to calculate the evolution of the intensities and spatial distributions of the excited modes under dynamic operation with reasonable computational effort. The VCSEL is divided into concentric cylindrical regions, where the structure only depends on z, and not r, in every region, see Fig. 9.6. Within each region, the electric field, $F(r, z, \phi, t)$, is separated into a z-dependent function $E_i(z)$ and a transverse and nearly harmonic time-dependent function, $E_{mn}(r, \phi, t)$,

$$F(r, z, \phi, t) \approx E_i(z) e^{-i\omega_0 t} \sum_{m,n=0,1}^{M,N} E_{mn}(r, \phi, t), \tag{9.6}$$

where the subscript i denotes a certain transverse region. The summation is taken over the different transverse modes, where the LP$_{mn}$ notation is used. The individual modes are described by,

$$E_{mn}(r, \phi, t) = A_{mn}\psi_{mn}(r, t)e^{\pm im\phi}e^{-i\Delta w_{mn}t}, \tag{9.7}$$

where A_{mn} is a complex constant and ψ_{mn} is the radial distribution. The deviation from the nominal angular oscillation frequency ω_0 is given by the real part of $\Delta\omega_{mn}$ and the imaginary part is the exponential growth or decay rate. If (9.6) and (9.7) are inserted into the wave equation, one obtains one equation in the longitudinal coordinate z, and one in the radial coordinate r. Starting with the longitudinal expression, one invokes the effective index approximation, assuming that the individual functions $E_i(z)$ are all approximately the same, yielding the one-dimensional wave equation

$$\frac{d^2 E_i}{dz^2} + k_0^2(1 - \xi_i)\epsilon_i(z)E_i = 0, \qquad (9.8)$$

where the real part of the eigenvalue ξ_i is related to the effective index and the imaginary part is related to the cavity loss. The relative permittivity is divided into a structural and a nonstructural component, $\epsilon(r,z) \equiv \epsilon_i(z) + \epsilon_g(r)$, where the latter takes into account radially dependent gain in the quantum wells, thermal lensing, and carrier-induced antiguiding effects. Equation (9.8) is solved once for each transverse region, to obtain the eigenvalue ξ_i and the longitudinal function E_i. To find the solution, we apply the finite difference method with a spatial resolution of 0.5 nm. Figure 9.7(a) shows a solution of the longitudinal intensity distribution.

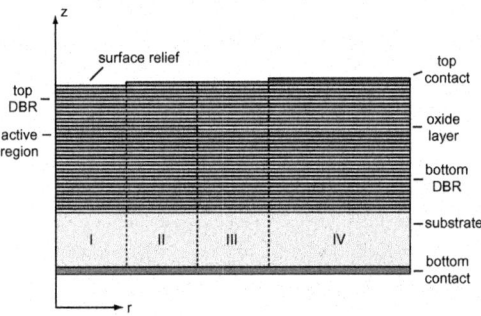

Fig. 9.6. Schematic cross-sectional view of a VCSEL geometry. In this example, a circular surface relief is etched in the center of the device. The different transverse regions, where the structure only depends on z, are indicated by Roman numbers.

From the radially dependent expression, a wave equation for the radial function of the electrical field, ψ_{mn}, can be derived under the slowly-varying envelope approximation,

$$\Delta\omega_{mn}\psi_{mn} = \frac{-c_0}{2k_0\langle\epsilon_i\rangle}\left[\frac{1}{r}\frac{\partial}{\partial r}r\frac{\partial}{\partial r} - \frac{m^2}{r^2} + k_0^2\Delta\epsilon_{eff}\right]\psi_{mn}, \qquad (9.9)$$

where $\langle\epsilon_i\rangle$ symbolizes the weighted value of the relative permittivity in the longitudinal direction, i.e., $\langle\epsilon_i\rangle \equiv \int_{-\infty}^{\infty} E_i^* \epsilon_i(z) E_i\, dz$ assuming $\int_{-\infty}^{\infty} E_i^* E_i\, dz =$

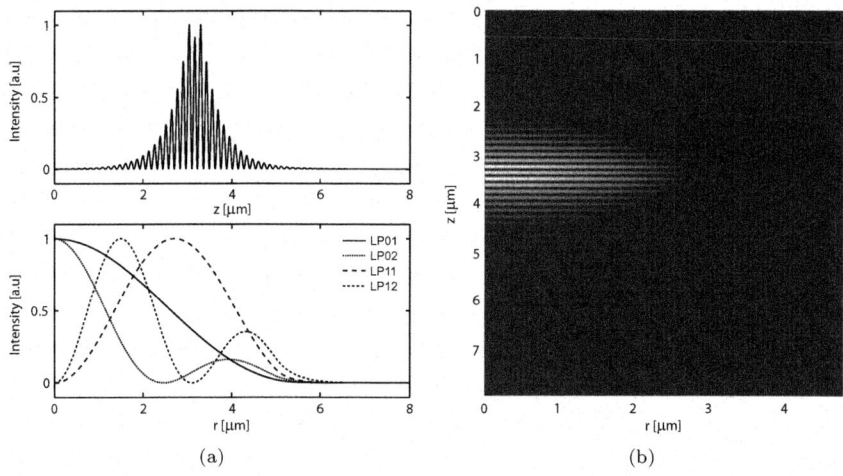

Fig. 9.7. Simulation of the optical fields, (a) top: longitudinal intensity distribution, bottom: transverse intensity distribution of four different modes for a VCSEL with a 10-μm oxide aperture, and (b) total intensity distribution of the LP$_{01}$-mode for a VCSEL with a 6-μm oxide aperture.

1. In the derivation, the time derivative of ψ_{mn} was neglected, which implies that we assume the electric field to settle momentarily from a perturbation in the refractive index. This is generally a good approximation as long as the refractive index fluctuations occur on a time scale longer than the photon lifetime. The variation in the effective dielectric constant was defined as $\Delta\epsilon_{\text{eff}} \equiv \langle\epsilon_g\rangle + \xi_i\langle\epsilon_i\rangle$. Including the changes in refractive index with temperature and the presence of carriers, we obtain $\langle\epsilon_g\rangle = 2\sqrt{\text{Re}(\langle\epsilon_i\rangle)}\, dn/dT\,(\langle T\rangle(r) - T_0) - gn_a\langle u\rangle(\beta_c + i)/k_0$, where dn/dT is the derivative of the refractive index with respect to temperature, T_0 is the ambient temperature, g and n_a are the material gain coefficient and refractive index of the quantum wells, u is a binary function with a value of unity in the quantum wells and zero elsewhere, and β_c is the anti-guiding factor. Equation (9.9) is solved at each point in time to obtain a new eigenvalue $\Delta\omega_{mn}$ and radial function ψ_{mn}. We again apply the finite difference method, but, in this, case with a spatial resolution of $\sim 0.1\,\mu$m. Figure 9.7(a) shows solutions of the transverse intensity distribution of the LP$_{01}$, LP$_{02}$, LP$_{11}$, and LP$_{12}$-modes, and Fig. 9.7(b) shows a solution of the total intensity distribution of the LP$_{01}$-mode. Typical evolutions of the oscillation wavelength and transverse intensity distribution of the LP$_{01}$ and LP$_{11}$-modes under digital modulation are displayed in Figs. 9.8(b)-(d). The time dependency is only caused by the anti-guiding effect, as the temperature distribution does not change at this high modulation frequency.

The dynamic evolution of the modal intensities, which we define by $I_{mn}(t) \equiv \int_0^{2\pi}\int_0^\infty |E_{mn}(r,\phi,t)|^2\,rdrd\phi$, is calculated by the following rate

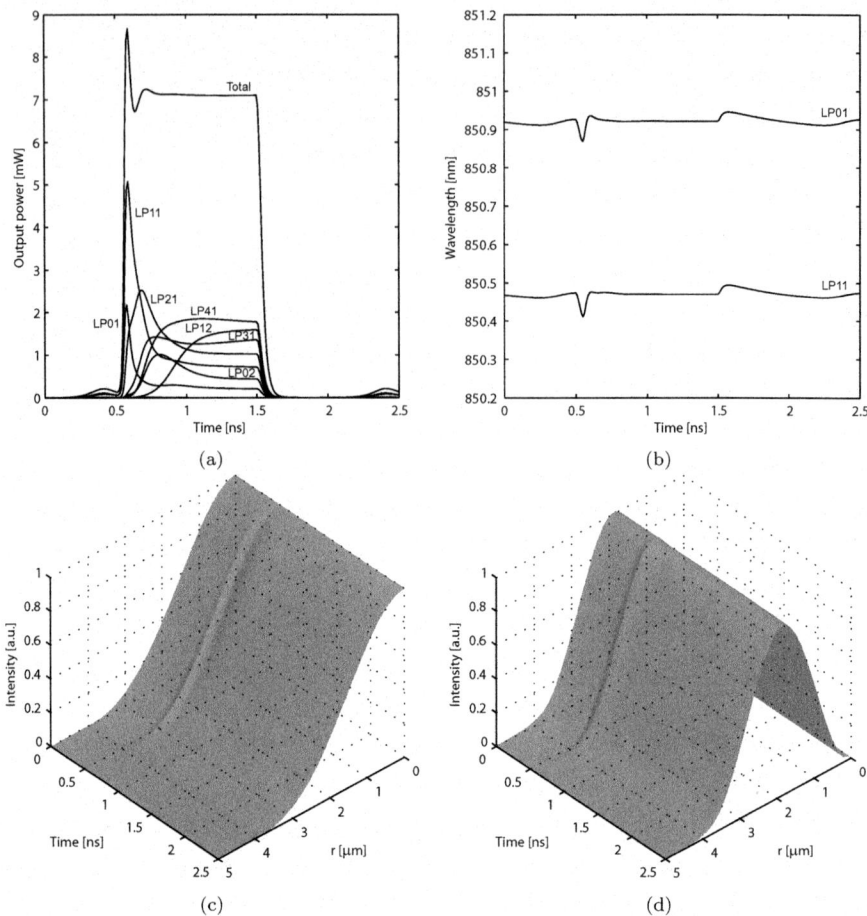

Fig. 9.8. Simulated dynamic behavior at 1 Gbit/s for a VCSEL with a 7-μm oxide aperture. (a) Total and individual mode output power, (b) oscillation wavelength of the LP_{01} and LP_{11}-modes, (c) transverse intensity distribution of the LP_{01}-mode, and (d) transverse intensity distribution of the LP_{11}-mode.

equation

$$\frac{dI_{mn}}{dt} = 2\text{Im}(\Delta\omega_{mn})I_{mn} + R_{sp}^{mn}. \tag{9.10}$$

The first term corresponds to the net effect of modal gain and modal losses, whereas the second term denotes the rate of spontaneously emitted photons that are coupled into the optical mode. A detailed expression for R_{sp}^{mn} can be found in [3]. The optical power in an individual mode has a temporal evolution during a modulation cycle that can strongly deviate from that of the total power, an example of which is given in Fig. 9.8(a) showing strong modal competition under digital modulation. As previously mentioned, the effective index method does not account for diffraction losses. The model can, therefore, not accurately handle VCSELs with very small oxide apertures (<3 μm), where the scattering from the oxide layer becomes substantial.

9.3.4 Material Gain

The gain in the quantum wells gives rise to the interaction between the carriers and the optical fields, and depends strongly on the material composition. Rigorous calculations of the material gain involves quantum mechanical considerations for the carriers in the form of transition and occupancy probabilities and density of available states. In our model, we use an empirical formula for the material gain coefficient, g_0, that accounts for the carrier density, wavelength, and temperature dependency

$$g_0(N_w, \lambda, T) = a_N \ln\left(\frac{N_w}{N_0}\right) - G_0\left(\lambda - \lambda_p\right)^2, \qquad (9.11)$$

where the temperature dependency of the parameters a_N, N_0, and λ_p are formulated as $a_N(T) = a_0 + a_1 T + a_2 T^2$, $N_0(T) = b_0 + b_1 T$, and $\lambda_p(T) = \lambda_0 + \lambda_1 T$, respectively [12]. The empirical gain model is approximately valid as long as the optical wavelength, λ, is within ±10 nm from the peak gain wavelength, λ_p. A phenomenological gain suppression factor, ε, is introduced to include the effects of hot carriers and spectral hole burning, $g = g_0/\left[1 + \varepsilon I_{den}\right]$, where I_{den} is the local photon density in the quantum wells.

9.3.5 Noise

No matter how comprehensive the models we use to simulate the behavior of a VCSEL, we will never be able to make an entirely correct prediction. This is because a real VCSEL always exhibits random fluctuations in all quantities that influence its performance. As a result, the laser output is also unpredictable to some degree. This shows as fluctuations, noise, in the intensity as well as the phase (or, equivalently, frequency of the lightwave emitted from the laser. Although the fluctuations are random, their statistical effects are quite predictable. The purpose of our noise model is, therefore, to obtain a truthful simulation of the laser output from a real VCSEL as a function of time, including fluctuations. Then statistical figures of merit are calculated based on time sequences long enough for a very large number of fluctuations to occur. These figures of merit can then be compared with measurements on actual devices or be compared with simulations on other VCSEL designs, trying to optimize the VCSEL structure. Common figures of merit are bit-error rate, relative intensity noise, and laser linewidth. Basically, the reason for the noise is that the processes occurring in the VCSEL are random processes, where the individual events occur with some probability. The most obvious random process is perhaps the spontaneous emission of photons. In fact, it is common to explain the noise in the output from lasers as a result of the uncertainty of the phase of the spontaneously emitted photon. However, it is also possible to interpret noise from the stimulated emission, which is also a random process with a temporal uncertainty. The two different viewpoints are not quite so different as one might intuitively feel, if one recognizes that, in fact, the spontaneous emission into a laser mode is closely related to the stimulated emission into the same mode [8]. Whichever viewpoint is taken, it should be remembered that it is a simplification of a more correct quantum mechanical approach [21]. Therefore, in both cases, it is necessary to make a few not-so-obvious assumptions to arrive at the correct quantitative results in accordance with the rigorous theory. In our spatially dependent noise model, we take the second approach, formally neglecting the contribution from spontaneous emission to the noise. Further, we view the noisy quantities — the photons in each optical mode and the carriers in the SCH and quantum wells — as particles in suitably defined reservoirs. For the photons in each mode, the whole cavity is one reservoir. For the carriers, the SCH and the quantum wells are divided into a large number of concentric ring-shaped reservoirs. Figure 9.9(a) shows a segment of the ith ring reservoir of the quantum wells, where the processes that lead to particle increase or decrease are indicated, and Fig. 9.9(b) shows an example of the fluctuating carriers in the quantum wells. In each reservoir, the rate equation for the quantity under consideration is expressed as usual. The only difference is that an extra term, a so-called Langevin term [5] denoted by $F_N(t)$, is formally added to the rates to emphasize that the momentary rates exhibit fluctuations

$$\frac{dN}{dt} = R_{incr} - R_{decr} + F_N(t) \,, \qquad (9.12)$$

where N is the particle number in the reservoir and R_{incr} and R_{decr} are the rates of the processes leading to particle number increase and decrease, respectively, in the reservoir. Transport of particles into the reservoir and particle generation within the reservoir contribute to R_{incr}, whereas R_{decr} is determined by annihilation and particle flux out of the reservoir. Evidently, the reservoirs are connected through carrier transport between neighboring reservoirs, but also through absorption and emission processes that link the carrier reservoirs to the photon reservoirs. This implies that a detailed balance must exist between any two connected reservoirs so that the same process is consistently described from either reservoir. Therefore, the noise terms for connected reservoirs are correlated. In the numerical simulation, the rate equation is converted into a time-stepping equation for the change of particle number ΔN during a timestep Δt. This is very simple, of course, the only problem is the conversion of $F_N(t)$ into its time-stepping counterpart. If all processes are random, so that they obey Poisson statistics, it can be shown that the time-stepping version of (9.12) becomes

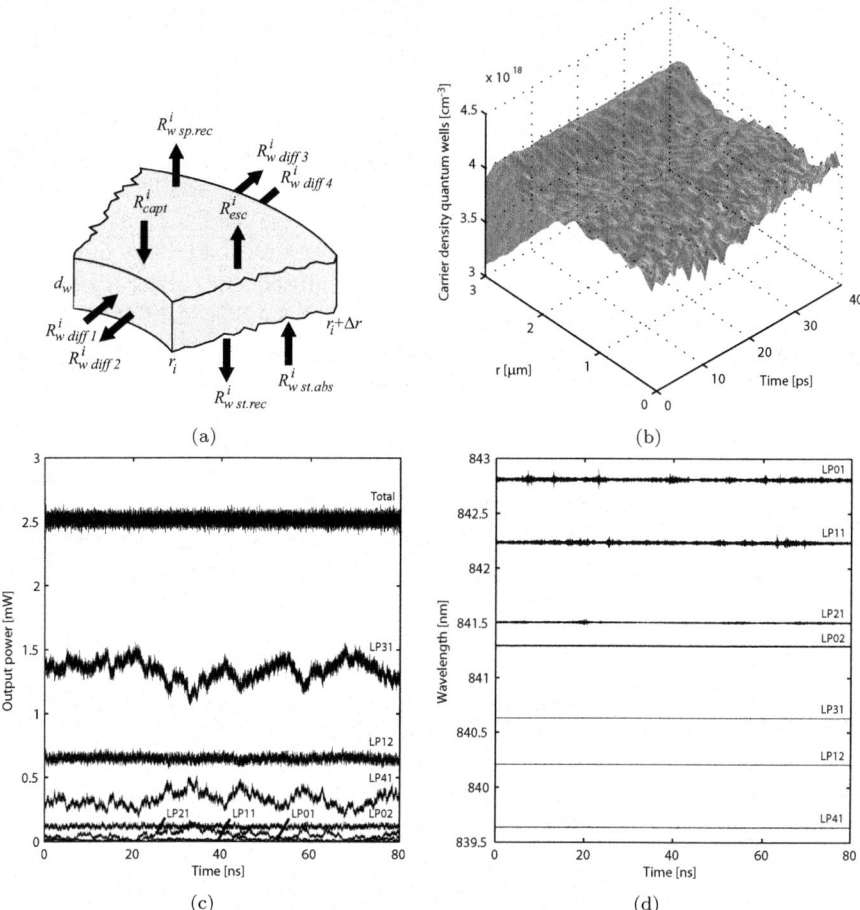

Fig. 9.9. (a) Schematic illustration of a segment of the i:th ring reservoir used for the modeling of the quantum wells. The processes that lead to carrier increase and decrease are indicated by the incoming and outgoing arrows, respectively. (b)-(d) Simulated time-fluctuating carrier density in the quantum wells, total and individual mode output power, and oscillation wavelengths, respectively, for a VCSEL with a 6-μm oxide aperture.

$$\Delta N = (R_{incr} - R_{decr})\, \Delta t + \sqrt{(R_{incr} + R_{decr})\, \Delta t}\, \xi \ , \quad (9.13)$$

where ξ is a random number with a normal distribution, zero mean, and a variance equal to one. Thus, ξ can be positive or negative and the actual value of ΔN may be slightly smaller or larger than without the noise contribution. It is important to note that, in the noise term, the total rate, not the net rate, appears. This means that from a noise point of view, rates do not cancel. Therefore, as an example, you can have noise from diffusion of carriers even if there is no concentration gradient for the carriers. This noise is caused by the

Brownian motion of carriers into or out of the reservoir in the radial direction. In addition to diffusion, the noisy processes that influence the carrier number in the SCH and quantum wells are the capture and escape from the quantum wells, recombination of carriers, and, for the quantum wells, the creation of a carrier by absorption of a photon. Conventionally, VCSEL noise models neglect the injection of carriers into the SCH by the externally connected current source. The reason is that there is a negative feedback mechanism built into the current supplying circuit, such that an unusually large injection of carriers temporarily lowers the probability for subsequent carrier injection. Therefore, the number of carriers injected into the entire SCH exhibits smaller fluctuations than expected from Poisson statistics. In fact, the fluctuations can be neglected provided the external current supplying circuit is arranged properly [7]. However, since our noise model is spatially dependent, the SCH is divided into a large number of concentric reservoirs. The distribution of injected carriers among the reservoirs is quite random and thus the number of carriers injected into any one of the reservoirs is a noisy quantity. It turns out that this noise can be calculated exactly like the fluctuations in the other noisy quantities in the VCSEL, only, a small correction is made in a post-processing step so that the sum of carriers injected into all reservoirs during δt is equal to its deterministic value, i.e., noise-free. Finally, the noisy processes that influence the photon numbers are, aside from the stimulated processes, the loss of photons by internal absorption and transmission through the bottom and top mirrors.

Including noise in the model is, thus, rather straightforward, identifying the relevant physical processes and for each process use appropriate expressions for the total rates. This procedure is outlined in detail in [22]. Figures 9.9(c) and (d) show an example of the noisy output from a VCSEL. Note, in Fig. 9.9(c), the anti-correlated power fluctuations of individual modes (MPN), which are particularly pronounced for modes with a high degree of spatial overlap.

9.3.6 Iterative Procedures

The nonlinear equations describing the optical, electrical, and thermal behavior of the VCSEL are strongly interdependent. A self-consistent solution is obtained by iteration until a common solution is found for each point in time. The time constants involved in the fluctuations of carriers and photons are orders of magnitude shorter than that of the temperature. To avoid numerical redundancies, we therefore use longer steps in the time discretization for the temperature. Figure 9.10 shows a flowchart of the iterative procedure for the carriers and photons, referred to as the inner loop. To incorporate temperature, we use two different approaches depending on the modulation. The two cases are illustrated in Fig. 9.11(a) and (b) and are referred to as outer loops. The procedure in Fig. 9.11(a) is used when the period of the modulation current is very short, so that the temperature distribution

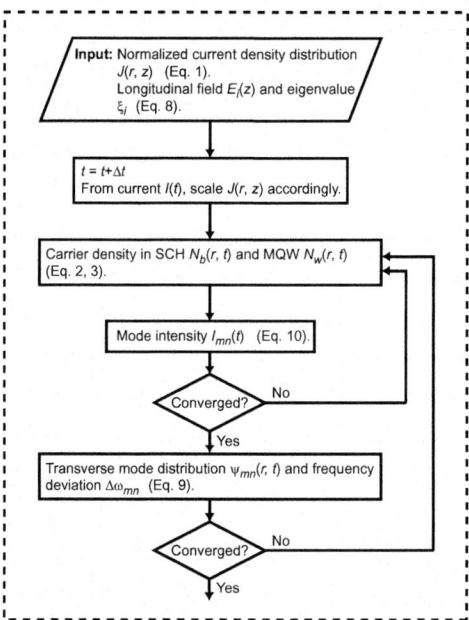

Fig. 9.10. Flowchart illustrating the iteration procedure for the inner loop, which iterates the carriers and photons and is used in the outer loops shown in Fig. 9.11.

becomes time-independent. Based on an initial guess of the temperature distribution, the behavior of the laser is traced during one period of the current modulation. In this way, the time average of the heat generated and lost by the different processes is found, and, consequently, a better approximation for the temperature distribution is calculated. This procedure is repeated for another period of the current modulation until convergence. We may remark that this method is also used to obtain the steady-state characteristics, such as the light-current curve, of the VCSEL, in which case, the modulation of the current should evidently be zero. The strategy in Fig. 9.11(a) fails if, for example, one wants to simulate intermodulation distortion, where the interference between the two modulation currents leads to the total modulation having a slowly varying envelope (typically with frequency $\leq 10\,\mathrm{MHz}$). As indicated in Fig. 9.11(b), the temperature distribution is assumed to be constant until a certain time has elapsed, of the order of the thermal time constant. Then, a correction of the temperature distribution might be necessary and is calculated from the time average of the heat sources since the preceding correction. Strictly, thus, the temperature calculation is not self-consistent in this case; but it is still reliable if the temperature update occurs frequently enough.

In order to reduce computational time, certain optimizations have been carried out. One example is the two-dimensional calculation of the local

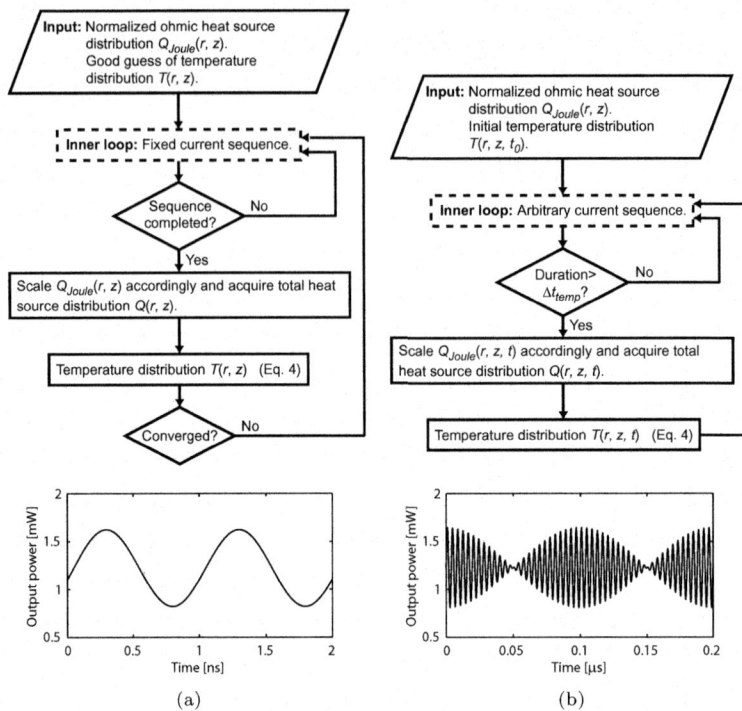

Fig. 9.11. Flowcharts illustrating the iteration procedures for the outer loop, which includes temperature. Algorithm (a) is applied when the modulation has a short period, whereas (b) is used for modulation with a slowly varying envelope. Examples of calculated output power for the two types of modulation are shown, note the different time scales.

electrostatic potential, which governs the current density distribution in the Bragg mirrors. As the Laplace equation is a linear equation, it can be normalized and solved initially. For a particular current, the current density distribution and, consequently, the Joule heating distribution are simply scaled accordingly, as pointed out in Figs. 9.10 and 9.11. Another example is the time-consuming calculation of the temperature distribution, which is done only in the outermost loop. To further reduce computational time, the one-dimensional calculations of the transverse mode distributions are not included in the innermost loop of Fig. 9.10.

9.4 Simulation Example: Fundamental-Mode-Stabilized VCSELs

The VCSEL has established itself as a light source of choice for cost-sensitive fiber-optic communication links and networks. Although most links and networks today use multimode VCSELs, more demanding communication applications will require single-mode VCSELs with good dynamic performance. Examples include high-speed links where chromatic dispersion in the fiber is a limitation, long-wavelength VCSEL-based links where single-mode fibers are used, and free-space optical interconnects where the beam quality is of importance for efficiency and cross-talk. Many of these future "single-mode" applications for VCSELs will also necessitate high output power. This has been a challenge because the VCSEL is by nature multimode, because of its relatively large transverse dimensions.

Considerable effort has been invested to develop VCSELs with high single-mode output powers. With "single-mode" operation, we implicitly assume "fundamental-mode" operation, i.e., lasing in the lowest-order mode. The fundamental-mode is preferred due to its low beam divergence. Several techniques have been developed with the objective to change the transverse guiding or introduce mode-selective losses or gain. To give a flair of the multitude of proposed and demonstrated techniques for increasing the VCSEL tendency to lase in the fundamental-mode only, Fig. 9.12 lists several designs for single-mode VCSELs. The most common technique for achieving single-mode emission is simply to decrease the diameter of the index-guiding aperture of a standard oxide-confined VCSEL, Fig. 9.12(a), until higher-order modes are no longer supported by the waveguide or prevented from lasing due to high diffraction losses. A single-mode power of 4.8 mW has been demonstrated using an oxide aperture of diameter 3.5 μm [23]. This method limits the output power due to a small current confining region, which increases the differential resistance and, thereby, the self-heating. The latter also results in a poor reliability. A further problem is the difficulty in reproducing a small oxide aperture, and thereby reproducing the performance.

9.4.1 Surface Relief Technique

In this chapter, we will instead describe a different method, the surface relief technique illustrated in Fig. 9.12(d), and use our simulation method to determine its performance. The surface relief technique represents an almost negligible increase of fabrication complexity, compared with a conventional multimode VCSEL, yet it has produced very high levels of single-mode power; 6.5 mW was reported in [24]. The idea is to etch a shallow surface relief into the top mirror, whereby lateral differences in the mirror loss can be achieved without disturbing the waveguide properties of the VCSEL. The modal loss will then be mode-selective and dependent on the design of the surface relief. In order to favor the fundamental mode, the surface relief technique can be applied in two different ways [25]. Both methods are used to provide higher mirror loss in the peripheral region to discriminate the higher-order modes. The first, "conventional", way is to etch a surface relief in the form of a depressed ring, in the top layer of a conventional VCSEL structure. The second,

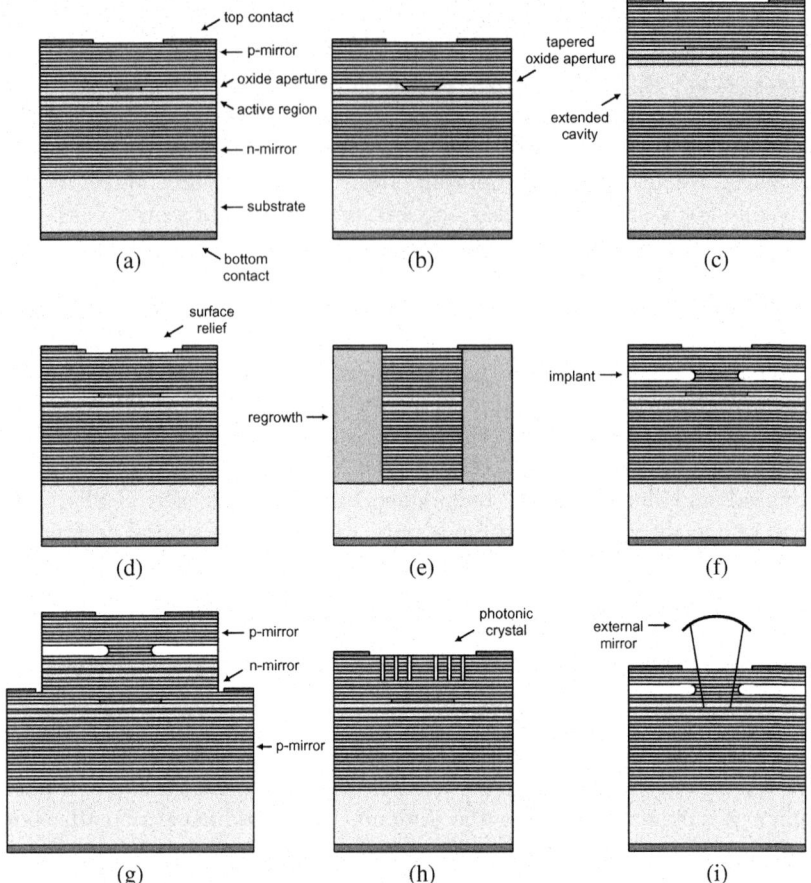

Fig. 9.12. Schematic illustration of different single-mode VCSEL designs; (a) small oxide aperture, (b) tapered oxide aperture, (c) extended cavity, (d) surface relief, (e) anti-guiding regrowth, (f) gain aperture, (g) coupled resonator, (h) photonic crystal, (i) external mirror.

"inverted" way is to add an extra $\lambda/4$-thick layer on top of a conventional VCSEL structure during epitaxial growth. This significantly increases the mirror loss by the anti-phase reflections from this layer with respect to the reflections further down in the mirror stack. A surface relief, in the form of a depressed circular disk, is then etched in the center of the device to remove the topmost layer and, thereby, locally restore the high reflectivity in the central region. By choosing an appropriate etch depth for the surface relief, a relatively high mirror loss contrast between the etched and unetched areas can be accomplished, with minor influence on the waveguide properties. As the DBR is a periodic structure with alternating layers of low- and high-index material, the mirror loss as a function of etch depth is also periodic. In the "conventional" case, the mirror loss has maxima at etch depths of odd numbers of quarter wavelengths, whereas the "inverted" case has maxima at zero and even numbers of quarter wavelengths. The maxima correspond to anti-phase reflections from the semiconductor/air interface with respect to the reflections further down in the mirror stack. Figure 9.13(a) plots the calculated mirror loss and resonance wavelength as a function of etch depth for a VCSEL structure, in which the "inverted" method is applied. The inset demonstrates the effect of the anti-phase reflections from the extra $\lambda/4$-thick layer on the longitudinal standing wave pattern in the VCSEL. The "inverted" method has a fabricational advantage. It utilizes the high thickness precision in the epitaxial growth, to reach a narrow local maximum in the mirror loss. This will then relax the required etch depth precision because a local minimum in the mirror loss is much broader, see Fig. 9.13(a). With the effective index method, the change in resonance wavelength can be translated to a change in effective index [14]. It turns out that the change in effective index, for these shallow etch depths, is order(s) of magnitude lower than 10^{-2}, which means that the index guiding properties of the structure are unchanged by the surface relief.

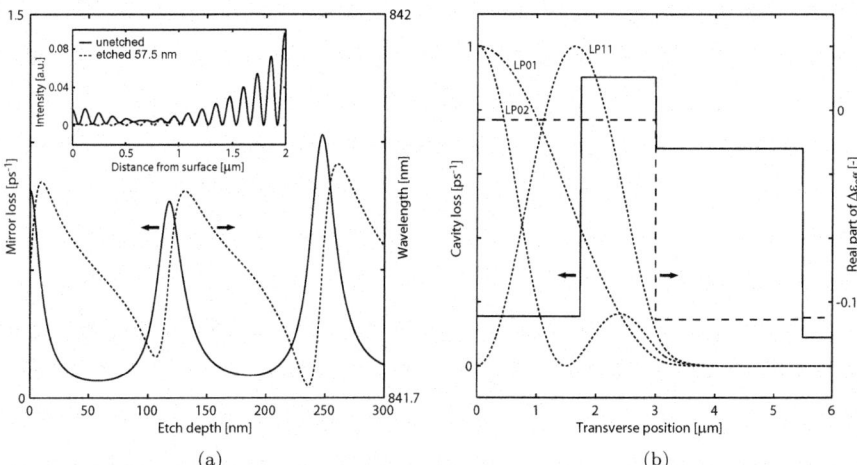

Fig. 9.13. (a) Mirror loss and wavelength as functions of etch depth. The inset shows the longitudinal standing wave pattern in the proximity of the surface. (b) Cavity loss and effective index distribution for a 6-μm oxide aperture and 3.5-μm surface relief diameter. The intensity distribution of the LP_{01}, LP_{11}, and LP_{02} modes are also displayed. For a comparison with a more commonly used unit, a loss rate of $1\,\mathrm{ps}^{-1}$ corresponds to a distributed loss of $\sim 110\,\mathrm{cm}^{-1}$.

As mentioned, the "inverted" surface relief is used to introduce lower mirror loss in the central region. Figure 9.13(b) shows the transverse cavity loss distribution for four different transverse regions, for a VCSEL with a surface relief having a depth of 57.5 nm and a diameter of 3.5 μm. The oxide aperture, diameter 6 μm, induces a guiding effective index distribution, which is also included in the figure together with the transverse distribution of the LP_{01}, LP_{11}, and LP_{02} modes. All the modes experience a higher loss in the periphery. However, the higher-order modes are discriminated because of their larger overlap with the high-loss region. The large modal losses will prevent the higher-order modes from lasing or delay their onset to larger drive currents, which will allow for higher fundamental-mode output powers. This is illustrated in Fig. 9.14, where the light-current characteristics for two VCSELs with surface relief diameters of 4 μm and 11 μm, respectively, are shown. The oxide aperture diameter is 6 μm. The former case corresponds to a fundamental-mode-stabilized VCSEL, and the latter case, with an effectively infinite surface relief diameter, corresponds to a conventional multimode VCSEL, where the modal loss for the different modes is lower and very similar to each other. As can be seen, the fundamental-mode-stabilized VCSEL has a higher threshold current and slope efficiency due to its higher modal loss.

When designing a VCSEL with a fundamental-mode stabilizing surface relief, there exists an optimal combination of relief depth, relief diameter, and oxide aperture diameter, to produce the highest single-mode output power [26]. Too shallow a relief will provide insufficient difference in the modal loss, and too deep a relief will obstruct the current injection and alter the waveguide properties. Moreover, a smaller relief diameter will postpone the onset of higher-order modes as they now have a large overlap with the unetched high-loss region. However, this will also increase the threshold current of the LP_{01}-mode and thereby reduce the maximum output power, which is often limited by the thermal rollover. For larger relief diameters, the higher-order modes will reach threshold at lower currents, and thus, limit the maximum single mode output power. Finally, a larger oxide aperture will, in general, increase the difference in modal loss by the decreased overlap between the mode distributions. On the other hand, the larger current-confining aperture leads to a more nonuniform current injection, which favors the onset of higher-order modes. Other effects that contribute to a decreased mode discrimination are SHB and thermal lensing, which become increasingly important as the drive current is increased.

The surface relief technique offers a simple solution to produce single-mode emission. It only involves a slight modification to standard VCSEL processing. The successful method is associated with a low differential resistance and self-heating by allowing a large current aperture. A single-mode output power of 5.7 mW at wavelength 980 nm has been achieved using the "conventional" surface relief technique [27] and a single-mode output power of 6.5 mW at wavelength 850 nm has been achieved using the "inverted" surface relief technique [24]. In the latter case, the surface relief diameter was 3 μm and the oxide aperture diameter was 6 μm, which resulted in a differ-

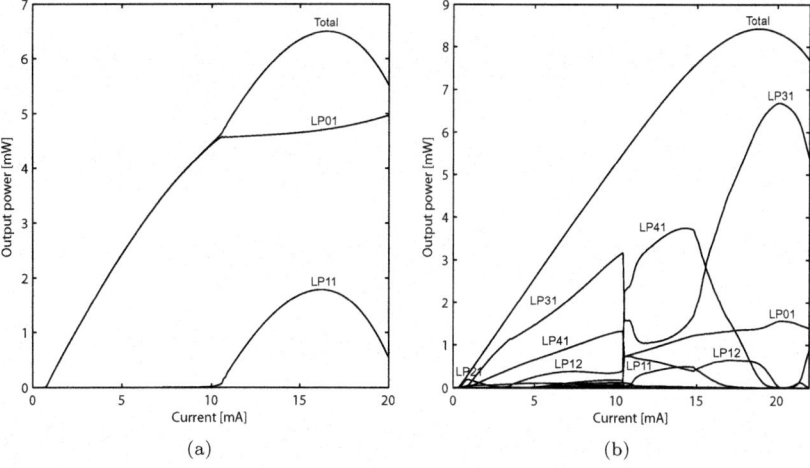

Fig. 9.14. Output power as a function of current for an oxide aperture and surface relief diameter of (a) 6 and 4 μm, (b) 6 and 11 μm, respectively. The contributions from individual modes are also indicated.

ential resistance of 64 Ω. The disadvantage with the method is the critical alignment between the surface relief and the oxide aperture, which normally requires electron beam lithography. However, a self-alignment technique has been developed to circumvent this problem [28]. Further, due to the increased out-coupling, the photon density in the cavity is lowered, which somewhat reduces the speed of the device. The single transverse mode causes severe SHB, which influences, for example, the linearity of the device. The possible degradation of the dynamic performance compared with conventional multimode VCSELs will be evaluated in Sect. 9.4.3. It should finally be noted that the surface relief technique offers the possibility of combing fundamental-mode stabilization with polarization stabilization. By using a nonsymmetric surface relief or incorporating a grating in the surface relief, the polarization can also be controlled [29].

9.4.2 Device Structure

A schematic cross-sectional view of the VCSEL geometry that is to be simulated is shown in Fig. 9.15, and the detailed epitaxial structure, which is designed for 850-nm emission, is listed in Table 9.1. To enable high output powers from VCSELs, it is important to optimize the epitaxial structure. The top mirror should have a relatively low reflectivity to increase the out-coupling. However, too high an out-coupling will result in large threshold and drive currents and, therefore, excessive device heating, which limits the output power. The SCH contains three GaAs quantum wells, and the p-doped top and n-doped bottom Bragg mirrors are built up by 22 and 34 mirror pairs, respectively, with graded interfaces. To the top mirror, a topmost $\lambda/4$-thick GaAs layer is added to obtain the above mentioned anti-phase reflection of the unetched surface. A 30-nm-thick $Al_{0.98}Ga_{0.02}As$ layer, positioned 30.5 nm above the SCH, is selectively oxidized to form the oxide aperture, which provides both optical and current confinement. The oxide layer is purposely located in a node of the longitudinal optical field. Further, a modulation doping scheme is applied to the top mirror in order to decrease the differential resistance, while maintaining a low free-carrier absorption loss. A low differential resistance is desirable as it reduces self-heating and thereby delays the thermal rollover, which allows for higher output powers. Finally, a circular surface relief of depth 57.5 nm is etched in the center of the device.

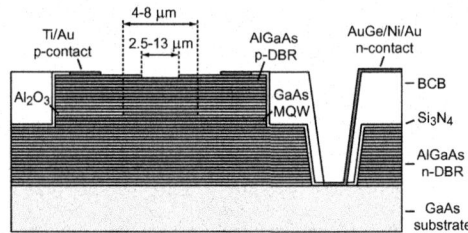

Fig. 9.15. Schematic cross-sectional view of the VCSEL geometry. A circular surface relief is etched in the center of the device.

9.4.3 Simulation Results

In the simulations of the VCSELs with a surface relief, we use the device and material parameters listed in Table 9.2, and the material gain parameters in Table 9.3. The free-carrier absorption coefficient for the doped mirror layers is crudely estimated by $\alpha_{fca} \approx 3 \times 10^{-18} N_d + 7 \times 10^{-18} N_a$ [cm^{-1}], where N_d [cm^{-3}] and N_a [cm^{-3}] are the n- and p-doping concentrations, respectively [30].

A numerical study was first performed to optimize the surface relief and oxide aperture diameters. From simulated static light-current characteristics, we found that the VCSELs with an oxide aperture between 4 and 8 µm produced the highest single fundamental-mode output powers. Figure 9.16 shows the maximal values when a side-mode-suppression ratio of 20 dB is required. For each oxide aperture diameter, there is an optimum surface relief diameter. Further, the optimal combination is obtained by oxide aperture and surface relief diameters of 5 and 3 µm, respectively, producing 6.1 mW. In the following, we will study three fundamental-mode stabilized VCSELs having oxide apertures of 4, 6, and 8 µm with an optimized surface relief diameter, referred to as single-mode VCSELs. We will also study three VCSELs with the same oxide aperture diameters but with surface relief diameters so large that they

Table 9.1. VCSEL Layer Structure.

Material	Composition (x)	Thickness [nm]	Doping (N_d/N_a) [cm^{-3}]	Repetition
GaAs		57.5	$-/1 \times 10^{19}$	
GaAs		5	$-/1 \times 10^{19}$	
Al$_x$Ga$_{1-x}$As	0.12	44	$-/1 \times 10^{19}$	
Al$_x$Ga$_{1-x}$As	0.9 − 0.12	20	$-/6 \times 10^{18}$	
Al$_x$Ga$_{1-x}$As	0.9	50	$-/3 \times 10^{18}$	20
Al$_x$Ga$_{1-x}$As	0.12 − 0.9	20	$-/4 \times 10^{18}$	20
Al$_x$Ga$_{1-x}$As	0.12	39	$-/3 \times 10^{18}$	20
Al$_x$Ga$_{1-x}$As	0.9 − 0.12	20	$-/6 \times 10^{18}$	20
Al$_x$Ga$_{1-x}$As	0.9	50	$-/2 \times 10^{18}$	2
Al$_x$Ga$_{1-x}$As	0.12 − 0.9	20	$-/2 \times 10^{18}$	2
Al$_x$Ga$_{1-x}$As	0.12	39	$-/2 \times 10^{18}$	2
Al$_x$Ga$_{1-x}$As	0.9 − 0.12	20	$-/2 \times 10^{18}$	2
Al$_x$Ga$_{1-x}$As	0.98	30	$-/2 \times 10^{18}$	
Al$_x$Ga$_{1-x}$As	0.9	30.5	$-/2 \times 10^{18}$	
Al$_x$Ga$_{1-x}$As	0.3 − 0.6	87		
Al$_x$Ga$_{1-x}$As	0.3	20		
GaAs		6		
Al$_x$Ga$_{1-x}$As	0.3	8		2
GaAs		6		2
Al$_x$Ga$_{1-x}$As	0.3	20		
Al$_x$Ga$_{1-x}$As	0.6 − 0.3	87		
Al$_x$Ga$_{1-x}$As	0.9	60	$2 \times 10^{18}/-$	
Al$_x$Ga$_{1-x}$As	0.12 − 0.9	20	$2 \times 10^{18}/-$	
Al$_x$Ga$_{1-x}$As	0.12	39	$2 \times 10^{18}/-$	34
Al$_x$Ga$_{1-x}$As	0.9 − 0.12	20	$2 \times 10^{18}/-$	34
Al$_x$Ga$_{1-x}$As	0.9	50	$2 \times 10^{18}/-$	34
Al$_x$Ga$_{1-x}$As	0.12 − 0.9	20	$2 \times 10^{18}/-$	34
GaAs		(substrate)	$3 \times 10^{18}/-$	

do not introduce any mode discrimination, referred to as multimode VCSELs (because of the top $\lambda/4$ epitaxial layer of GaAs, also the multimode VCSELs must have a surface relief to remove this layer, but the relief diameter is so large that it is effectively infinite). The important device dimensions are listed in Table 9.4.

The simulated light-current characteristics for the single- and multimode VCSELs are shown in Fig. 9.17. The single-mode VCSELs are fundamental-mode-stabilized over the entire operational range and produce powers above 5 mW at the thermal rollover. The threshold current and initial slope efficiency for the three single-mode and three multimode devices are 0.5, 1.1, 2.5 and 0.2, 0.3, 0.5 mA and 0.78, 0.64, 0.55 and 0.55, 0.55, 0.55 W/A, respectively. The higher modal loss for the single-mode VCSELs is responsible for the higher threshold currents and slope efficiencies. Also, the larger difference in the modal loss for the single-mode VCSELs contributes to the larger variation in the threshold current and slope efficiency with oxide aperture diameter.

Table 9.2. Device and Material Parameters.

Parameter	Symbol	Value
Nominal oscillation wavelength	λ_0	842 nm
Refractive index temp. coeff.	dn/dT	$2.3 \times 10^{-4}\,\text{K}^{-1}$
Anti-guiding factor	β_c	3
Electrical conduct. p-mirror	σ_r, σ_z	$2000, 100\,\Omega^{-1}\,\text{m}^{-1}$
Ambipolar diffusion coeff. SCH	D_b	$10\,\text{cm}^2\,\text{s}^{-1}$
Ambipolar diffusion coeff. QW	D_w	$10\,\text{cm}^2\,\text{s}^{-1}$
Diffusion and capture time SCH	τ_{bw}	20 ps
Thermionic emission time QW	τ_{wb}	100 ps
Nonradiative recomb. time SCH	τ_b	5 ns
Nonradiative recomb. time QW	τ_w	5 ns
Bimolecular recomb. coeff.	B	$10^{-10}\,\text{cm}^3\,\text{s}^{-1}$
Auger recomb. coeff.	C	$3.5 \times 10^{-30}\,\text{cm}^6\,\text{s}^{-1}$
Ambient temperature	T_0	295 K
Thermal conduct. mirror	κ_r, κ_z	$12, 10\,\text{W}\,\text{m}^{-1}\,\text{K}^{-1}$
Thermal conduct. substrate	κ_r, κ_z	$45, 45\,\text{W}\,\text{m}^{-1}\,\text{K}^{-1}$
Thermal conduct. contact	κ_r, κ_z	$315, 315\,\text{W}\,\text{m}^{-1}\,\text{K}^{-1}$
Gain suppression factor	ε	$10^{-17}\,\text{cm}^3$

Table 9.3. Optical Gain Parameters.

Symbol	Value
a_0	$2478.3\,\text{cm}^{-1}$
a_1	$-7.6369\,\text{cm}^{-1}\,\text{K}^{-1}$
a_2	$1.3177 \times 10^{-2}\,\text{cm}^{-1}\,\text{K}^{-2}$
b_0	$8.4226 \times 10^{-17}\,\text{cm}^{-3}$
b_1	$5.0679 \times 10^{15}\,\text{cm}^{-3}\,\text{K}^{-1}$
λ_0	$0.765\,\mu\text{m}$
λ_1	$2.501 \times 10^{-4}\,\mu\text{m}\,\text{K}^{-1}$
G_0	$6.25 \times 10^6\,\text{cm}^{-1}\,\mu\text{m}^2\,(\lambda > \lambda_p)$
G_0	$1.254 \times 10^6\,\text{cm}^{-1}\,\mu\text{m}^2\,(\lambda \leq \lambda_p)$

Table 9.4. Device Geometry.

Device	Surface relief diameter [μm]	Oxide aperture diameter [μm]	Contact aperture diameter [μm]
SR25OA4[1]	2.5	4	9
SR35OA6[1]	3.5	6	11
SR4OA8[1]	4	8	13
SR9OA4	9	4	9
SR11OA6	11	6	11
SR13OA8	13	8	13

The surface relief depth is 57.5 nm for all devices.
[1] Fundamental-mode-stabilized VCSEL.

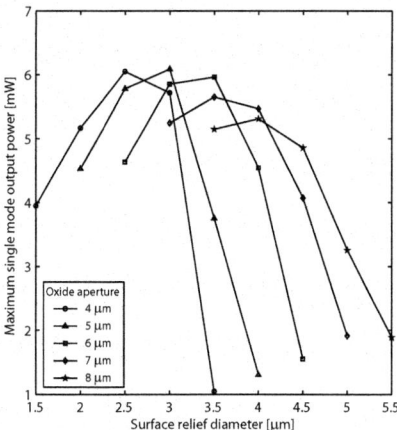

Fig. 9.16. Calculated maximum single fundamental-mode output power as a function of surface relief diameter for different oxide aperture diameters. Single-mode operation is defined by a side-mode-suppression ratio of >20 dB. The surface relief depth is 57.5 nm.

A larger oxide aperture reduces the mode confinement, which increases the overlap between the mode and the high-mirror-loss region. The cold-cavity losses for the three respective single-mode VCSELs are 0.400, 0.405, and 0.479 ps^{-1}. Moreover, at low currents, the output power of the single-mode devices exhibits a more nonlinear behavior, which is aggravated for larger oxide aperture diameters. This is mainly caused by the enhanced nonlinear effects of SHB. The effect of SHB is to depress the carrier density in the center of the active region by the intense stimulated recombination in this region. At low currents, the stimulated recombination rate is small and the carrier distribution resembles the injected current density distribution, but with higher currents, SHB becomes increasingly effective. As a consequence, the carrier density in the periphery instead increases strongly because the modal gain must still be equal to the modal loss even though the local gain in the center has decreased due to SHB. An increasing number of injected carriers will thus be lost to the peripheral reservoir of carriers where they are unlikely to cause stimulated recombination because of the low photon density. This leads to a slight sublinear light-current characteristic. Further, for lasers with large oxide diameters, the injected current is larger at the periphery, which enhances the SHB effect. At moderate currents, the enhanced nonlinear effects of thermal lensing adds to the effect of SHB in the single-mode devices. It is caused by the heating in the central part of the laser, leading to an increase of the refractive index in this region and, therefore, a higher confinement of the mode to the center. This effect is most outspoken for devices with a large oxide aperture. In this way, the optical mode abandons the peripheral region, where the losses are high, and therefore the modal loss is reduced, as well as the slope of the light-current characteristic as a smaller fraction of the generated photons are lost through the top mirror. At high currents, the elevated temperature leads to a wavelength mismatch between the gain peak and the cavity resonance. The lower gain is responsible for the rollover, which limits the output power. As a rule, the multimode VCSELs normally reach higher output powers by the coexistence of several transverse modes, which can more effectively utilize the laterally distributed carriers in the active region for stimulated emission. However, in some cases, the single-mode devices can produce a higher output power by their higher out-coupling. This is the case for the single-mode VCSEL with an oxide aperture of 4 μm, which has also been shown experimentally [31].

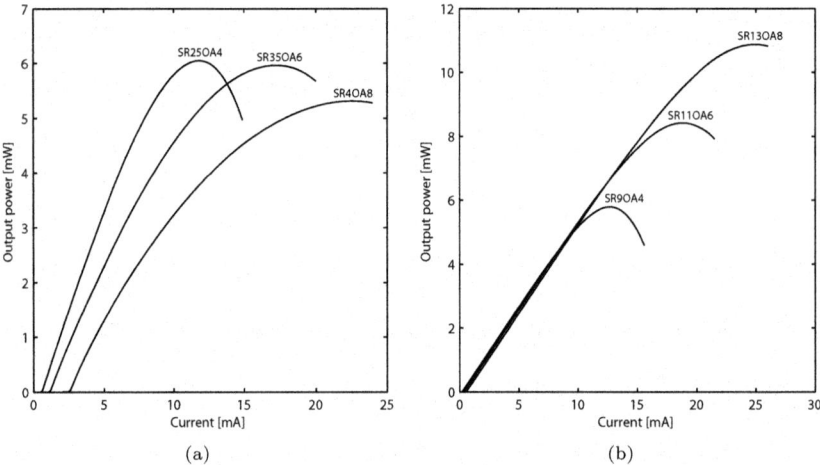

Fig. 9.17. Simulated total output power as a function of current. (a) Fundamental-mode-stabilized VCSELs and (b) multimode VCSELs.

9.4.3.1 Modulation Response

The modulation response was computed from a small-signal analysis, where the equations for the fluctuating quantities in the VCSEL are linearized. Figure 9.18 shows the modulation response for the single- and multimode VCSELs at five different bias currents, which correspond to a steady-state output power of 0.25, 0.5, 1, 2.5, and 5 mW. The RO frequency of the three single-mode devices is lower than that of the three respective multimode devices. This is mainly caused by the lower photon density in the cavity, compelled by the higher modal loss. The RO frequency decreases with increasing oxide aperture diameter due to the reduced photon density, recognizing that the RO frequency is roughly proportional to the square root of the photon density. The damping of the RO normally depends linearly on the photon density. However, the single-mode VCSELs deviate from this dependency, especially at low output powers. This is most clearly seen in Fig. 9.18(a) for the VCSEL with an oxide aperture of 8 µm, where the RO peak decreases at the lowest output power, indicating increased damping. Also, the single-mode VCSELs show a clear low-frequency roll-off in the modulation response. The increased damping and the low-frequency roll-off are caused by the dynamic effects of SHB. The influence of SHB on the modulation response of single-mode VCSELs has been studied in [32] and [6]. Also there, SHB was found to induce a low-frequency roll-off and strong damping of the RO. These effects were more pronounced in a smaller device, where the spatial distribution of the photons varies more abruptly.

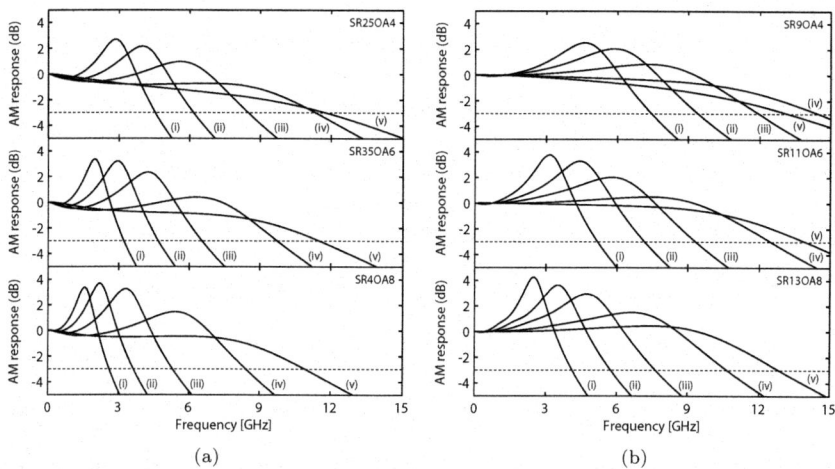

Fig. 9.18. Calculated modulation response. The indices (i)-(v) correspond to a steady-state output power of 0.25, 0.5, 1, 2.5, and 5 mW, respectively. (a) Fundamental-mode-stabilized VCSELs and (b) multimode VCSELs. The dotted line indicates the -3-dB limit for bandwidth determination.

To explain the influence of SHB, it is important to distinguish between the static and dynamic effects. At DC operation, the carrier density decreases slightly in the central parts and increases strongly in the periphery as the current is increased. This is an obvious effect of the higher photon density, which depresses the carrier density in the center, whereas the higher carrier injection increases the carrier number in the periphery where stimulated recombination is negligible. Only at high currents, stimulated recombination becomes significant also in the periphery, which leads to a more uniform increase in the carrier density as the current is further increased. This is also reflected in the dynamic behavior at quasi-static modulation frequencies — small carrier density fluctuations in the center and large in the periphery, especially at low and moderate bias. However, as will be explained shortly, the long lifetime of the carriers in the periphery leads to a dramatic reduction of the fluctuations in this region as the modulation frequency is increased. Above a few GHz, the carrier density fluctuations are of roughly equal magnitude irrespective of the lateral position. Therefore, the dynamic effects of SHB are most pronounced at low frequencies.

With these observations, the SHB-induced low-frequency roll-off can be rather simply understood. In the central region of the VCSEL, the fluctuations in the carrier density are small and their amplitude is independent of the modulation frequency. This is because the high photon density causes intense stimulated recombination, which makes the carrier lifetime much shorter than the modulation period. The small-signal carrier density is therefore determined by the instantaneous value of the small-signal injection current. The situation is quite different in the peripheral region. As a result of the low photon density, stimulated recombination is small and carrier lifetime longer than the modulation period. The small-signal carrier density is, therefore, solely determined by the small-signal injection current, which adds carriers during the positive part of the modulation cycle and removes an equal amount of carriers during the negative part. The carrier density is, therefore, not dependent on the instantaneous value of the injection current but by the time that has elapsed from the start of the last modulation cycle. Obviously, with increasing frequency, the modulation period becomes shorter, which means less time to add and remove carriers, and, thus, the fluctuations in carrier density become smaller. Although the intensity is low in this region of the VCSEL, the reduced fluctuations in carrier density do reduce the modal gain fluctuations somewhat. At these moderate frequencies, the photon number of the mode directly follows the fluctuations in the modal gain. Therefore, we observe a slightly reduced modulation of the photon number with the modulation frequency, which explains the roll-off of the modulation response at low frequencies. At frequencies above the cut-off frequency for SHB, which is typically 1–3 GHz, the fluctuations in the peripheral region are down to the same low level as those in the central region. Instead of SHB, the small-signal modulation is now increasingly influenced by RO effects.

The increased damping of the RO by SHB effects is understood as follows. Far below the RO frequency, the modulation is so slow that the negative feedback mechanism, the stimulated recombination, that determines the carrier density and photon number have had time to settle at each point in time. Intuitively then, the RO frequency could be described as the modulation frequency at which the period is still long enough for the photon number to rise significantly in the direct response to an increase in the carrier density, but short enough for any negative feedback effects (the consecutive lowering of the carrier density followed by a lower photon number, resulting from the increased stimulated recombination) not to occur within that period. We may thus say that the ROs occur at a modulation period roughly equal to the feedback time. However, for a VCSEL strongly affected by SHB, the feedback time is not a single well-defined entity: As the feedback time is longer, the lower the rate of stimulated recombination, the feedback time in the periphery is considerably longer than in the center. Thus, the onset of ROs is not sharply defined, which leads to a less pronounced resonance peak, i.e., an increased SHB-induced damping [32].

9.4.3.2 Eye Diagram

The digital modulation characteristics are commonly evaluated by studying eye diagram and bit-error rate (BER) at various modulation conditions. Here, we present examples of the large-signal performance at 2.5 and 10 Gbit/s. The eye diagrams are generated by first recording the response in the output power from a $2^7 - 1$ pseudo random bit sequence modulation current with non-return-to-zero data format. Secondly, the response is divided into 2-bit intervals and overlayed on each other to create an eye diagram. In the simulations, we used a timestep of 0.4 and 0.1 ps for the 2.5- and 10-Gbit/s modulation, respectively.

It is often desirable to reduce the off-state output power, for instance, to improve the extinction ratio. However, this is accompanied by increased timing jitter and influence of the previous bit pattern, particularly at higher bit rates [4]. Figure 9.19 shows simulated eye diagrams at 2.5 and 10 Gbit/s for the single- and multimode VCSELs. For each device, two eye diagrams are displayed, where the off and on currents correspond to a steady-state output power of 1 and 5 mW, and 0.25 and 5 mW, respectively. Note that there is a discrepancy in the actual output power, especially in the on state, due to a different self-heating in the modulated case. At 2.5 Gbit/s, the multimode VCSELs exhibit more well defined on and off states and less timing jitter. This is caused by the larger number of possible ways for the laser to reach a certain state, by the coexistence of multiple transverse modes. The multimode devices have a higher damping of the RO, which results in a shorter settling time. When the off-state output power is reduced from 1 to 0.25 mW, the RO frequency and damping in the off state decrease. For the VCSELs with a larger oxide aperture diameter, the previous bit pattern starts to influence the timing jitter. The settling time of the RO is then on the order of the bit period. Depending on when the laser is switched from the off to the on state, the initial conditions will be different and the laser responds accordingly. This often results in a turn-on transition in the form of multiple, almost parallel, discrete paths, each corresponding to a certain previous bit pattern. Two such paths can be barely distinguished for the multimode VCSEL with an oxide aperture diameter of $8\,\mu$m at 2.5 Gbit/s, the inset in the bottom right eye diagram of Fig. 9.19(c). Naturally, at 10 Gbit/s, the influence of the previous bit pattern on the timing jitter is even stronger. The number of paths is increased and the on and off states are further broadened. The lower RO frequency and larger timing jitter for the single-mode VCSELs result in a somewhat higher degree of eye closure. When the off-state output power is reduced to 0.25 mW, the single-mode devices show only a small degradation of the timing jitter, whereas the multimode devices are more influenced. It should be noted that the multimode devices have a somewhat lower actual output power in the off state, especially the devices with a larger oxide aperture diameter, which increases the timing jitter.

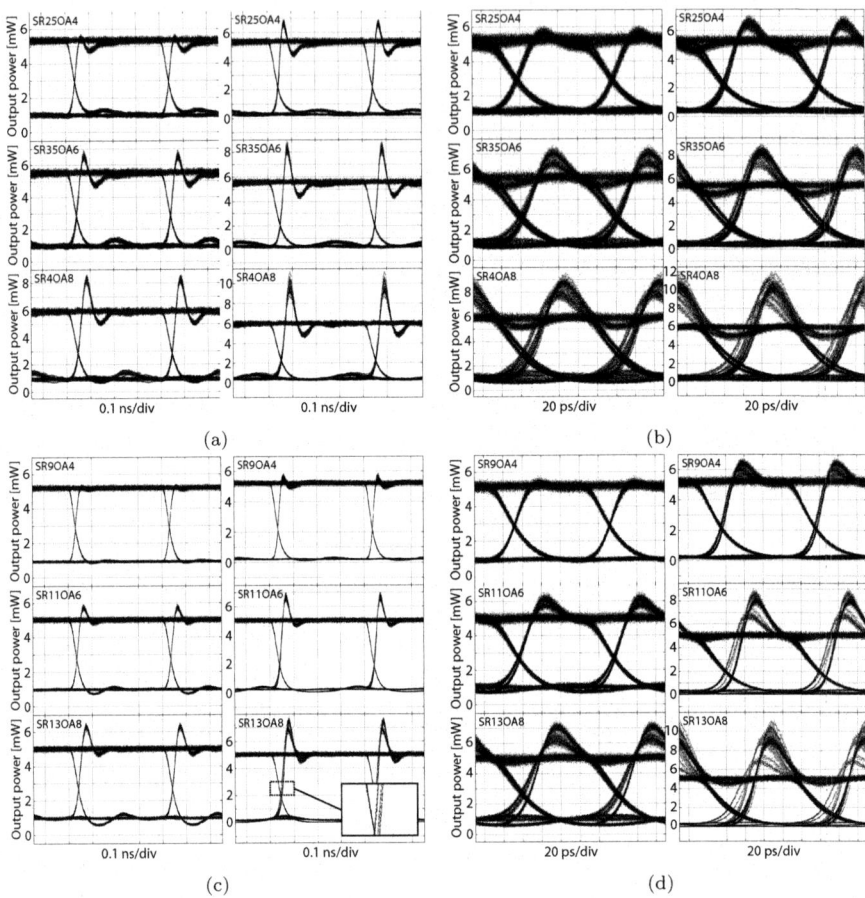

Fig. 9.19. Simulated eye diagrams. (a) 2.5 Gbit/s fundamental-mode-stabilized VCSELs, (b) 10 Gbit/s fundamental-mode-stabilized VCSELs, (c) 2.5 Gbit/s multimode VCSELs, (d) 10 Gbit/s multimode VCSELs. For each VCSEL, two eye diagrams are displayed, where the diagram to the right corresponds to a lower off-state output power.

In Fig. 9.20, the eye diagrams of the total output power, together with the individual eye diagrams of the two most dominant modes are shown for the multimode VCSEL with a 6-μm oxide aperture at 2.5 and 10 Gbit/s. The eye qualities of the individual modes are appreciably deteriorated compared with the eye diagrams of the total output power. The degradation is a consequence of anti-correlated power fluctuations of the individual modes due to the strong competition for carriers from a common reservoir (MPN). The effect becomes particularly pronounced for modes with a high degree of spatial overlap due to increased carrier competition. This example demonstrates the importance of collecting the power from all existing modes in optical links using multimode VCSELs.

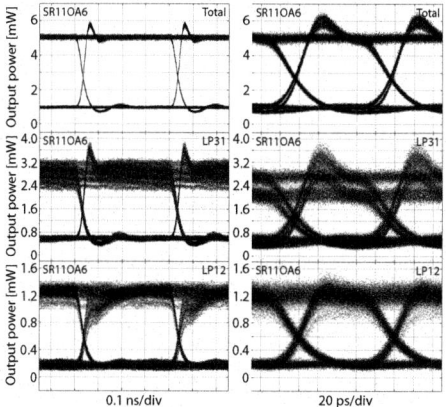

Fig. 9.20. Modally resolved eye diagrams for the multimode VCSEL with a 6-μm oxide aperture. The eye diagrams to the left and right correspond to 2.5 and 10 Gbit/s, respectively.

9.4.3.3 Bit-Error Rate (BER)

Studying the BER versus average received optical power is a standard procedure for evaluating the transmission performance of an optical link. Here, we study the BER for back-to-back transmission. The receiver consists of a photodetector, low-pass filter, and decision circuitry. The photodetector is assumed to have a responsivity of 0.4 A/W, output impedance of 50 Ω, and dominating thermal noise with a noise-equivalent power of 50 pW/√Hz. The low-pass filter is assumed to be a fifth-order Bessel filter with a 3-dB bandwidth of $f_{3dB} = 0.7B$, where B is the bit rate. This corresponds approximately to the optimum bandwidth for the non-return-to-zero data format [33]. The fifth order Bessel filter is described in the frequency domain by its transfer function

$$H_{Bessel}^{5th}(\omega) = \frac{a_0}{\left(\frac{i\omega}{\omega_0} + a_1\right)\left[\left(\frac{i\omega}{\omega_0} + a_2\right)^2 + a_3^2\right]\left[\left(\frac{i\omega}{\omega_0} + a_4\right)^2 + a_5^2\right]}, \quad (9.14)$$

where $\omega_0 = 2\pi f_{3dB}/\sqrt{9\ln 2}$ [34]. The constants are $a_0 = 945$, $a_1 = 3.6467$, $a_2 = 3.3520$, $a_3 = 1.7427$, $a_4 = 2.3247$, and $a_5 = 3.5710$. A semi-analytical approach described in [33] is used to calculate the BER,

$$BER(I_d, t_d) = \frac{1}{2(2^n - 1)} \left[\sum_{"0"s} \text{erfc}\left(\frac{I_d - I(t_d)}{\sqrt{2}\,\sigma(t_d)}\right) + \sum_{"1"s} \text{erfc}\left(\frac{I(t_d) - I_d}{\sqrt{2}\,\sigma(t_d)}\right) \right], \quad (9.15)$$

where $2^n - 1$ is the word length, I is the low-pass filtered photocurrent, and σ is the standard deviation of the superimposed noise current from the receiver. The BER depends on the decision level and time of the decision circuitry, I_d and t_d, respectively. A minimum BER is obtained by optimizing I_d and t_d.

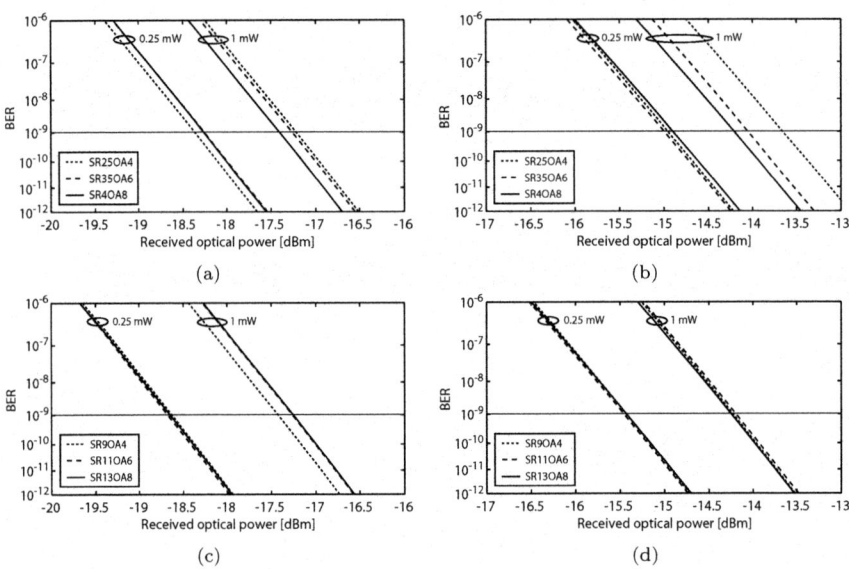

Fig. 9.21. Calculated minimum BER as a function of average received optical power. (a) 2.5 Gbit/s fundamental-mode-stabilized VCSELs, (b) 10 Gbit/s fundamental-mode-stabilized VCSELs, (c) 2.5 Gbit/s multimode VCSELs, (d) 10 Gbit/s multimode VCSELs. The BER is plotted for two different off-state output powers. The horizontal line indicates a BER of 10^{-9}, the limit for receiver sensitivity determination.

Figure 9.21 shows the calculated minimum BER versus averaged received optical power at 2.5 and 10 Gbit/s for the single- and multimode VCSELs. The BER is calculated for the output power responses generating the eye diagrams in Fig. 9.19. At 2.5 Gbit/s, the single- and multimode devices have similar receiver sensitivities (received optical power at a BER of 10^{-9}) that are better than -17 dBm for an off-state power of 1 mW. The small variation in receiver sensitivity between devices with different oxide apertures is mainly due to the previously mentioned discrepancy in actual output power in the on and off states, especially for the single mode devices. This results in a small variation in the extinction ratio between the devices. The BER value cannot be attributed to a particular feature of the eye diagram. For instance, the receiver sensitivity is strongly influenced by both the extinction ratio and pulse overshoot. The latter is particular pronounced for the single-mode VCSELs at 10 Gbit/s, which can partly explain the larger variation in receiver sensitivity for these devices.

When the off-state power is reduced to 0.25 mW at 2.5 Gbit/s, the receiver sensitivity is improved by about 1 and 1.5 dB for the single- and multimode VCSELs, respectively. This is caused by the increased extinction ratio. The more well defined on and off states for the multimode VCSELs result in a somewhat higher quality of the eye. Going to 10 Gbit/s, the receiver sensitivity degrades by about 3 to 3.5 dB for both the single and multimode VCSELs. The main part of this degradation, 3 dB, is a result of the necessarily increased bandwidth of the receiver, which increases the thermal receiver noise. The additional degradation, ~ 0.5 dB, is caused by the increased eye closure.

9.4.3.4 Relative Intensity Noise (RIN)

The RIN is of great importance for optical links because it ultimately limits the signal-to-noise ratio and the dynamic range. The RIN is defined by $\text{RIN} = 2S_P/\langle P \rangle^2$ in the frequency domain, where $S_P(\omega)$ is the double-sided spectral density of the output power and $\langle P \rangle$ is the average output power. In general, the double-sided spectral density, $S_f(\omega)$, of a complex fluctuating variable, $f(t)$, is equal to the Fourier transform (F) of its auto correlation function, $S_f(\omega) = F\{\langle f(t)f^*(t-\tau)\rangle\} = F\{f(t)\}F\{f^*(t)\} = |F\{f(t)\}|^2$. The RIN spectrum is thus obtained by Fourier transforming the time-fluctuating output power, which was simulated using a temporal resolution of 1 ps. By averaging over 50 trajectories, each of duration 40 ns, we obtain a spectral resolution of 25 MHz.

Figure 9.22 shows the RIN spectra for the single- and multimode VCSELs at three different bias currents, which correspond to a steady-state output power of 0.5, 2.5, and 5 mW. The peak RIN occurs at the RO frequency and is strongly dependent on the damping of the RO. For the single-mode devices, the peak RIN varies from -140 dB/Hz, for a device with an oxide aperture diameter of 4 μm at an output power of 5 mW, to -105 dB/Hz for a device with an oxide aperture diameter of 8 μm at an output power of 0.5 mW. For the corresponding multimode devices, the peak RIN is 10 to 15 dB lower, due to significantly higher damping of the RO. The damping of the RO is strongly influenced by SHB, carrier diffusion, and nonlinear gain suppression. However, the main reason for the lower damping of the RO for the single-mode VCSELs is the significantly lower photon density in the cavity, induced by the higher modal loss. Below the RO frequency, the RIN declines and reaches a low-frequency RIN that is ∼ 10 to 20 dB lower than the peak RIN, depending on the damping of the RO. In this frequency region, very faint multiple peaks can be distinguished for the multimode VCSELs, especially for the devices with a larger oxide aperture diameter. The peaks result from anti-correlated power fluctuations of individual modes due to MPN. At these "mode partition frequencies," the competition is particularly pronounced, so that a small noise-induced disturbance has a large effect on the instantaneous intensity relation between the modes.

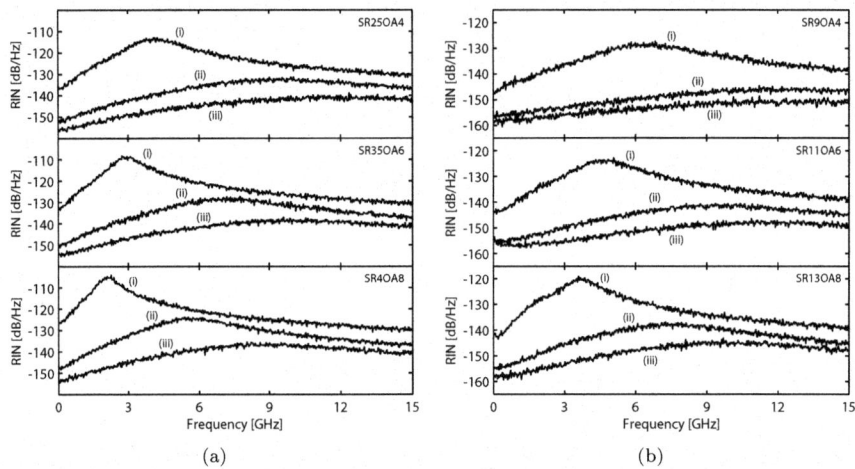

Fig. 9.22. Simulated RIN spectra. The indices (i)-(iii) correspond to a steady-state output power of 0.5, 2.5, and 5 mW, respectively. (a) Fundamental-mode-stabilized VCSELs and (b) multimode VCSELs.

Figure 9.23 shows modally resolved RIN spectra for the multimode VCSEL with an 8-μm oxide aperture at an output power of 0.5 mW. The RIN spectra of the individual modes demonstrate the substantially higher low-frequency RIN, in this case, by more than 40 dB/Hz, due to MPN. The peaks at the "mode partition frequencies" are also more pronounced. This is a result of having only one carrier reservoir for all modes. The feedback mechanism in the laser efficiently reduces the fluctuations in the total power, but much less so for the individual modes. Consequently, if a multimode VCSEL is to be used in an optical link, it is important that all modes are equally coupled to the detector in order to ensure low-noise operation. As a final remark, it has been shown theoretically that an efficient single-mode semiconductor laser can exhibit sub-Poissonian noise in the output at low frequencies, provided that a quiet current supply is used [35]. As mentioned in Sect. 9.2.3, this phenomenon is referred to as amplitude squeezing and occurs at high output powers. Our single-mode devices did not exhibit squeezing though, which is probably due to a too-high free-carrier absorption loss and strong SHB.

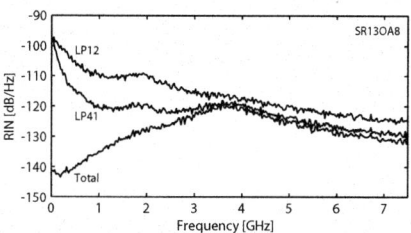

Fig. 9.23. Simulated RIN spectrum for the multimode VCSEL with an 8-μm oxide aperture at an output power of 0.5 mW. The individual RIN spectra of the two most dominant modes are also included.

9.4.3.5 Frequency Noise (FN) and Linewidth

The FN is defined by FN = $2S_\nu$ in the frequency domain, where $S_\nu(\omega)$ is the double-sided spectral density of the frequency of the emitted field. The FN spectrum is thus obtained by Fourier transforming the time fluctuating frequency. The same averaging procedure is performed as for calculating the RIN, giving the same spectral resolution.

Figure 9.24 displays the FN spectra for the single- and multimode VCSELs at the same three bias currents as for the RIN spectra in Fig. 9.22. The FN of the multimode devices are defined by the FN of the most dominant mode. Similar to the peak RIN, the peak FN occurs at the RO frequency and is also strongly dependent on the damping of the RO. The single-mode devices have a peak FN that varies from 5 MHz*Hz/Hz, for a device with an oxide aperture diameter of 4 μm at an output power of 5 mW, to 450 MHz*Hz/Hz for a device with an oxide aperture diameter of 8 μm at an output power of 0.5 mW. For the corresponding multimode devices, the peak FN is about five to ten times lower, because of the higher damping of the RO. Below the RO frequency, the FN declines, however, for frequencies well below the RO, the FN saturates and can be considered as white noise. Irregularities are observed for the multimode VCSELs and are probably induced by the MPN.

The linewidth of the emitted field can be estimated by $\Delta\nu = S_\nu(\omega = 0)$, provided that the linewidth is considerably smaller than the RO frequency [8]. In Fig. 9.25, the so estimated linewidth as a function of inverse output power is plotted for the single- and multimode VCSELs. The inset shows the intensity spectrum for the single-mode VCSEL with a 6-μm oxide aperture at an output power of 0.5 mW. The intensity spectrum is equal to the double-sided spectral density of the complex electric field and, thus, obtained by Fourier transforming the time fluctuating electric field. Characteristic of the

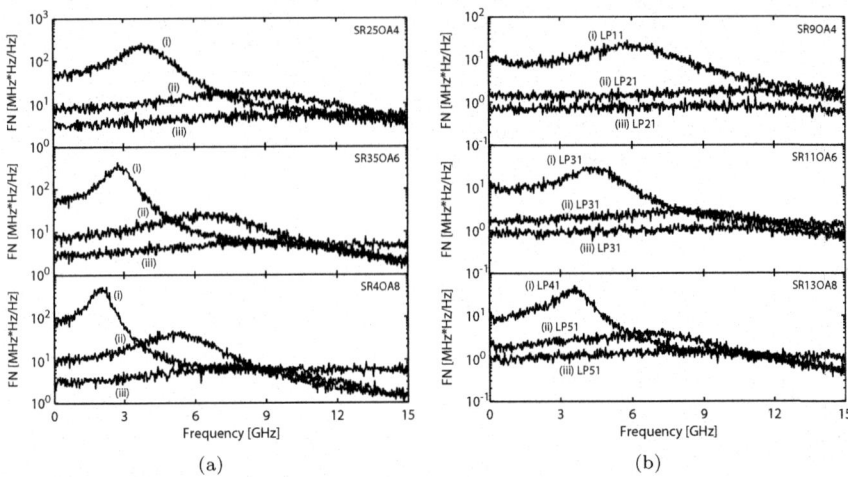

Fig. 9.24. Simulated FN spectra. The indices (i)-(iii) correspond to a steady-state output power of 0.5, 2.5, and 5 mW, respectively. (a) Fundamental-mode-stabilized VCSELs and (b) multimode VCSELs.

intensity spectrum is the central Lorentzian-shaped spectrum with a full-width half-maximum that corresponds to the linewidth. Thus, this presents an alternative way to estimate the VCSEL linewidth, but, because of the limited spectral resolution, the accuracy of such a determination would be quite poor. Another characteristic of the intensity spectrum is the satellite peaks that are separated from the main peak by the RO frequency, and multiples thereof if the damping of the RO is low [8]. The linewidth of the single-mode devices is appreciably broader than that of the multimode devices. This is, again, a result of the higher mirror loss, which causes a lower photon density and, consequently, a lower damping of the field fluctuations. As a rule, the linewidth is inversely proportional to the output power and may saturate or rebroaden at higher output powers [36]. At an output power of 5 mW, we observe linewidths of the single- and multimode devices as low as 1.5 and 0.5 MHz, respectively. It should be noted that these estimated narrow linewidths will probably be somewhat broader in practice, due to effects not accounted for in the model, such as noise in the current supply and $1/f$ noise.

9.5 Conclusion

In this chapter, we have described important quantities that characterize the performance of high-speed VCSELs, and pointed out the need for a comprehensive numerical model if these quantities are to be accurately obtained by computer simulation. The model presented in this chapter is an effort to truthfully mimic the numerous important, and interrelated, physical

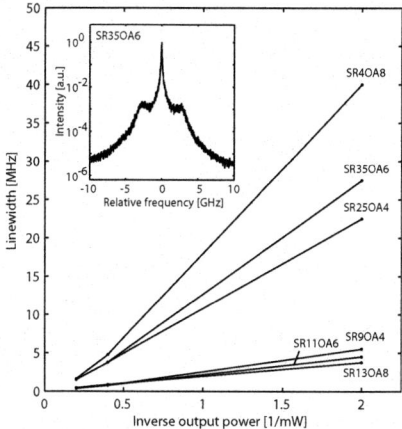

Fig. 9.25. Estimated linewidth as function of inverse output power. The linewidth of the multimode VCSELs is defined by the linewidth of the most dominant mode. The inset shows the intensity spectrum for the fundamental-mode-stabilized VCSEL with a 6-μm oxide aperture at an output power of 0.5 mW.

processes that take place in a real VCSEL, mainly governed by the dynamically changing distributions of the optical field, carrier density, and temperature. The model is quasi-three-dimensional (cylindrically symmetric) and includes the detailed epitaxial layer structure and device geometry in the calculation of the optical field. In particular, the model can predict the detailed evolution in both time and space of the individual modes in the VCSEL.

As an example that is both a challenging modeling task and represents a VCSEL of considerable practical interest, we simulated fundamental-mode-stabilized VCSELs of wavelength 850 nm, which are identical to conventional VCSELs except for an etched, shallow surface relief. The relief induces a mode-selective loss that suppresses the higher-order modes and thus enables high-power single-mode operation. From the simulations, quantities such as modulation bandwidth, relative intensity noise (RIN), bit-error rate (BER), and laser linewidth were estimated. It was found that the surface relief technique should have negligible detrimental effect on the high-speed performance of VCSELs.

References

1. K. L. Lear, A. Mar, K. D. Choquette, et al.: Electron. Lett. **32**, 457 (1996)
2. J. S. Gustavsson, Å. Haglund, C. Carlsson, et al.: IEEE J. Quantum Electron. **39**, 941 (2003)
3. J. S. Gustavsson, Å. Haglund, J. Bengtsson, et al.: IEEE J. Quantum Electron. **40**, 607 (2004)

4. J. S. Gustavsson, Å. Haglund, J. Bengtsson, et al.: IEEE J. Quantum Electron. **38**, 1089 (2002)
5. L. A. Coldren and S. W. Corzine: *Diode Lasers and Photonic Integrated Circuits*, 1st edn (Wiley, New York 1995)
6. Y. Liu, W.-C. Ng, B. Klein, et al.: IEEE J. Quantum Electron. **39**, 99 (2003)
7. Y. Yamamoto ed.: *Coherence, Amplification, and Quantum Effects in Semiconductor Lasers.* 1st edn (Wiley, New York 1991)
8. K. Petermann: *Laser Diode Modulation and Noise.* 1st edn (Kluwer Academic Publishers, Dordrecht 1991)
9. D. C. Kilper, P. A. Roos, J. L. Carlsten, et al.: Phys. Rev. A **55**, 3323 (1997)
10. C. Degen, J.-L. Vej, W. Elsäßer, et al.: Electron. Lett. **34**, 1585 (1998)
11. D. Wiedenmann, P. Schnitzer, C. Jung, et al.: Appl. Phys. Lett. **73**, 717 (1998)
12. J. S. Gustavsson, J. A. Vukušić, J. Bengtsson, et al.: IEEE J. Quantum Electron. **38**, 203 (2002)
13. S. F. Yu: *Analysis and Design of Vertical Cavity Surface Emitting Lasers.* 1st edn (Wiley, New Jersey 2003)
14. G. R. Hadley: Opt. Lett. **20**, 1483 (1995)
15. M. J. Noble, J. P. Lohr, and J. A. Lott: IEEE J. Quantum Electron. **34**, 1890 (1998)
16. J. P. Zhang and K. Petermann: IEEE J. Quantum Electron. **30**, 1529 (1994)
17. H. K. Bissessur, F. Koyama, and K. Iga: IEEE J. Select. Topics Quantum Electron. **3**, 344 (1997)
18. B. Demeulenaere, P. Bienstman, B. Dhoedt, et al.: IEEE J. Quantum Electron. **36**, 358 (1999)
19. G. P. Bava, P. Debernardi, and L. Fratta: Phys. Rev. A **63**, 023816-1 (2001)
20. G. R. Hadley, K. L. Lear, M. E. Warren, et al.: IEEE J. Quantum Electron. **32**, 607 (1996)
21. Y. Yamamoto and N. Imoto: IEEE J. Quantum Electron. **22**, 2032 (1986)
22. J. S. Gustavsson, J. Bengtsson, and A. Larsson: to be published in IEEE J. Quantum Electron. (Sept. 2004)
23. C. Jung, R. Jäger, M. Grabherr, et al.: Electron. Lett. **33**, 1790 (1997)
24. Å. Haglund, J. S. Gustavsson, J. Vukušić, et al.: IEEE Photon. Technol. Lett. **16**, 368 (2004)
25. H. J. Unold, S. W. Z. Mahmoud, R. Jäger, et al.: IEEE J. Select. Topics Quantum Electron. **7**, 386 (2001)
26. J. Vukušić, H. Martinsson, J. S. Gustavsson, et al.: IEEE J. Quantum Electron. **37**, 108 (2001)
27. H. J. Unold, M. Golling, F. Mederer, et al.: Electron. Lett. **37**, 570 (2001)
28. H. J. Unold, M. Grabherr, F. Eberhard, et al.: Electron. Lett. **35**, 1340 (1999)
29. H. J. Unold, M. C. Riedl, R. Michalzik, et al.: Electron. Lett. **38**, 77 (2002)
30. H. C. Casey and M. B. Panish: *Heterostructure Lasers.* 1st edn (Academic Press, New York 1978)
31. H. J. Unold, S. W. Z. Mahmoud, R. Jäger, et al.: Proc. SPIE, 218 (2002)
32. R. Schatz, M. Peeters, and H. Isik: Proc. LFNM, 108 (2002)
33. G. P. Agrawal: *Fiber-Optic Communication Systems.* 1st edn (Wiley, New York 1992)
34. M. C. Jeruchim, P. Balaban, and K. S. Shanmugan: *Simulation of Communication Systems.* 1st edn (Plenum Press, New York 1992)
35. Y. Yamamoto and S. Machida: Phys. Rev. A **35**, 5114 (1987)
36. G. Morthier and P. Vankwikelberge: *Handbook of Distributed Feedback Laser Diodes.* 1st edn (Artech House, Boston 1997)

10 GaN-based Light-Emitting Diodes

J. Piprek[1] and S. Li[2]

[1] University of California, Santa Barbara, CA 93106-9560, U.S.A.,
piprek@ieee.org
[2] Crosslight Software, 202-3855 Henning Dr., Burnaby, BC V5C 6N3 Canada,
simon@crosslight.com

10.1 Introduction

Light-emitting diodes (LEDs) offer several advantages over traditional light sources, such as smaller size, longer lifetime, higher efficiency, and greater mechanical ruggedness. Continuing developments in LED technology are producing devices with increased output power and efficiency as well as a wider range of colors [1]. Recent progress in the fabrication of GaN-based compound semiconductors enabled the practical breakthrough of short-wavelength LEDs that emit green, blue, or ultraviolet light [2]. In particular, compact ultraviolet (UV) light sources are currently of high interest for applications in white-light generation, short-range communication, water purification, and biochemical detection. Prime candidates are nitride LEDs with AlGaN quantum wells. However, their performance is still below the requirement for practical applications.

We present here a self-consistent physics-based three-dimensional (3D) simulation of an AlGaN/GaN LED and study performance limiting internal mechanisms. Good agreement with measured device characteristics [3] is achieved by refinement of the physical model and by calibration of material parameters. Based on this agreement, we are able to analyze the practical impact of microscale and nanoscale physical effects such as current crowding, carrier leakage, nonradiative recombination, and built-in polarization.

The device structure is given in the next section. Section 10.3 describes the theoretical models and the material parameters used. The main simulation results are presented and discussed in Sect. 10.4.

10.2 Device Structure

Our example device was grown on c-face sapphire by metal organic chemical vapor deposition (MOCVD) [3]. The layer structure is given in Table 10.1. It includes an AlGaN multi-quantum well (MQW) active region that is covered by a p-$Al_{0.3}Ga_{0.7}N$ electron blocker layer. MQW and blocker layer are sandwiched between two 42-period AlGaN superlattice (SL) cladding layers. A quadratic mesa with 300-μm edge length is etched down to the n-GaN contact layer. The U-shaped n-side contact covers three of the four sides of the mesa (Fig. 10.1). The top p-contact layer is semi-transparent for top emission. More details on device design, fabrication, and performance are given in [3, 4].

Table 10.1. Layer Structure and Room-Temperature Parameters of the AlGaN/GaN LED (d, layer thickness; N_{dop}, doped carrier density; μ, majority carrier mobility (low field); n_{r}, refractive index at wavelength 340 nm; κ_{L}, lattice thermal conductivity).

Parameter Unit	d (nm)	N_{dop} (1/cm^3)	μ (cm^2/Vs)	n_{r} —	κ_{L} (W/cmK)
p-GaN	5	1×10^{18}	10	2.77	1.3
p-AlGaN SL cladding	126	4×10^{17}	0.5	2.48	0.2
p-Al$_{0.3}$Ga$_{0.7}$N blocker	15	1×10^{17}	5	2.02	0.1
i-Al$_{0.10}$Ga$_{0.90}$N well	5	—	300	2.79	0.2
n-Al$_{0.16}$Ga$_{0.84}$N barrier	13	2×10^{18}	185	2.48	0.2
i-Al$_{0.10}$Ga$_{0.90}$N well	5	—	300	2.79	0.2
n-Al$_{0.16}$Ga$_{0.84}$N barrier	13	2×10^{18}	185	2.48	0.2
i-Al$_{0.10}$Ga$_{0.90}$N well	5	—	300	2.79	0.2
n-Al$_{0.16}$Ga$_{0.84}$N barrier	13	2×10^{18}	185	2.48	0.2
i-Al$_{0.10}$Ga$_{0.90}$N well	5	—	300	2.79	0.2
n-Al$_{0.16}$Ga$_{0.84}$N barrier	13	2×10^{18}	185	2.48	0.2
n-AlGaN SL cladding	126	2×10^{18}	10	2.48	0.2
n-GaN contact layer	500	2×10^{18}	200	2.77	1.3

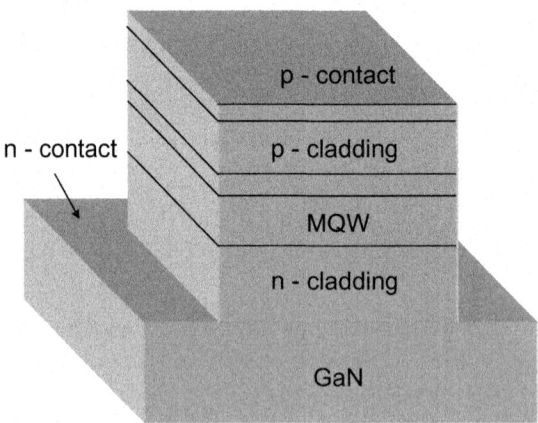

Fig. 10.1. Schematic 3D view of the light-emitting diode structure.

10.3 Models and Parameters

We employ the simulation software APSYS [5], which self-consistently combines the 3D simulation of carrier transport, self-heating, spontaneous photon emission, and optical ray tracing. The code was significantly improved during our investigation to provide more realistic results. The wurtzite energy band structure is considered for all GaN-based semiconductors. The main features of the models are summarized in the following, and more details can be found in [6].

The inclusion of correct material parameters is of paramount importance for realistic device simulations. These parameters depend on the material composition and may be different for every layer in the device. Published values sometimes spread over a wide range, and it is difficult to select the number most appropriate for a given device. We therefore include a detailed discussion of those material parameters that are crucial for our simulation. Hereby, we mainly rely on the recent review of III-nitride parameters in [7]. Binary material parameters are listed in Table 10.2, and they are interpolated linearly for AlGaN unless noted otherwise in the following.

10.3.1 Wurtzite Energy Band Structure

Most GaN-based semiconductor compounds are grown as wurtzite (hexagonal) crystals. Their energy band structure is different from that of traditional zinc blende III-V semiconductors. The three valence bands of wurtzite semiconductors are referred to as heavy-hole (hh), light-hole (lh), and crystal-field split-hole (ch) band. Spin–orbit interaction leads to only slight separations between the three band edges. We here briefly summarize the 6×6 $\mathbf{k} \cdot \mathbf{p}$ model for the band structure of strained wurtzite semiconductors as developed by Chuang and Chang [8, 9]. Their material parameters are replaced by the more recent data listed in Table 10.2.

The epitaxial growth of $Al_xGa_{1-x}N$ on GaN is typically along the c axis of the wurtzite crystal, which is parallel to the z axis in our coordinate system. The natural lattice constant $a_0(x)$ is enlarged to the one of the GaN substrate, a_s, imposing biaxial tensile strain in the transverse plane

$$\epsilon_t = \frac{a_s - a_0}{a_0} \tag{10.1}$$

and compressive strain in the growth direction

$$\epsilon_z = -2\frac{C_{13}}{C_{33}}\epsilon_t. \tag{10.2}$$

The nondiagonal elements of the strain tensor are zero. The valence band edge energies are

Table 10.2. Material Parameters used for Wurtzite Semiconductors GaN and AlN at Room Temperature [7] ($\Delta_{\rm cr} = \Delta_1$, $\Delta_{\rm so} = 3\Delta_2 = 3\Delta_3$).

Parameter	Symbol	Unit	GaN	AlN
Electron eff. mass (c axis)	m_c^z	m_0	0.20	0.32
Electron eff. mass (transverse)	m_c^t	m_0	0.20	0.30
Hole eff. mass parameter	A_1	—	-7.21	-3.86
Hole eff. mass parameter	A_2	—	-0.44	-0.25
Hole eff. mass parameter	A_3	—	6.68	3.58
Hole eff. mass parameter	A_4	—	-3.46	-1.32
Hole eff. mass parameter	A_5	—	-3.40	-1.47
Hole eff. mass parameter	A_6	—	-4.90	-2.64
Direct band gap (unstrained)	E_g^0	eV	3.438	6.158
Thermal band gap shrinkage	dE_g^0/dT	meV/K	-0.42	-0.56
Spin–orbit split energy	$\Delta_{\rm so}$	eV	0.017	0.019
Crystal–field split energy	$\Delta_{\rm cr}$	eV	0.01	-0.169
Lattice constant	a_0	Å	3.189	3.112
Elastic constant	C_{33}	GPa	398	373
Elastic constant	C_{13}	GPa	106	108
Hydrost. deform. potential (c axis)	a_z	eV	-4.9	-3.4
Hydrost. deform. potential (transverse)	a_t	eV	-11.3	-11.8
Hydrost. deform. potential (cond. band)	a_c	eV	-6.8	-7.6
Shear deform. potential	D_1	eV	-3.7	-17.1
Shear deform. potential	D_2	eV	4.5	7.9
Shear deform. potential	D_3	eV	8.2	8.8
Shear deform. potential	D_4	eV	-4.1	-3.9
dielectric constant	ε	—	9.5	8.5

$$E_{\rm hh} = E_{\rm v} + \Delta_1 + \Delta_2 + \theta_\epsilon + \lambda_\epsilon \tag{10.3}$$

$$E_{\rm lh} = E_{\rm v} + \frac{\Delta_1 - \Delta_2 + \theta_\epsilon}{2} + \lambda_\epsilon + \sqrt{\left(\frac{\Delta_1 - \Delta_2 + \theta_\epsilon}{2}\right)^2 + 2\Delta_3^2} \tag{10.4}$$

$$E_{\rm ch} = E_{\rm v} + \frac{\Delta_1 - \Delta_2 + \theta_\epsilon}{2} + \lambda_\epsilon - \sqrt{\left(\frac{\Delta_1 - \Delta_2 + \theta_\epsilon}{2}\right)^2 + 2\Delta_3^2} \tag{10.5}$$

with the average valence band edge $E_{\rm v}$ and

$$\theta_\epsilon = D_3 \epsilon_z + 2 D_4 \epsilon_t \tag{10.6}$$

$$\lambda_\epsilon = D_1 \epsilon_z + 2 D_2 \epsilon_t. \tag{10.7}$$

As a result of the negative crystal-field split energy $\Delta_{\rm cr} = -0.169$ eV in AlN, the light-hole band edge $E_{\rm lh}$ is above the heavy-hole band edge $E_{\rm hh}$ in all of our AlGaN layers. We therefore use the unstrained band edge $E_{\rm lh}^0$ as reference in the calculation of the conduction band edge [9]

$$E_{\rm c} = E_{\rm lh}^0 + E_g^0 + P_{c\epsilon} , \tag{10.8}$$

with the hydrostatic energy shift

$$P_{c\epsilon} = a_{cz}\epsilon_z + 2a_{ct}\epsilon_t. \tag{10.9}$$

The hydrostatic deformation potential is anisotropic (a_z, a_t), and half of the deformation is assumed to affect the conduction band (a_{cz}, a_{ct}). For a given material, a_{cz} and a_{ct} can be translated into the isotropic APSYS input parameter

$$a_c = \frac{a_{ct} - a_{cz}\frac{C_{13}}{C_{33}}}{1 - \frac{C_{13}}{C_{33}}}. \tag{10.10}$$

The AlGaN band gap is known to deviate from the linear Vegard law and a wide range of bowing parameters C_g has been reported [7]. We adopt an average value of $C_g = 0.7$ eV and approximate the unstrained $Al_xGa_{1-x}N$ band gap by

$$E_g^0(x) = xE_g^0(\text{AlN}) + (1-x)E_g^0(\text{GaN}) - x(1-x)C_g. \tag{10.11}$$

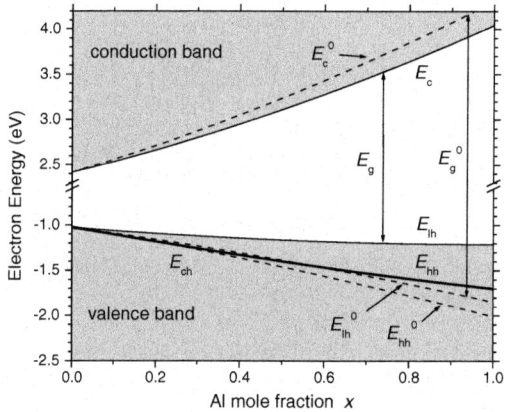

Fig. 10.2. Band edges of $Al_xGa_{1-x}N$ grown on GaN (solid — strained, dashed — unstrained).

Figure 10.2 plots the $Al_xGa_{1-x}N$ band edges with strain (solid lines) and without strain (dashed lines). For GaN, E_{hh} is the top valence band edge. For $x > 0.02$, E_{lh} becomes the top valence band edge, which increasingly reduces the effective band gap $E_g(x)$ compared with the unstrained band gap $E_g^0(x)$. E_{ch} is always slightly below E_{hh}. As $E_{lh}(x)$ varies little with increasing x, the valence band offset between different AlGaN layers is small, leading to poor hole confinement in the quantum wells. We here consider the valence band offset ratio $\Delta E_\mathrm{v}/\Delta E_\mathrm{g} = 0.3$ for unstrained material, using $E_\mathrm{v}(x) = -0.3\, E_\mathrm{g}^0(x)$. However, this ratio is an uncertain parameter that corresponds to the 0.85 eV extracted from the literature as a most reliable result for the GaN/AlN valence band offset [7].

The dispersion $E_\mathrm{c}(\boldsymbol{k})$ of the conduction band can be characterized by a parabolic band model with electron effective masses m_c^t and m_c^z perpendicular and parallel to the c-growth direction, respectively. The three valence bands are nonparabolic. Near the Γ point, the bulk hole effective masses can be approximated as

$$m_\mathrm{hh}^z = -m_0(A_1 + A_3)^{-1} \tag{10.12}$$

$$m_\mathrm{hh}^t = -m_0(A_2 + A_4)^{-1} \tag{10.13}$$

$$m_\mathrm{lh}^z = -m_0\left[A_1 + \left(\frac{E_\mathrm{lh} - \lambda_\epsilon}{E_\mathrm{lh} - E_\mathrm{ch}}\right)A_3\right]^{-1} \tag{10.14}$$

$$m_\mathrm{lh}^t = -m_0\left[A_2 + \left(\frac{E_\mathrm{lh} - \lambda_\epsilon}{E_\mathrm{lh} - E_\mathrm{ch}}\right)A_4\right]^{-1} \tag{10.15}$$

$$m_\mathrm{ch}^z = -m_0\left[A_1 + \left(\frac{E_\mathrm{ch} - \lambda_\epsilon}{E_\mathrm{ch} - E_\mathrm{lh}}\right)A_3\right]^{-1} \tag{10.16}$$

$$m_\mathrm{ch}^t = -m_0\left[A_2 + \left(\frac{E_\mathrm{ch} - \lambda_\epsilon}{E_\mathrm{ch} - E_\mathrm{lh}}\right)A_4\right]^{-1}, \tag{10.17}$$

using the hole effective mass parameters A_i given in Table 10.2. Details on the numerical calculation procedure for quantum well valence bands are given in [9].

10.3.2 Carrier Transport

APSYS employs the traditional drift-diffusion model for semiconductors. The current density of electrons $\boldsymbol{j}_n$ and holes $\boldsymbol{j}_p$ is caused by the electrostatic field $\boldsymbol{F}$ (drift) and by the concentration gradient of electrons and holes, ∇n and ∇p, respectively,

$$\boldsymbol{j}_n = q\mu_n n \boldsymbol{F} + qD_n \nabla n \tag{10.18}$$

$$\boldsymbol{j}_p = q\mu_p p \boldsymbol{F} - qD_p \nabla p, \tag{10.19}$$

with the elementary charge q, the mobilities μ_n and μ_p, and the carrier densities n and p. The diffusion constants D_n and D_p are replaced by mobilities using the Einstein relation $D = \mu k_\mathrm{B} T/q$ with the Boltzmann constant k_B and the temperature T. The electric field is affected by the charge distribution, which includes electrons n and holes p as wells as dopant ions $(p_\mathrm{D}, n_\mathrm{A})$ and other fixed charges N_f (the latter are of special importance in GaN-based devices to account for built-in polarization). This relationship is described by the Poisson equation

$$\nabla \cdot (\varepsilon\varepsilon_0 \boldsymbol{F}) = q(p - n + p_D - n_A \pm N_f). \tag{10.20}$$

Changes in the local carrier concentration are accompanied by a spatial change in current flow $\nabla \boldsymbol{j}$ and/or by the generation G or recombination R of electron–hole pairs. This relation is expressed by the continuity equations

$$q\frac{\partial n}{\partial t} = \nabla \cdot \boldsymbol{j}_n - q(R - G) \tag{10.21}$$

$$q\frac{\partial p}{\partial t} = -\nabla \cdot \boldsymbol{j}_p - q(R - G). \tag{10.22}$$

Generation of electron-hole pairs by reabsorption of photons is not considered in our simulation. The relevant carrier recombination mechanisms in our device are spontaneous recombination and Shockley–Read–Hall (SRH) recombination. Spontaneous (radiative) recombination is discussed below. The defect-related nonradiative SRH recombination rate is given by

$$R_{SRH} = \frac{np - n_i^2}{\tau_p^{SRH}\left(n + N_c \exp\left[\frac{E_t - E_c}{k_B T}\right]\right) + \tau_n^{SRH}\left(p + N_v \exp\left[\frac{E_v - E_t}{k_B T}\right]\right)}, \tag{10.23}$$

and it is governed by the SRH lifetimes τ_n^{SRH} and τ_p^{SRH} ($N_{c,v}$ — density of states of conduction, valence band; E_t — mid-gap defect energy). SRH lifetimes are different for electrons and holes, but the SRH recombination rate is usually dominated by the minority carrier lifetime so that $\tau_{nr} = \tau_n^{SRH} = \tau_p^{SRH}$ is assumed in the following. The nonradiative carrier lifetime τ_{nr} is a crucial material parameter for GaN-based LEDs. In fact, the low output power of LEDs is often attributed to the high defect density and the correspondingly short nonradiative lifetime of carriers in AlGaN and GaN epitaxial layers. Defect density and nonradiative lifetime depend on the substrate used and on the growth quality, and they are hard to predict. As SRH lifetime studies have not yet been performed on our example LEDs, we assume the uniform value of 1 ns in our simulations. The SRH lifetime in quantum wells is of particular importance, and it may be used as a fit parameter to find agreement with experimental LED characteristics [4].

Our model includes Fermi statistics and thermionic emission of carriers at hetero-interfaces [6]. The doping densities given in Table 10.1 represent actual densities of free carriers. Although the Si donor exhibits a low ionization energy, the Mg acceptor is known for its high activation energy so that the Mg density is significantly above the hole density. The hole mobility is hardly investigated for AlGaN, and Table 10.1 lists empirical estimations. In contrast, significant attention has been paid to electron mobility modeling and we use approximate values extracted from Monte-Carlo simulations [10]. However, the metal-semiconductor contact resistance contributes significantly to the device bias. Figure 10.3 compares the calculated current - voltage (IV) characteristics with the measurement. Without contact resistance, the calculated IV curve shows a steep turn-on at 3.6 V. Better agreement with the measurement can be obtained by adding an ohmic p-contact resistance of 20 Ω to the simulation. This results in a linear IV slope that slightly deviates from the superlinear experimental curve. The superlinearity is possibly caused by a Schottky-type contact, which is hard to model [6].

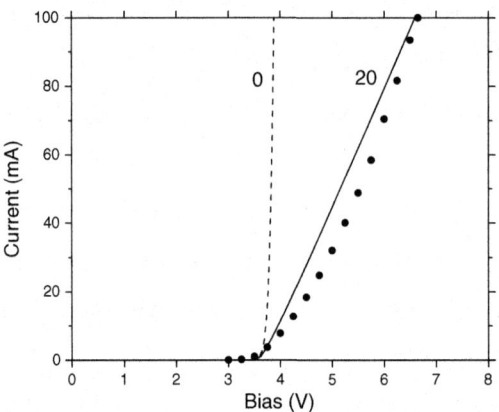

Fig. 10.3. Calculated current-voltage characteristics with (solid) and without (dashed) contact resistance (Ω); the dots give the measured curve [3].

Built-in polarization is another important issue for GaN LEDs. Spontaneous and piezoelectric polarization of nitride compounds is larger than in other III-V semiconductors. It depends on the compound's composition so that net charges remain at hetero-interfaces. Much theoretical effort has been invested in the prediction of these polarization charges, leading to relatively simple nonlinear interpolation formulas that are in close agreement with experimental observations [11]. For our $Al_xGa_{1-x}N$ layers, the spontaneous polarization P_{sp} [C/m^2] is calculated as

$$P_{sp} = -0.09x - 0.034(1-x) + 0.019x(1-x). \tag{10.24}$$

The piezoelectric polarization P_{pz} [C/m^2] is given by

$$P_{pz} = x[-1.808\epsilon_t(x) - 7.888\epsilon_t^2(x)] + (1-x)[-0.918\epsilon_t(x) + 9.541\epsilon_t(x)^2], \tag{10.25}$$

and it is linearly interpolated between the binary polarizations, which are nonlinear functions of the positive (tensile) transverse strain ϵ_t.[3] Spontaneous and piezoelectric polarization add up to the surface charge density. At each interface, the difference of the surface charge densities gives the net polarization charge density, which is listed in Table 10.3 for all types of interfaces in our device. The $Al_{0.16}Ga_{0.84}N/Al_{0.20}Ga_{0.80}N$ superlattices are represented by a uniform $Al_{0.16}Ga_{0.84}N$ layer in our simulation as the internal SL interface charges compensate each other.

Table 10.3. Fixed Interface Charge Densities for Different Types of LED Interfaces.

Interface	Built-in Charge Density
$GaN/Al_{0.16}Ga_{0.84}N$	$+6.88 \times 10^{12} cm^{-2}$
$Al_{0.16}Ga_{0.84}N/Al_{0.10}Ga_{0.90}N$	$-2.73 \times 10^{12} cm^{-2}$
$Al_{0.10}Ga_{0.90}N/Al_{0.16}Ga_{0.84}N$	$+2.73 \times 10^{12} cm^{-2}$
$Al_{0.10}Ga_{0.90}N/Al_{0.30}Ga_{0.70}N$	$+9.89 \times 10^{12} cm^{-2}$
$Al_{0.30}Ga_{0.70}N/Al_{0.16}Ga_{0.84}N$	$-7.16 \times 10^{12} cm^{-2}$
$Al_{0.16}Ga_{0.84}N/GaN$	$-6.88 \times 10^{12} cm^{-2}$

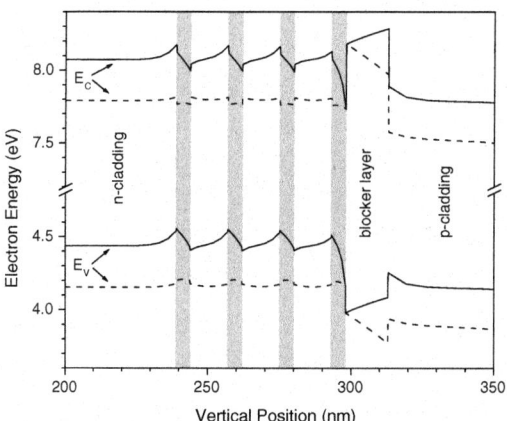

Fig. 10.4. Energy band diagram of the active region with (solid) and without (dashed) built-in polarization; the quantum wells are marked gray (E_c — conduction band edge, E_v — valence band edge).

---

[3] The strain is compressive (negative) when AlN is used as substrate, resulting in a different formula (10.25) [11].

Figure 10.4 shows the energy band diagram of the MQW region with and without interface polarization charges. Built-in polarization causes a strong deformation of the quantum wells accompanied by a strong electrostatic field. Consequently, electrons and holes are separated within the wells and the spontaneous emission rate is reduced, limiting the LED output power. In addition, polarization affects the electron blocking by the $Al_{0.3}Ga_{0.7}N$ layer. Our comparison with measurements in the next section will show that (10.11) somewhat underestimates the band gaps in our device, in particular for the blocker layer. We therefore increase the band gap E_g^0 of $Al_{0.3}Ga_{0.7}N$ from 4.1 eV to 4.5 eV. This unstrained blocker band gap is reduced by strain to the actual band gap $E_g = 4.2$ eV (cf. Fig. 10.2).

10.3.3 Heat Generation and Dissipation

Self-heating often limits the performance of LEDs. Rising internal temperature reduces the band gap and leads to a red-shift of the emission wavelength. Device heating is generated when carriers transfer part of their energy to the crystal lattice. In our device, main heat sources are the electrical resistance, resulting in the Joule heat density

$$H_J = \frac{j_n^2}{q\mu_n n} + \frac{j_p^2}{q\mu_p p} , \qquad (10.26)$$

and to a lesser extent, nonradiative carrier recombination, which gives the recombination heat density

$$H_R = R_{SRH} (E_{Fn} - E_{Fp}) , \qquad (10.27)$$

with the quasi-Fermi levels E_{Fn} and E_{Fp} for electrons and holes, respectively. The total heat power density $H_{heat}(x, y, z)$ enters the steady-state heat flux equation

$$-\nabla(\kappa_L \nabla T) = H_{heat} , \qquad (10.28)$$

which is used to calculate the internal device temperature $T(x, y, z)$. The thermal conductivity κ_L is 130 W/Km for GaN and 285 W/Km for AlN. It is a strong function of composition for AlGaN due to alloy scattering [12]. The approximate numbers given in Table 10.1 also consider phonon mean free path restrictions by interface scattering [13].

Earlier investigations reveal a relatively low temperature slope near the active region of GaN devices due to the relatively high thermal conductivity [14]. Most of the temperature rise is caused by the thermal resistance of the sapphire substrate and the mounting, which are outside the simulated device region. We therefore add an empirical external resistance of 100 K/W to the heat sink at the bottom of our LED.

10.3.4 Spontaneous Photon Emission

The local spontaneous emission rate in bulk AlGaN is approximated by

$$R_{\rm sp} = B(np - n_{\rm i}^2) \tag{10.29}$$

using the bimolecular recombination coefficient $B = 2 \times 10^{-10} {\rm cm}^3/{\rm s}$ ($n_{\rm i}$ - intrinsic density). This simple equation includes the full spectrum of photons generated by spontaneous band-to-band recombination processes. The spontaneous emission spectrum of our quantum wells is calculated as a function of the photon energy $h\nu$ by

$$r_{\rm sp}(h\nu) = \left(\frac{q^2 h}{2m_0^2 \varepsilon \varepsilon_0}\right)\left(\frac{1}{h\nu}\right) D_{\rm opt} D_{\rm r} |M|^2 f_{\rm c}(1 - f_{\rm v}), \tag{10.30}$$

with Planck's constant h, the free electron mass m_0, and the photon frequency ν. The photon emission rate is proportional to the density of photon states

$$D_{\rm opt}(h\nu) = \frac{\varepsilon n_{\rm r}}{\pi^2 \hbar^3 c^3}(h\nu)^2, \tag{10.31}$$

with the reduced Planck constant $\hbar = h/2\pi$ and the light velocity c. For quantum wells of thickness d_z, the reduced density of states in each subband is

$$D_{\rm r} = \frac{m_{\rm r}}{\pi \hbar^2 d_z}, \tag{10.32}$$

and all subbands are added up in (10.30).

The transition strength is given by $|M|^2$ (transition matrix element), and its computation is based on the $\boldsymbol{k} \cdot \boldsymbol{p}$ electron band structure model outlined in Sec. 10.3.1. $|M|^2$ is averaged over all photon polarization directions. For a quantum well grown in the hexagonal c direction, the transition matrix elements for heavy (hh), light (lh), and crystal-field holes (ch), respectively, are

$$|M_{\rm hh}^{\rm TE}|^2 = \frac{3}{2} O_{ij}(M_{\rm b}^{\rm TE})^2 \tag{10.33}$$

$$|M_{\rm lh}^{\rm TE}|^2 = \frac{3}{2} \cos^2(\theta_{\rm e}) O_{ij}(M_{\rm b}^{\rm TE})^2 \tag{10.34}$$

$$|M_{\rm ch}^{\rm TE}|^2 = 0 \tag{10.35}$$

$$|M_{\rm hh}^{\rm TM}|^2 = 0 \tag{10.36}$$

$$|M_{\rm lh}^{\rm TM}|^2 = 3 \sin^2(\theta_{\rm e}) O_{ij}(M_{\rm b}^{\rm TM})^2 \tag{10.37}$$

$$|M_{\rm ch}^{\rm TM}|^2 = 3 O_{ij}(M_{\rm b}^{\rm TM})^2. \tag{10.38}$$

These equations consider the angle $\theta_{\rm e}$ of the electron $\boldsymbol{k}$ vector with the k_z direction

$$k_z = |\boldsymbol{k}| \cos(\theta_e) , \qquad (10.39)$$

with $\cos(\theta_e) = 1$ at the Γ point of the quantum well subband. For transverse electric (TE) polarization, the photon electric field vector lies within the quantum well plane, whereas the photon magnetic field vector lies within the quantum well plane for transverse magnetic (TM) polarization. The matrix element also depends on the photon energy through the quantum well dispersion functions $E_m(\boldsymbol{k})$. It is different for each subband m. The overlap integral O_{ij} of the electron and hole wave functions can assume values between 0 and 1. At the Γ point, O_{ij} is nonzero only for subbands with the same quantum number m. Away from the Γ point, O_{ij} may be nonzero for any transition. Thus, at higher photon energies, summation over all possible subband combinations is included in the calculation. The anisotropic bulk momentum matrix elements are given by [15]

$$(M_b^{TM})^2 = \frac{m_0}{6}\left(\frac{m_0}{m_c^z} - 1\right)\frac{(E_g + \Delta_1 + \Delta_2)(E_g + 2\Delta_2) - 2\Delta_3^2}{E_g + 2\Delta_2} \qquad (10.40)$$

$$(M_b^{TE})^2 = \frac{m_0}{6}\left(\frac{m_0}{m_c^t} - 1\right)\frac{E_g[(E_g + \Delta_1 + \Delta_2)(E_g + 2\Delta_2) - 2\Delta_3^2]}{(E_g + \Delta_1 + \Delta_2)(E_g + \Delta_2) - \Delta_3^2}. \qquad (10.41)$$

Note that the bulk electron mass is different in transversal (m_c^t) and in parallel directions (m_c^z) relative to the hexagonal c axis. The material parameters are given in Table 10.2.

The Fermi factor $f_c(1 - f_v)$ in (10.30) gives the probability that the conduction band level is occupied and the valence band level is empty at the same time. The final spontaneous emission spectrum $r_{spon}(h\nu)$ is obtained by including the transition energy broadening according to

$$r_{spon}(h\nu) = \frac{1}{\pi}\int dE\, r_{sp}(E)\frac{\Gamma_s}{(h\nu - E)^2 + \Gamma_s^2} \qquad (10.42)$$

using a Lorentzian line shape with the half-width $\Gamma_s = 6.6$ meV in our simulation.

10.3.5 Ray Tracing

Only a small fraction of generated photons is able to escape from the LED. This is attributed to total internal reflection as well as to internal absorption. Calculation of the external light power requires 3D ray tracing from every emission point within the device, weighted by the local emission rate. The ray tracing model is based on simple geometrical optics. Assuming equal numbers of TE and TM polarized photons, Fresnel's formulas are employed to account for reflection

$$r_{12}^{TE} = \frac{E_r^{TE}}{E_i^{TE}} = \frac{n_{r1}\cos\vartheta_i - n_{r2}\cos\vartheta_t}{n_{r1}\cos\vartheta_i + n_{r2}\cos\vartheta_t} \tag{10.43}$$

$$r_{12}^{TM} = \frac{E_r^{TM}}{E_i^{TM}} = \frac{n_{r2}\cos\vartheta_i - n_{r1}\cos\vartheta_t}{n_{r2}\cos\vartheta_i + n_{r1}\cos\vartheta_t} \tag{10.44}$$

and transmission

$$t_{12}^{TE} = \frac{E_t^{TE}}{E_i^{TE}} = \frac{2n_{r1}\cos\vartheta_i}{n_{r1}\cos\vartheta_i + n_{r2}\cos\vartheta_t} \tag{10.45}$$

$$t_{12}^{TM} = \frac{E_t^{TM}}{E_i^{TM}} = \frac{2n_{r1}\cos\vartheta_i}{n_{r2}\cos\vartheta_i + n_{r1}\cos\vartheta_t} \tag{10.46}$$

of the optical field at each light ray transition from material 1 to material 2 (E_i — incident field, E_r — reflected field, E_t — transmitted field, ϑ_i — incident angle, ϑ_t — angle of refraction as given by Snell's law). The multitude of reflections and the variety of possible light paths in our LED elongates the computation time. We therefore limit the number of initial rays to 6000.

Material parameters are the refractive index and the absorption coefficient, which are both a function of photon energy and alloy composition. We adopt a physics-based model developed by Adachi for photon energies close to the semiconductor band gap [16]. For nitride III–V compounds, the valence band splitting is very small and Adachi's model for the transparency region can be approximated by only one interband transition giving the refractive index

$$n_r^2(h\nu) = A\left(\frac{h\nu}{E_g}\right)^{-2}\left\{2 - \sqrt{1 + \left(\frac{h\nu}{E_g}\right)} - \sqrt{1 - \left(\frac{h\nu}{E_g}\right)}\right\} + B. \tag{10.47}$$

This approximation shows good agreement with measurements on GaN, AlN, and InN [17]. For $Al_xGa_{1-x}N$ with $x < 0.38$, the material parameters

$$A(x) = 9.827 - 8.216x - 31.59x^2 \tag{10.48}$$
$$B(x) = 2.736 + 0.842x - 6.293x^2 \tag{10.49}$$

have been extracted from measurements [18]. The resulting data for our device are listed in Table 10.1. For 340-nm wavelength, the photon energy is larger than the GaN band gap and the overall absorption in our LED is dominated by the GaN layers with an absorption coefficient of $\alpha = 11 \times 10^4$ cm^{-1}. The ray tracing also considers the semi-transparent p-contact, which comprises a 3-nm-thick palladium layer ($n_r = 1.1, \alpha = 18 \times 10^4$ cm^{-1}) and a 5-nm-thick gold layer ($n_r = 1.4, \alpha = 59 \times 10^4$ cm^{-1}). Background loss of $\alpha = 20$ cm^{-1} is assumed for all other layers.

Fig. 10.5. 3D plots of the LED giving the radiative recombination rate on the left and the vertical current density on the right at 100-mA injection current (the discontinuity near the MQW is related to strong lateral current).

10.4 Results and Discussion

10.4.1 Internal Device Analysis

Figure 10.5 shows on the left a 3D plot of the LED radiative recombination rate, which is strongest in the four quantum wells and which decays toward the device center, in agreement with experimental observations. This lateral nonuniformity is attributed to current crowding along the sides with an adjacent n-contact (right-hand side of Fig. 10.5). The current density is highest in the two corners of the U-shaped contact. This corner position is therefore chosen in the next few graphs to show vertical profiles of different physical properties, all at 100-mA injection current.

Figure 10.6 plots the vertical profile of electron and hole density. As a result of the quantum well deformation by the polarization charges shown in Fig. 10.4, the hole density peaks on the n-side of each well and the electron density on the p-side. This separation of electron and hole wavefunctions reduces the radiative emission rate. The quantum well electron density is higher than the hole density, and it is highest in the p-side asymmetric quantum well because of the better electron confinement. The highest hole density occurs above the electron blocker layer due to the valence band edge maximum caused by the negative interface charges. Figure 10.7 shows vertical profiles of the electron-hole recombination rates. The nonradiative Shockley–Read–Hall recombination peaks within the quantum wells. It is about two orders of magnitude stronger than the radiative recombination. In other words, not more than 1% of the injected carriers contribute to the light emission. Figure 10.7 also indicates significant carrier leakage from the quantum wells, which leads to additional carrier recombination outside the wells. Non-radiative recombi-

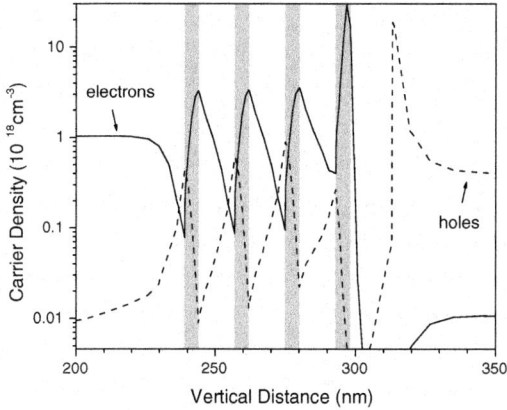

Fig. 10.6. Vertical profile of electron and hole density

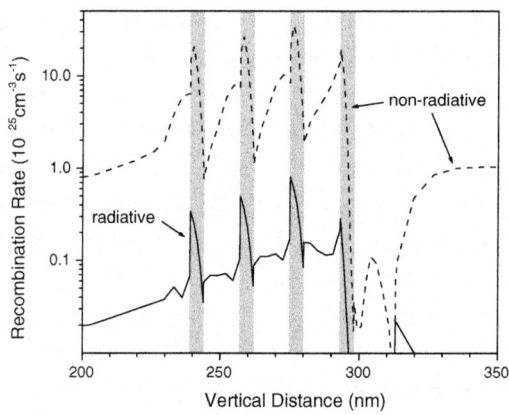

Fig. 10.7. Vertical profile of radiative and nonradiative recombination rate.

nation occurs even beyond the electron blocker layer, where leaking electrons meet injected holes.

Vertical components of the current density are plotted in Fig. 10.8 (the current is negative because it flows from the top to the bottom). Ideally, electrons and holes meet in the quantum wells and recombine completely. However, some holes leave the MQW region and leak into the lower n-cladding where they recombine with electrons. More severe is the electron leakage in the opposite direction. A large part of the electrons injected from the n-side

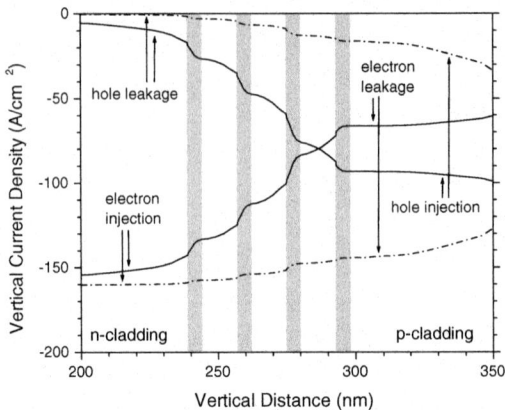

Fig. 10.8. Vertical current density j_y of electrons and holes (dashed — original blocker band gap, solid — adjusted blocker band gap).

into the MQW leaks into the p-cladding layer. This electron leakage strongly depends on the conduction band offset between the top quantum well and the blocker layer (cf. Fig. 10.4). According to our default material parameters from [7], the offset is 153 meV, resulting in very strong electron leakage and in light emission below the measured light power (dashed lines). Considering the large variation of band gap and offset data extracted from the literature [7], we therefore use it as fit parameter to find better agreement with the measured light output. The fit of the light-current characteristic below is obtained using a 403-meV conduction band offset between quantum well and blocker layer (solid lines in Fig. 10.8).

10.4.2 External Device Characteristics

The calculated emission spectrum is compared with the measurement in Fig. 10.9. The theoretical spectrum exhibits multiple peaks and shoulders attributed to different transition energies within the quantum wells. Without the electrostatic field, the asymmetric top well exhibits a lower emission energy than the other three symmetric quantum wells. The built-in polarization field leads to a shift of the existing quantum levels and to the creation of additional levels, especially in the top quantum well. However, the measured spectrum is smoother, probably due to statistical variations of the quantum well structure. There may be less polarization charges in the real device, or the shape of the quantum well may deviate otherwise from the theoretical assumption. The experimental emission peak is at slightly higher photon energy than the calculated one, which may also be attributed to nonideal quantum well growth or to an overestimation of the band gap by (10.11). The experimental peak hardly shifts with increasing current, indicating little self-heating. In our simulation, the internal temperature rise of up to 63 K leads to a slight shift of the emission peak by about 1 nm. Thus, the assumed thermal resistance of 100 K/W seems to slightly overestimate the influence of substrate and heat sinking on our LED self-heating. The temperature difference within the simulation region is less than 1 K due to the relatively high thermal conductivity of nitride alloys.

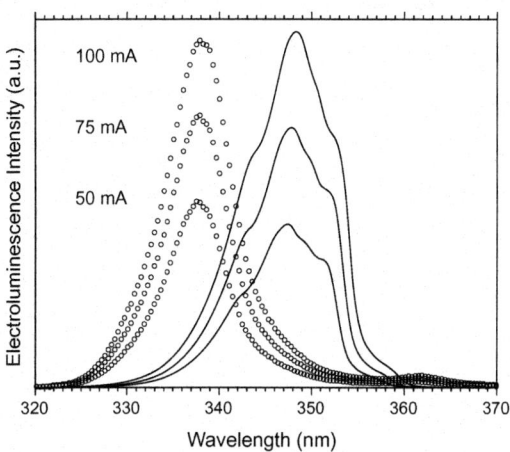

Fig. 10.9. LED emission spectra at three different injection currents (dots — measurement [3], lines — simulation).

Most photons are unable to escape from the LED, due to total internal reflection, mainly at the device surfaces, and to photon absorption, primarily in the bottom GaN layer. The detected external quantum efficiency η_{det} gives the ratio of detected photons to the injected electron-hole pairs. It is calculated from

$$\eta_{det} = \eta_{int}\eta_{opt}\eta_{cap}, \qquad (10.50)$$

with the internal quantum efficiency η_{int} (fraction of photons generated inside the LED per electron-hole pair injected), the photon extraction efficiency η_{opt} (fraction of escaped photons per photon generated inside the LED), and the detector capture efficiency η_{cap} (fraction of detected photons per escaped photons). The last number cannot be exactly determined in this study. A large-area detector (100 mm^2) was used to measure the top emission, and we estimate $\eta_{cap} = 82\%$. The strong absorption in GaN prevents any bottom emission. Only $\eta_{opt} = 4.5\%$ of all internally generated photons escape from the LED. About 78% of all photons are absorbed within the GaN layers and the remaining 17% are absorbed by the semi-transparent Pd/Au contact.

But the main limitation of the emitted light power originates from the internal quantum efficiency, which is $\eta_{\text{int}} = 0.96\%$ for 100-mA injection current. As indicated by the strong nonradiative recombination in Fig. 10.7, less than 1 out of 100 injected electron-hole pairs generate a photon within the quantum wells. Nonradiative recombination and carrier leakage are key causes for the low external quantum efficiency of $\eta_{\text{det}} = 0.035\%$, which is close to the measured value of 0.032% [3].

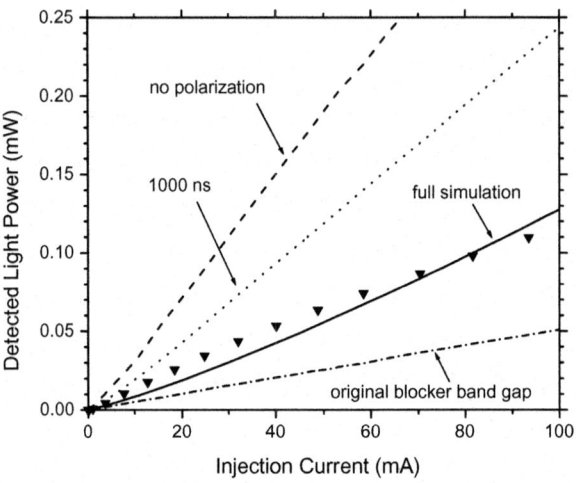

Fig. 10.10. LED emission light vs. current characteristics (triangles — measurement [3], lines — simulation).

Figure 10.10 plots the detected light power as a function of injection current (LI characteristic). The simulation is in good agreement with the measurement, which indicates that model and parameters are fairly accurate in our simulation. The solid line in Fig. 10.10 is fitted to the measurement by increasing the blocker layer band gap from $E_g^0 = 4.1$ eV to 4.5 eV. For comparison, Fig. 10.10 also shows the much lower output power calculated with the original blocker band gap, which is attributed to stronger carrier leakage (cf. Fig. 10.8). Without built-in polarization, electrons and holes are less separated within each quantum well and the radiative emission rate is significantly enhanced, as shown by the dashed curve in Fig. 10.10 ($\eta_{\text{int}} = 2.9\%$). Finally, the dotted curve represents a simulation with extremely long carrier lifetime of $\tau_{\text{nr}} = 1000$ ns within the quantum wells. Due to the suppressed nonradiative recombination in the wells, the internal quantum efficiency is doubled to $\eta_{\text{int}} = 1.8\%$. This surprisingly low improvement underlines the detrimental effect of carrier leakage from the MQW; i.e., more than 98% of all carriers recombine outside the quantum wells.

10.5 Summary

We have presented a self-consistent three-dimensional simulation of electronic, optical, and thermal processes in AlGaN/GaN ultraviolet light-emitting diodes. Good agreement with measurements is achieved underlining the accuracy of models and parameters. We find that the low output power of practical devices is mainly restricted by the low internal quantum efficiency of less than 1%, which is primarily attributed to carrier leakage from the quantum wells.

Acknowledgment

This work was partially supported by the Solid-State Lighting and Display Center (Director: Prof. Shuji Nakamura) at the University of California at Santa Barbara. J. P. is grateful to Dr. Thomas Katona and Dr. Stacia Keller for valuable discussions and measurements. S. L. acknowledges Dr. Oleksiy Shmatov's contribution to the initial phase of the 3D ray-tracing project.

References

1. E. F. Schubert: *Light-Emitting Diodes* (Cambridge Univ. Press, Cambridge 2003)
2. S. Nakamura: Development and future prospects of GaN-based LEDs and LDs. In: *Introduction to Nitride Semiconductor Blue Lasers and Light-Emitting Diodes*, ed by S. Nakamura and S. F. Chichibu (Taylor & Francis, London 2000)
3. T. Katona: Development of Ultraviolet Nitride-based Light-Emitting Diodes. PhD Thesis, University of California, Santa Barbara (2003)
4. J. Piprek, T. Katona, S. P. DenBaars, and S. Li: 3D Simulation and Analysis of AlGaN/GaN Ultraviolet Light–Emitting Diodes. In: *Light-Emitting Diodes: Research, Manufacturing and Applications VIII* SPIE Proceedings **5366** (The International Society for Optical Engineering, Bellingham 2004)
5. APSYS Version 2003.12.01 by Crosslight Software, Inc., Burnaby, Canada (http://www.crosslight.com)
6. J. Piprek: *Semiconductor Optoelectronic Devices: Introduction to Physics and Simulation* (Academic Press, San Diego 2003)
7. I. Vurgaftman and J. R. Meyer: J. Appl. Phys. **94**, 3675 (2003)
8. S. L. Chuang and C. S. Chang: Phys. Rev. B **54**, 2491 (1996)
9. S. L. Chuang and C. S. Chang: Semicond. Sci. Technol. **12**, 252 (1997)
10. M. Farahmand, C. Garetto, E. Bellotti, K. F. Brennan, M. Goano, E. Ghillino, G. Ghione, J. D. Albrecht, and P. P. Ruden: J. Appl. Phys. **48**, 535 (2001)
11. V. Fiorentini, F. Bernardini, and O. Ambacher: Appl. Phys. Lett. **80**, 1204 (2002)
12. B. C. Daly, H. J. Maris, A. V. Nurmikko, M. Kuball, and J. Han: J. Appl. Phys. **92**, 3820 (2002)

13. J. Piprek, T. Troger, B. Schroter, J. Kolodzey, and C. S. Ih: IEEE Photon. Technol. Lett. **10**, 81 (1998)
14. J. Piprek and S. Nakamura: IEE Proceedings, Optoelectronics **149**, 145 (2002)
15. S. L. Chuang: IEEE J. Quantum Electron. **32**, 1791 (1996)
16. S. Adachi: *Physical Properties of III-V Semiconductor Compounds* (Wiley, New York 1992)
17. T. Peng and J. Piprek: Electron. Lett. **32**, 2285 (1996)
18. G. M. Laws, E. C. Larkins, I. Harrison, C. Molloy, and D. Somerford: J. Appl. Phys. **89**, 1108 (2001)

11 Silicon Solar Cells

P. P. Altermatt

Australian National University, Engineering, Canberra ACT 0200, Australia,
pietro.altermatt@anu.edu.au

11.1 Operating Principles of Solar Cells

A solar cell can deliver electrical power $P = I \cdot V$ because it generates *both* a charge current I and a voltage V at the terminal contacts. The cell can do so because the photogenerated electrons and holes thermalize in separate Fermi–Dirac distributions with their quasi-Fermi energies E_{Fn} and E_{Fp}, respectively. To deliver I, the electrons and holes need to be separated; i.e., we need to design the two terminal contacts such that there occurs mainly electron extraction at one contact and mainly electron injection at the other [1]. This is achieved by having a highly doped n-type region in front of one contact and having a highly doped p-type region in front of the other. For this reason, most solar cells have a p-n junction and a characteristic diode behavior depicted in Fig. 11.1. The source of I is not the p-n junction, but the gradients ∇E_{Fn} and ∇E_{Fp} that drive the free carriers toward the contacts. Only small gradients are required, as the resistivity of the silicon material is selected to be low. Therefore, the quasi-Fermi energies of the majority carriers are relatively constant in the whole device, and the delivered V is the difference between the photo-induced E_{Fn} and E_{Fp} near the p-n junction.

In contrast to many other electronic devices, p-n junction solar cells operate in line with their *minority* carrier properties [2] and, hence, are commonly classified as "minority carrier devices." This is an important feature that can be understood as follows. The free carriers distribute themselves throughout the device in the most probable way, which is the Fermi–Dirac (FD) distribution. A carrier type is the minority carrier (with density c_{min}) at one edge of the p-n junction, whereas at the other edge, it is the majority carrier (with density c_{maj}). Due to the FD distribution, c_{min} and c_{maj} are related to each other by the Boltzmann factor, i.e., by:

$$c_{min} = c_{maj} e^{-q\psi/kT}, \qquad (11.1)$$

because the potential barrier ψ across the p-n junction is larger than the thermal voltage kT/q (here, c stands for either electron density n or hole density p). In thermal equilibrium, ψ is the built-in voltage ψ_0, and (11.1) becomes $c_{min,0} = c_{maj,0} e^{-q\psi_0/kT}$, where $c_{min,0}$ and $c_{maj,0}$ denote the minority and majority carrier density in thermal equilibrium, respectively. If additional carriers are injected, ψ_0 is lowered by the separation of the quasi-Fermi levels, i.e., by $E_{Fn} - E_{Fp} \approx V$, and we have $c_{min} = c_{maj}e^{-q(\psi_0 - V)/kT}$. Inserting (11.1) finally yields

$$c_{min} = c_{min,0} e^{qV/kT}. \tag{11.2}$$

This means that V is directly related to the excess minority carrier density $\Delta c = c_{min} - c_{min,0}$ at either edge of the depletion region. As V determines the cell's power output $P = I \cdot V$, most optimization work aims in some way or other to maximize Δc near the p-n junction depletion region.

How can this be done? Let us consider the general expression $\Delta c = G\tau$, where G is the photogeneration rate of electron-hole pairs, and τ is the excess carrier lifetime. This expression reveals that there are two main categories for maximizing Δc, via either G or τ.

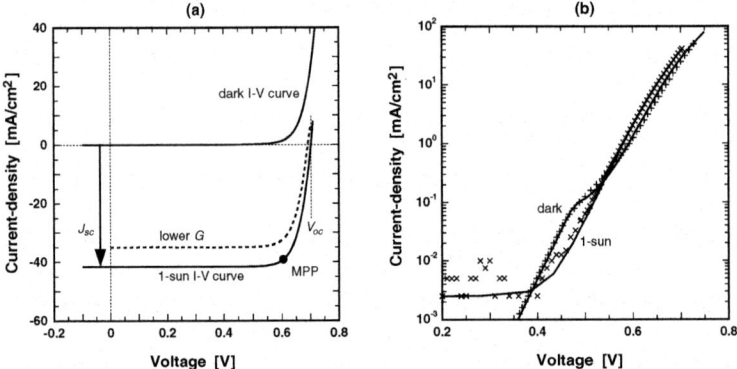

Fig. 11.1. The I-V characteristics of solar cells have a diode behavior. In (a), the I-V curve is plotted in the dark and under terrestrial, nonconcentrated (i.e., "1-sun") illumination (solid lines). According to the superposition principle, the illuminated curve can be approximated with the dark curve shifted by the short-circuit current density J_{sc}. The operating point with maximum power output (MPP) is where $P = I \cdot V$ is maximal. In (b), the same I-V curves are plotted (as symbols) in logarithmic representation; to do so, the illuminated I-V curve is shifted by J_{sc}, revealing both the violation of the superposition principle and fluctuations of the light intensity during the measurements at low voltages. The lines are numerical simulations.

Commonly, G is increased by two kinds of surface treatments: first, by covering the surface with an anti-reflection (AR) coating [3, 4], for example, SiO_2; and second by texturing it, for example, with etching solutions [5, 6]. The second procedure not only reduces the reflection losses, it also deflects the incoming sun rays sideways, so that they are likely to undergo total internal reflection. Such a path-length enhancement leads to better absorption and is called light-trapping. Figure 11.1 shows that enhancing G acts mainly on I and only to a lesser extent on V.

In the other category of optimization, the excess-carrier lifetime τ is increased, and this procedure acts mainly on V. In general terms, τ is the mean time that elapses between the generation of an excess electron-hole pair and its subsequent recombination. If there were only one excess electron-hole pair (per cm^3), then the rate R of recombination events (per cm^3 per second) would be $1/\tau$; but as there are $\triangle c$ pairs in excess, we have $R = \triangle c/\tau$. Hence, τ is a measure of the driving force that restores thermal equilibrium in the solar cell, and we want to minimize this driving force so the photogenerated carriers can reach the terminal contacts and deliver power.

Equilibrium is attained in different device parts in different ways. In lowly doped device regions, R is usually dominated by recombination via defects, which is quantified using the Shockley–Read–Hall (SRH) formalism via τ_{SRH} [7, 8]. In highly doped or injected regions, R is dominated by Auger recombination, i.e., by τ_{Au} [9]. At the surfaces and the metal contacts, interface recombination occurs, quantified using the SRH formalism via the surface recombination velocity S (in units of cm/s) [10]. Mainly because the excess carriers can move to various device parts, they have a range of recombination options. In this case, their lifetime is called effective lifetime τ_{eff}. This is the quantity that is normally measurable [11]. Hence, numerical modeling is a helpful tool to quantify the contributions to τ_{eff} from various device parts and from various recombination mechanisms. Such investigations provide concrete strategies of how to minimize recombination losses and of how to optimize the device.

Although this may sound like a straightforward task, optimization is intricate, mainly because the carriers move around, i.e., because their transport properties are heavily involved. From each generated electron-hole pair, the minority carrier needs to diffuse to the junction, where it drifts across, and subsequently is driven as a majority carrier to a terminal contact. Its survival probability is influenced by the various sources of recombination from different device parts. On top of this, G is very inhomogeneously dispersed in the device: blue light from the solar spectrum is strongly absorbed and hence creates free carriers only near the surface, whereas red light is weakly absorbed and creates electron-hole pairs all over the device. In the following sections, we will follow various optimization procedures and will encounter various tradeoffs in delivered power.

11.2 Basic Modeling Technique

Most commercial and prototype solar cells are p-n junction devices with a large area (usually above 100 cm^2). Nowadays, the majority of silicon cells are fabricated on monocrystalline wafers or on wafers made from cast multicrystalline silicon, both having typical thickness between 180 and 350 μm.

The wafer is the base, usually p-type. An n-type emitter is formed, usually by diffusing phosphorus into the front surface (junction depth $\approx 0.5 - 1.2\,\mu\text{m}$). The front metal contacts are minimized to lines having a width of a few micrometers to avoid shading losses. Hence, the smallest features, relevant in terms of electrical output, are smaller than one micron — whereas the surface area of solar cells reaches many square centimeters.

In numerical modeling, it has been a challenge to spatially discretize the device by means of a "grid" (in 2D) or a "mesh" (in 3D). A mesh with a uniformly high point density cannot be handled because it would contain millions of mesh points. Thus, it is necessary to restrict a dense mesh to regions where any input or output variable changes within small distances; in the remaining regions, a much coarser grid suffices. A good choice of the grid decides whether the applied numerical algorithms converge fast and to a correct solution or not. In addition, the grid must be chosen such that the fluctuations of the output current I are kept well below 1 %. Compared with integrated circuits, where I may fluctuate by 10 % without affecting the binary logic, solar cells are highly sensitive to the photogenerated current. As a check, the mesh density is increased until the differences from the previous results are acceptably small.

To model the electrical output precisely in most cell types, it is sufficient to solve the basic semiconductor equations [12] with the drift-diffusion approximation:

$$-\nabla \boldsymbol{E} = \nabla \cdot (\epsilon \nabla \Psi) = -q(p - n + N_d^+ - N_a^-) \quad (11.3)$$

$$\frac{\partial n}{\partial t} = \frac{1}{q} \nabla \cdot \boldsymbol{J_n} + G - R_n \quad (11.4)$$

$$\frac{\partial p}{\partial t} = -\frac{1}{q} \nabla \cdot \boldsymbol{J_p} + G - R_p \quad (11.5)$$

$$\boldsymbol{J_n} = -q\mu_n n \nabla \Psi + q D_n \nabla n \quad (11.6)$$

$$\boldsymbol{J_p} = -q\mu_p p \nabla \Psi - q D_p \nabla p. \quad (11.7)$$

The Poisson (11.3), continuity (11.4 and 11.5), and drift-diffusion equations (11.6 and 11.7) form a set of three coupled nonlinear second-order partial differential equations in three dependent variables n, p, and the electrical potential Ψ. N_d^+ and N_a^- denote ionized donors and acceptors, respectively, μ denotes the mobility, and $D = (kT/q)\mu$ denotes the diffusion constant. In oxide regions, where no currents are considered, the three equations are replaced by the Laplace equation:

$$\nabla \cdot (\epsilon_{ox} \nabla \Psi) = 0. \quad (11.8)$$

These equations can only be solved if the boundary conditions are given. There are two types of boundary conditions in solar cells. On the terminal contacts, the unknowns are directly given (Dirichlet type of boundary condition). For example, in the case of ohmic contacts, both E_{Fn} and E_{Fp}

coincide with the Fermi level of the metal, which is directly related to the external voltage. This is in fact a thermal equilibrium condition, but it has proven useful because ohmic contacts influence the overall device behavior only insignificantly. Elsewhere in the device, derivatives of the unknowns are involved in the boundary conditions (Neumann type). For example, at the Si/SiO$_2$ interface, the quasi-Fermi level of the minority carriers is related to the amount of occurring recombination. For a discussion on boundary conditions, see [13].

The optical generation G is calculated with a commercially available ray-tracing software such as Optik [14] or Sunrays [15]. Silicon is not a particularly good absorber — photons beyond the red part of the solar spectrum soon reach an absorption length of millimeters [16]. In order to boost the light response of solar cells, sophisticated anti-reflection coating and texturing methods have been developed in the photovoltaic community [5, 6]. With current light trapping schemes, a path-length enhancement of about 16 is commonly achieved [17]. This means that a ray-tracing program needs to account for a multitude of internal reflections to deliver precise results. Also, care needs to be taken that the optical functions of the individual coating materials are accurately known; deviations, especially in the UV range where many coatings are weakly absorbent, lead to imprecise results when simulating the spectral response of devices. An other common source of imprecision is the widespread assumption that metal contacts, evaporated on the polished rear side of the cell, lead to very high internal reflection. Ray tracing shows that up to 35% of photons, reaching the rear metal interface, are absorbed due to the disparity of the optical constants between Si and the metal. Hence, choosing a high, wavelength-independent value of the internal reflection causes considerable error if the rear is fully metalized.

The equations (11.3) – (11.8) are discretized into a set of $3X$ discrete coupled equations, where X is the number of grid points [18]. In this chapter, the software DESSIS [19] is used to solve the equation system. DESSIS discretizes the device domain with the "box method" [20] (also called "finite volumes"), and linearizes the equations with the method of Bank and Rose [21]. The three equations are solved by starting with the initial value Ψ_0, n_0, and p_0, and by applying the iterative procedure to the whole $3X$ dimensional system. This is called the "coupled" iteration approach. Alternatively, one can first solve only the Poisson equation for Ψ (starting with Ψ_0, n_0 and p_0), then use the new Ψ (with n_0 and p_0) to solve the continuity equation for n, and finally use the new values of Ψ and n (with p_0) to solve the drift-diffusion equation for p. This must then be iterated until a self-consistent solution is achieved. This method is called "plugin" or "Gummel" iteration.

The entire I-V curve of Fig. 11.1 is calculated in the voltage-dependent mode, sweeping in selected steps from $V = 0$ toward the open-circuit voltage V_{oc}; only at the end is the current forced to zero in current-dependent mode in order to determine V_{oc} precisely. It is most time-saving to solve voltage-driven boundary conditions with the plugin procedure, and the current-driven ones with the coupled procedure.

At present, there are no useful theoretical error bounds known for the solution of the basic semiconductor equations if they are discretized with the box method. It is often a matter of intuition to enable and to accelerate the convergence to a meaningful physical solution by choosing an appropriate grid and reasonable initial values of Ψ, n, and p. Although there are error bounds known for another discretization method—called finite elements—the application of finite elements to computation had turned out less satisfactorily than the usage of the box method.

11.3 Techniques for Full-Scale Modeling

When the number of mesh points is minimized by adapting the density of grid points to the local conditions, a 3D simulation of a complete cell requires about 500 Gbytes of computer memory and is therefore infeasible. However, the semiconductor structure of most cells is highly symmetrical, and the full cell can be reduced to a geometrically irreducible "standard domain," shown in Fig. 11.2. Its horizontal size is defined by half the (front and rear) contact spacings. As the whole domain is electronically active, any further restrictions of the domain would cause severe errors in the solutions.

The standard domain contains only a small part of the front metal grid (and completely discounts its resistive losses), and the recombination losses in the perimeter region of the cell are also dismissed. These two losses may be included in the simulations with the following approach [22]. The cell can be thought of as being "tiled" with standard domains, as indicated in Fig. 11.2. Then, a typical cell consists of a few thousand standard domains connected in a circuit by ohmic resistances that represent the front metal fingers. For symmetry reasons, only half the front metal grid must be modeled.

The boundary conditions of the standard domains are chosen such that current flows only through the metal contacts. Thus, "tiling" works only if, in real cells, the currents through silicon across these artificially introduced boundaries are negligible. This condition is usually fulfilled, because these boundaries are chosen considering the symmetries of the device. We only need to take care that the voltage drop along the front finger within each standard domain is negligible, and that the voltage difference is small between two neighboring fingers. Thus, it is possible to treat the standard domains in isolation, i.e., to first simulate one single 3D domain, and afterward feed its I-V curve multiple times into the circuit simulation.

The number of devices (nodes) in the circuit simulation can be drastically reduced, for two reasons. First, at each node, several devices can be connected in parallel. For example, there is one standard domain at each side of the finger, corresponding to two standard devices in the circuit model (cf. Fig. 11.2). As these devices are identical, it is sufficient to use just one

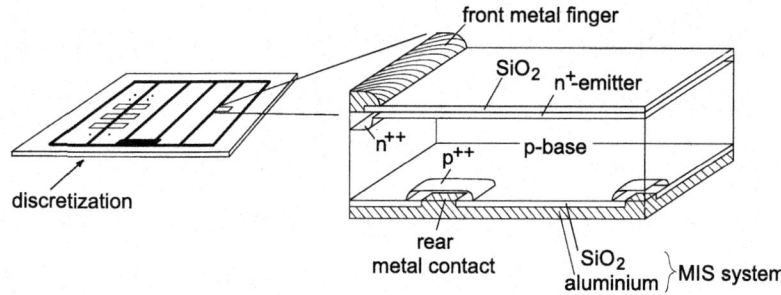

Fig. 11.2. For symmetry reasons, the simulation domain of most cells can be reduced to a geometrically irreducible "standard domain." Shown here is a schematic representation (not to scale) of the standard domain of the PERL cell [23]. It has a moderately doped n^+ emitter with local n^{++} diffusions under the front finger contacts, and local p^{++} diffusions at the rear point-contacts. Both surfaces are passivated with a thermally grown oxide.

standard domain and double its current. Second, the resulting circuit can be discretized, as indicated in Fig. 11.2. The discretization error can be adjusted by varying the coarseness of the discretization.

Losses in the cell's perimeter region may be significant and need to be included in full-scale modeling. Some cells are cut out of the wafer at some distance from the emitter diffusion in order to minimize losses in the p-n junction, whereas laboratory cells often stay embedded in the wafer. In the latter case, a perimeter regions of 5 mm or more needs to be simulated. This is not a problem, as the perimeter region is essentially two-dimensional. This poses the question of how 2D simulations of perimeter effects can be combined with 3D simulations of the illuminated cell areas in a consistent fashion. It was shown in [24] that the perimeter region can be treated as a perturbation of the outermost standard domains, i.e., that the perimeter effect can be superimposed onto the behavior of a cell with ideal borders.

This procedure reduces the required computer memory from about 500 gigabytes to 260 megabytes [22], while keeping the discretization errors sufficiently small. Most of the simulations discussed below were obtained with this technique so that all relevant performance losses can be incorporated and quantified.

11.4 Derivation of Silicon Material Parameters

As a first application of the above-described simulation technique, we model specific experiments to improve the precision of silicon material parameters. Solar cells provide excellent opportunities for this, as they are highly optimized large-area p-n junction devices.

The determination of the intrinsic carrier density n_i in crystalline silicon is an example. The exact quantification of n_i is necessary for an improved understanding and design of semiconductor devices in general, because this parameter enters almost all calculations that relate responses to excitations. In the case of solar cells, n_i influences the minority carrier density and, hence, the voltage across the p-n junction via (11.2). Prior to 1990, $n_i = 1.45 \times 10^{10}$ cm^{-3} was commonly used, leading to significant deviations between the theoretically predicted and the measured behavior of devices [25]. A new value of $n_i = 1.08(8) \times 10^{10}$ cm^{-3} was suggested [25] in 1990 (the number in the parentheses represents the estimated one-standard-deviation uncertainty in the preceding digit). Shortly thereafter, $n_i = 1.00(3) \times 10^{10}$ cm^{-3} was experimentally confirmed by Sproul and Green [26, 27]. In their experiment, Sproul and Green fabricated highly optimized cells on high-purity float-zone (FZ) wafers. After intentionally increasing the rear surface recombination velocity S, the external current of these cells was limited by the electron flow through the boron-doped base. Thus, the current-voltage (I-V) curves of these cells depended only on very few parameters: n_i, the mobility of minority electrons $\mu_{e,min}$ in the base, the dopant density N_{acc} in the base, and the base thickness W. As $\mu_{e,min}$, N_{acc} and W were known from separate measurements of the same material, only n_i was unknown. Sproul and Green obtained n_i by reproducing their measured I-V curves with an analytical model, where n_i was adjusted as a free input parameter. Hence, their technique is among the most direct methods [25] of determining n_i. The symbols in Fig. 11.3(a) show the measured dark I-V curves, and the circles in Fig. 11.3(b) depict n_i extracted from these cells using the analytical model [26].

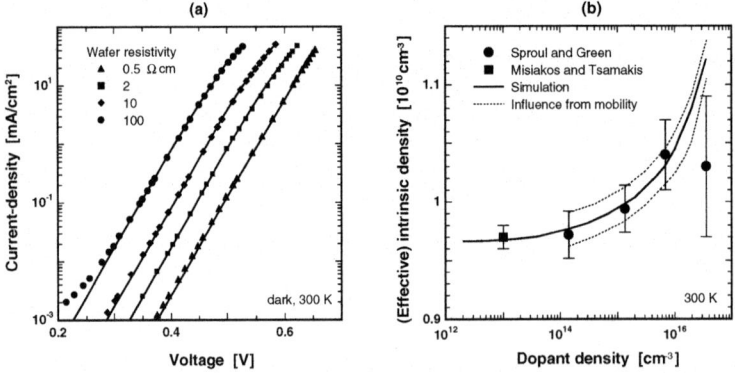

Fig. 11.3. In (a), the dark I-V curves of specially designed silicon solar cells [26] (symbols) are reproduced by numerical simulations [28] (lines). This permits the extraction of the effective intrinsic carrier density $n_{i,eff}$ [lines in (b)] because these I-V curves depend on only a very few material and device parameters, which were separately measured. The intrinsic carrier density n_i is the asymptotic value approached in the left part of (b).

Figure 11.3(b) suggests that Sproul and Green's n_i values increase with rising dopant density and, hence, are influenced by band gap narrowing (BGN). This would imply that the data points in Fig. 11.3(b) do not represent n_i, but they actually represent the effective intrinsic carrier density $n_{i,eff}$ according to the formula:

$$n_{i,eff} = n_i\, e^{\Delta E_g/2kT}, \tag{11.9}$$

where $\Delta E_g > 0$ is the BGN. In this interpretation, n_i is given by the asymptotic value approached in the left part of Fig. 11.3(b). However, there were no measurements available in 1991 indicating BGN at $N_{dop} < 2 \times 10^{17}\,\text{cm}^{-3}$. Hence, Sproul and Green neglected BGN and interpreted the scattering of their data as inaccuracies of their measurement or data evaluation technique. Thus, they took the average of their n_i values, which is $n_i = 1.00(3) \times 10^{10}\,\text{cm}^{-3}$. Today, there are BGN measurements available at low N_{dop}; however, they scatter considerably [29]. Hence, ΔE_g is computed here by means of Schenk's BGN model [30], which is derived with quantum mechanical principles. As described in the previous section, the $2 \times 2\,\text{cm}^2$ large cells of Sproul and Green are simulated here, including 1 cm of wafer material surrounding the cells (which remained embedded in the wafer after fabrication).

The lines in Figs. 11.3(a) and 11.3(b) show the simulation results [28]. All measured I-V curves in Fig. 11.3(a) are reproduced very precisely using the (single) value of $n_i = 9.65 \times 10^9\,\text{cm}^{-3}$, which is, in terms of measurements precision, significantly lower than Sproul and Green's value. The simulated $n_{i,eff}$ [solid line in Fig. 11.3(b)] reproduces the apparent n_i values of Sproul and Green very well, except at $N_{dop} = 3.6 \times 10^{16}\,\text{cm}^{-3}$ (the dashed curves in Fig. 11.3(b) are obtained by using the error bounds of the measured mobility). The revised value of $n_i = 9.65 \times 10^9\,\text{cm}^{-3}$ is the asymptotic value toward low dopant densities of the solid line in Fig. 11.3(b). It has the same error bound as Sproul and Green's experiment (3 %). It is consistent with capacitance measurements of Misiakos and Tsamakis [31], which is included in Fig. 11.3(b) (square). In this way, long-prevailing inconsistencies between independent measurement techniques for the determination of n_i were resolved [28].

The above reinterpretation of the n_i measurements makes it apparent that a sophisticated band gap narrowing model is necessary for precise simulations. Usually, empirical BGN parameterizations have been used, which were derived from measurements in doped silicon and were given as functions of the dopant density. There has been disagreement about which measuring method best yields BGN data relevant to device simulation. The values obtained by absorption measurements are significantly smaller than those obtained by photoluminescence or electronic techniques [28]. It is commonly experienced that solar cells can be most precisely simulated using the electronic BGN data. As these BGN values were extracted from transport measurements, they are influenced by the transport model employed in the data evaluation.

It has been theoretically shown that the transport properties of free carriers are mainly influenced by degeneracy [32, 33] at high doping levels, but to a minor extent also by the change in the density-of-states (DOS) due to the formation of an impurity band [34] at medium to high dopant densities, and by the asymmetry of the gap shrinkage [33, 34, 35], i.e., by $\Delta E_c \neq \Delta E_v$. As all of these influences are neglected in the data evaluation of the empirical BGN values, the obtained values do not generally reflect the band gap narrowing ΔE_g of (11.9), but they are instead a conglomeration of various effects. Hence, they are referred to as an "apparent band gap narrowing" ΔE_g^{app}. For a more detailed treatment of this subject, please refer to [28].

Compared with empirical models, Schenk's model improves the precision of simulations considerably. This quantum mechanical model differentiates between BGN caused by interactions among free carriers (regardless of whether they are induced by doping or by injection) and BGN caused by interactions between the dopant ions and free carriers. Hence, Schenk's BGN is expressed as a function of n, p, N_{acc} and N_{don}, as well as of T. For a comparison with experimental data in low-injection, see [28], and in high-injection, see [36]. Using this model rather than the common empirical expressions has three major advantages: (1) Devices with medium dopant densities are modeled considerably more precisely [37], whereas the empirical models are imprecise; (2) Schenk's model can be used together with Fermi–Dirac statistics [38], as we will see below; and (3) Highly injected devices can be simulated [36], whereas there are no empirical models known to the author that precisely quantify BGN under such conditions.

To demonstrate the advantage of Schenk's BGN model over empirical models, we look at the simulation of highly doped device parts. It was established in the early days of semiconductor physics that Fermi–Dirac statistics need to be applied for N_{dop} above $\approx 1 \times 10^{19}$ cm^{-3}, because Pauli-blocking reduces the pn product significantly. Despite this well-established fact, highly doped device regions are usually modeled using Boltzmann statistics, for historical reasons: FD statistics cannot be applied to analytical models in a feasible way. When using Boltzmann statistics, it is necessary to compensate for the overrated pn product. This is usually done by means of the apparent ΔE_g^{app}, although it has been shown on mathematically rigorous grounds that such adjustments lead to various inconsistencies [35, 34].

FD statistics can be applied in numerical modeling without difficulties. However, they cannot be used together with the ΔE_g^{app} values, as this would cause an overestimation of the degeneracy effects. Adapting the ΔE_g^{app} values to FD statistics is a difficult task, because the ΔE_g^{app} values were obtained using various transport models, and we cannot correct them for FD statistics with one well-recognized procedure.

To avoid these difficulties, BGN and other experiments are reproduced here with numerical modeling based entirely on FD statistics [38]. The FD statistics is implemented as follows. Because the well-known relation $n_i^2 = pn$

Table 11.1. Physical Device Parameters and Models Used in the Numerical Simulations.

Cell geometry	As determined by the manufacturer
Perimeter region	1 cm from cell's edge [39]
Doping profile	As determined with SIMS measurements
Temperature	300 K
Carrier statistics	Fermi–Dirac
Intrinsic carrier density	9.65×10^9 cm^{-3} [28]
Band gap narrowing	Schenk's model [30]
Carrier mobility	Klaassen's model [40], adapted in FZ material to $\mu_{L,e} = 1520$ and $\mu_{L,h} = 500$ cm^2/Vs [28]
SRH recombination in bulk	Mid-gap defects with equal capture cross-sections, excess carrier lifetime τ as given in [9]
SRH recombination at surface	Mid-gap defects with equal capture cross-sections, recombination velocity S as given in [38]
SRH recombination at contacts	Ohmic, i.e., $E_{Fn} = E_{Fp}$ [13, 19], equivalent to $S = \infty$ cm/s
Auger recombination	Injection dependent model [9]
Contact resistance	1×10^{-6} Ω cm [39]
Optical properties	Silicon: [16]; SiO$_2$ and metal: DESSIS [19]; antireflection coatings: Sunrays [15] and unpublished measurements

holds only if the carriers do not interact strongly with each other (analogous to the law of mass action for ideal gases), n is calculated in heavily doped n-type silicon with:

$$n = N_c F_{1/2} \left[\frac{E_{fn} - E_c^{(0)} + \Delta E_c}{kT} \right]. \tag{11.10}$$

N_c is the effective density of states in the conduction band, $F_{1/2}$ is the Fermi integral of order 1/2, $E_c^{(0)}$ is the energy of the intrinsic conduction band edge, and ΔE_c the shift of the conduction band edge due to BGN. In the simulator, p is expressed in an analogous way. However, to keep the following discussion simple, we restrict ourselves here to n-type emitters, where the minority holes are nondegenerate and can be calculated with the simple Boltzmann expression:

$$p = N_v exp \left[-\frac{E_{fp} - E_v^{(0)} - \Delta E_v}{kT} \right]. \tag{11.11}$$

The symbols have equivalent meaning as in (11.10) but for holes and the valence band. In order to clarify the influence of FD statistics on the simulations, we write the pn product in such a way that n_i^2, degeneracy, BGN, and deviations from thermal equilibrium are separated in different factors:

$$pn = N_c N_v exp\left[-\frac{E_c^{(0)} - E_v^{(0)}}{kT}\right] \cdot F_{1/2}\left[\frac{E_{fn} - E_c^{(0)} + \Delta E_c}{kT}\right] \cdot$$

$$exp\left[-\frac{E_{fp} - E_v^{(0)} - \Delta E_v}{kT}\right] exp\left[\frac{E_c^{(0)} - E_v^{(0)}}{kT}\right]$$

$$= n_i^2 \times \frac{F_{1/2}\left[\frac{E_{fn} - E_c^{(0)}}{kT}\right]}{exp\left[-\frac{E_c^{(0)} - E_{fn}}{kT}\right]} \times \frac{F_{1/2}\left[\frac{E_{fn} - E_c^{(0)} + \Delta E_c}{kT}\right]}{F_{1/2}\left[\frac{E_{fn} - E_c^{(0)}}{kT}\right]} exp[\Delta E_v] \times$$

$$exp\left[\frac{E_{fn} - E_{fp}}{kT}\right]$$

$$\equiv n_i^2 \times \gamma_{deg} \times \gamma_{BGN} \times \gamma_{neq}. \tag{11.12}$$

The degeneracy factor γ_{deg} is a measure of how much the pn product deviates from its value as an ideal gas. Only if the electrons are also nondegenerate, as is the case in lowly doped n-type emitters, do we experience $\gamma_{deg} \to 1$ and $\gamma_{BGN} \to 1$, so that $pn \to n_{i,eff}^2 = n_i^2 exp[(\Delta E_c + \Delta E_v)/kT]$, i.e., only then do Boltzmann statistics describe the situation well. To illustrate the difference between Boltzmann and FD statistics, (11.12) is plotted in Fig. 11.4(a), as simulated in a highly doped emitter. FD statistics with BGN of [30] is used, or Boltzmann statistics with BGN of [41], respectively. Although BGN increases the pn product toward the surface of this diffused emitter, degeneracy tends to decrease it, leading to a maximum value of pn within the bulk. Such counteracting effects between BGN and carrier degeneracy cannot be adequately quantified using Boltzmann statistics as well as "apparent" BGN ΔE_g^{app} [35, 34]. For example, as Boltzmann statistics overrate p, ΔE_g^{app} is made smaller than the actual shifts of the band edges in order to adjust the pn product, i.e., $\Delta E_g^{app} < \Delta E_g \equiv \Delta E_c + \Delta E_v$.

In the simulation of highly doped n-type emitters, silicon parameters and device models other than BGN come into play as well. These are mainly Auger recombination, the minority carrier mobility, and the surface recombination velocity parameter of holes, S_{p0}. While Auger recombination and mobility are known from independent measurements, there is no method available to independently measure S_{p0}, except when the emitter is covered with a thin metal layer: S_{p0} of such surfaces is then limited by the thermal velocity of free carriers v_{th}, i.e., $S_{p0} = 1.562 \times 10^7$ cm/s and $S_{n0} = 2.042 \times 10^7$ cm/s at 300 K, independently of the crystal orientation [19, 42, 43]. Hence, the metal-coated samples of Cuevas et al. [41] were a precious opportunity to verify the consistency of the applied model and material parameters. Indeed, the measured emitter saturation-current densities J_{oe} of all of these devices (i.e., at $N_{dop} < 2 \times 10^{20}$ cm^{-3}) were reproduced without any further adjustments [38]. With previous models, such consistency has not been achieved at high N_{dop}, mainly because Boltzmann statistics were applied.

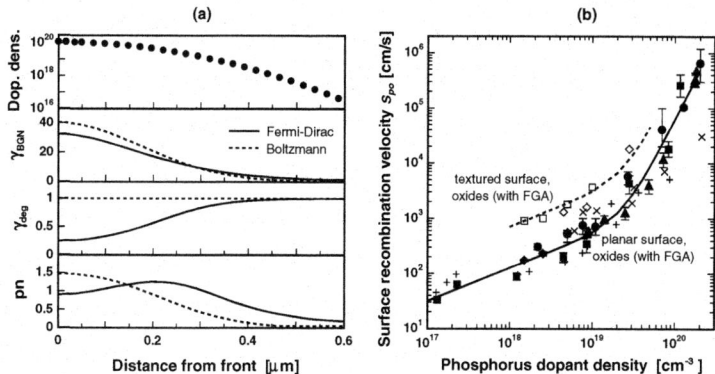

Fig. 11.4. Simulation of a heavily doped emitter [41] with Boltzmann and with Fermi–Dirac statistics, respectively [38]. Shown are, in (a), the phosphorus dopant profile in units of cm^{-3}, band gap narrowing γ_{BGN}, degeneracy γ_{deg}, and the pn product in units of 10^{33} cm^{-6}, all using (11.12). In (b), the surface recombination velocity S_{p0} is shown, as extracted by numerical simulations of saturation current-density measurements, made by various groups [38] on planar (filled symbols) and textured surfaces (empty symbols). The data obtained by using Boltzmann statistics are shown as crosses.

With this improved model, S_{p0} of nonmetalized emitters was determined by re-evaluating J_{oe} measurements made by various research groups using a broad range of samples [38]. Figure 11.4(b) summarizes the S_{p0} values of oxidized samples that acquired a forming gas anneal (FGA). The S_{po} values of planar devices are approximated by the following parameterization:

$$S_{p0} = S_{p1} \left(\frac{N_{dop}}{N_{p1}}\right)^{\gamma_{p1}} + S_{p2} \left(\frac{N_{dop}}{N_{p2}}\right)^{\gamma_{p2}}, \qquad (11.13)$$

where $S_{p1} = 500$ cm/s, $S_{p2} = 60$ cm/s, $\gamma_{p1} = 0.6$, $\gamma_{p2} = 3$, and $N_{p1} = N_{p2} = 1 \times 10^{19}$ cm^{-3}. For comparison, S_{p0} values that were obtained with Boltzmann statistics are plotted in Fig. 11.4(b) as well (crosses). They are considerably lower, in order to compensate for the incorrect statistics.

BGN is important not only in highly *doped* device parts, but also in devices that are highly *injected* with free carriers as, for example, in solar cells that operate under concentrated sunlight [36] (as well as in power devices and some optoelectronic devices). In such devices, BGN is caused mainly by carrier–carrier interactions, whereas in highly doped devices, carrier–dopant interactions come into play as well. Therefore, empirical BGN models that have been given as a function of N_{dop} are not suitable for the simulation of such cells (which are lowly doped). BGN — together with n_i — is important in concentrator cells because it affects the carrier density, which, in turn, affects the Auger recombination rate:

$$R_{au} = (C_n n + C_p p)(np - n_{i,eff}^2). \tag{11.14}$$

These three parameters (n_i, BGN, and the Auger coefficients C_n and C_p) form the foundation for the modeling of highly injected devices and have a sensitive effect on the open-circuit voltage V_{oc} of concentrator cells.

Figure 11.5 shows the measured short-circuit current-density J_{sc} as well as V_{oc} of a concentrator cell made by Sinton and Swanson [44] (symbols). This experiment is very valuable for demonstrating consistency among n_i, BGN, and the Auger coefficients. It can be reproduced very precisely over a wide injection range by using the revised n_i and Schenk's BGN model (lines in Fig. 11.5). If such devices are simulated with the usual empirical BGN models, V_{oc} is significantly overestimated (see dashed line in Fig. 11.5).

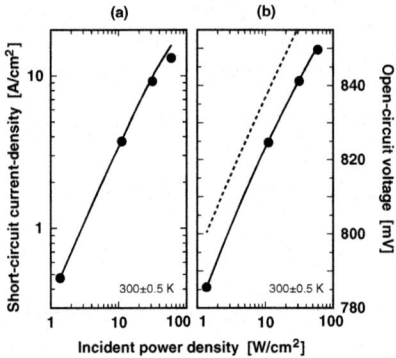

Fig. 11.5. Measured short-circuit current density (a) and open-circuit voltage (b) as a function of incident illumination power density [44] (symbols). The simulated values are obtained using an advanced model for free-carrier-induced band gap narrowing (solid lines) and without such a model (dashed line).

It is important to note that in highly injected devices, the carrier density is also affected by the applied transport model, which should ideally quantify the free-carrier mobility using a tensor [45]. Because DESSIS (like most simulators) does not have such a "complete" transport model, one needs to test the approximations involved. A suitable test is the simulation of J_{sc} as a function of injection density. In Fig. 11.5, it coincides with the measured values except at the highest illumination intensities. There, J_{sc} is very sensitive to the dimensions of the doped regions of the metal contacts, because these regions were minimized to gain high V_{oc} values. The results of Fig. 11.5 show that the approximations in the transport model are sufficiently accurate, and they are even better under open-circuit conditions, where no external current-flow exists.

V_{oc} of the cells made by Sinton and Swanson is limited by band-to-band Auger recombination. In this recombination process, an electron recombines with a hole, transferring its excess energy and momentum to another electron or hole. This process is an intrinsic property of silicon because no crystal dislocations or impurity atoms are involved. In the simulations shown in Fig. 11.5, the ambipolar Auger coefficient $C_a = C_n + C_p$ had to be chosen as injection-dependent. This is so because Auger recombination is affected by Coulomb interactions between the free carriers: The electron density is increased in the vicinity of a hole, whereas it is decreased in the vicinity of another electron. As the Auger recombination rate (11.14) strongly depends on the particle densities, recombination is increased by the nonuniform distribution of the charge carriers. To account for these effects, the Auger coefficients in (11.14) are multiplied by the enhancement factors g_{eeh} and g_{ehh}, respectively. They decrease with increasing injection density [9, 46, 47, 48]. At low-injection conditions, g_{eeh} and g_{ehh} are extracted from measurements of the effective excess carrier lifetime τ_{eff} as a function of N_{dop} [9]. Although τ_{eff} measurements directly yield the Auger lifetime $\tau_{au} = \Delta n / R_{au}$ at high N_{dop}, various other recombination paths contribute significantly to τ_{eff} at lower N_{dop}, such as SRH recombination in the bulk (τ_{SRH}) and at the surfaces, as well as radiative recombination (τ_r). Therefore, contributions from surface recombination need to be subtracted from τ_{eff} to obtain the bulk excess carrier lifetime τ_b, as explained in [47], and then τ_{au} is obtained from:

$$\frac{1}{\tau_{au}} = \frac{1}{\tau_b} - \frac{1}{\tau_{SRH}} - \frac{1}{\tau_r}. \tag{11.15}$$

The enhancement is obtained by comparing the measured τ_{au} with (11.14), using C_n and C_p of Dziewior and Schmid [49]. The same procedure is applied at medium- and high-injection conditions, but C_n and C_p cannot be determined separately due to $n \approx p$. At all injection conditions, the scatter of the data has lead to various parameterizations [48], which differ by relatively small amounts. Most recently, it has become apparent that some of the experiments were influenced by a large amount of photon-recycling, because some experiments were carried out on electrically rather than optically optimized samples. Hence, the amount of photon-recycling is determined with ray tracing, leading to an effective $\tau_{r,eff}$, which has to be inserted in (11.15) instead of τ_r. In this way, τ_{au} can be determined more precisely at lower N_{dop} [50].

11.5 Evaluating Recombination Losses

We now apply the above-described models and parameters to the characterization and optimization of PERL cells, whose structure is depicted in Fig. 11.2. PERL stands for "passivated emitter and rear locally diffused," meaning that the surface recombination velocity S of the moderately doped n$^+$ emitter is reduced (i.e., the surface is passivated) by an oxide and an aluminium anneal, and that the 10×10 μm^2 large rear metal contacts are located at an array of local p^{++} diffusions. These features make this type of cell highly efficient for energy conversion. The PERL cells made at the University of New South Wales, Sydney, Australia [23] reach an efficiency level of 24.7% [51], the highest value for any silicon-based solar cell under terrestrial illumination. Commercial cells usually have a highly doped emitter without considerable surface passivation, and the rear contacts are made to a continuous p^+ diffusion. Their efficiency is in the range between 14% and 16%, although commercial cells reaching 20% are in pilot-line production [52].

The efficiency of solar cells is mainly limited by recombinative and resistive losses (apart from optical losses). Hence, these two types of losses need to be evaluated and optimized. When doing so, the first main step is to distinguish between these two types of losses in existing cells. This is not a straightforward task, as both affect the I-V curve. Resistive losses shift the dark I-V curve to the left of Fig. 11.1(b), and the illuminated one to the right. Recombination losses shift both I-V curves upward in Fig. 11.1(b), i.e., they reduce V_{oc}. The following procedure has proven to be very useful for distinguishing between these two losses. The I-V curve of a cell is measured at two slightly different light intensities to evaluate the total lumped series resistance R_s, which is the sum of all resistive losses in the different device parts [39]. As we will see in Sect. 11.7, R_s depends on J, V, and the light intensity. Then the measured I-V curve is corrected for $R_s(J)$ by plotting it with $V_{corr} = V_{meas} - J \cdot R_s(J)$. This corrected I-V curve results from recombination alone. The same is done in the simulations. Hence, to assess and optimize the recombination losses, it is necessary to work exclusively with such corrected I-V curves, both in experiment and in the simulations (for a detailed description of how to address resistive losses, see Sect. 11.7).

As a second step in evaluating and optimizing devices, we need to know how much recombination occurs in the various device regions. This is important, because only then can we target specific device regions for improving the output power, and this saves time and costs. Without such an assessment, optimization is left to trial and error, which is not effective because one may miss out on the dominating losses, and on top of this, different losses counteract each other due to the transport properties of the free carriers.

To assess the recombination losses in various device regions, one needs to know the relevant input parameters from independent measurements. The S_{p0} values at the emitter surface, shown in Fig. 11.4(b), are an example. Other parameters include τ_{SRH} in the base, the losses at the rear surface, dopant profiles (which affect the amount of Auger recombination losses), and others. Apart from these, the optical properties need to be known as well. The cell may also suffer from shunt leakage paths or other additional losses. If these parameters are known from independent measurements, one should be able to reproduce the experimental I-V curves by modeling without varying the input parameters by more than their experimental precision. Some examples are shown in Fig. 11.3(a). In the process of adjusting these input parameters, it is often advantageous to start with the dark I-V curve, where the optical properties are irrelevant. In this case, though, one has to consider injection-dependent recombination parameters. An example is the recombination at the rear metal-insulator-semiconductor (MIS) structure of the PERL cells, shown in Fig. 11.2 [53, 54].

Once the experimental R_s-corrected I-V curves are reproduced, the relative importance of the recombination in each region is determined as follows. The recombination rate R (in e-h pairs per second and a cell area of $1\,\text{cm}^2$) is separately determined by integration in each device region. These integrated rates can be plotted like I-V curves, because multiplying R by the electron charge q yields the current produced in each device region. An example of a PERL cell is shown in Fig. 11.6. The uncertainties originating from the experimental error bounds of the input parameters are given as shaded areas. Figure 11.6 can be understood as follows. In the dark, the cell's forward current at low voltages ($< 500\,\text{mV}$) is completely dominated by recombination at the rear Si-SiO$_2$ interface. There, the recombination rate is so high that virtually every electron arriving at the rear surface recombines instantaneously; i.e., the recombination rate is limited by the diffusion of carriers to the rear surface. Between 500 and 600 mV, the recombination rate at the rear oxidized surface saturates, as will be explained in the next section. Due to this, SRH recombination in the base dominates above 580 mV. However, the situation is different under 1-sun illumination. In the 0–500-mV voltage range, the carrier densities throughout the cell do not vary strongly (being dominated by photogeneration), and hence, recombination rates stay approximately constant. The simulations predict a total recombination loss of $2.4 \times 10^{15}\,\text{s}^{-1}\text{cm}^{-2}$ electron–hole pairs or, equivalently, a J_{sc} loss of $0.38\,\text{mA/cm}^2$. This is only about 1 % of J_{sc}. Above 550 mV, the forward-injected current across the p-n junction starts to become comparable with J_{sc}, increasing the electron and hole density throughout the cell and hence the recombination losses. As the recombination rate at the rear oxidized surface saturates slightly with increasing electron density in the base, rear surface losses increase at a slower rate than losses in the base or at the front surface. Figure 11.6(b) shows that at MPP, the losses in the base are therefore larger than at the rear surface or in the emitter. The plot also shows that, compared with these losses, recombination in other areas of the device are negligible (such as in the heavily doped regions below the metal contacts).

Once the assessment of state-of-the-art cells is finished, the optimized device design is explored by varying the input parameters in the simulations. Nowadays, sophisticated software is available for this task [55]. Various procedures are used, such as the "Design of Experiments" (which reduces the number of required simulations considerably), the "Response Surface Models" (which builds and evaluates empirical models iteratively), "Generic Optimization" (which allows us to combine multiple optimization goals for different responses), as well as sensitivity and uncertainty analysis. For example, as

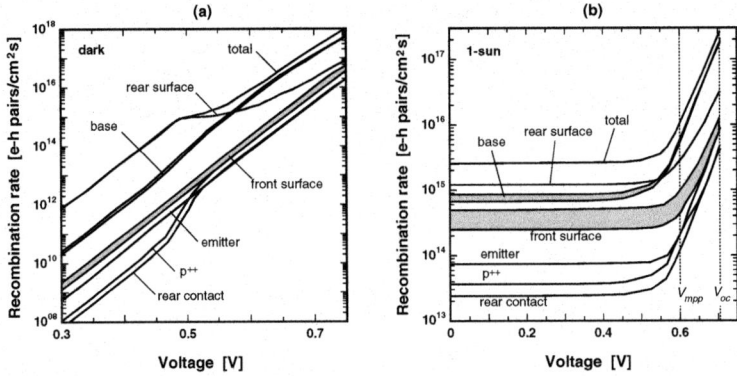

Fig. 11.6. Simulated recombination rates in a PERL cell as a function of the cell voltage in the dark (a) and under 1-sun illumination (b). These simulations are based on independently measured device parameters [such as in Fig. 11.4(b)], and they are varied within their uncertainties by fitting the I-V curves in Fig. 11.1(b), also taking resistive losses of Fig. 11.10 into account. The shaded regions indicate uncertainties.

recombination in the base is dominating the losses in these cells at MPP, the output power relies heavily on the base (wafer) thickness [56].

Once an optimized design is found, a feedback loop needs to be established among the simulations, the fabrication of the cells, and their characterization. Depending on how feasible the suggestions are from the modeling person to the manufacturer, and depending on the precision of the input model parameters, it may take a few loops between modeling and fabrication until considerable progress is achieved. Still, in the author's experience, the combination of fabrication with modeling has greatly reduced time and costs for optimization.

11.6 Modeling the Internal Operation of Cells

Before we proceed to the optimization of resistive losses, we will first discuss an application of the above analysis. In complex devices, it is often difficult to elucidate the internal operation from the I-V characteristics (e.g., the behavior of carrier densities, the p-n junction voltages, etc.). Accordingly, it is often hard to evaluate how the device can be improved or how certain features of the I-V characteristics can be changed. This is where numerical modeling becomes a very valuable tool.

For example, the passivation of the rear surface of PERL cells is obtained by a thermally grown oxide that forms an MIS structure, as shown in Fig. 11.2. Si–SiO$_2$ (and Si–SiN$_x$) interfaces contain immobile positive charges with density Q_s. These charges influence n and p at the interface and, hence, its recombination properties. At low voltages in the dark, n is too small to compensate for Q_s; hence, the silicon region near the interface is depleted of holes. At higher V, more electrons are injected by the external bias, and Q_s is compensated for mainly by electron accumulation rather than by hole depletion. This happens to the extent that an inversion occurs near the interface ($n > p$ in p-type). The region near the interface undergoes a transition from depletion to inversion near a V value that depends on Q_s [53, 54]. Under illumination, there is an abundance of photogenerated electrons at all voltages. Hence, the transition from depletion to inversion occurs only if Q_s is very low — below about 2×10^{10} charges/cm^2. Usually, higher Q_s values are achieved, so the region near the surfaces is inverted at all forward biases. This explains the losses depicted in Fig. 11.6. Although the dark I-V curve has a pronounced shoulder where the rear surface goes from depletion to inversion, the illuminated I-V curve is only moderately influenced, because under illumination, the recombination in the inversion layer stays always hole-limited and therefore saturates only slightly. The injection-dependence of the rear surface passivation also has another effect on the illuminated I-V curve in Fig. 11.1: It bends below 0.4 V because the superposition principle is violated. This means that the illuminated I-V curve, shifted by J_{sc}, does not closely approximate the dark I-V curve (for more details, see [57]).

In an attempt to improve the rear surface passivation, a floating (i.e., a noncontacted) p-n junction was added at the rear surface, resulting in the "passivated emitter, rear floating p-n junction" (PERF) cell design. Although these cells exhibited record 1-sun open-circuit voltages of up to 720 mV [58], their efficiency was degraded by nonlinearities (shoulders) in the logarithmic I-V curves. In order to understand and manipulate such nonlinearities, a detailed investigation of the internal operation of PERF cells was made by means of numerical modeling, based on experimentally determined device parameters [59].

Floating junctions reduce the rear surface recombination by decreasing the hole density at the surface with phosphorus doping. See Fig. 11.7 for a schematic representation of the energy bands. The hole density in the floating junction depends, according to (11.2), on the induced floating voltage V_{fl} across the floating p-n junction. Equation (11.2) also implies that V_{fl} is induced by the excess minority carrier (electron) density at the rear of the p-type base. This excess density varies with the operating conditions, and hence caused the nonlinearities in the I-V curves. Therefore, we need an understanding of how this density behaves as a function of the voltage across the emitter p-n junction V_{em} (which is essentially the external cell voltage minus the resistive voltage losses).

Figure 11.7 plots the simulated V_{fl} as a function of V_{em} under various conditions. It can be understood as follows. In the dark, the electrons are injected from the emitter into the base. The electron density in the base is essentially uniform (due to the large diffusion length in good quality silicon). Thus, according to (11.2), V_{fl} is equal to V_{em} (solid line), and as a result, the

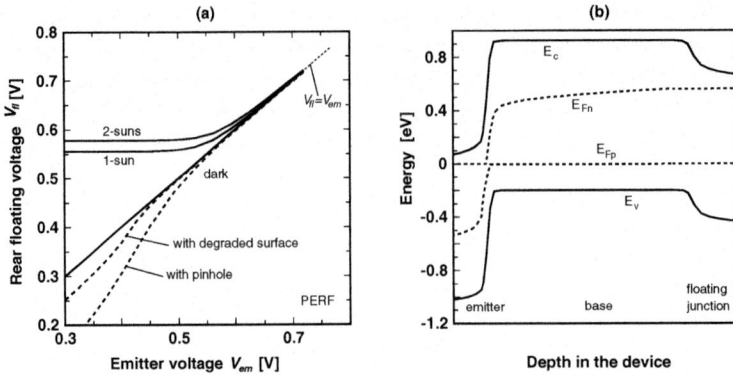

Fig. 11.7. (a) Simulated floating voltage V_{fl} at the rear of a PERF cell [59], as a function of the voltage V_{em} across the p-n junction of the emitter. The solid lines represent the ideal case of no recombination at the rear surface. The dashed lines include the effects of pinholes and of recombination at the rear surface. (b) is a schematic representation of the energy bands of a PERF cell at short-circuit condition.

rear floating junction is forward-biased due to the collection of electrons from the base. Under illumination, the electrons are photogenerated throughout the base. At low voltage, the electron density at the front of the base is lowered because the electrons are collected by the emitter junction and transferred to the emitter metal contacts. Thus, relative to the density at the front, the rear of the base has an increased electron density, and therefore, $V_{fl} > V_{em}$. When the cell voltage approaches V_{oc}, fewer electrons flow to the external load. Thus, most of them spread and recombine in the base, leading to a more uniform distribution of electrons in the base and to $V_{fl} = V_{em}$.

The dashed line in Fig. 11.7(a) shows that recombination at the surface of the floating junction is able to lower V_{fl}, but only at low bias in the dark. This is so because only relatively small numbers of holes can recombine at the rear surface, and this affects the hole density only where it is very small, such as at low voltages in the dark. Under illumination, a large number of holes is photogenerated, which is considerably larger than the number lost by recombination. Similar arguments hold for shunt leakage paths between the floating junction and the rear metal contacts: They can draw only a limited amount of current that is collected by the floating junction. Hence, to understand PERF cells, we do not need to consider possible changes in V_{fl}.

In contrast to V_{fl}, the I-V curves of PERF cells respond very sensitively to surface recombination and shunt leakage paths in the floating junction [59]. This is so because the I-V curve is determined primarily by the amount of recombination (or shunt current), and this amount also changes with V_{em} in cases where V_{fl} is undisturbed. Simulations are compared with the

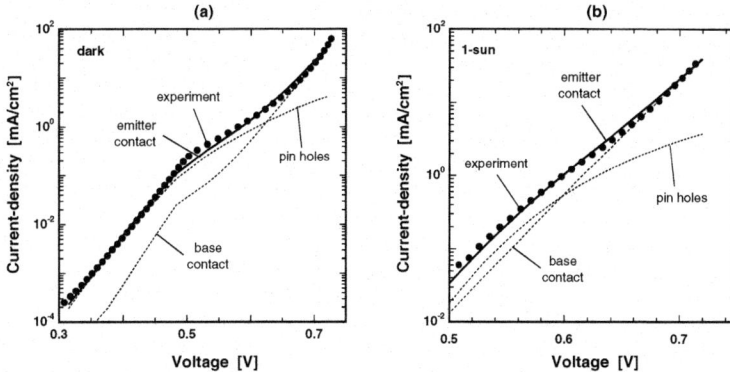

Fig. 11.8. Measured (symbols) and simulated (lines) I-V curves of a PERF cell whose floating junction has a Schottky-type pinhole, in the dark (a) and under 1-sun illumination (b).

experiment in Fig. 11.8. As the PERL and PERF cells differ only in the introduction of an n-diffusion at the rear surface, one would expect that the I-V curves of PERF cells are modeled very precisely by simply using the same device parameters as in PERL cells and by adding a rear floating junction. However, it turned out that the experimentally observed shoulders of PERF cells, shown in Fig. 11.8 as symbols, can only be reproduced if a Schottky-type shunt between the floating junction and the rear metal contacts is introduced. These findings lead to specific experimental investigations. They revealed that, during the manufacturing of the PERF cells, pinholes occur due to holes in the photolithographic film that cover the rear surface. Initially, a thin native oxide layer between the aluminium and the silicon in the pinhole area causes the contact to be ohmic. However, during the p^{++} diffusion, the pinholes are sintered and, hence, of the Schottky type. The shoulders appear because the current through the pinhole is determined by V_{fl} and by the Schottky diode characteristics. The simulated curve describes the experiment very accurately. It deviates most from the experimental one at the shoulder, because there the I-V curve is most sensitive to the specific shape of the Schottky characteristics. It is important to note that, under illumination, the shunt current is small compared with the photogenerated current. This has two main effects: (1) the shoulder in the I-V curve is effectively reduced under illumination; and (2) the pinhole does not change the electron density at the rear of the base substantially, and therefore, the shunting losses are essentially determined by the undisturbed V_{fl} (shown in Fig. 11.8) and by the Schottky characteristics. This is the reason why the shoulder in the I-V curve shifts toward higher voltages with increasing illumination levels, as was experimentally observed [59].

Numerical simulations have also been useful in modeling the internal operation of devices containing grain boundaries, such as polycrystalline solar cells [60, 61].

11.7 Deriving Design Rules for Minimizing Resistive Losses

Solar cells, operating under nonconcentrated sunlight, are typically manufactured on p-type wafers with a resistivity of around $1\,\Omega$cm ($N_{dop} \approx 1.5\times10^{16}$cm^3). In high-efficiency cells, an n-type emitter is formed with a sheet resistivity of $\rho \approx 200\,\Omega$/square. In addition, the front metal contacts are minimized to micron size to avoid shading losses. From these figures, it becomes apparent that the resistive losses are large if they are not carefully minimized.

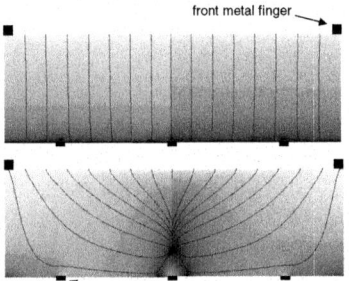

Fig. 11.9. Electron-current flow through the p-type base at short-circuit condition (top) and maximum power point (bottom). The electrons are injected mainly at the midpoint between the front metal fingers, due to the voltage drop along the emitter (top horizontal layer of the device). This causes the electrons to flow along curved paths (lines). The shading represents the current density, from low (dark) to high values (bright).

In the following, we outline the behavior of some resistive losses. The emitter sheet resistance is a *distributed* resistance. This is so because at short-circuit, photogenerated electrons flow vertically through the base and are swept over the p-n junction into the emitter, where they flow laterally to the front metal finger contacts (as shown in Fig. 11.9). This lateral flow causes a voltage drop along the emitter due to the emitter sheet resistance. When the cell is operating at higher external voltages — say, near the maximum power point — the electrostatic potential barrier across the p-n junction decreases, and electrons are thermally injected from the emitter into the base. Because there is a voltage drop along the emitter, this forward injection is strongest at the midpoint between the front metal fingers. The resulting electron density gradient in the base causes the electrons to flow along curved paths through the base, until most of them are finally collected in the emitter regions near the metal fingers. If the emitter sheet resistance is larger than the base sheet resistance, the total lumped R_s of the illuminated cell decreases with increasing external voltage, as is shown in Fig. 11.10 in the case of PERL cells. With this example, we see that the magnitude of distributed resistances depend on the operating conditions, i.e., on the external current density, external voltage, and illumination intensity. In optimization studies, it is therefore important that the I-V curves of manufactured cells can be precisely simulated.

Before being minimized, the resistive losses must be assessed in manufactured cells. In other words, we need to quantify the contributions of different device parts to the total resistive losses given in Fig. 11.10. First, we determine the losses in the emitter. To do this, both the voltage drop toward the finger contacts and the current-density distribution are spatially discretized to the values J_{x_i,y_i} and V_{x_i,y_i}, respectively (2D simulations are sufficient if the emitter has mainly two-dimensional features). The resistance originating from the emitter, R_{em}, is then calculated using

$$R_{em} = \frac{P_{em}}{J^2} \tag{11.16}$$

$$P_{em} = d \int \int (V_{x_{i+1},y_i} - V_{x_i,y_i}) \left(\frac{J_{x_{i+1},y_i} + J_{x_i,y_i}}{2} \right) \Delta y \Delta x \ . \tag{11.17}$$

The total current density J has the same dimensions as J_{x_i,y_i}, and d is a geometric normalization factor (i.e., d = emitter depth/finger spacing). In our example, we have $R_{em} = 0.088\,\Omega\mathrm{cm}^2$ at MPP, which causes 17.9 % of the total resistive losses of this cell.

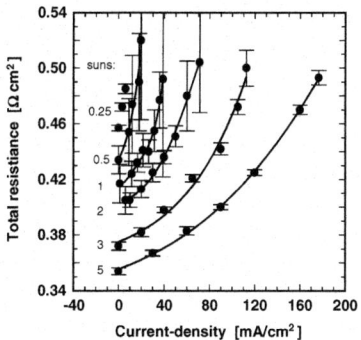

Fig. 11.10. The measured total series resistance R_s of a PERL cell (symbols) compared with simulations (lines) at various illumination intensities. As R_s is a distributed resistance, it depends on the operating conditions.

The base contacts in PERL cells are made with a point-contact metalization scheme to minimize recombination losses. Both the resulting (three-dimensional) current flow and the voltage drop are more complex than in emitters with continuous line contacts. Thus, (11.17) is unsuitable for determining the lumped resistance of the base, R_{base}. However, the resistivity of the metal/semiconductor interface may be set to zero in simulations using the 3D standard domain (cf. Sect. 11.3); the simulated total R_s is then the sum of R_{em} and R_{base}. Thus, R_{base} is obtained by subtracting the above-determined R_{em} from R_s.

The hole-flow pattern in the p-type base is governed by spreading resistance and not by injection effects. Thus, R_{base} is essentially independent of the external current density and voltage, so long as the illumination intensity is about one sun. This was verified by the 3D numerical simulations. Under 1-sun illumination, R_{base} of our example cell is 0.072 Ωcm^2, contributing 14.7 % to the total resistive losses at MPP.

The next contribution to resistive losses comes from the front metal fingers. They are included in the full-size simulations described in Sect. 11.3. The magnitude of the resistance in the metal fingers is adjusted such that the simulated I-V curves, the total R_s, and the voltage drops along the fingers can be fitted very accurately to the experimental values, as shown in Figs. 11.1(b), 11.10, and 11.11. Please note that the resulting I-V curves coincide with the measurements only if the recombination losses are accurately reproduced, as is the case in Fig. 11.6. The voltage drop along the front fingers of Fig. 11.11 can be understood as follows. The collected electrons flow through the fingers toward the contact pad and produce the voltage profile due to finger resistance. Because there is a voltage drop along the busbar as well ($\approx$10 mV), the electrons that are photogenerated at the far end of the cell do not flow along the nearest metal finger. Rather, they tend to flow first through the redundant line (for terminology, see Fig. 11.11). This results in an increased current flow through the central fingers, causing a larger voltage drop ($\approx$11 mV) there, compared with 5 mV in the fingers along the cell's edge. This nonuniform current distribution is the reason why the lumped grid resistance, R_{grid}, cannot be determined simply by measuring the voltage profile, by assuming a spatially homogeneous current generation and by applying $R = V/I$. R_{grid} is, instead, quantified by means of the full-size simulations: R_{grid} equals the difference in total R_s between a simulation that describes the experimentally determined voltage profile along the grid, and another full-scale simulation with zero grid resistance.

In the optimization that followed the assessment of resistive losses, the resistance of the busbar at its maximum width was reduced from 0.087 Ω/cm to 0.0314 Ω/cm (i.e., by 36 %), and R_{grid} fell from 0.330 to 0.174 Ωcm^2 (i.e., by 53 %). In order to understand why the busbar resistance has such a large impact on R_{grid}, we must be aware that the resistive losses in the front grid are only partly due to Joule dissipation, which is linear in R_s ($P_{loss} = R_s I^2$).

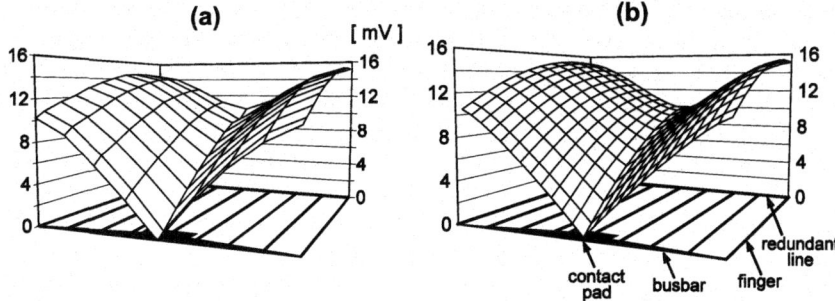

Fig. 11.11. Measured (a) and simulated (b) voltage profile along the front metal fingers of a PERL cell, operated at the maximum power point (MPP) at 1-sun illumination.

Part of the grid losses is caused by the exponential shape of the I-V curve: As a result of the voltage drop along the metal grid, not all regions of the cell are in the optimum local MPP condition when the cell has the terminal MPP bias. The exponential shape of the I-V curve causes the regions far from the contact pad to produce a smaller current density, because these regions operate at higher voltages than the terminal voltage. The power loss due to this effect is called nongeneration loss [62].

Both losses — the one due to nongeneration and the other to Joule dissipation — shift the I-V curves, and thus they cannot be distinguished by R_s measurements. However, it is advantageous for optimization studies to quantify these two types of resistive losses separately, because their behavior is so different. This can again be done by means of the full-scale simulations and by computing the locally produced current density (J_{loc}). Before the optimization, 1 mA/cm^2 less current was locally produced at the far end regions of the fingers than was produced near the contact pad (at 1-sun MPP). Multiplying the local nongenerated current density by the terminal voltage at the contact pad yields the nongeneration loss:

$$P_{nongen} = \text{average}[(J_{cp} - J_{loc})V_{cp}], \qquad (11.18)$$

where J_{cp} and V_{cp} are the current density and the voltage at the contact pad, respectively (for details, see [39]).

The resistance of the fingers reduces when the finger width is increased. However, this increases the shading losses as well. To optimize the finger width w, one needs first to evaluate the finger resistance as a function of w. This is determined by fitting the simulated voltage profile along the metal grid to measurements in fabricated cells that have various finger widths [symbols in Fig. 11.12(a)]. The theoretical curve in Fig. 11.12(a) (line) is explained in [39]. Using Fig. 11.12(a) as input, the cell efficiency can be simulated, taking optical losses into account, as shown in Fig. 11.12(b). The finger spacing is varied from 800 to 1750 μm. Figure 11.12(b) shows that the optimum finger width is in this case 23 μm, if the distance between the fingers is 1.25 mm.

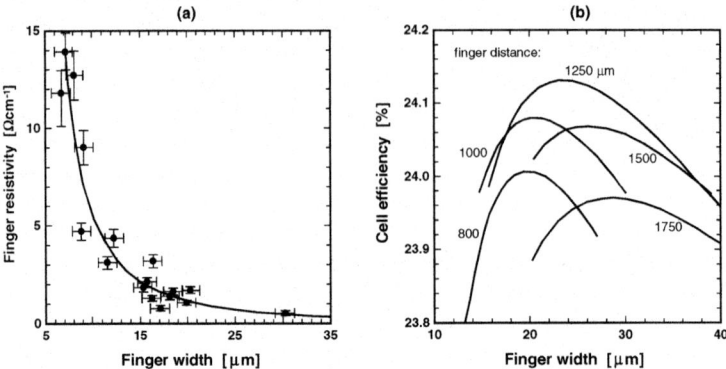

Fig. 11.12. (a) Finger resistance of PERL cells as a function of finger width, obtained from a combination of measurements and simulations (symbols). The line represents a model for fingers with circular edges [39]. (b) Simulated cell efficiency as a function of finger width, for various finger spacings.

Optimizing the resistive losses in the emitter is more complex, because the choice of its sheet resistivity has a large impact on other device parameters. In general, there is a tradeoff between resistive and recombination losses. A higher phosphorus doping density at the surface strongly increases the surface recombination velocity, as shown in Fig. 11.4(b). Lowering the sheet resistivity by increasing the doping density throughout the emitter profile lowers the free-carrier mobility and increases Auger recombination losses. This, in turn, decreases V_{oc} and, at high doping levels, the collection efficiency (and hence J_{sc}). For these reasons, the emitter is doped only moderately in PERL cells (where the surface phosphorus density is 5×10^{18} cm^{-3}), but the junction depth is relatively large (0.7 to 1.1 μm) [63, 64].

The remaining device part to be optimized is the base. The theory for spreading resistance predicts that R_{base} in PERL cells depends mainly on the geometrical configuration of the rear point-contacts and only marginally on the base thickness. The reason is that the current crowds around the point-contacts, producing a much larger voltage drop in these regions compared with the voltage drop across the bulk base. This is confirmed by the simulations even when the contact spacing is increased above 500 μm — the level at which additional resistive losses originate in the bulk due to lateral hole-current flow.

As the current-flow in the base has strong three-dimensional features, the spacing of the rear point-contacts is optimized by means of 3D modeling; apart from this, the 2D simulations are an accurate approximation of PERL cells. The simulations indicate that the cell efficiency depends only minimally on the rear contact spacing so long as the latter is chosen to be between 100 and 300 µm. A spacing larger than 300 µm increases R_{base}, mainly due to current-crowding effects at the contacts and, to a lesser extent, to an increased amount of lateral hole-current flow in the base. A spacing smaller than 100 µm substantially increases the amount of recombination at the contact regions due to both a larger metalization fraction and a larger total volume of the local p^{++} diffusions. The 3D simulations predict an optimum contact spacing of around 160 µm. In experiment, boron atoms in the p^{++} diffused regions and stress in the silicon oxide near the metal contacts may induce dislocations in the silicon crystal, enhancing recombination losses and thus increasing the optimum contact spacing. State-of-the-art PERL cells therefore have a rear contact spacing of 250 µm.

The above optimization study helped to achieve a new world record in Si solar cell efficiency in 1994: It was the first world record to which computer modeling contributed significantly [65, 66]. Since then, modeling has played its part in further enhancing the cell's power output.

References

1. P. Würfel: *Proc. 3rd World Conf. of Photovolt. Energy Conversion*, 2672–2675 (Osaka, Japan 2003)
2. M. A. Green: *Silicon Solar Cells: Advanced Principles and Practice* (University of New South Wales, Sydney, Australia 1995)
3. G. E. Jellison, J. Wood, and R. F. Wood: Solar Cells 18, 93–114 (1986)
4. H. Nagel, A. G. Aberle, and R. Hezel: Progress in Photovoltaics 7, 245–260 (1999)
5. P. Campbell and M. A. Green: J. Appl. Phys. 62, 243–249 (1987)
6. G. Kuchler and R. Brendel: Progress in Photovoltaics 11, 89–95 (2003)
7. W. Shockley and W. Read: Phys. Rev. 87, 835–842 (1952)
8. R. Hall: Phys. Rev. 87, 387 (1952)
9. P. P. Altermatt, J. Schmidt, G. Heiser, and A. G. Aberle: J. Appl. Phys. 82, 4938–4944 (1997)
10. A. G. Aberle: *Crystalline silicon solar cells – Advanced surface passivation and analysis* (University of New South Wales, Sydney, Australia 1999)
11. R. A. Sinton and A. Cuevas: Appl. Phys. Lett. 69, 2510–2512 (1996)
12. W. van Roosbroeck: Bell System Technical Journal 29, 560–607 (1950)
13. D. Schröder: *Modelling of Interface Carrier Transport for Device Simulation* (Springer, Vienna 1994)
14. *Optik version 8.0, Manual* (Integrated Systems Engineering AG, Zurich, Switzerland 2002)
15. R. Brendel: *Proc. 12th European Photovoltaic Solar Energy Conference*, pp. 1339–1342 (Amsterdam, 1994)
16. M. A. Green and M. J. Keevers: Optical properties of intrinsic silicon at 300 K. Progress in Photovoltaics 3, 189–192 (1995)
17. T. Trupke, E. Daub, and P. Würfel: Solar Energy Mat. Solar Cells 53, 103–114 (1998)

18. E. M. Buturla, P. E. Cottrell, B. M. Grossman, and K. A. Salsburg: IBM Journal for Research Development 25, 218–239 (1981)
19. *DESSIS version 8.0, Manual* (Integrated Systems Engineering AG, Zurich, Switzerland 2002)
20. R. S. Varga: *Matrix iterative analysis* (Springer, Vienna 2000)
21. R. E. Bank and D. J. Rose: Numerical Mathematics 37, 279–295 (1985)
22. G. Heiser, P. P. Altermatt, and J. Litsios:*Proc. 6th Int. Conf. Simulation of Semiconductor Devices and Processes*, pp. 348–351, (Erlangen, Germany 1995)
23. A. Wang, J. Zhao, and M. A. Green: Appl. Phys. Lett. 57, 602–604 (1990)
24. P. P. Altermatt, G. Heiser, and M. A. Green: Progress in Photovoltaics 4, 355–367 (1996)
25. M. A. Green: J. Appl. Phys. 67, 2944–2954 (1990)
26. A. B. Sproul and M. A. Green: J. Appl. Phys. 70, 846–854 (1991)
27. A. B. Sproul and M. A. Green: J. Appl. Phys. 73, 1214–1225 (1993)
28. P. P. Altermatt, A. Schenk, F. Geelhaar, G. Heiser, and M. A. Green: J. Appl. Phys. 93, 1598–1604 (2002)
29. D. B. M. Klaassen, J. W. Slotboom, and H. C. de Graaff: Solid-State Electron. 35, 125–129 (1992)
30. A. Schenk: J. Appl. Phys. 84, 3684–3695 (1998)
31. K. Misiakos and D. Tsamakis: J. Appl. Phys. 74, 3293–3297 (1993)
32. F. A. Lindholm and J. G. Fossum: IEEE Electron Devices Lett. 2, 230–234 (1981)
33. A. H. Marshak, M. Ayman-Shibib, J. G. Fossum, and F. A. Lindholm: IEEE Trans. Electron Devices 28, 293–298 (1981)
34. M. S. Lundstrom, R. J. Schwartz, and J. L. Gray: Solid-State Electron. 24, 195–202 (1981)
35. H. Kroemer: RCA Review 28, 332–342 (1957)
36. P. P. Altermatt, R. A. Sinton, and G. Heiser: Solar Energy Materials and Solar Cells 65, 149–155 (2001)
37. S. W. Glunz, J. Dicker, and P. P. Altermatt: *Proc. 17th EU Photovoltaic Energy Conference*, pp. 1391–1395 (Munich, Germany 2001)
38. P. P. Altermatt, J. O. Schumacher, A.Cuevas, S. W. Glunz, R. R. King, G. Heiser, and A. Schenk: J. Appl. Phys. 92, 3187–3197 (2002)
39. P. P. Altermatt, G. Heiser, A. G. Aberle, A. Wang, J. Zhao, S. J. Robinson, S. Bowden, and M. A. Green: Progress in Photovoltaics 4, 399–414 (1996)
40. D. B. M. Klaassen: Solid-State Electron. 35, 953–959 (1992)
41. A. Cuevas, P. A. Basore, G. Giroult-Matlakowski, and C. Dubois: J. Appl. Phys. 80, 3370–3375 (1996)
42. G. Ottaviani, L. Reggiani, C. Canali, F. Nava, and A. Alberigi-Quaranta: Phys. Rev. B12, 3318–3329 (1975)
43. C. Canali, C. Jacoboni, F. Nava, G. Ottaviani, and A. Alberigi-Quaranta: Phys. Rev. B12, 2265–2284 (1975)
44. R. Sinton and R. Swanson: IEEE Trans. Electron Devices 34, 1380–1389 (1987)
45. D. E. Kane and R. M. Swanson: J. Appl. Phys. 72, 5294–5304 (1992)
46. A. Hangleiter and R. Häcker: Phys. Rev. Lett. 65, 215–218 (1990)
47. J. Schmidt, M. Kerr, and P. P. Altermatt: J. Appl. Phys. 88, 1494–1497 (2000)
48. M. J. Kerr and A. Cuevas: J. Appl. Phys. 91, 2473–2480 (2002)
49. J. Dziewior and W. Schmid: Appl. Phys. Lett. 31, 346–348 (1977)
50. To be published.

51. M. A. Green, K. Emery, D. L. King, S. Igari, and W. Warta: Progress in Photovoltaics 11, 347–352 (2003)
52. R. Hezel: Solar Energy Mat. Solar Cells 74, 25–33 (2002)
53. A. G. Aberle, S. Glunz, and W. Warta: J. Appl. Phys. 71, 4422–4431 (1992)
54. S. J. Robinson, S. R. Wenham, P. P. Altermatt, A. G. Aberle, G. Heiser, and M. A. Green: J. Appl. Phys. 78, 4740–4754 (1995)
55. *OptimISE version 8.0, Manual* (Integrated Systems Engineering AG, Zurich, Switzerland 2002)
56. A. G. Aberle, P. P. Altermatt, G. Heiser, S. J. Robinson, A. Wang, J. Zhao, U. Krumbein, and M. A. Green: J. Appl. Phys. 77, 3491–3504 (1995)
57. S. J. Robinson, A. G. Aberle, and M. A. Green: J. Appl. Phys. 76, 7920–7930 (1994)
58. S. R. Wenham, S. J. Robinson, X. Dai, J. Zhao, A. Wang, Y. H. Tang, A. Ebong, C. B. Honsberg, and M. A. Green: *Proc. 1st IEEE World Conf. Solar Energy Conversion*, pp. 1278–1282 (Waikoloa, HI 1994)
59. P. P. Altermatt, G. Heiser, X. Dai, J. Jürgens, A. G. Aberle, S. J. Robinson, T. Young, S. S. Wenham, and M. A. Green: J. Appl. Phys. 80, 3574–3586 (1996)
60. P. P. Altermatt and G. Heiser: J. Appl. Phys. 91, 4271–4274 (2002)
61. P. P. Altermatt and G. Heiser: J. Appl. Phys. 92, 2561–2574 (2002)
62. J. E. Mahan and G. H. Smirnov: *Proc. 14th IEEE Photovoltaic Specialists Conference*, pp. 612–618 (San Diego, 1980)
63. J. Zhao, A. Wang, and M. A. Green: Progress in Photovoltaics 1, 193–202 (1993)
64. A. Cuevas and D. A. Russell: Progress in Photovoltaics 8, 603–616 (2000)
65. J. Zhao, A. Wang, P. P. Altermatt, S. R. Wenham, and M. A. Green: *Proc. 1st IEEE World Conference on Solar Energy Conversion*, pp. 1277–1281 (Waikoloa, HI 1994)
66. J. Zhao, A. Wang, P. Altermatt, and M. A. Green: Appl. Phys. Lett. 66, 3636–3638 (1995)

12 Charge-Coupled Devices

C. J. Wordelman[1] and E. K. Banghart[2]

[1] Synopsys Corporation, TCAD Research & Development Division, 700 E. Middlefield Rd. Mountain View, CA 94043-4033, U.S.A.,
Carl.Wordelman@synopsys.com
[2] Eastman Kodak Company, Image Sensor Solutions Division, Kodak Research & Development Laboratories, 1999 Lake Avenue, Rochester, NY 14650-2008, U.S.A., Edmund.Banghart@kodak.com

12.1 Introduction

Charge-coupled devices (CCDs) are the solid-state photosensitive and charge-transferring elements used in most electronic imaging systems. Shortly after their invention in 1969 at Bell Laboratories [1], CCDs were also used in memory and in signal-processing applications [2]. The early success of CCDs was due largely to their highly modular construction and similarity to metal-oxide-semiconductor (MOS) capacitors. Today, CCDs are used almost exclusively for image sensing, and the continuous development of CCDs for this purpose has contributed to today's digital imaging revolution [3].

Recently, the demand for high-resolution imagers with millions of pixels (or picture elements) has brought about the rapid scaling down of the pixel size. Today's aggressive designs typically call for pixels with dimensions of five micrometers or less [4, 5, 6]. The small pixels and the complicated physical processes used to manufacture them demand extensive use of computer modeling tools.

Technology computer-aided design (TCAD) tools are used to predict the structure of CCDs, based on the manufacturing processes, and their electrical and optical behaviors. TCAD tools are also used to study the fundamental physical mechanisms that occur in CCDs and control their performance. When used in conjunction with theory, TCAD tools provide a means to extract key characterization parameters and to improve the device design. For these reasons, TCAD modeling has become a critical part of the development of CCDs.

In this chapter, the basic aspects of CCD design are studied with an emphasis on device modeling using Synopsys's TCAD modeling tools [7]. Section 12.2 discusses the principles of operation of CCDs and the architectures used for CCD-based image sensors. Section 12.3 describes the process, device, and optical models used in this study and the methods used to implement them. Sections 12.4 through 12.8 focus on several basic aspects of CCD design: charge capacity, charge transfer, charge blooming, dark current, and charge trapping. In each section, examples of TCAD modeling are presented. The examples are primarily based on the full-frame architecture, but the same models and methods may be used to study other architectures. The chapter concludes with a summary, information about how to obtain the examples, acknowledgments, and references.

12.2 Background

12.2.1 Principles of Operation of CCDs

CCDs, or charge-coupled devices, are typically arranged in an array and operated in deep depletion [8, 9, 10, 11, 12, 13]. A linear array of CCDs (see Fig. 12.1) may be described as a shift register, because signal packets (or bits of information) can be moved along the array by sending appropriate signals to the controlling gates. The signal packets are ultimately sent to an output amplifier for charge-to-voltage conversion. In Fig. 12.1(a), two pixels of a shift register using two-phase CCD technology [14] are shown. Each pixel is composed of regions that normally store charge (storage regions) and regions that normally confine charge (barrier regions). Above the storage and barrier regions is a gate. Together, these three regions may be referred to as a phase or stage. Pixels based on two, three, and four phases [9] are commonly used, and pixels based on just a single phase [15] are also possible. Pixels with three or more phases typically do not have barrier regions, because, in these cases, alternate phases can be used as barriers.

CCDs are normally fabricated using silicon (see Table 12.1). It is also possible to fabricate CCDs using other elemental and compound semiconductors [16, 17, 18]. The storage regions in CCDs are normally formed by implanting an n-type layer near the surface of the p-type substrate, just below the insulator and the gates. This structure is described as a buried-channel CCD [19], because the storage regions of these devices lie below the semiconductor-insulator interface. CCDs designed with a buried channel suffer less charge trapping than those designed with a surface channel because carriers are transported away from the semiconductor-insulator interface where traps are often present. N-type buried-channel devices also benefit from the higher mobility of electrons relative to that of holes. In the design shown in Fig. 12.1(a), the barriers are formed by implanting p-type impurities. The gates are typically made of transparent materials such as polysilicon or indium tin oxide (ITO). For nonimaging applications, metallic light shields are typically used [10].

When light representing a scene is incident on the pixels of the sensor, photons can excite electrons from the valence band to the conduction band. In the process, holes are created in the valence band, that is, an electron-hole pair is generated. While the generated holes move to the substrate, the generated electrons are collected as a signal charge in the storage wells of the pixel (see Fig. 12.1). A diagram of the conduction band during the charge-integration mode is shown in Fig. 12.1(b), where every phase is assumed

Table 12.1. Material Properties for Intrinsic Silicon at T=300 K.

Property	Value	Unit
band gap	1.12	eV
relative permittivity	11.9	-
electron mobility	1430	$cm^2V^{-1}s^{-1}$
hole mobility	480	$cm^2V^{-1}s^{-1}$
conduction band state density	2.86×10^{19}	cm^{-3}
valence band state density	3.10×10^{19}	cm^{-3}

to be exposed to light. During this mode, both gates are biased to a large negative value to accumulate holes at the semiconductor surface and, thereby, to suppress dark current [20]. Next, the charge in phase 1 must be combined with that in phase 2. This is accomplished by applying a voltage on phase 2 that is greater than the voltage applied to phase 1 [see Fig. 12.1(c)]. To read out the integrated charge, the biases on the gates are alternated in a sequential fashion. This is known as the charge transfer mode. In all cases, as the barrier to the right of the charge packet is lowered, the packet spills to the right into the next phase. Meanwhile, the barrier to the left of the packet remains high and blocks the flow of charge in the opposite direction. Thus, under the combined influence of the clocks and the barriers, the signal charges (electrons) are transported along the shift register to the right [see Fig. 12.1(c)]. Modeling examples that depict charge transfer in CCDs are presented in Sect. 12.5.

12.2.2 CCD Architectures

In addition to an array of CCDs, a complete image sensor contains an array of color filters for color processing and an array of lenses to focus the light on the CCDs [10]. The entire assembly is mounted in a package. Several types of arrays, or architectures, exist for the CCD-based image sensors. These include: linear arrays, area arrays, full-frame arrays, frame-transfer arrays, interline arrays, and CMOS arrays [2, 11, 12, 10].

12.2.2.1 Linear Arrays

A linear array is the simplest architecture and typically consists of thousands of pixels that feed into an output amplifier. In the marketplace, linear arrays are normally used for document-scanning applications. Such an array is just an extension of the simple shift register (see Fig. 12.1).

12.2.2.2 Area Arrays

An area array combines multiple linear arrays and is typically used for two-dimensional (2D) imaging. The three most common types of area arrays are

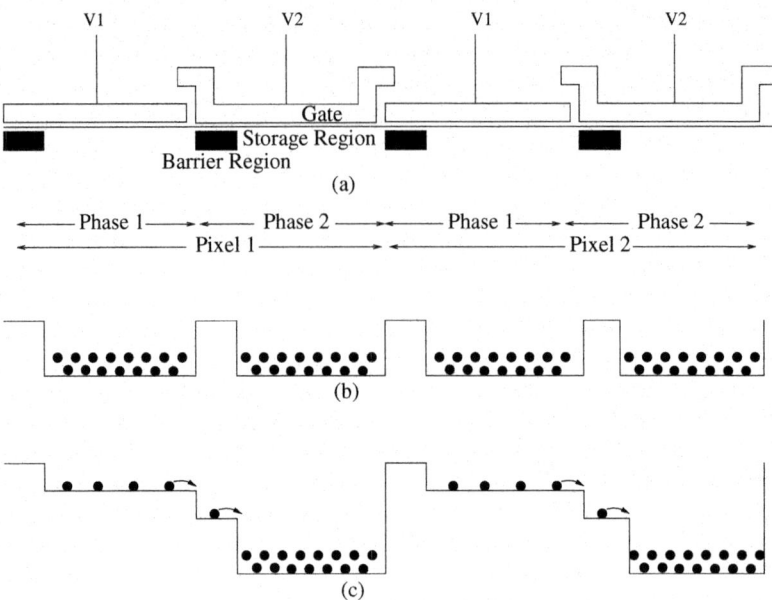

Fig. 12.1. (a) Two-phase CCD structure, (b) conduction band along the channel in the charge-integration mode, and (c) conduction band along the channel in the charge-transfer mode.

full-frame, frame-transfer, and interline (described in the following subsections). As each column in an area array can contain thousands of pixels and thousands of columns may be used, the total number of pixels can be in the millions.

12.2.2.3 Full-Frame and Frame-Transfer Arrays

A full-frame array is shown in Fig. 12.2. When light is incident on the pixels within the array, electrons are photogenerated in the storage wells. This process of generating and collecting charge was introduced in Sect. 12.2.1 and is known as charge integration. After a scene has been exposed, a mechanical shutter is used to cover the device and stop the photogeneration of charge. At this point, a record of the image is stored in the pixels. The charge is transferred out of the vertical columns of the array, one row at a time, to a horizontal register. Before reading the next row, the horizontal register shifts that row to the output amplifier. The process is repeated until all rows have been shifted to the output, at which point the array is ready for the next exposure. In full-frame arrays, the CCDs both integrate and transfer the charge. Full-frame arrays also offer a large photoactive area, which is useful for applications that require high sensitivity or high dynamic range [12, 13].

A variation of the full-frame array is the frame-transfer array [8, 10]. The frame-transfer array contains a second CCD array that is adjacent to the first and covered with a light shield. The frame-transfer array uses the first array for imaging and the second array for storage such that simultaneous imaging and read-out are possible.

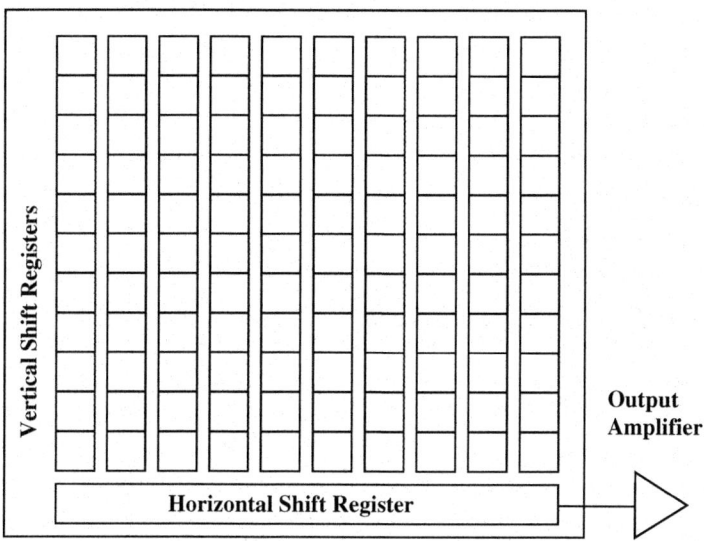

Fig. 12.2. Schematic diagram of a full-frame array.

12.2.2.4 Interline Arrays

An interline array is shown in Fig. 12.3. Contrary to a full-frame or a frame-transfer array, an interline array has pixels with separate regions for the photosensing and charge-transferring operations. The photosensing elements are typically n-type photodiodes with their surface potential pinned to ground by a highly doped p-type implant [21]. The charge-transfer elements are CCD shift registers (see Sect. 12.2.1), which are normally covered with a metal light shield. In this configuration, charge is integrated continuously in the photodiodes. A large pulse is periodically applied to the transfer gate to move charge captured from a scene in the photodiodes into the vertical shift registers. From there, the charges are moved to the horizontal shift register and, ultimately, to the output amplifier (see Fig. 12.3). As a light shield covers the vertical CCD shift registers, the signal charges can be moved as a new scene is integrated and no mechanical shutter is necessary. For this reason,

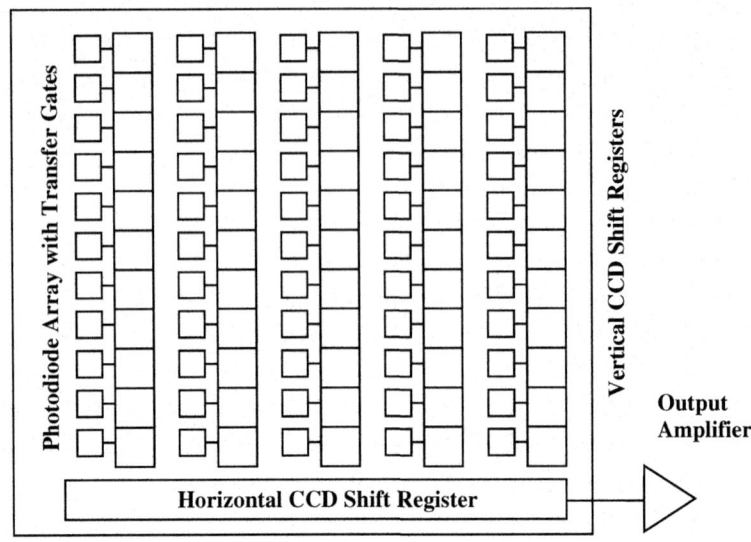

Fig. 12.3. Schematic diagram of an interline array.

interline arrays are ideally suited for motion imaging. Due to their complexity, interline arrays are more costly to fabricate than full-frame arrays. Also, interline arrays have lower sensitivity than full-frame arrays because they have a smaller photosensitive area.

12.2.2.5 CMOS Arrays

A CMOS array, or active-pixel MOS array, consists of pixels in which each photodiode is attached to a CMOS transistor circuit [22]. CMOS arrays have recently received considerable interest because of their ability to route signals from individual pixels. The photodiodes and transfer gates used in CMOS arrays share many features with those used in interline arrays.

12.3 Models and Methods

Computer modeling of electronic devices including CCDs can be divided into process and device modeling [23]. Process modeling predicts the structure of a device based on the manufacturing processes, while device modeling predicts the behavior of a device based on its operating conditions. The basic equations that describe these models and the methods by which they are solved is described in the following sections.

12.3.1 Process Models

Different process models are used to predict the outcome of each step of the manufacturing process. A comprehensive treatment of process modeling can be found in [23]. For the CCD examples presented in this chapter, the process modeling was limited to ion implantation, impurity diffusion, oxidation, and geometrical deposition and etching of the photoresist, insulators, and contacts.

Ion implantation was modeled using the dual Pearson model [24]. Each Pearson distribution used in the dual form is defined by a differential equation of the form

$$\frac{df(v)}{dv} = \frac{(v-a)f(v)}{b_o + av + b_2 v^2}, \tag{12.1}$$

where f is the normalized Pearson distribution, v is the vertical distance, and the other quantities are defined in terms of the moments of $f(v)$ [24].

Impurity diffusion was treated by solving the diffusion equation [23] for each type of impurity, namely

$$\frac{\partial C}{\partial t} = -\nabla \cdot (\boldsymbol{J_m} + \boldsymbol{J_n}), \tag{12.2}$$

where C is the dopant concentration, t is time, $\boldsymbol{J_m}$ and $\boldsymbol{J_n}$ are the flux vectors of vacancies and interstitials, respectively, and are given by

$$\boldsymbol{J_m} = -D_m(\boldsymbol{\nabla}(C_m \frac{M}{M'}) - z_s(C_m \frac{M}{M'})\frac{q\boldsymbol{E}}{kT}) \tag{12.3}$$

$$\boldsymbol{J_n} = -D_n(\boldsymbol{\nabla}(C_m \frac{N}{N'}) - z_s(C_m \frac{N}{N'})\frac{q\boldsymbol{E}}{kT}), \tag{12.4}$$

where D_m and D_n are the diffusivities of vacancies and interstitials, respectively, z_s is the charge of the ionized impurity, C_m is the mobile impurity concentration, q is the electronic charge, k is the Boltzmann constant, T is the temperature, M/M' and N/N' are the diffusion-enhancement factors of vacancies and interstitials [25], respectively, and $\boldsymbol{E}$ is the electric field. Both ion implantation and diffusion can also be treated with Monte Carlo methods [26, 27, 28], although these methods were not used for the results shown here.

The gate oxides were created using models for wet and dry oxidation [29]. The electrodes were created by geometric deposition and etching using the masks [27].

12.3.2 Device Models

Device modeling of CCDs typically consists of modeling the electrical behavior, the optical behavior, and the coupling between them. The models used for each of these domains are described in the sections below.

12.3.2.1 Electronic Models

Most electronic modeling of CCDs is done with the drift-diffusion model, although the simplified hydrodynamic or energy-balance models may also be solved to account for carrier heating [30]. The drift-diffusion model consists of Poisson's equation and the continuity equations [9]. Poisson's equation

$$\epsilon \nabla^2 \psi = -q(p - n + N_d^+ - N_a^-) \tag{12.5}$$

is solved for the electrostatic potential, ψ, where ϵ is the permittivity, q is the electron charge, p and n are the hole and electron densities, respectively, and N_d^+ and N_a^- are the ionized donor and acceptor densities, respectively. The ionized donor and acceptor densities are determined either directly from the solution of the process modeling equations (see Sect. 12.3.1) or are defined using analytic functions.

The electron and hole continuity equations

$$\frac{\partial n}{\partial t} = \frac{1}{q} \nabla \cdot \boldsymbol{J_n} - (U_n - G_n) \tag{12.6}$$

$$\frac{\partial p}{\partial t} = \frac{-1}{q} \nabla \cdot \boldsymbol{J_p} - (U_p - G_p) \tag{12.7}$$

are normally solved for the electron and hole concentrations, n and p, respectively, where the electron and hole current densities, J_n and J_p, respectively, are given by

$$\boldsymbol{J_n} = q\mu_n n \boldsymbol{E} + q D_n \nabla n \tag{12.8}$$

$$\boldsymbol{J_p} = q\mu_p p \boldsymbol{E} - q D_p \nabla p \tag{12.9}$$

where t is the time, U_n and U_p are the electron and hole recombination rates, respectively, G_n and G_p are the electron and hole generation rates, respectively, $\boldsymbol{E}$ is the electric field, D_n and D_p are the electron and hole diffusion coefficients, respectively, and μ_n and μ_p are the electron and hole mobilities, respectively [9]. In the device models above, the subscript n refers to electrons and not to interstitials as used in the process models. Also, in the device models above, G_n and G_p refer to generation due to external stimuli (such as light) and not due to internal processes [see (12.10)].

The electron and hole recombination rates, U_n and U_p, are modeled using the Shockley–Read–Hall (SRH) model [31, 32]

$$U_n = U_p = \frac{pn - n_i^2}{\tau_p(n + n_i e^{\frac{E_t - E_i}{kT}}) + \tau_n(p + n_i e^{\frac{E_i - E_t}{kT}})}, \tag{12.10}$$

where τ_n and τ_p are the electron and hole minority carrier lifetimes, n_i is the intrinsic carrier concentration, E_t is the energy level of the SRH center, and E_i

is the intrinsic energy level. Similar terms describe the surface recombination [32]. When U_n and U_p are positive, they refer to recombination and when they are negative, they refer to internal (thermal) generation.

Three different boundary conditions are employed in the CCD examples presented here. Dirichlet boundary conditions [30] are used for Ohmic contacts. Neumann boundary conditions [30] are used at reflecting domain boundaries. Periodic boundary conditions are used to model periodic behavior in CCD arrays.

12.3.2.2 Optical Models

Optical modeling of CCDs may be performed using a ray-tracing model [33, 34, 35], although models based on Maxwell's equations [36] may also be used. Models based on Maxwell's equations can account for the interference and diffraction of light, whereas models based on ray tracing cannot. Models based on ray tracing are, however, more flexible in terms of the meshing and may be used more quickly than models based on Maxwell's equations.

Ray tracing is based on Snell's law [37]

$$n_1 \sin(\theta_1) = n_2 \sin(\theta_2) , \qquad (12.11)$$

where n_1 and n_2 are the indices of refraction of the optical media and θ_1, θ_r ($= \theta_1$), and θ_2 are the angles of the incident, reflected, and transmitted rays with respect to the surface normal. Ray tracing also uses the Fresnel formulae for the reflected and transmitted waves [37, 33].

The ray-tracing model accounts for the absorption of light as the light propagates using a decaying exponential function for the intensity, I, namely

$$I(x, \lambda) = I_o(\lambda) e^{-\alpha(\lambda)x} , \qquad (12.12)$$

where x is the position along the ray, λ is the wavelength, I_o is the intensity at $x = 0$ (the start of the ray), and α is the absorption coefficient.

12.3.2.3 Optoelectronic Coupling

The coupling between the electronic and optical models is accounted for through the use of a model for the photogeneration of charge. The photogeneration of charge carriers at point x by light of wavelength, λ, is taken to have the form [33]

$$G_\lambda(x, \lambda) = -Q(\frac{\lambda}{hc})\frac{dI(x,\lambda)}{dx} , \qquad (12.13)$$

where Q is the quantum efficiency, h is Planck's constant, c is the speed of light, and $I(x, \lambda)$ is the intensity of light. The photogeneration of carriers at point x by light of all wavelengths is determined by integrating over the light spectrum, that is

$$G_n = G_p = G_{\text{opt}}(x) = \int G_\lambda(x, \lambda) d\lambda. \tag{12.14}$$

As an alternative to optical modeling, the optical charge-carrier generation rates needed in electronic simulation of optoelectronic devices can also be directly provided using analytic functions [33].

12.3.3 Solution Methods

In order to solve the equations described in the models above, it is common to use a mesh to capture the geometry and to solve discretized equations on the nodes of the mesh via the finite difference and box integration methods [30]. The nonlinearity of the equations may be treated by fully coupling them and basing the update of the solution on the full Jacobian matrix (by Newton's method) or by iteratively solving the coupled system (by Gummel's method) [30, 33, 38]. Newton's method is preferred for convergence, and Gummel's method is preferred for memory conservation.

When using Newton's method, which is the more common approach, each step of the fully coupled nonlinear calculation requires the solution of a linear subproblem for the step. The subproblem is solved with a direct method (LU decomposition [30]) or an iterative method (conjugate gradient [39]). Direct methods are robust and converge in fewer iterations than iterative methods because more information is retained during the solution process. As a result of this, the memory requirements of direct methods are larger and direct methods are normally not used for problems with a large numbers of nodes. For the simulations presented here, direct methods were used for the 2D simulations (where the number of nodes is typically lower), and iterative methods were used for the 3D simulations (where the number of nodes is typically higher). In either case, sparse matrix methods [30, 39] are used to limit the storage of the matrix elements.

When modeling CCDs and other unipolar devices, a specialized approach, known as the constant (or fixed) quasi-Fermi level method [40], is often used. The method allows for the study of the electrostatic and electrodynamic response to specified solutions to the continuity equations (or a specified distribution of the charge carriers). This is carried out either by fixing one of the Fermi levels and solving the continuity equation for the unspecified carrier density along with Poisson's equation or by fixing both Fermi levels and solving for the Poisson equation alone. The method of fixing both Fermi levels and solving the Poisson equation alone can only be applied in situations where there is no current because neither continuity equation is solved. This situation occurs in full-frame CCDs at the end of the charge integration when charge is created in the storage regions but remains within these regions until the biases are changed. Generally, the constant quasi-Fermi level method is useful for modeling CCDs because it allows one to decouple the optical and electronic problems in the situation described above and to solve them separately. The approach is used in Sect. 12.4 on charge capacity to obtain electrostatic solutions and in Sect. 12.5 on charge transfer to establish the initial conditions before the transfer. This method is not used in the remaining sections of this chapter where the optoelectronic model equations are solved in a fully self-consistent manner.

12.4 Charge Capacity

A pixel representative of those found in a full-frame array is shown in Fig. 12.4. The pixel is square, measuring 9 μm on a side and rests on a grounded p-type silicon substrate 8-μm thick and with a uniform doping concentration of 2.0×10^{15} cm^{-3}. The pixel consists of two phases of equal length, each with a barrier and storage region similar to those described in Sect. 12.1. As the electric field has almost no component in the x direction in the planes parallel to $x = 0$, which pass through the center of the barrier and storage regions, respectively, it is possible to split the domain in either of these planes and use reflective boundary conditions. This approach is shown in Fig. 12.4, where the domain is split in the plane that passes through the center of the phase-1 barrier. In this pixel, electrons are transferred in the positive x direction and $y = 0$ corresponds to the semiconductor-insulator interface. The device contains an n-type buried channel that is uniform in x and z and is modeled using a Gaussian profile with a peak doping at $y = 0$ of 2.5×10^{17} cm^{-3} and a characteristic length along y of 0.15 μm. To form the barrier regions, a compensating p-type Gaussian profile is used with peak doping at the surface equal to 8.0×10^{16} cm^{-3} and a characteristic length along y of 0.25 μm. The barrier regions measure 1.5 μm in length along x and are uniform along z. Isolation regions are formed along the sides of the device using heavily doped p-type Gaussian profiles with a peak doping at the surface of 1.0×10^{18} cm^{-3} and a characteristic y-depth of 0.3 μm. A dielectric layer consisting of silicon dioxide with a thickness of 500 Å is placed uniformly over the silicon surface. Two electrodes, one for each phase, are defined for the structure and are typically made of polysilicon or indium tin oxide (ITO).

In Fig. 12.5, electrostatic potential contours are shown based on a 3D solution of Poisson's equation with the Fermi levels chosen to deplete the buried channel of the device of both electrons and holes. In order to make the wells empty of charge, the applied biases V_1 and V_2 are set to 0 V, the electron Fermi levels in phases 1 and 2 (ϕ_{n1} and ϕ_{n2}) are set to 20 V, and the hole Fermi levels in phases 1 and 2 (ϕ_{p1} and ϕ_{p2}) are set to 0 V. As a result of symmetry in z, only half of the structure is simulated. In Fig. 12.6, the electrostatic potential is shown along y in the barrier region (at $x = z = 0$ μm) and in the storage region (at $x = 2.25$ μm, $z = 0$ μm). From these plots, it can be seen that a maximum in the electrostatic potential along the y-axis is produced in the buried channel region at about 0.15 μm below the Si/SiO$_2$ interface. This potential is known as the channel potential and is shown in

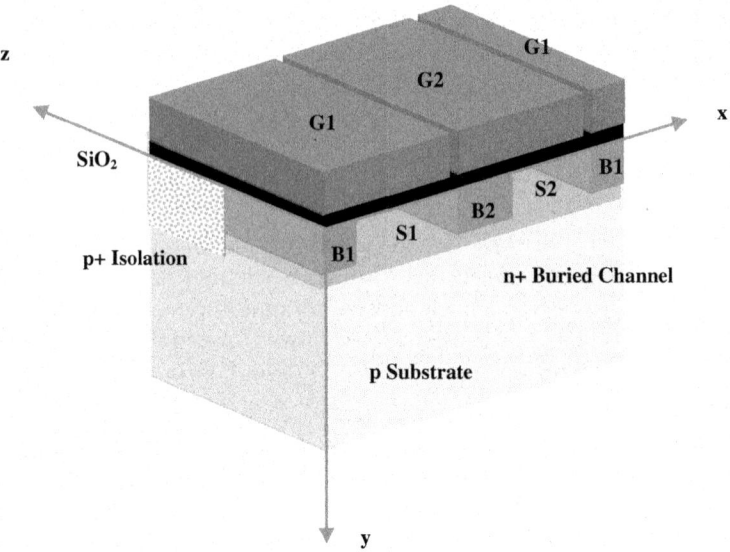

Fig. 12.4. Two-phase full-frame CCD pixel where G, B, and S refer to the gate, barrier, and storage regions, respectively, and the numbers denote the phases.

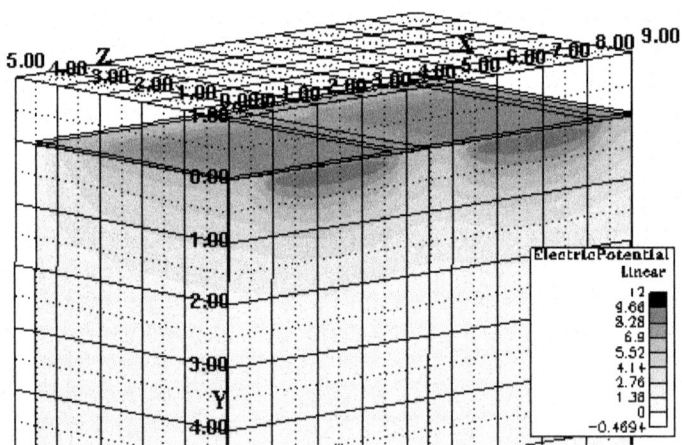

Fig. 12.5. Electrostatic potential in V for the pixel shown in Fig. 12.4 for empty-well conditions in which $V_1 = V_2 = 0$ V, $\phi_{n1}=\phi_{n2} = 20$ V, and $\phi_{p1}=\phi_{p2} = 0$ V. Note that the vertical axis is directed downward.

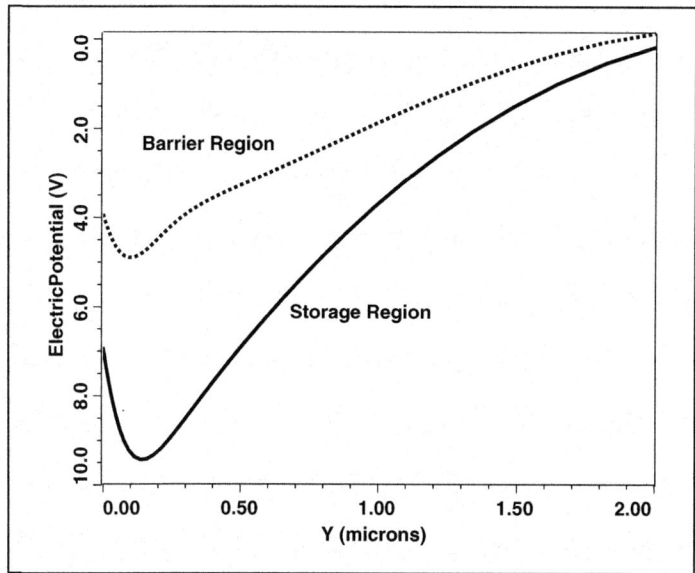

Fig. 12.6. Electrostatic potential along y in the barrier region (at $x = z = 0$ μm) and in the storage region (at $x = 2.25$ μm and $z = 0$ μm) for empty-well conditions. Note that the vertical axis is directed downward.

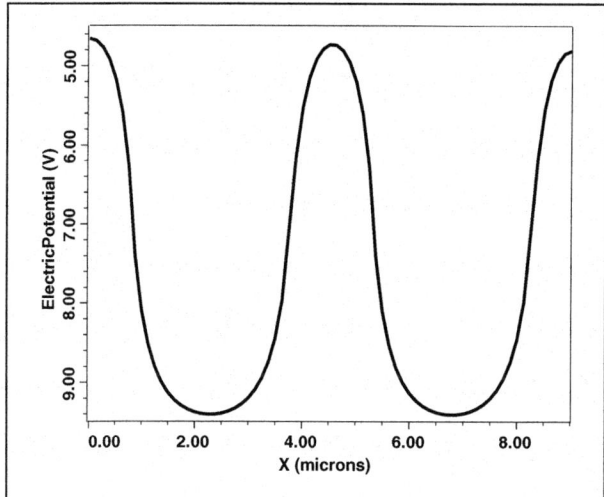

Fig. 12.7. Electrostatic potential along a line through the center of the channel from $(x, y, z) = (0, 0.15, 0)$ μm to $(9, 0.15, 0)$ μm for empty-well conditions. This is known as the channel potential.

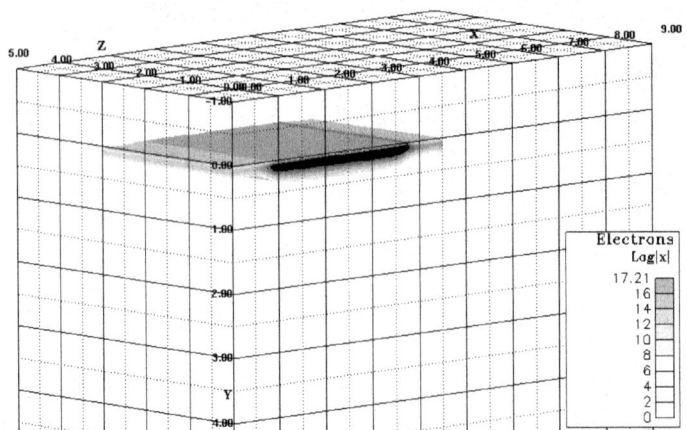

Fig. 12.8. 3D electron density in cm^{-3} for the pixel in Fig. 12.4 for full-well conditions in which $V_1 = V_2 = 0$ V with $\phi_{n1} = 4.5$ V, $\phi_{n2} = 20$ V, and $\phi_{p1} = \phi_{p2} = 0$ V.

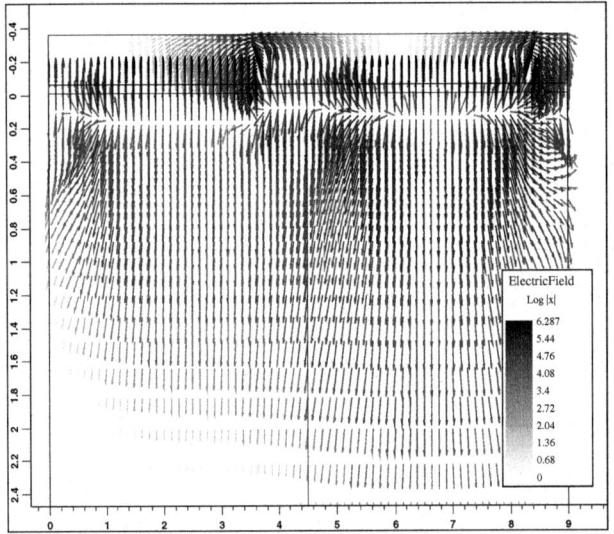

Fig. 12.9. Vector electric field in Vcm^{-1} in the $z = 0$ plane when the device is biased in the transfer mode with $V_1 = -10$ V, $V_2 = 0$ V, $\phi_{n1} = \phi_{n2} = 20$ V, and $\phi_{p1} = \phi_{p2} = 0$ V. The x-component of the field represents the fringing field. The vertical axis shows y in μm, and the horizontal axis shows x in μm.

the direction of charge transfer (i.e., along the x-axis) in Fig. 12.7 for $z = 0$. From this figure, it can be seen that the maximum channel potential in the storage region is 9.40 V, and the minimum channel potential in the barrier regions is 4.65 V when the buried channel is completely empty of charge. The difference between the storage and barrier potentials is known as the storage voltage, which, in this example, is 4.75 V.

To determine the charge capacity, the electron Fermi level in the storage region of phase 1, ϕ_{n1}, is manually adjusted until the storage potential is approximately 0.5 V higher than the lowest available barrier potential. This value is chosen to minimize losses to electron emission over the barrier and has been determined experimentally and confirmed through numerical simulation [41]. At these full-well conditions, the charge capacity is determined to be 1.43×10^5 electrons by integrating the electron density throughout the storage region. (The charge capacity of the full pixel is twice this value.)

The electric field in the $z = 0$ plane is shown in Fig. 12.9 for a new set of bias conditions with $V_1 = -10$ V, $V_2 = 0$ V, $\phi_{n1} = \phi_{n2} = 20$ V, and $\phi_{p1} = \phi_{p2} = 0$ V to transfer electrons from phase 1 to phase 2. The component of this field in the direction of transfer (that is, along x) is known in the literature as the fringing electric field [42]. The fringing field can be used to estimate the time required for carriers to transit the device. By examining the figure closely, it can be seen that the fringing field reaches a minimum approximately in the center of the storage region of phase 1. It is in this vicinity that the last electrons to transfer out of the phase reside. Although the fringing field is a useful analytical tool, a complete description of the charge-transfer problem, which accounts for the presence of charge and the clocking dynamics, requires a self-consistent solution of the Poisson equation with the electron and hole continuity equations. Simulations of this type are described in the next section.

12.5 Charge Transfer

A key measure of CCD performance is the charge-transfer efficiency, η [8, 9, 13]. The charge-transfer efficiency for a CCD is defined as the ratio of the total charge successfully transferred out of a phase to the initial amount of charge present in the phase before the transfer. Alternately, the charge-transfer inefficiency, ϵ, is defined as the ratio of the total charge remaining in the phase to the initial amount of charge in the phase before the transfer, where $\eta + \epsilon = 1$. As several thousand transfers are required in a large sensor, designs typically require a charge-transfer efficiency of 99.999% or higher per phase.

Many factors determine the charge-transfer efficiency, including the length of the phase, the strength of the fringing fields, the time allowed for the transfer, the size and shape of the clock waveforms, and the presence of obstacles such as barriers and wells [43, 44] along the charge-transfer path. In some cases, the charge can become trapped in either bulk or interface states. Thus, determining the dynamics of capture and emission from these traps is important for understanding the extent to which these trapped electrons alter the device performance [32, 45]. This topic will be explored in greater detail in Sect. 12.8.

In this section, two example TCAD simulations of charge transfer in a two-phase full-frame CCD are presented. In the first example, a structure is obtained by 2D process modeling (see Sect. 12.3.1) of a single pixel. The processed structure is designed to match the ideal structure shown in Fig. 12.4. In order to study pixels under conditions of periodicity, the single-pixel structure obtained from process modeling is replicated to form a 3-pixel structure. The second pixel (along with portions of the first and third pixels) is shown in Fig. 12.10. Finally, the 2D three-pixel structure is extruded to 3D and a $p+$ channel stop is added with an analytic profile.

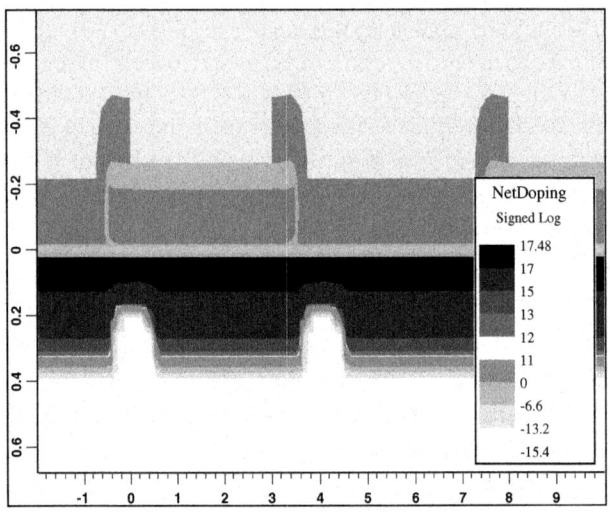

Fig. 12.10. Net doping in cm^{-3} for the structure obtained by 2D process modeling. The center pixel of the three-pixel structure (along with portions of the first and third pixels) is shown. The vertical axis shows y in μm, and the horizontal axis shows x in μm.

Using this structure, the Poisson equation is solved for a static charge configuration with the electron and hole Fermi levels fixed, as described in Sect. 12.3.3. Transient solutions of the charge transfer are obtained using this initial distribution of the charge. This is accomplished by solving the time-dependent drift-diffusion equations [see (12.6) through (12.9)]. In this structure, charge traps are ignored and the influence of the barriers and wells on the transport is minimized by using large voltage swings on the gates.

The electron densities along the center of the channel during the transfer are shown, respectively, in Fig. 12.11. Electrons are initially held in phase 1 with the gate voltages set at $V_1 = 0$ V and $V_2 = -10$ V. The gates are linearly ramped over 0.1 ns to $V_1 = -10$ V and $V_2 = 0$ V. At the end of the allowed period for the transfer, 1 ns, the charge-transfer efficiency is computed to be 99.67%. In Fig. 12.12, the integrated charges in the left and right phases are shown as a function of time. The effect of the total voltage swing on the charge-transfer efficiency is presented in Fig. 12.13. The plot shows that an increased high clock swing can reduce the charge-transfer inefficiency by about 30%. The improvement in the CCD performance can be explained in terms of an increase in the fringing electric field (see Sect. 12.4). A higher clock swing is also useful in pulling charge over small barriers and out of small wells that are produced at the edges of gates because of lateral diffusion and misalignment [44].

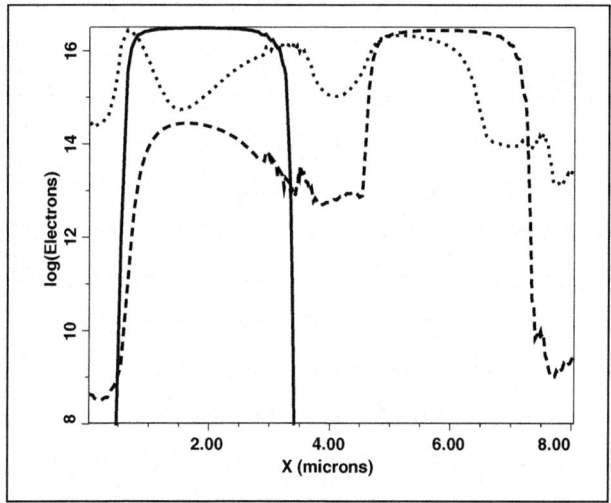

Fig. 12.11. Electron density along the channel through the center of the storage regions ($y = 0.15$ μm and $z = 0$ μm) at the times: $t = 0.00$ ns (solid line), $t = 0.07$ ns (dotted line), and $t = 1.00$ ns (dashed line) for the center pixel of the three-pixel structure (see Fig. 12.10).

12.5.1 Charge Transport Mechanisms

Historically, three mechanisms have been identified for charge transfer in CCDs: self-field, drift, and diffusion [8, 9]. The self-field is a repulsive force that is active during the initial period of charge transfer. The field rapidly dissipates as the charge moves out of the phase. The drift mechanism is attributable to the fringing fields coupling adjacent phases (see Sect. 12.4).

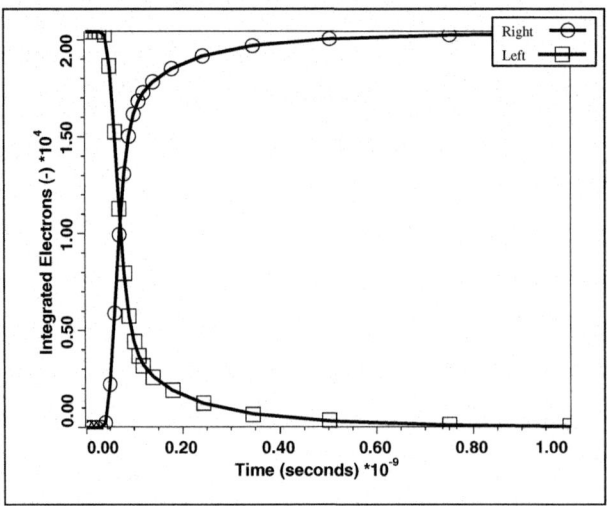

Fig. 12.12. Integrated electrons in the left phase (phase 1) and right phase (phase 2) during the charge transfer for the example shown in Fig. 12.11.

The drift mechanism is normally the most important for removing the final portion of the charge from the phase. To enhance the drift (or increase the strength of the fringing fields), shorter phase lengths, deeper buried channels, and higher clock swings are used, as demonstrated in the previous section. When the fringing fields are weak, Brownian motion or diffusion [46] dominates the transport. This situation commonly occurs for pixels with very long phases, such as those used in astronomy in which high sensitivity is required [13].

When diffusion dominates the transport, it is possible to use simple analytic models to study and design CCDs. In a second example of charge transport, the accuracy of these analytical models is explored for a series of two-phase full-frame designs. In this case, the length of the storage regions in the structures is increased from 3 to 15 μm in increments of 3 μm, whereas the length of the barrier regions remains fixed at 1.5 μm. The initial charge condition and the clock waveforms are similar to those used for the previous example in this section. Transient simulations are performed in 2D using periodic boundary conditions along the x direction, and the integrated charge remaining in the left phase (phase 1) is shown as a function of time in Fig. 12.14.

The instantaneous decay time of the charge is determined from the reciprocal of the slope in Fig. 12.14 and shown as a function of time (see Fig. 12.15). It can be observed in Fig. 12.15 that at large times a characteristic decay time, τ, is reached for each storage length studied. In particular, the characteristic decay time matches the instantaneous decay time at the right-most points of zero slope in Fig. 12.15. The characteristic decay time

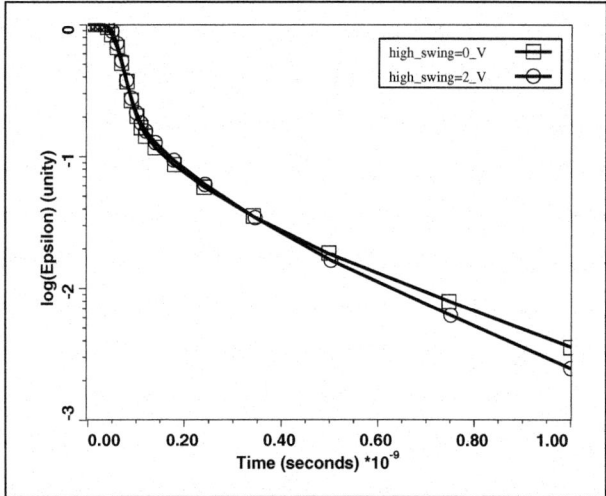

Fig. 12.13. The charge-transfer inefficiency for two different settings of the high clock swing (0 V and 2 V) in which the low clock swing is -10 V in both cases. The case with the high clock swing of 2 V shows reduced inefficiency and higher performance.

represents the time dependence of the first-order term of an infinite series solution to the time-dependent diffusion equation [11, 46]. The characteristic decay times versus the storage length are shown in Fig. 12.16. In the limit of large storage lengths, the curve approaches a straight line, which is in accordance with the diffusion limit [9, 11], namely

$$\tau = \frac{4L^2}{\pi^2 D_n}, \qquad (12.15)$$

where L is the storage length and D_n is the electron diffusion coefficient.

The characteristic decay time τ is plotted versus storage length squared in Fig. 12.17. From the slope of this curve at large storage lengths, the diffusion coefficient associated with transport in the buried channel is estimated to be 16.6 cm^{2}V^{-1}s^{-1} (the diffusion coefficient appears in (12.8) in which both drift and diffusion are considered). By using the Einstein relation, $D_n = \mu_n kT/q$, and assuming room temperature operation, the electron mobility in the channel is calculated to be about 600 cm^{2}V^{-1}s^{-1}. This estimate is consistent with the value given by the table-based concentration-dependent mobility model [47] used in the simulations in the storage regions where the total doping concentration is about 1.5×10^{17}cm^{-3}.

It can also be seen in Fig. 12.16 that the diffusion limit for CCDs with this type of design is valid only for storage lengths in excess of 8 or 9 μm. Ignoring the barrier regions, this indicates that the diffusion limit only applies for two-phase pixel sizes in excess of 16 to 18 μm. However, state-of-the-art devices

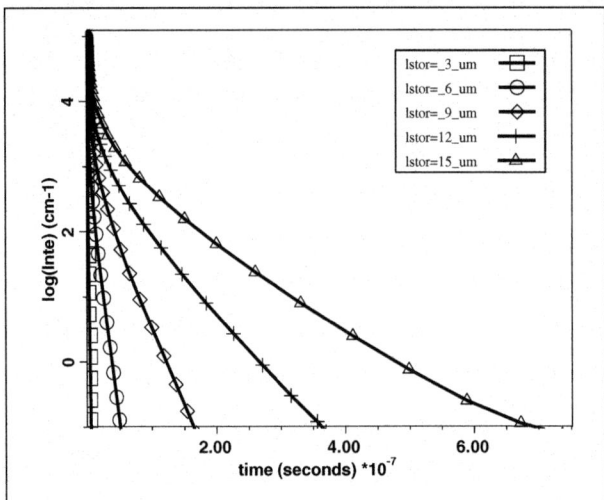

Fig. 12.14. Charge remaining in the left phase (phase 1) versus time for various storage lengths.

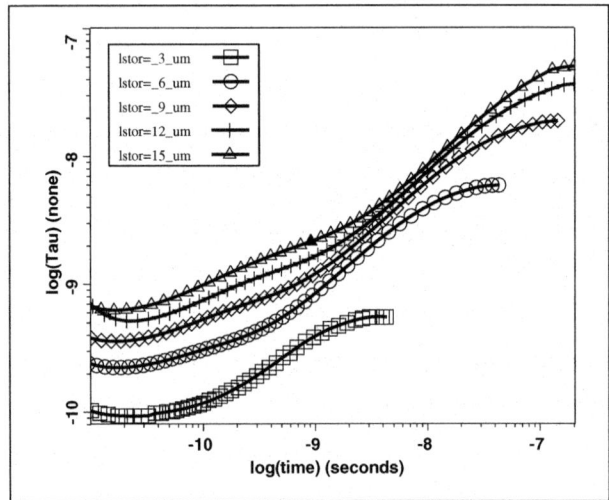

Fig. 12.15. Instantaneous decay time of the integrated charge versus time for the cases shown in Fig. 12.14.

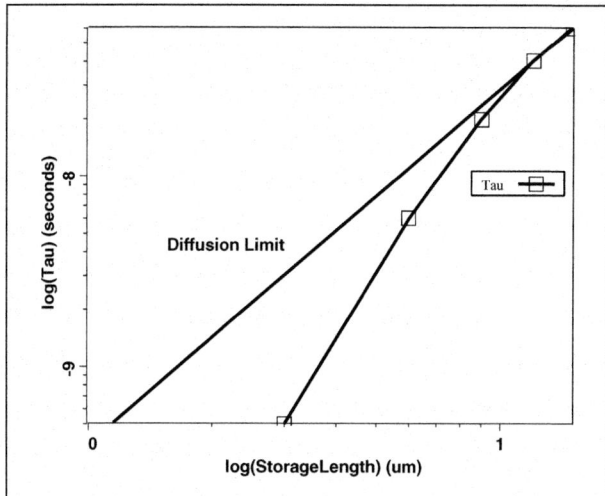

Fig. 12.16. Characteristic decay time versus storage length obtained from Fig. 12.15. In the limit of large storage lengths, the characteristic decay time approaches a straight line associated with the diffusion limit.

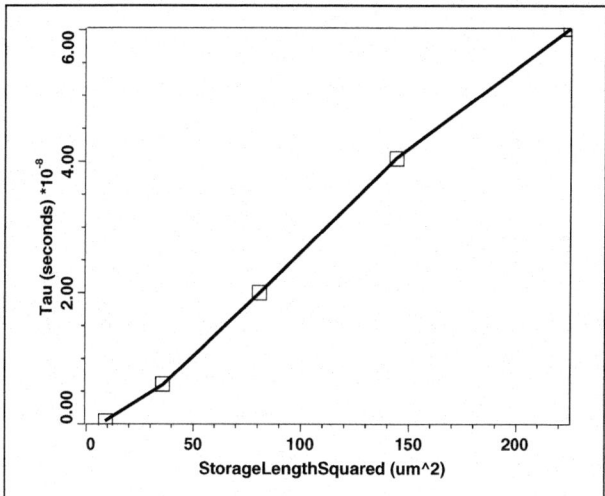

Fig. 12.17. Characteristic decay time versus storage length squared, from which the mobility in the channel can be estimated in the limit of large storage length.

now have pixel sizes of just 5 to 6 μm [14, 48], and CCDs with pixel sizes as small as 2 to 4 μm have been reported [4, 5, 6]. Therefore, this example demonstrates the importance of drift in the electronic transport modeling of present and future CCDs. Furthermore, the drift is expected to be strongly dependent on the 2D and 3D electrostatic properties of these devices.

12.6 Charge Blooming

When the photogenerated signal exceeds the charge capacity in the storage region of the imager pixel, the excess charge spills into the adjacent pixels in the vertical columns. For particularly bright scenes, the number of pixels that are needed to contain the excess charge in the column can be large and will appear as a bright line in the image. This effect is called blooming [2, 8, 10, 12, 13]. The intensity of the illumination, the duration of the charge integration, and the charge capacity of the CCD are critical factors in determining when blooming will occur.

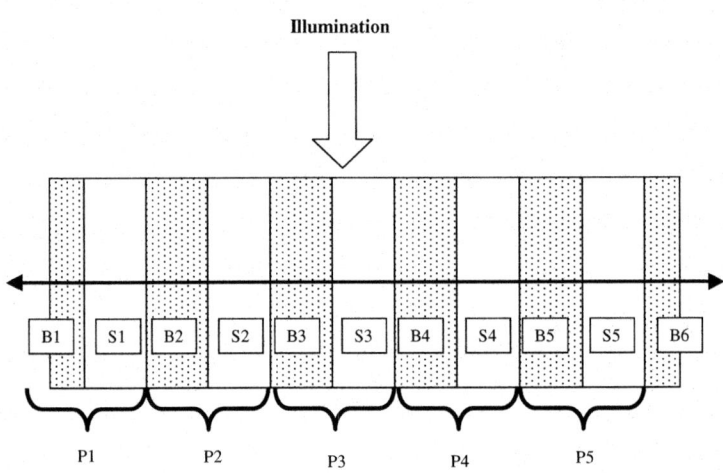

Fig. 12.18. Five-phase CCD with light incident only on the center phase (P3).

By using TCAD modeling tools, it is possible to create a blooming event and to study its behavior. In this example, a structure composed of five identical phases (similar to those described in Sect. 12.4) is modeled in 2D (see Fig. 12.18). The substrate is chosen to be 10-μm thick and to have a uniform p-type doping concentration of $4 \times 10^{15} \text{cm}^{-3}$. The length of the storage and barrier regions (which appear in each phase) are taken to be 2 μm. The n-type buried-channel profile has a peak density at the surface of

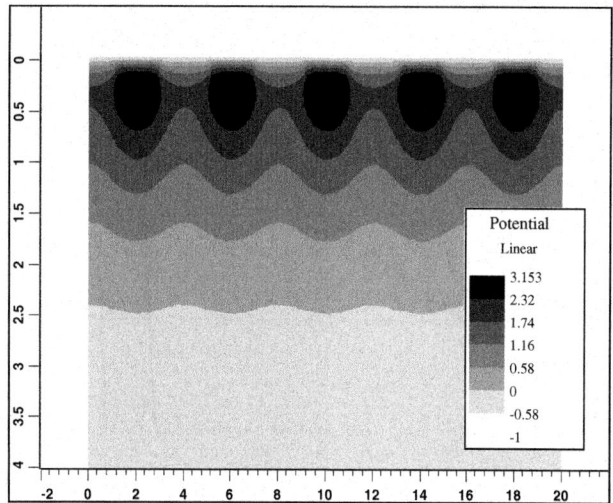

Fig. 12.19. Electrostatic potential for structure in Fig. 12.18 with all gates set to -10 V and the buried channels fully depleted The vertical axis shows y in μm, and the horizontal axis shows x in μm.

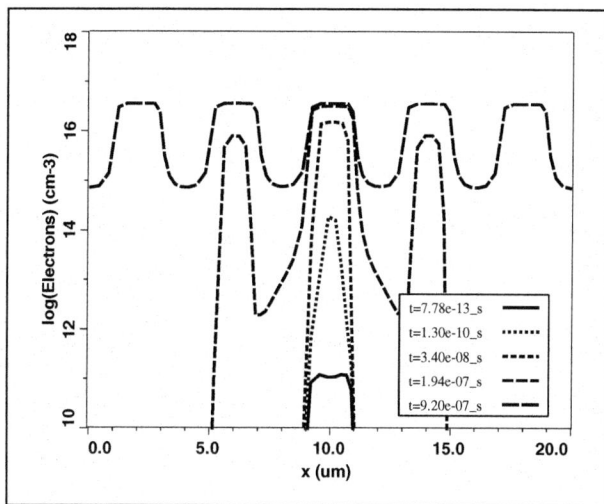

Fig. 12.20. Electron density along x at $y = 0.15$ μm at various times during the blooming process.

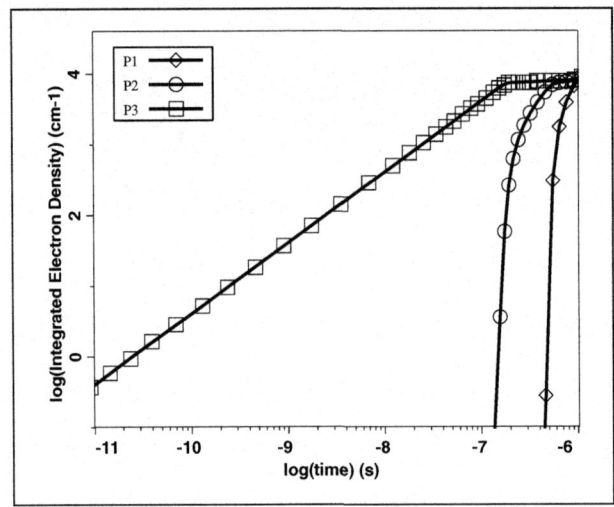

Fig. 12.21. Integrated electron density in phases P1, P2, and P3 versus time for the blooming case. Curves are not shown for phases P4 and P5 because they coincide with those for P2 and P1, respectively.

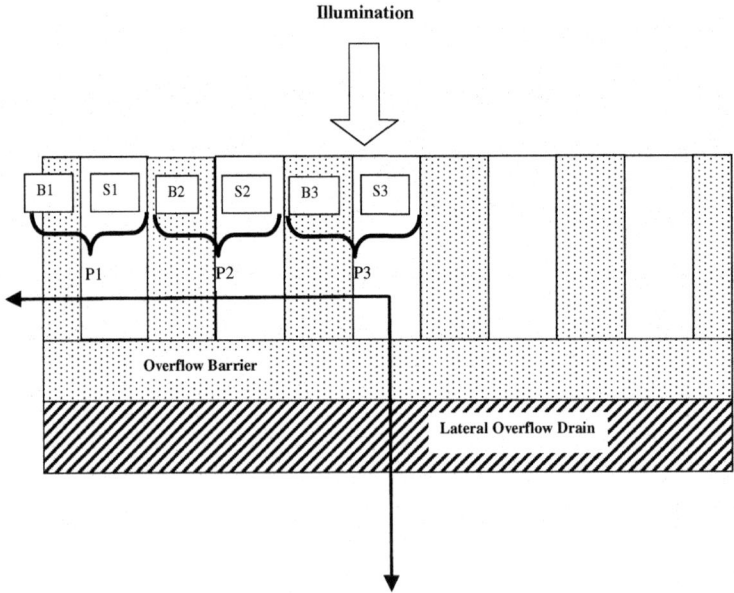

Fig. 12.22. Five-phase CCD with lateral overflow drain (LOD) with light incident only on the center phase (P3). A 2D simulation is performed along the solid line as an approximation to the 3D geometry.

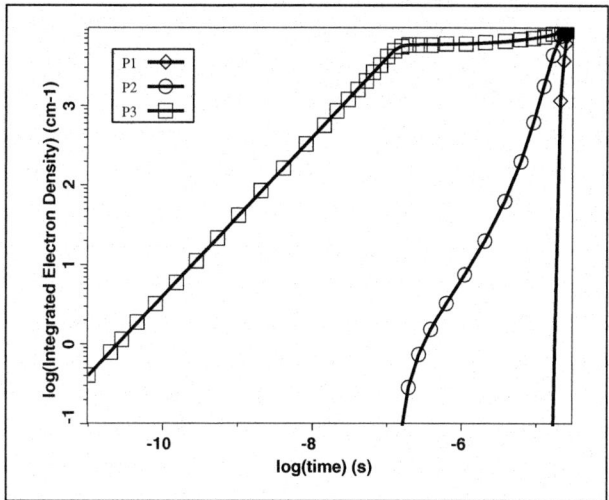

Fig. 12.23. Integrated electron density in phases P1, P2, and P3 versus time for the anti-blooming (LOD) case.

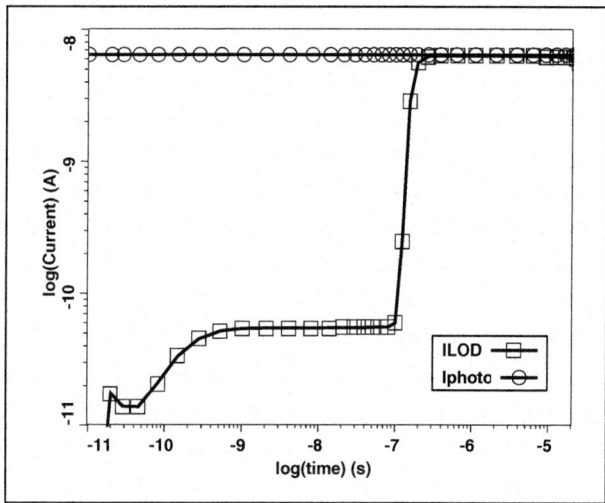

Fig. 12.24. Photocurrent and LOD current versus time for the antiblooming case shown in Fig. 12.22.

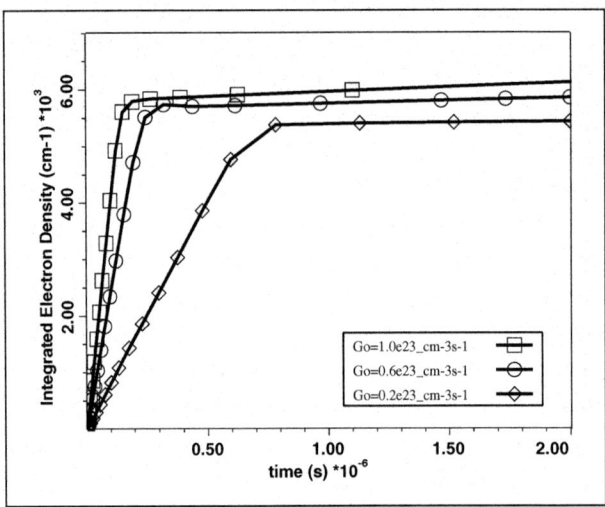

Fig. 12.25. Integrated electron density in phase P3 versus time for the anti-blooming case for three surface-photogeneration rates corresponding to three intensities of light.

2×10^{17} cm^{-3} and a characteristic length along y of 0.25 μm. The barrier regions are formed with a p-type profile with a peak density at the surface of 8×10^{17} cm^{-3} and a characteristic length along y of 0.35 μm. The electrodes are assumed to be fully transparent and are set to -10 V. The channel potentials corresponding to this bias are 3.2 V in the storage regions and 1.8 V in the barrier regions. The electrostatic potential for this structure when the wells are empty is shown in Fig. 12.19. In this case, periodic boundary conditions are applied along x, although the choice of boundary conditions is not critical as the simulation ends before charge reaches the boundary.

Next, a scene is imaged with the CCD in which light is assumed to be normally incident only on the center phase (P3) (see Fig. 12.18). The photogeneration rate is modeled with a decaying exponential function in y with a surface value of 1.0×10^{23} electron-hole pairs cm^{-3} s^{-1} and with an absorption depth, α^{-1}, of 0.2 μm. The absorption coefficient, α, is chosen to concentrate the photogeneration near the center of the channel (the maximum of the potential in the storage region in Fig. 12.6). The photogeneration is assumed to occur from $t = 0$ μs to $t = 1$ μs, and the thermal generation of carriers via the SRH mechanism is ignored. The electron density along x at $y = 0.15$ μm is shown at various times during the photogeneration process in Fig. 12.20. From this figure, it can be seen that, once the maximum charge capacity of the center storage phase is reached, the excess charge spills over into the two adjacent phases (P2 and P4). Over time, these two phases also reach their maximum capacity and the charge again overflows (this time into

P1 and P5). The process will repeat until the light source is removed or until all unfilled neighboring phases have been exhausted.

A convenient summary of the blooming process is obtained by integrating the electron density in each phase and plotting the result versus time (see Fig. 12.21). In Fig. 12.21, it can be seen that the integrated electron density in the center phase (P3) increases linearly with time and abruptly stops increasing at 200 ns, at its maximum capacity. At this time, the two adjacent phases (P2 and P4) begin to fill and reach their maximum capacity at 600 ns and so on.

To limit the spread of blooming in CCD imaging arrays, it is common practice to include a means for conducting the excess signal charge away from the storage wells to a drain. One such structure is known as the lateral overflow drain (LOD) [49] (see Fig. 12.22). Another type of structure for control of blooming is the vertical overflow drain (VOD) [10, 50], which is not discussed here.

As an LOD adds lateral geometry to a CCD, it is accurate to model an array of CCDs with LODs in 3D (see Fig. 12.22). However, a useful approximation to the behavior of the 3D structure is to perform 2D modeling along an L-shaped cut plane that passes through a portion of the CCD array and through the LOD (see Fig. 12.22). Using this approach, the five-phase structure used in the previous example is modified to include a highly doped n-type LOD in storage region S4. In this example, storage regions S1 and S2 are still included in the shift register, but storage region S5 is effectively disconnected from the register and may be ignored. In this configuration, barrier region B4 serves as the overflow barrier. The gates for all phases are set, as before, to -10 V, and the LOD is contacted to an internal electrode set to 10 V. As the donor doping in the drain is significantly higher than the donor doping in the channel, the electrostatic potential in B4 is increased with respect to the potential in barrier regions B1, B2, and B3. This implies that electrons in S3 face a barrier to emission to the LOD, which is relatively lower than the barrier to emission to S2.

As in the prior example for blooming, a light source, incident on phase P3 is turned on at $t = 0$ s, and the electron densities are integrated during a transient simulation. In Fig. 12.23, it can be observed that the integrated electron density in phase P3 increases linearly with time and saturates. Unlike the prior example in which excess charge in P3 moves to adjacent pixels (see Fig. 12.21), in this example, the excess charge is shunted to the LOD. Although phases P2 and P1 do eventually bloom at about 10 μs (see Fig. 12.23), this occurs much later than it did in the prior example in which phase P2 bloomed at about 0.6 μs and phase P1 bloomed shortly thereafter. The shunting operation of the LOD may also be observed in Fig. 12.24, where the LOD current becomes significant at 0.1 μs and rapidly increases to match the photocurrent. In Fig. 12.25, results for the integrated electron density versus time in phase P3 are shown for three intensities of light. It is observed that saturation occurs earliest for the highest intensity. Studies of this type are useful to determine the photoresponse linearity [51].

12.7 Dark Current

In order to create the storage potentials for charge integration and charge transfer, it is necessary to operate CCDs in the deep depletion mode [8]. The deep depletion mode is produced by rapidly applying a large positive bias to the gates to remove all free carriers. While in this mode, ideally only photogenerated electrons will collect in the CCD storage wells. However, as a result of the principle of detailed balance [32], thermally generated electron-hole pairs are produced continuously in a depleted semiconductor in an effort to restore equilibrium. The source of these thermally generated carriers is normally the recombination-generation (RG), or SRH centers [see (12.10)], which have energy near the mid-gap of the semiconductor [31]. The presence of RG centers in semiconductors is generally attributable to defects in the crystalline structure and to the presence of impurities [32, 52].

In CCDs, the dark current is primarily composed of three components: (1) generation at the semiconductor surface (semiconductor-insulator interface), (2) generation in the depletion region associated with the buried channels, and (3) generation in the quasi-neutral region (also known as the diffusion current [52]). These components are illustrated in Fig. 12.26, where a 2D simulation has been performed and the magnitude of the electron recombination rate, U_n in (12.10), is plotted along y through the CCD storage region for 0 V applied to the gate. As U_n is negative in these regions, it corresponds to generation. In this example, the simulation uses a structure similar to those found in Sect. 12.5. The silicon substrate is considered to be of high quality and the electron and hole minority carrier lifetimes are, therefore, set to long times (100 μs) throughout the device. The silicon interface, to the contrary, is considered to be of poor quality with the electron and hole surface recombination velocities, S_n and S_p, both set to 1.0×10^7 cm/s. In Fig. 12.26, a very large peak in the generation rate is observed near the silicon surface. Distinct regions in which the quasi-neutral and depletion currents dominate are also observed.

As it is more difficult to control the quality of the lattice in the surface regions than in bulk regions, depleted surfaces must be avoided in image sensor design. For this reason, negative biases are applied to the gates during the charge-integration mode to accumulate holes near the surface and suppress surface generation [20]. In Fig. 12.27, the total integrated electron density collected in the storage well after 1 ps is plotted as the applied bias on the gates is decreased from 0 to -10 V. An abrupt decrease in the dark current is observed at about -7 V, at which point the hole accumulation layer forms. For this reason, accumulation-mode-clocking schemes are generally preferred

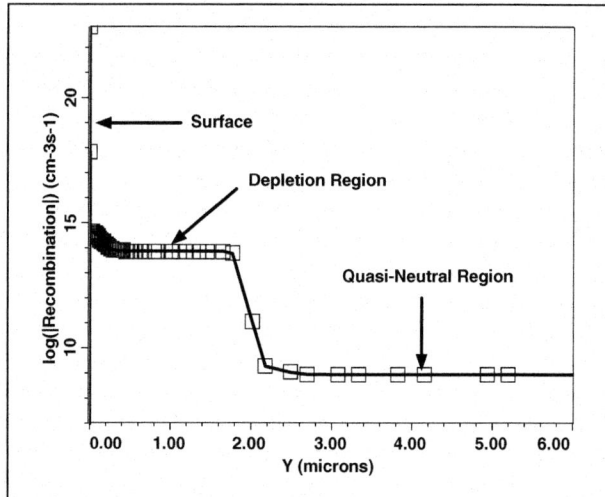

Fig. 12.26. Magnitude of the electron recombination rate, U_n, along y in the storage region. As U_n is negative in this case, it corresponds to generation. All gate potentials are set to 0 V, and S_n and S_p are set to 10^7 cm/s.

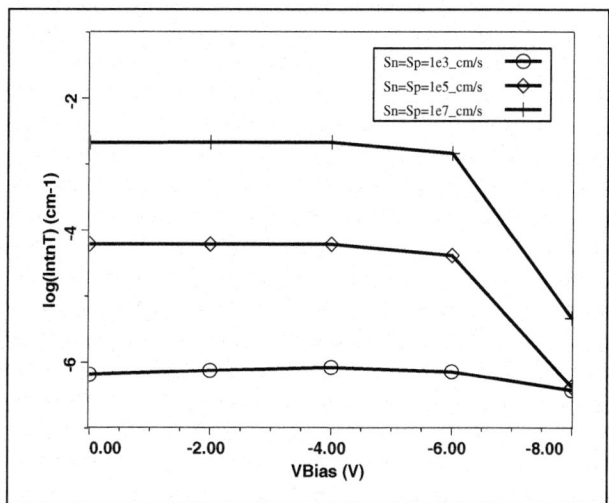

Fig. 12.27. Integrated electron density caused by dark current sources versus the gate bias after 1 ps.

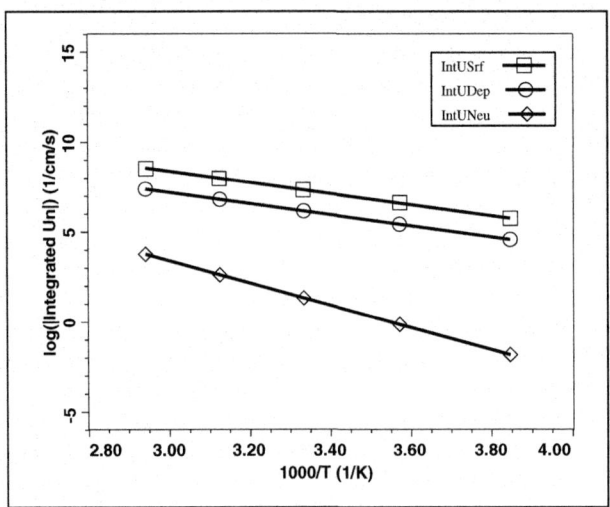

Fig. 12.28. Magnitude of the integrated recombination rate, U_n, versus $1000/T$ at the surface, in the depletion region, and in the quasi-neutral region.

for image sensor operation [14, 49]. In Fig. 12.27, curves are shown for surfaces with varying quality by setting S_n and S_p to 10^3, 10^5, and 10^7 cm/s. As shown in Fig. 12.27, accumulation-mode clocking does not offer significant improvement when the surface quality is very high ($S_n = S_p = 10^3$ cm/s). Furthermore, it is seen that the depletion and quasi-neutral components persist regardless of the applied voltage.

As the presence of dark current greatly reduces the dynamic range of image sensors [12, 13], the identification of dark current sources is an important topic. One frequently used experimental technique for the study of dark current in image sensors is temperature variation [53, 54, 55]. As an example, the dark current for the $L_{store}=3$ μm structure described in Sect. 12.5 is modeled at temperatures ranging from 260 K to 340 K. The magnitude of the individual electron recombination rates shown in Fig. 12.26 are integrated over their respective regions of the 2D device, and the results of these integrations are shown in Fig. 12.28. These integrated recombination (or "generation") rates are plotted versus $1000/T$ in Fig. 12.28, where T is the temperature in Kelvin.

When viewing the temperature dependence shown in Fig. 12.28, it is useful to consider analytical expressions for the components of the total generation current

$$J_{n,\text{dark}} = J_{n,\text{surface}} + J_{n,\text{depletion}} + J_{n,\text{quasi-neutral}}. \tag{12.16}$$

The surface, depletion, and quasi-neutral components, derived in 1D [9, 32, 52], are

$$J_{n,\text{surface}} = \frac{qn_i S_n S_p}{S_n + S_p} \qquad (12.17)$$

$$J_{n,\text{depletion}} = \frac{qn_i W_D}{\tau_n + \tau_p} \qquad (12.18)$$

$$J_{n,\text{quasi-neutral}} = \frac{qn_i^2 \sqrt{D_n/\tau_n}}{N_a}, \qquad (12.19)$$

where S_n and S_p are the electron and hole surface recombination velocities, respectively, τ_n and τ_p are the electron and hole minority carrier lifetimes, respectively, D_n is the electron diffusion coefficient, N_a is the substrate acceptor doping, n_i is the intrinsic carrier concentration, and W_D is the width of the depletion region. The intrinsic carrier concentration that appears in the current components above, has a temperature dependence dominated by the exponential term $exp(\frac{-E_g}{2k_B T})$, where E_g is the energy band gap. This dependence relates to the slopes of the curves in Fig. 12.28 and shows that the surface and depletion components have an activation energy equal to half the silicon band gap, whereas the diffusion component in the quasi-neutral region has an activation energy equal to the full silicon band gap.

12.8 Charge Trapping

When defects in semiconductors are located near mid-gap, they serve as highly efficient RG centers. However, when these centers have energies that occur closer to the conduction and valence band edges, it is more likely they will trap and re-emit carriers into the same band than they will transfer electrons between the bands. In this situation, the centers are referred to as traps, and a generalized form of SRH statistics that includes time dependence is necessary. The generalized electron and hole recombination rates and the time rate of change of electrons and holes in the traps are, respectively [32],

$$U_{n,\text{general}} = c_n p_T n - e_n n_T \qquad (12.20)$$

$$U_{p,\text{general}} = c_p n_T p - e_p p_T \qquad (12.21)$$

$$\frac{dn_T}{dt} = U_{n,\text{general}} - U_{p,\text{general}}, \qquad (12.22)$$

where c_n and c_p are the electron and hole capture coefficients, respectively, e_n and e_p are the electron and hole emission coefficients, respectively, n_T and p_T are the filled and empty occupation densities, respectively, and $N_T = n_T + p_T$ is the total density of the traps. From detailed balance [32] and using Boltzmann statistics [9], it is found that $e_n = c_n n_i e^{(Et-Ei)/kT}$ and

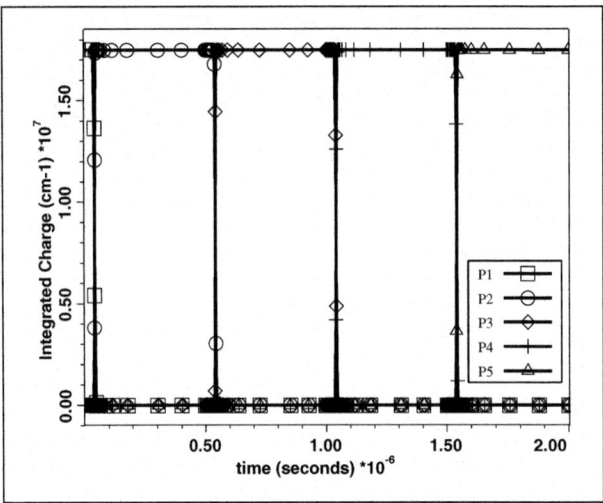

Fig. 12.29. Integrated free-electron density versus time in each phase of a five-phase CCD with no traps.

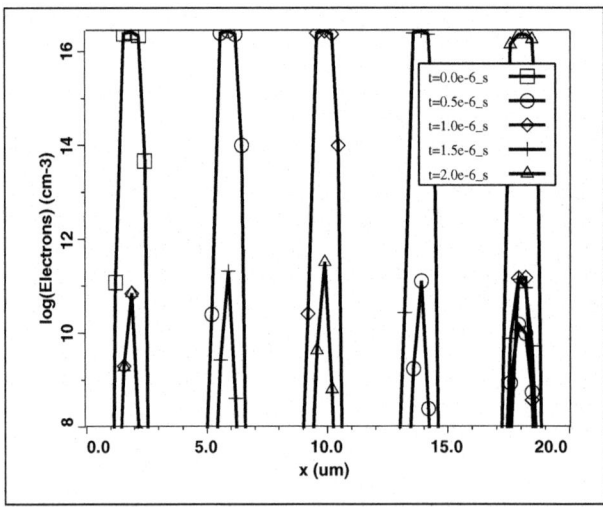

Fig. 12.30. Maximum free-electron density in the interval $0~\mu\mathrm{m} < y < 2~\mu\mathrm{m}$ versus x in a five-phase CCD with no traps. The curves show the maximum free-electron densities at the end of each clock cycle.

$e_p = c_p n_i e^{(E_i - E_t)/kT}$. Also, the capture coefficients are related to the carrier lifetimes through, $\tau_n = 1/(c_n N_T)$ and $\tau_p = 1/(c_p N_T)$.

When a single trap level is considered, the generalized electron and hole recombination rates, $U_{n,\text{general}}$ and $U_{p,\text{general}}$, replace the standard electron and hole recombination rates, U_n and U_p [see (12.10)], in (12.6) and (12.7), and (12.22) is solved for the occupation density. When multiple trap levels are considered, the generalized electron and hole recombination rates for each trap level [see (12.20) and (12.21)] are summed separately and the sums are added to (12.6) and (12.7), respectively. Also, for each trap level, the appropriate forms of (12.22) are solved [56]. Furthermore, Poisson's equation [see (12.5)] is modified to include traps using

$$\epsilon \nabla^2 \psi = -q(p - n + N_d^+ - N_a^- + \Sigma_i(N_{T_i} - n_{T_i}) - \Sigma_j n_{T_j}), \quad (12.23)$$

where i refers to the donor-like traps and j refers to the acceptor-like traps. This model is particulary useful for studying CCDs operated at low temperature [45].

The behavior of charge trapping is studied in a structure similar to the one used for the blooming analysis (see Sect. 12.6). Results are obtained first for the case without traps and are shown in Fig. 12.29. Initially, a signal charge is placed in the storage well of the first phase of the shift register by fixing the electron Fermi level of this phase to 6.3 V and solving Poisson's equation. The clocks are arranged so that the first, third, and fifth phases are initially in depletion ($V_g = 0$ V), and the second and fourth phases are in accumulation ($V_g = -10$ V). The period chosen for each clock cycle is 100 ns and the period chosen to ramp the signals is 0.1 ns. A 2D transient drift-diffusion simulation using (12.5) through (12.10) is performed to determine the charge-transfer efficiency after four clock cycles. To suppress dark current through mid-gap centers, SRH recombination is not included. The results in Fig. 12.29 show that the charge-transfer efficiency after four cycles is almost 100%. The same information is conveyed in Fig. 12.30, where the maximum electron density is shown along x at the end of each cycle. In each phase, the same peak electron density is achieved with the amount of residual charge in the phases several orders of magnitude below the peak.

A second simulation is performed with traps present (see Figs. 12.31 and 12.32). In this example, the transient drift-diffusion model includes the generalized electron and hole recombination rates [(12.20) through (12.23)]. A single donor-like trap level is considered with an energy 0.3 eV above mid-gap and with a density, N_T, of 1.0×10^{15} cm^{-3} distributed uniformly throughout the device. The electron and hole lifetimes chosen for the trap are 1.0 μs and 0.1 s, respectively. From the definitions given above, it can be seen that the electron and capture coefficient, c_n, and the electron emission coefficient, e_n, are relatively large, whereas the hole capture and emission coefficients, c_p and e_p, are relatively small. Thus, electrons will easily be trapped but will

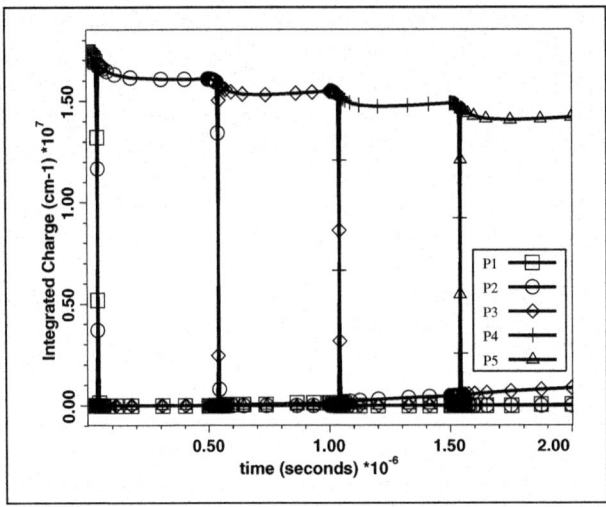

Fig. 12.31. Integrated free-electron density versus time in each phase of a five-phase CCD with traps.

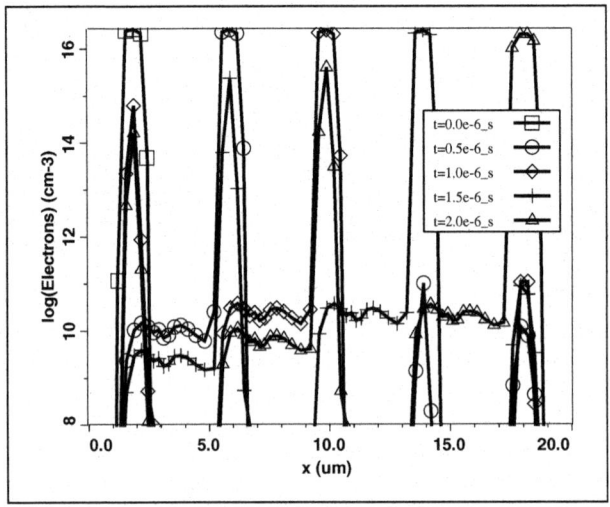

Fig. 12.32. Maximum free-electron density in the interval $0\ \mu\text{m} < y < 2\ \mu\text{m}$ versus x in a five-phase CCD with traps. The curves show the maximum free-electron densities at the end of each clock cycle.

not recombine with holes, because it is not likely that holes will be captured or emitted. Instead, provided there is sufficient time, the trapped electrons will be re-emitted to the conduction band. In Fig. 12.31, it can be observed that the number of free electrons in the lead phase decreases from cycle to cycle. Also, the trapping of free carriers in the lead phase can be observed by the exponential tail at the beginning of each cycle. Also, it is observed at the bottom of Fig. 12.31 that a portion of the orginally trapped electrons are re-emitted during later cycles. An alternative view of the trapping process is provided in Fig. 12.32, where the maximum electron density in the interval $0\ \mu m < y < 2\ \mu m$ is shown versus x. In Fig. 12.32, a non-negligible density of free electrons appears in the barrier regions ($3\ \mu m < x < 5\ \mu m$, $7\ \mu m < x < 9\ \mu m$, etc.) as a result of the steady emission of electrons from the traps. In the previous example without traps, no such density is observed (see Fig. 12.30).

12.9 Summary

In this chapter, an introduction to CCDs was given, and a description of the models used to simulate them was included. Several essential topics for CCD performance and design were examined by using TCAD modeling tools in 2- and 3D. These included: charge capacity, charge transfer, charge blooming, dark current, and charge trapping. Some other important topics not treated here include quantum efficiency, cross-talk, the modulation transfer function, smear, and noise [2, 8, 10, 11, 12, 13]. Although the simulations presented here consider only full-frame architectures, interline and CMOS architectures can also be studied using the same techniques.

A Example Distribution

Input files for all of the TCAD modeling examples presented here are available at http://solvnet.synopsys.com/retrieve/010719.html. The CCD structures used in the examples and presented here have been modified in the interest of confidentiality. Information on Synopsys modeling tools is available at http://www.synopsys.com. Information on Kodak image sensors is available at http://www.kodak.com/go/imagers.

Acknowledgment

C. J. Wordelman acknowledges J. G. Rollins for prior work on CCD modeling [57] helpful in this study, A. H. Gencer for help in developing the 3D charge-transfer example shown here, and S. J. Motzny for helpful discussions.

E. K. Banghart acknowledges J. P. Lavine and R. D. McGrath for helpful discussions regarding CCD theory and for their review of this chapter, and E. G. Stevens for helpful discussions on LOD operation.

References

1. W. S. Boyle and G. E. Smith: Bell Syst. Techn. J. 49, 587–593 (1970)
2. C. H. Sequin and M. F. Tompsett, *Charge Transfer Devices* (Academic Press, New York 1975)
3. M. G. Kang, *Selected Papers on CCD and CMOS Imagers* (SPIE Press, Bellingham, WA 2003)
4. T. Yamada, H. Tanaka, K. Henmi, M. Kobayashi, H. Mori, Y. Katoh, and Y. Miyata: *2003 IEEE Workshop on CCDs and Advanced Image Sensors* (2003)
5. K. Itakura, T. Nobusada, N. Kokusenya, R. Nayayoshi, and R. Ozaki: IEEE Trans. Elec. Dev. 47, 65–70 (2000)
6. A. Tanab: IEEE Trans. Elec. Dev. 47, 1700–1706 (2000)
7. http://www.synopsys.com
8. D. K. Schroder: *Advanced MOS Devices*. (Addison-Wesley, Reading, MA 1987)
9. S. M. Sze: *Physics of Semiconductor Devices*. (Wiley, New York 1981)
10. A. Theuwissen: *Solid-State Imaging with Charge-Coupled Devices*. Boston, MA: Kluwer Academic Publishers, 1995.
11. M. J. Howes and D. V. Morgan: *Charged-Coupled Devices and Systems*. (Wiley, New York 1979)
12. G. C. Holst: *CCD Arrays, Cameras and Displays*. (SPIE Press, Bellingham, WA 1998)
13. J. R. Janesick: *Scientific Charge-Coupled Devices*. (SPIE Press, Bellingham, WA 2001)
14. E. J. Meisenzahl, W. C. Chang, W. DesJardin, S. L. Kosmen, J. Shepard, E. G. Stevens, and K. Y. Wong: Proc. SPIE 3764, 261–268, (1999)
15. J. Hynecek: IEEE Trans. Elec. Dev. 30, 941–948 (1983)
16. S. T. Sheppard, M. R. Melloch, and J. A. Cooper, Jr.: Tech. Digest, IEEE Electron Device Meeting, 721–723 (1996)
17. K. Konuma, S. Tohyama, N. Teranishi, K. Masubuchi, T. Saito, and T. Muramatsu: IEEE Trans. Elec. Dev. 39, 1633–1637 (1992)
18. M. V. Wadsworth, S. R. Borrello, J. Dodge, R. Gooch, W. McCardel, G. Nado, and M. D. Shilhanek: IEEE Trans. Elec. Dev. 42, 244–250 (1995)
19. R. H. Walden, R. H. Krambeck, R. J. Strain, J. McKenna, N. L. Schryer, and G. E. Smith: Bell Syst. Techn. J. 51, 1635 (1972)
20. N. S. Saks: IEEE Electron Device Lett. 1, 131–133 (1980)
21. B. C. Burkey, W. C. Chang, J. Littlehale, T. H. Lee, T. J. Tredwell, J. P. Lavine, and E. A. Trabka: IEEE Electron Device Meeting, 28–31 (1984)
22. E. R. Fossum: IEEE Trans. Elec. Dev. 44, 1689–1698 (1997)
23. R. W. Dutton and Z. Yu, *Technology CAD: Computer Simulation of IC Processes and Devices*. (Kluwer, Boston, MA 1993)
24. W. K. Hofker: Phillips Res. Reports 8, 1–121 (1975)
25. N. E. B. Cowern: J. Appl. Phys. 64, 4484–4490 (1988)
26. B. J. Mulvaney, W. B. Richardson, and T. L. Crandle: IEEE Trans. Computer-Aided Design 8, 336–349 (1989)

27. Taurus-Process Manual, Synopsys Inc. (http://www.synopsys.com) (2004)
28. S. J. Morris, B. Obradovic, S.-H. Yang, and A. F. Tasch: Tech. Digest, IEEE Electron Device Meeting, 721–723 (1996)
29. B. E. Deal and A. S. Grove: J. Appl. Phys. 36, 3770, (1965)
30. S. Selberherr: *Analysis and Simulation of Semiconductor Devices*. (Springer, Wien 1984)
31. W. Shockley and J. W. T. Read: Phys. Rev. 87, 835–842 (1952)
32. R. F. Pierret: *Advanced Semiconductor Fundamentals*. (Addison-Wesley, Reading, MA 1987)
33. Taurus-Device Manual, Synopsys Inc. (http://www.synopsys.com) (2004)
34. H. Mutoh: IEEE Trans. Elec. Dev. 44, 1604–1610 (1997)
35. H. Mutoh: IEEE Trans. Elec. Dev. 50, 19–25 (2003)
36. T. O. Koerner and R. Gull: IEEE Trans. Elec. Dev. 47, 931–938 (2000)
37. F. A. Jenkins and H. E. White: *Fundamentals of Optics*. (McGraw-Hill, New York 1976)
38. Medici Manual, Synopsys Inc. (http://www.synopsys.com) (2004)
39. Y. Saad: *Iterative methods for sparse linear systems*. (SIAM, Philadelphia 2003)
40. B. C. Burkey, G. Lubberts, E. A. Trabka, and T. J. Tredwell: IEEE Trans. Elec. Dev. 31, 423–429 (1984)
41. N. M. S. R. Kawai and N. Teranishi: IEEE Trans. Elec. Dev. 44, 1588–1592 (1997)
42. J. E. Carnes, W. F. Kosonocky, and E. G. Ramberg: IEEE Trans. Elec. Dev. 19, 798–808 (1972)
43. J. P. Lavine and E. K. Banghart: IEEE Trans. Elec. Dev. 44, 1593–1598 (1997)
44. E. K. Banghart, J. P. Lavine, J. M. Pimbley, and B. C. Burkey: COMPEL J. Computation and Mathematics in Electrical and Electronic Eng. 10, 203–213 (1991)
45. E. K. Banghart, J. P. Lavine, E. A. Trabka, E. T. Nelson, and B. C. Burkey: IEEE Trans. Elec. Dev. 38, 1162–1174 (1991)
46. J. Crank: *The Mathematics of Diffusion*. (Oxford University Press, Oxford 1980)
47. N. D. Arora, J. R. Hauser, and D. J. Roulston: IEEE Trans. Elec. Dev. 29, 292–295 (1982)
48. J. T. Bosiers, B. G. M. Dillen, C. Draijer, A. C. Kleimann, F. J. Polderdijk, M. A. R. C. de Wolf, W. Klaassens, A. J. P. Theuwissen, H. L. Peek, and H. O. Folkerts: IEEE Trans. Elec. Dev. 50, 254–265 (2003)
49. E. G. Stevens, S. L. Kosman, J. C. Cassidy, W. C. Chang, and W. A. Miller: Proc. SPIE 1147, 274–282 (1991)
50. D. N. Nichols, W. C. Chang, B. C. Burkey, E. G. Stevens, E. A. Trabka, D. L. Losee, T. J. Tredwell, C. V. Stancampiano, T. M. Kelley, R. P. Khosla, and T. H. Lee: Tech. Digest, IEEE Electron Device Meeting, 120–123 (1987)
51. E. G. Stevens: IEEE Trans. Elec. Dev. 38, 299–302 (1991)
52. A. S. Grove: *Physics and Technology of Semiconductor Devices*. (Wiley, New York 1967)
53. R. D. McGrath, J. Doty, G. Lupino, G. Ricker, and J. Vallerga: IEEE Trans. Elec. Dev. 34, 2555–2557 (1987)
54. W. C. McColgin, J. P. Lavine, and C. V. Stancampiano: Mat. Res. Soc. Symp. Proc. 378, 713–724 (1995)
55. R. Widenhorn, M. M. Blouke, A. Weber, A. Rest, and E. Bodegon: *IEEE Workshop on Charge-Coupled Devices and Advanced Image Sensors* (2003)

56. J. R. F. McMacken and S. G. Chamberlain: IEEE Trans. CAD of Integrated Circuits and Systems 11, 629–637 (1992)
57. J. G. Rollins: "Simulation of CCDs with TMA tools," *unpublished* (1996)

13 Infrared HgCdTe Optical Detectors

G. R. Jones, R. J. Jones, and W. French

Silvaco International, 4701 Patrick Henry Drive, Santa Clara, CA 95054, U.S.A., www.silvaco.com

13.1 Introduction

The history of optoelectronics dates back to the early 1800s when in 1839 Edmund Becqueral discovered the photovoltaic effect. The fundamental factors resulting in this observation then are the same today as dictated by quantum mechanics embedded in the characteristics of photon–electron interaction. This interaction ideally results in a detectable current, which forms the basis of photon detection. This chapter will be primarily based on discussing and investigating photon detection simulated using Silvaco International's technology computer aided design (TCAD) software. The type of photodetector chosen is a HgCdTe n-type doped semiconductor material typically used in the infrared spectrum. A brief description of photodetection is first given. The relevant properties of Silvaco simulation tools are then summarized with the remaining part of the chapter focusing on the device mentioned.

13.2 Photon Detection

Photon detection can be achieved through the fabrication of a photodiode using semiconductor materials. A basic photodiode consists of a pn junction of doped semiconductors. An n-doped semiconductor refers to a semiconductor that has been doped with impurities of higher atomic number relative to the host, yielding an excess of free electrons in the conduction band. A p-type semiconductor refers to a semiconductor that is doped with impurities of lower atomic number relative to the host, resulting in an excess of holes in the valence bands. At the pn junction, this disparity in dopant causes a concentration gradient that causes electrons to diffuse into the p-type region and holes to diffuse into the n-type region. This diffusion results in a region depleted of free carriers with remaining ionized atoms, resulting in an opposing electrical potential that inhibits further diffusion. Within this depletion region, this electrical potential causes any free charge carriers to be rapidly swept through drift to the appropriate layer.

In a generic photodiode, light enters the device through a thin layer whose absorption typically cause the light intensity to exponentially drop with penetration depth. For enhanced performance it is often necessary for the device

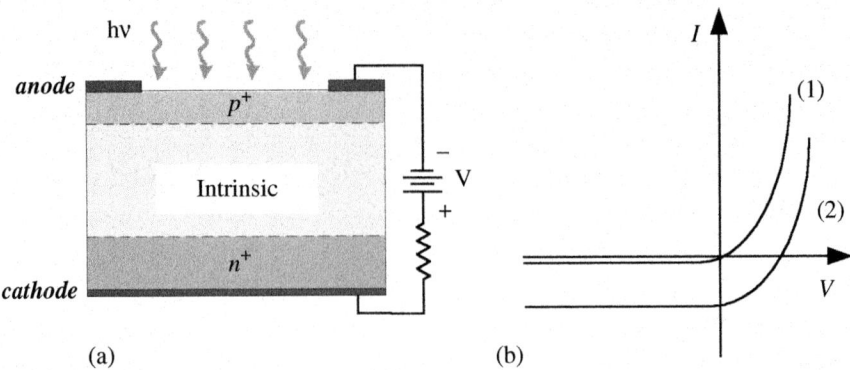

Fig. 13.1. (a) Pin photodiode operated in reverse bias. $h\nu$ is the energy of radiation, V is the bias, and R_L is the load resistance. p^+ signifies high acceptor doping, and n^+ signifies high donor concentration. (b) I-V characteristics for a pin photodiode: (1) with no light, (2) with light.

to have a shallow junction followed by a wide depletion region where most of the photon absorption and electron–hole generation should take place. Consequently, the basic pn junction is modified to incorporate intrinsic (i) or lightly doped material (π) that is sandwiched between the p- and the n- material resulting in a pin diode. An example of a photodiode device together with device electrodes is shown in Fig. 13.1(a). Photons that have sufficient energy and reach the depletion region will produce electron–hole pairs in that vicinity. Due to the presence of the electric field within the depletion region, these charge carriers are immediately separated and are swept across the junction. The movement of charge carriers across the junction perturbs the electrical balance and results in a small detectable current at the device electrodes. The magnitude of the generated current is proportional to the intensity of the incident light. It can be used as a photon detector by operating it in the third quadrant of its electrical current-voltage (I-V) characteristics, as shown in Fig. 13.1(b) under both dark (curve 1) and light (curve 2) conditions. The detection capability stems from the sharp increase in the reverse current I_0 to $I_0 + I_\mathrm{ph}$ with incident photons of energy $h\nu$. A small reverse voltage is therefore usually sufficient to extend the depletion region across the whole intrinsic region, resulting in improved device performance and enhanced quantum efficiency. The quantum efficiency is then a measure of how many detectable electrons are produced for every incident photon.

Advances in technology continued with the advent of sophisticated materials. Of popular choice for infrared (IR) radiation detection was mercury cadmium telluride (HgCdTe), which offered an attractive property of being able to alter its intrinsic material properties through the alteration of its constitutive molar fractions. Pioneering this work was a research group led by Lawson [1] at the Royal radar establishment in the United Kingdom who first synthesized HgCdTe material in 1958. This work was the successful outcome of a deliberate effort to engineer a direct band gap intrinsic semiconductor for the long wavelength infrared (LWIR) spectrum region (8–14-μm wavelength). Advancements in crystal growth technology continued and several useful properties now qualify this material for infrared detection, which include an adjustable band gap wavelength from 0.7 to 25 μm, direct band gap with high absorption coefficient, moderate dielectric constant and index of refraction, moderate thermal coefficient of expansion, and the availability of a wide band gap lattice matched substrate for excellent epitaxial growth. HgCdTe has found extensive use in optical detection and indeed wide use in infrared photodetectors over the past few decades. It is particularly important in electronic focal plane arrays, which have use in military applications for infrared sensing equipment. Applications in this area has been the main driving force for research on this material [2].

13.3 Summary of Simulation Tools

13.3.1 Introduction

Silvaco International provides several electrical simulation tools for semiconductor materials and advanced materials such as HgCdTe. The simulation tools of great interest and relevance for optical simulation include Atlas, S-Pisces, Blaze, and Luminous together with the C-interpreter. The simulation syntax is written in an interactive virtual wafer fab environment using a tool called Deckbuild, and the results are plotted using a graphics package called Tonyplot.

Atlas is a modular and extensive framework for one-, two- or three-dimensional semiconductor device simulation. It predicts the electrical behavior of specified semiconductor devices and provides an insight into the physical mechanisms associated with device simulation. Atlas can be used as a stand-alone device simulator for silicon using S-Pisces or can be implemented with additional simulators including Blaze and Luminous. Blaze is a 2D device simulator for II–IV and III–V semiconductor materials. Blaze accounts for the effects of positional-dependent band structures by modifying charge transport equations associated with Atlas. Luminous is a ray trace and light absorption simulator that calculates optical intensity profiles and converts these profiles into photogeneration rates.

The C-interpreter is a versatile ANSI C-compatible interpreter and debugging environment that permits the user to define specific code to be developed and substituted for the default models within a number of Silvaco applications. Upon execution of an input deck, the C-interpreter is simply invoked and the user-defined syntax is executed in real time modifying the existing models within the original code. This versatility of the C-interpreter adds significant power to gain accurate and personal simulations. This coupling of simulation tools permits the user to simulate the electrical response to optical signals for a broad range of devices. For recent examples, see [3, 4].

13.3.2 Fundamentals of Device Simulation

Many years of research into device physics has resulted in a mathematical model describing the operation of many semiconductor devices. This model consists of a set of fundamental equations that link together the electrostatic potential and the carrier densities within a simulation domain. These equations consist of Poisson's equation, current continuity equations, and transport equations. Poisson's equation relates variations in electrostatic potential to local charge densities where as the continuity and transport equations help calculate these local charge densities by describing the way electron and hole densities evolve as a result of transport, generation, and recombination processes. These formulas ultimately help predict quantitatively the electrical behavior of numerous devices forming the basis of semiconductor modeling today.

To predict quantitatively the electrical behavior of a semiconductor device, the carrier distributions must first be known and expressed in appropriate equations. These equations are based on fundamental electrostatic properties, two of which are summarized in (13.1) and (13.2), where:

$$\frac{\mathrm{d}E}{\mathrm{d}x} = \frac{q}{\varepsilon_\mathrm{s}}(p - n + N_\mathrm{d} - N_\mathrm{a}) \tag{13.1}$$

and

$$E = -\frac{\mathrm{d}\psi}{\mathrm{d}x}. \tag{13.2}$$

The derivative of the electrical field E with respect to the distance x is equal to the net charge concentration at that location divided by the dielectric constant of the semiconductor material ε_s and is given in (13.1). The net charge is the electron charge q multiplied by the carrier concentration. The carrier concentration is equal to the sum of the free hole concentration (p) and the donor doping density (N_d) minus the sum of the free electron concentration (n) and the acceptor doping density (N_a). The electrical potential ψ is defined as a scalar quantity such that the negative of its gradient is equal to the electric field. Its physical significance relates to the work done in carrying a charge from one point to another. Poisson's equation is obtained by merging (13.1) and (13.2), thus giving a relationship between electrical potential and the net charge concentration. The carrier concentrations n and p are established according to:

$$\frac{\partial n}{\partial t} = \frac{1}{q}\nabla \cdot \boldsymbol{J}_n + G_n - R_n \tag{13.3}$$

and

$$\frac{\partial p}{\partial t} = \frac{1}{q}\nabla \cdot \boldsymbol{J}_p + G_p - R_p . \tag{13.4}$$

Here $\boldsymbol{J}_n$ and $\boldsymbol{J}_p$ are the electron and hole current densities, respectively; G_n and G_p are the generation rates for electrons and holes, respectively; and R_n and R_p are the recombination rates for electron and holes, respectively. Up to now these expressions provide a general framework for device simulation. However, additional expressions are needed and, in particular, models for $\boldsymbol{J}_n$, $\boldsymbol{J}_p$, G_n, G_p, R_n, and R_p.

The current density equations or charge transport models are obtained by applying approximations and simplifications to Boltzmann's transport equation. These assumptions result in a number of transport models which Atlas uses and include the drift-diffusion model, the energy balance model and the hydrodynamic model.

The simplest model of charge transport is the drift-diffusion model that proves adequate for most applications and as such is the default transport model used within Atlas. The current densities in the continuity equations (13.3) and (13.4) can be expressed according to the drift-diffusion model as [5]:

$$\boldsymbol{J}_n = -q\mu_n n \nabla \varphi_n \tag{13.5}$$

and

$$\boldsymbol{J}_p = -q\mu_p p \nabla \varphi_p . \tag{13.6}$$

Here μ_n and μ_p are the electron and the hole mobilities, respectively, and φ_n and φ_p are the quasi-Fermi levels of the electron and holes, respectively. The quasi-Fermi levels are then linked to the carrier concentrations and the potential through the two Boltzmann approximations, where:

$$n = n_{\text{ie}} \exp\left(\frac{q(\psi - \varphi_n)}{kT_{\text{L}}}\right) \tag{13.7}$$

and

$$p = n_{\text{ie}} \exp\left(\frac{-q(\psi - \varphi_p)}{kT_{\text{L}}}\right) . \tag{13.8}$$

Here n_{ie} is the effective intrinsic concentration and T_{L} is the lattice temperature. By substitution, this yields:

$$\boldsymbol{J}_n = qn\mu_n \boldsymbol{E}_n + qD_n \nabla n \tag{13.9}$$

and

$$\boldsymbol{J}_p = qp\mu_p \boldsymbol{E}_p - qD_p \nabla n . \tag{13.10}$$

It should be noted that this derivation of the drift-diffusion equations assumes that the Einstein relationship holds, where D_n and D_p represent the electron and hole diffusivities such that:

$$D_n = \frac{kT_{\text{L}}}{q}\mu_n \tag{13.11}$$

and
$$D_p = \frac{kT_L}{q}\mu_p. \qquad (13.12)$$

If Fermi–Dirac statistics is assumed, the diffusivity D_n becomes:

$$D_n = \frac{\frac{kT_L}{q}\mu_n F_{1/2}\left(\frac{1}{kT_L}(E_{Fn}-E_C)\right)}{F_{-1/2}\left(\frac{1}{kT_L(E_{Fn}-E_C)}\right)}, \qquad (13.13)$$

where $F_{1/2}$ is the Fermi–Dirac integral of order $\alpha = \frac{1}{2}$ expressed as:

$$F_\alpha(\eta_s) = \frac{2}{\sqrt{\pi}} \int_0^\infty \frac{\eta^\alpha}{1+\exp(\eta+\eta_s)} d\eta. \qquad (13.14)$$

Here η_s can take the form $\eta_n = (E_{Fn} - E_c)//kT)$ for electrons or $\eta_v = (E_{Fn} - E_v)/(kT)$ for holes and η is the electron energy.

The conventional drift-diffusion model as described here neglects nonlocal transport effects such as velocity overshoot diffusion associated with the carrier temperature and the dependence of impact ionisation rates on carrier energy distributions. These phenomena can have a significant effect on the terminal properties of submicron devices. As such, Atlas offers two nonlocal models of charge transport termed the energy balance model and the hydrodynamic model. These models add continuity equations for the carrier temperatures and treat mobilities and impact ionisation coefficients as functions of the carrier temperatures rather than functions of the local electric field as the drift-diffusion model does. These models are based on the work in [6], which the interested reader is encouraged to study for further discussion.

Having briefly discussed the current density expressions, attention should also be addressed to modeling heterojunction devices where discontinuities in band alignment develop. An abrupt heterojunction has discontinuities in both the conduction and the valence bands. The discontinuity creates a spike in the potential profile that obtrudes out of the otherwise smooth conduction band level (Fig. 13.2). When the electrons leave one side for the other, they must first climb over the barrier. This spike, therefore, impedes current flow and can be undesirable. How the band gap difference is distributed between the conductance and valence bands has a large impact on the charge transport in these devices. Atlas has three methods for defining the conduction band alignment for a hetero interface, which include the affinity rule, the align parameter, and by manually adjusting the material affinities using the affinity parameter. The align parameter allows the user to specify the fraction of the band gap difference that will appear as the conduction band discontinuity. It is this approach that has been adopted in the simulations presented here. Having briefly discussed the current density expresions and modifications to band alignemnt, attention is now given to generation and recombination mechanisms that contribute to the net charge concentration.

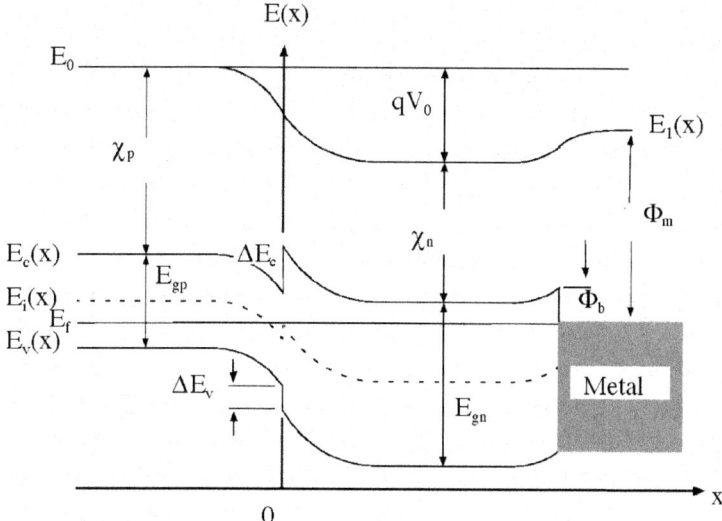

Fig. 13.2. Basic heterostructure with Schottky contact. $E_{c(x)}$, $E_{v(x)}$, and $E_{i(x)}$ are the specially dependent conduction band, valence band, and intrinsic energy levels, respectively. E_f is the Fermi energy level, E_0 is the vacuum energy level. E_{gp} and E_{gn} are the p-type and n-type material band gaps respectively. ΔE_v is the valence band offset, ΔE_c is the conduction band offset, and qV_0 is the build in potential of the heterojunction. χ_p and χ_n are the electron affinities in the p- and n-type materials, respectively. Φ_m is the work function of the metal. Φ_b is the Schottky barrier height.

13.3.3 Carrier Generation and Recombination Mechanisms

If we consider a homogeneously doped semiconductor with equilibrium concentrations n_0 and p_0, then at equilibrium, a steady-state balance exists where $n_0 p_0 = n_i^2$, where n_i is the intrinsic carrier concentration. However, under excitation, n_0 and p_0 are disturbed from their equilibrium states. A net recombination results that attempts to return the semiconductor to equilibrium. The processes responsible for recombination are well known and fall into several categories, which include photon transitions, phonon transitions, Auger transitions, surface recombination, and tunneling. Atlas attempts the simulation of all of these processes; however, we only describe the main ones here, including Shockley–Read–Hall (SRH) recombination, Auger recombination, and spontaneous photon emission.

13.3.4 Shockley–Read–Hall Recombination

One of the most important physical processes resulting in the generation and recombination of carriers is the capture and emission of carriers through localized energy states, generally termed traps, located in the energy band gap.

These traps typically occur due to the presence of lattice defects or impurity atoms that introduce energy levels near the center of the energy band gap. In this process, these localized states act as stepping stones for an electron from the conduction band to the valence band. A model that formulates this process adequately is the Shockley–Read–Hall theory [7] where the recombination rate R_{SRH} becomes:

$$R_{\text{SRH}} = \frac{pn - n_{\text{ie}}^2}{\tau_p \left[n + n_{\text{ie}} \exp\left(\frac{E_{\text{tr}}}{kT_{\text{L}}}\right)\right] + \tau_n \left[p + n_{\text{ie}} \exp\left(\frac{-E_{\text{tr}}}{kT_{\text{L}}}\right)\right]} . \quad (13.15)$$

Here E_{tr} is the difference between the trap energy level and the intrinsic Fermi level, T_{L} is the lattice temperature, and τ_n and τ_p are the electron and hole lifetimes, respectively. The constant carrier lifetimes that are used in the SRH model above may be a function of impurity concentration [8].

13.3.5 Auger Recombination

Auger recombination is a nonradiative recombination event, which involves three particles. Typically, an electron and a hole will recombine in a band-to-band transition and will give off the resulting energy to another electron or hole. Eventually, this energized particle will lose energy through phonon emission within the atomic lattice. The involvement of a third particle affects the recombination rate such that Auger recombination needs to be treated differently from typical band-to-band recombination. Auger recombination is modeled within Atlas using the equation:

$$R_{\text{Auger}} = C_n(pn^2 - nn_{\text{ie}}^2) + C_p(np^2 - pn_{\text{ie}}^2) . \quad (13.16)$$

C_n and C_p are empirical Auger parameters (see Sect. 13.5.1).

13.3.6 Recombination Through Photon Emission

Another physical mechanism we have to consider is radiative carrier recombination by spontaneous photon emission. This mechanism takes place in one step: An electron loses energy on the order of the band gap and moves from the conduction band to the valence band. This effect is important for narrow gap semiconductors and semiconductors whose specific bandstructure allows direct transitions. The total spontaneous emission rate equates to:

$$R^{\text{OPT}} = B(np - n_{\text{ie}}^2) , \quad (13.17)$$

with the empirical parameter B (see Sect. 13.5.1).

13.4 Optoelectronic Simulation

Optoelectronic device simulation is provided by Silvaco's device simulator Luminous, which has been designed to model light absorption and photogeneration in both planar and nonplanar devices. Exact solutions for general optical sources are obtained using geometric ray tracing. This feature enables Luminous to account for arbitrary topologies, internal and external reflection and refraction, polarization dependencies, and dispersion. In practice, Luminous can simulate up to ten monochromatic or multispectral optical sources, and provides special parameter extraction capabilities unique to optoelectronics. DC, AC, transient, and spectral responses of general device structures can be simulated in the presence of arbitrary optical sources.

Optical electronic device simulation is split into two distinct models that are calculated simultaneously at each DC bias point or transient timestep. The two models are the optical ray trace, which uses the real component of refractive index to calculate the optical intensity, and second, the absorption or photogeneration model, which uses the imaginary component of the refractive index to calculate a new carrier concentration at each grid point. Once these parameters are resolved, an electrical simulation is then performed by S-Spices or Blaze to calculate terminal currents.

13.4.1 Optical Beam Characteristics

An optical beam is modeled as a collimated source. The origin and positional properties of the beam are defined using several parameters as summarized in Fig. 13.3(a). To perform accurate simulations of planar and nonplanar geometry, the beam is automatically split into a series of rays such that the sum of the rays covers the entire width of the illumination window. When the beam is split, Atlas automatically resolves discontinuities along the region boundaries of the device.

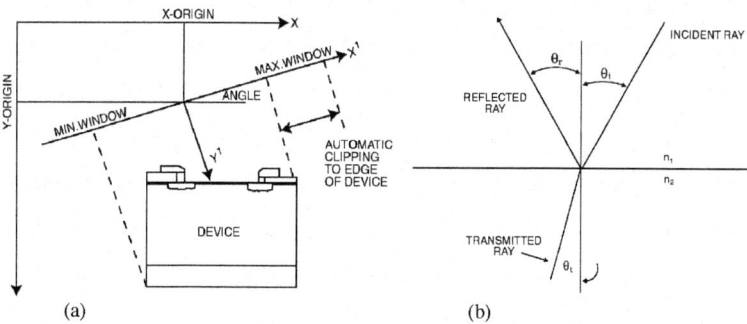

Fig. 13.3. (a) Device geometry and beam specification. (b) Summary of incident and reflected and transmitted rays.

Rays are split into a transmitted ray and a reflected ray at interfaces between two regions where there is a change in the refractive index. Figure 13.3(b) shows the relationship among the angles of incidence (θ_i), reflection (θ_r), and transmission (θ_t). The reflection and transmission coefficients of the light for parallel and perpendicular polarisation are calculated according to the Fresnel equations:

$$E_r = \frac{n_1 \cos\theta_t - n_2 \cos\theta_i}{n_1 \cos\theta_t + n_2 \cos\theta_i} E_i \quad \text{(parallel polarisation)} \tag{13.18}$$

$$E_t = \frac{2n_1 \cos\theta_i}{n_1 \cos\theta_t + n_2 \cos\theta_i} E_i \quad \text{(parallel polarisation)} \tag{13.19}$$

$$E_r = \frac{n_1 \cos\theta_i - n_2 \cos\theta_t}{n_1 \cos\theta_i + n_2 \cos\theta_t} E_i \quad \text{(perpendicular polarisation)} \tag{13.20}$$

$$E_t = \frac{2n_1 \cos\theta_i}{n_1 \cos\theta_i + n_2 \cos\theta_t} E_i \quad \text{(perpendicular polarisation)} \tag{13.21}$$

$$R = \left(\frac{E_r}{E_i}\right)^2 \tag{13.22}$$

$$T = \left(\frac{E_t}{E_i}\right) \frac{n_2}{n_1}, \tag{13.23}$$

where $\theta_r = \theta_i$ and $n_1 \sin\theta_1 = n_2 \sin\theta_t$. Here E_i is the incident field, E_r is the reflected field, E_t is the transmitted field, R is the reflection coefficient, and T is the transmission coefficient. n_1 and n_2 are the refractive indices on the incident and refractive transmission sides respectively.

13.4.2 Light Absorption and Photogeneration

The cumulative effects of the reflection coefficients, the transmission coefficients, and the integrated loss due to absorption over the ray path are saved for each ray. The generation associated with each grid point can be calculated by integration of the generation rate over the area of intersection between the ray and the polygon associated with that grid point. The photogeneration rate is given by:

$$G = \eta_0 \frac{P^* \lambda}{hc} \alpha e^{-\alpha y}. \tag{13.24}$$

Here, the optical intensity P^* represents the cumulative effects of reflections, transmissions, and loss due to absorption over the ray path; η_0 is the internal quantum efficiency; y is the relative distance for the ray in question; h is Planck's constant; λ is the wavelength; c is the speed of light; and α is the absorption coefficient expressed as:

$$\alpha = \frac{4\pi}{\lambda} k_i, \tag{13.25}$$

where k_i is the imaginary component of the complex index of refraction.

13.5 Device Simulation

13.5.1 Material Parameters

The adjustable band gap of HgCdTe stems from the ability to alter the stociometric ratio of Hg and Cd in the form $Hg_{(1-x)}Cd_xTe$, thus resulting in the ability to detect multispectral sources through the creation of multispectral infrared detectors [9]. Multispectral infrared detectors are highly beneficial for a variety of applications and have been implemented in applications such as missile warning and guidance systems, surveillance, target detection, and tracking [9, 10]. Dual band detection in the medium wavelength infrared (MWIR) and LWIR atmospheric windows has been performed using HgCdTe photodiodes [11, 12]. The structure proposed and simulated here is based on [12] and consists of a monolithic $Hg_{(1-x)}Cd_xTe$ photoconductive device that is suitable for dual band on pixel registered infrared photodetector arrays in the atmospheric transmission window of 3–5 μm and 8–12 μm.

The band gap of HgCdTe is a function of the molar fraction of Cd in the composite material as previously discussed. A number of equations have been developed to summarize the measured relationship and of popular choice is the expression developed by Hansen et al. [15], which describes the energy bands in a parabolic form where:

$$E_g = -0.302 + 1.93x - 0.810x^2 + 0.832x^3 + 5.35 \times 10^{-4}(1-2x)T . \quad (13.26)$$

Here T is the temperature in degrees Kelvin and x is the molar fraction, E_g is the material band gap in electronvolts, and x is the fractional composition value. With varying the value of x, the spectral response can be tailored to detect varying wavelengths. Consequently, in order to detect long wavelength radiation, x must be altered accordingly, resulting in a semiconductor having a very narrow band gap. This narrow band gap causes a significant problem in that a large number of intrinsic electron–hole pairs are thermally generated when the device is operated at room temperature ($T = 300$ K). To circumvent this problem, the device is normally operated at low temperatures near $T = 70$ K [16]. Consequently, much study has been performed on temperature characteristics of HgCdTe material, resulting in several expressions describing its properties. For this work, the applied formulas describing effective electron and hole masses are given in (13.27) and (13.28), respectively. The electron and hole mobilities are given by (13.29) and (13.30), respectively. The static dielectric constant is given in (13.31):

$$\frac{m_e^*}{m_0} = m_e^{*\prime} = \left[-0.6 + 6.333\left(\frac{2}{E_g} + \frac{1}{E_g+1}\right)\right]^{-1} \quad (13.27)$$

$$\frac{m_h^*}{m_0} = m_h^{*\prime} = 0.55 \quad (13.28)$$

$$\mu_e = 9 \times 10^4 \left(\frac{0.2}{x}\right)^{7.5} T^{-2(0.2/x)^{0.6}} \qquad (13.29)$$

$$\mu_h = 0.01\,\mu_e \qquad (13.30)$$

$$\varepsilon_s = 20.5 - 15.5x + 5.7x^2 \,. \qquad (13.31)$$

Here $m_e^*/m_0 = m_e^{*\prime}$ is the effective electron mass, $m_h^*/m_0 = m_h^{*\prime}$ is the effective hole mass, μ_e is the electron mobility in m²/Vs, μ_h is the hole mobility in m²/Vs, and ε_s is the static dielectric constant. The remaining symbols are as defined previously.

Recombination models are also important to consider in any simulation in order to determine how changes to any perturbation away from equilibrium are finalized. As the device studied here is doped nondegenerative n-type, it can be assumed that limited Shockley–Read defects are present. Consequently, only Auger and radiative recombination will be considered. Expressions for Auger and radiative recombination in silicon are well documented. However, detailed expressions for HgCdTe that include temperature variation and material band gap fraction dependencies must now be taken into account. Such expressions have been determined by Wenus et al. [14] and have been used for the simulations presented in this work and are shown here:

$$C_n = \left[2n_{ie}^2 \frac{3.8 \times 10^{-18} \varepsilon_s^2 \left(1 + m_e^{*\prime}/m_h^{*\prime}\right)^{0.5} \left(1 + 2m_e^{*\prime}/m_h^{*\prime}\right)}{m_e^{*\prime}(0.2)^2 (kT/E_g)^{1.5}}\right.$$
$$\left.\times \exp\left(\frac{1 + 2(m_e^{*\prime}/m_h^{*\prime})}{1 + m_e^{*\prime}/m_h^{*\prime}} \frac{E_g}{kT}\right)\right]^{-1} \qquad (13.32)$$

$$C_p = C_n \left[\frac{6\left(1 - \frac{5E_g}{4kT}\right)}{1 - \frac{3E_g}{2kT}}\right]^{-1} \qquad (13.33)$$

$$B = 5.8 \times 10^{-19} \varepsilon_s^{1/2} (m_e^{*\prime} + m_h^{*\prime})^{-1.5} (1 + m_e^{*\prime -1})$$
$$\times \left(\frac{300}{T}\right)^{1.5} \left(\frac{E_g^2 + 3kTE_g}{q} + \frac{3.75k^2T^2}{q^2}\right). \qquad (13.34)$$

The Auger recombination rate for electron and holes is given in (13.32) and (13.33), respectively. Optical recombination is expressed using (13.34). Here C_n is the Auger electron recombination coefficient in m⁶/s, C_p is the Auger hole recombination coefficient in m⁶/s, B is the radiative recombination coefficient in m³/s, k is Boltzmann's constant, q is the electron charge, and n_i is the intrinsic carrier concentration.

Of critical importance to modeling the absorption of incident radiation is the absorption coefficient. In general, this property should be wavelength dependent in order to give realistic behavior similar to a real device. Direct band gap semiconductors, such as HgCdTe, have a sharp onset of optical absorption as the photon energy increases above the band gap of the material. In contrast, indirect semiconductors, such as silicon or germanium, have softer absorption curves and will therefore not be as sensitive to a specific wavelength. The absorption coefficient used should therefore have similar properties to the material of choice. A popular expression representing the absorption coefficient is [17]:

$$\alpha = \frac{\sqrt{2c}}{\tau}\sqrt{\left[1 - \frac{\lambda}{\lambda_g}\right]}\left(\sqrt{\frac{m_e^{*\prime}\lambda}{h}}\right)^3. \quad (13.35)$$

Here α is the absorption coefficient in m^{-1}, λ is the incident wavelength in meters, h is Planck's constant, τ is the electron lifetime in seconds, c is the speed of light, and λ_g is the cutoff wavelength determined by the band gap of the material. For the studies in this work, the absorption coefficient based on (13.35) has been evaluated independently and has been found to give similar values to experimental data as measured by Scott [18] who investigated the absorption coefficient for varying alloy composition.

13.5.2 Device Structure

The device is shown schematically in Fig. 13.4(a). The device consists of three layers of HgCdTe material with varying x and hence material band gap situated on top of an CdTe IR transparent substrate. All HgCdTe layers are doped n-type. Several factors favor n-type doped HgCdTe photodetectors

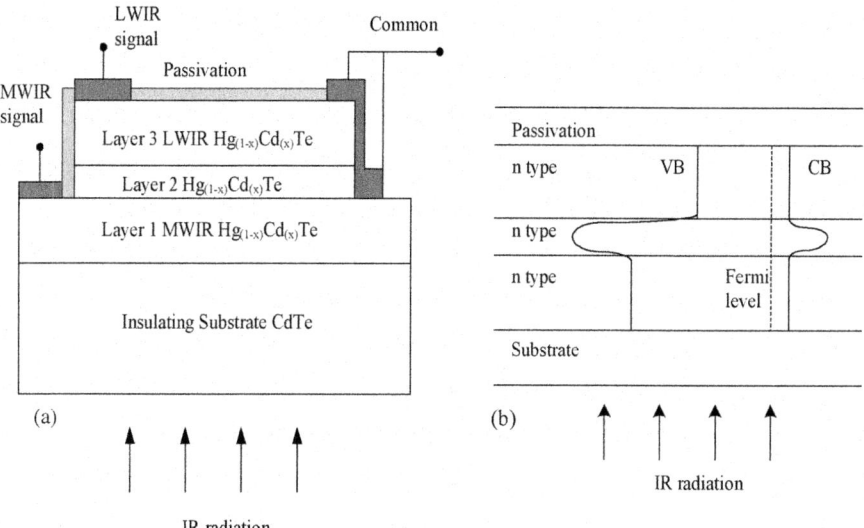

Fig. 13.4. (a) Schematic diagram of dual band monolithic HgCdTe photodetector. (b) Energy band diagram for layer 1, layer 2, and layer 3.

over p-type doped detectors. The n-type HgCdTe at low carrier concentrations is easier to control and passivate as well as is comparatively free of Shockley–Read defects that limit the lifetime of p-type material. Radiation between 2 and 5 μm now referred to as MWIR will pass through CdTe substrate and will not be absorbed due to the high band gap energy of the CdTe material but will be absorbed by the HgCdTe material in layer 1. Radiation between 6 and 12 μm now referred to as LWIR will also pass through the CdTe and in this case pass through layer 1 and layer 2 as it has insufficient energy to excite any electrons in these materials. However, radiation will be absorbed in layer 3 due to its smaller energy band gap. As such, it is intended that no LWIR or MWIR radiation be absorbed in layer 2, which has the widest band gap. Ideally, the layer 2 material acts as an electrically isolating layer between the two different absorbing layers, i.e., as a highly resistive layer for majority carriers and as a blocking layer for minority carriers [13, 14]. A simplified band diagram for the structure is shown in Fig. 13.4(b). As shown here, the presence of heterojunction barriers in the valence band will prohibit the flow of photogenerated minority carriers between layers 1 and layer 3, thus confining them to their respective layers. The barrier in the conduction band will provide a high resistance between the two absorbing layers for majority carrier flow.

Electrodes are placed at either edge of the device. The LWIR electrode extends from 0 μm to 10 μm in the vertical y direction. The MWIR electrode extends from 12 μm to 22 μm. The common electrode is placed at the other edge and extends down from 0 μm to 22 μm. The fractional composition value x for the LWIR layer, the insulating layer, and the MWIR layer are set to 0.21, 0.7, and 0.29, respectively, i.e., LWIR = $Hg_{0.79}Cd_{0.21}Te$, MWIR = $Hg_{0.71}Cd_{0.29}Te$, and the insulating layer = $Hg_{0.3}Cd_{0.7}Te$. The optical dimensions (length × width × thickness) of the LWIR and MWIR detector are $425 \times 300 \times 10$ μm^3. The optical dimensions for the insulating layer are $425 \times 300 \times 2$ μm^3.

All computations were performed using Fermi–Dirac statistics for nondegenerate semiconductor models with parabolic energy bands ($T = 77$ K). The photodiode was backside illuminated through the transparent CdTe substrate for all simulations. Surface power density of incident radiation was set to 0.1 W/cm^2. Each simulation was performed with the LWIR and MWIR electrode held at 0.1 V with respect to ground.

Figure 13.5(a) shows a one-dimensional cut line displaying the x compositional value through the device. Figure 13.5(b) shows the corresponding energy band diagram in electronvolts for the x compositional values. It is clear that three distinct energy bands are present. Layer 1, which has $x = 0.21$, has the smallest band gap and is suitable for LWIR detection. Layer 3, which has $x = 0.29$, has a larger band gap and is suitable for MWIR detection. Layer 2 has $x = 0.7$ and as such has a significantly larger band gap and consequently should act as an electrically and optically isolating layer.

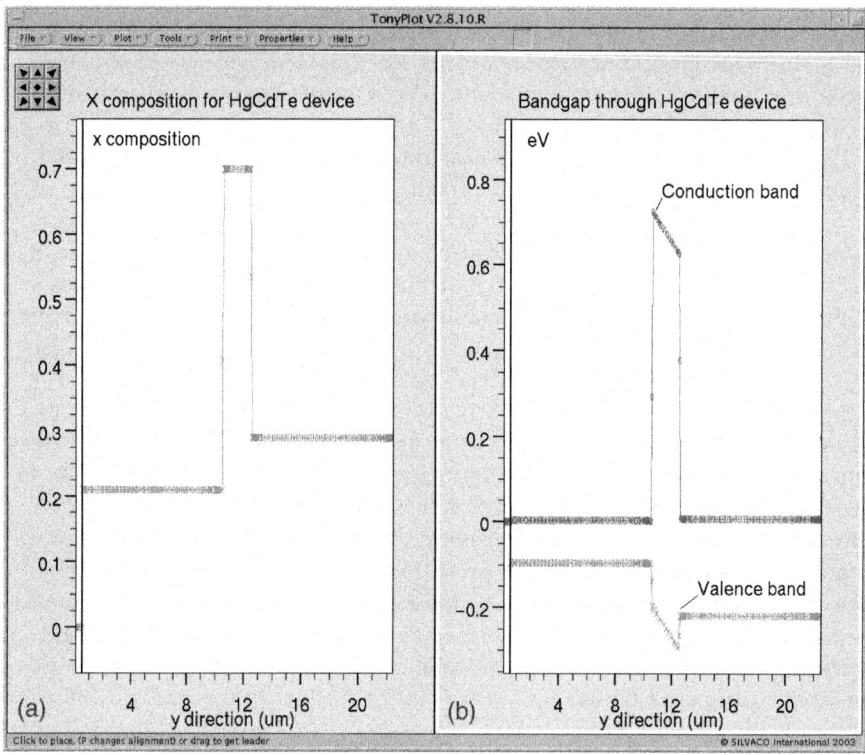

Fig. 13.5. (a) One-dimensional cutline showing fractional composition x through device. (b) Corresponding one-dimensional cutline display for the energy band diagram through the device.

13.5.3 Cross Talk Considerations

In practice, there will be a degree of cross talk between the layers affecting the electrical and optical performance. For example, a relevant characteristic of monolithic dual band detectors is the amount of signal cross talk occurring within the device structure between two detection bands, which should be kept to a minimum. That is, when detecting MWIR (LWIR), the amount of signal in the MWIR (LWIR) detector due to LWIR (MWIR) should be minimized. For example, if all incident mid-wave radiation is not absorbed by the MWIR layer, some will be transmitted and eventually be absorbed by the LWIR material, thus resulting in mid-wave radiation contributing to the LWIR signal current. This may be minimized by having sufficient MWIR material to absorb all of the mid-wave radiation. In effect, the MWIR layer acts as a filter of mid-wave radiation for the LWIR detector. The opposite effect of long wave radiation being absorbed by the MWIR material will not occur because the cutoff wavelength of the MWIR material is far shorter than the impinging LWIR radiation. For layer 2 that has the highest band gap, the impinging radiation will have insufficient energy and so will not be absorbed in this layer. For devices of this sort [19], calculations have shown that approximately 1% of the incident MWIR radiation is not absorbed by the 10-μm-thick MWIR layer.

The electrical cross talk between each device is also important to consider. Intradevice electrical cross talk is defined as an output signal appearing at the terminals of one device due to the absorption and modulation occurring in another device. The three HgCdTe layers shown in Fig. 13.4(a) can be considered as an interconnected distributed resistor network. As a consequence of this configuration, any modulation of the resistance of one layer due to impinging radiation will result in a change in the overall resistance and, correspondingly, a change in the resistance of the other devices. For example if the resistance of layer 3 is altered due to impinging MWIR radiation, the resistance and hence output signal at the LWIR terminals will change unless layer 2 is perfectly insulating [19]. To ensure that this is the case, the resistance of the insulating layer can be altered using such options as to decrease the doping in the insulating layer, increase the band gap of the insulating layer, or decrease its area. The optimum situation thus occurs when the middle layer is assumed to have an infinitely high resistance and the two LWIR and MWIR detectors act completely independently, resulting in no intradevice electrical cross talk. This situation is assumed in the simulations presented here.

13.5.4 Photogeneration and Spectral Response

The photogeneration of electrons within the device for two different wavelengths are shown in Fig. 13.6. This photogeneration rate is determined using the absorption coefficient described in (13.35). It is clear that two distinct cases are present. At a wavelength of 4.5 μm, the photogeneration rate is approximately three orders of magnitude higher in the MWIR detector compared with the LWIR detector. As the wavelength is increased to 9.5 μm, the energy of the incident radiation is reduced and so the photogeneration rate in the MWIR detector is reduced. In contrast to this, the photogeneration rate in the LWIR detector with the lowest fractional compositional of x is seen to increase dramatically. In both cases of incident radiation, the photogeneration rate in the insulating layer with the highest compositional fraction of x is negligible. Also of interest is the slope of the photogeneration rate. It can be seen that the photogeneration rate is highest with respect to the MWIR and LWIR detectors at the extremities of their layers, i.e., at approximately 22 μm and 11 μm, respectively. This is because the device is backside illuminated and the intensity of the light is decreasing as it penetrates the device. As described in (13.24), the photogeneration rate decreases exponentially with penetration depth, which is shown here. A continuation of this study is to illuminate the device with wavelengths covering the entire designed spectrum to determine the spectral response of the device. Figure 13.7 details

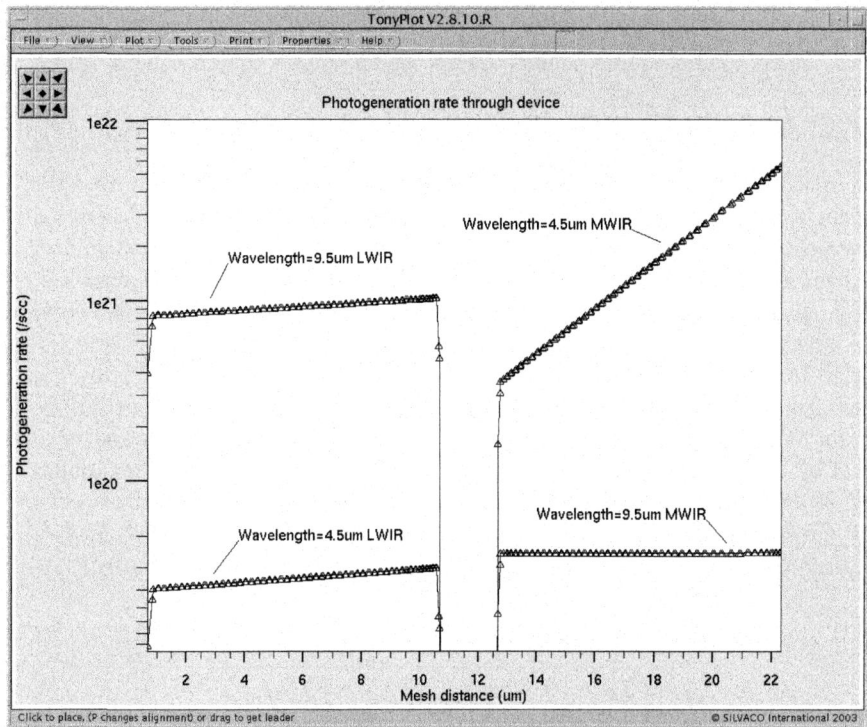

Fig. 13.6. One-dimensional cutline showing photogenerated carriers within the LWIR and MWIR detector for 4.5 μm and 9.5 μm of incident radiation.

the spectral response of the device over the wavelength range 1–11 μm. The LWIR and MWIR current is plotted against the wavelength. It is clear that two distinct areas of device responsivity exist. As the wavelength is increased from 1 μm, the current in the MWIR detector increases. This current reaches a maximum at approximately 5 μm then falls sharply to a smaller value. The nature of this sharp reduction stems from the direct band gap nature of the device and is represented in the second term of (13.35). Here it can be seen that as the wavelength λ approaches the cutoff wavelength λ_g, which is determined by the band gap of the material, the quotient will tend to unity. This will result in the sum tending to zero, which will give the sharp cutoff response characteristic of such materials. In contrast to the MWIR response, the LWIR response is negligible in the wavelength range 0 to 5 μm. As the wavelength is increased further, the LWIR detectors response increases and reaches a maximum at approximately 9.5 μm. The LWIR current then reduces in a similar fashion as the MWIR detector. This illustrates the effect of the MWIR layer acting as a filter of medium wave radiation for the LWIR detector. It is therefore clear that two distinct bands of radiation, medium wavelength infrared 2–5 μm and long wavelength infrared 5–12 μm, are detectable using the device described here.

13.5.5 Recombination Studies

Of particular importance in device simulation is having the ability to alter parameters or include or exclude effects that can ultimately lead to improved device characteristics or improved match of simulation results to benchmark experimental data. Within Atlas, once a suitable device has been simulated, many parameters are easily alterable to aid such investigations. Demonstrated here is what effect recombination properties have on the device performance. Figure 13.8 shows three situations of varying recombination within the device. Situation one is to include no recombination. It is clearly seen that the current is increased in both the MWIR and LWIR regimes compared with Fig. 13.7 that used both radiative and Auger recombination in the simulation. With respect to the MWIR current in Fig. 13.8, the introduction of only radiative recombination appears to have a greater effect on the MWIR response current compared with the LWIR response. Clearly the magnitude

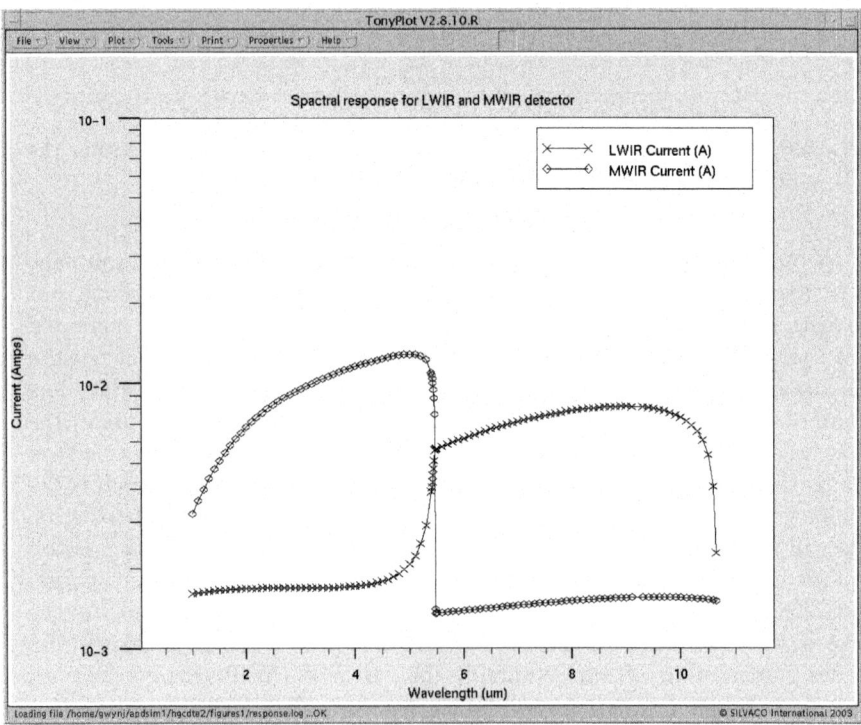

Fig. 13.7. Spectral response of LWIR and MWIR current as a function of wavelength.

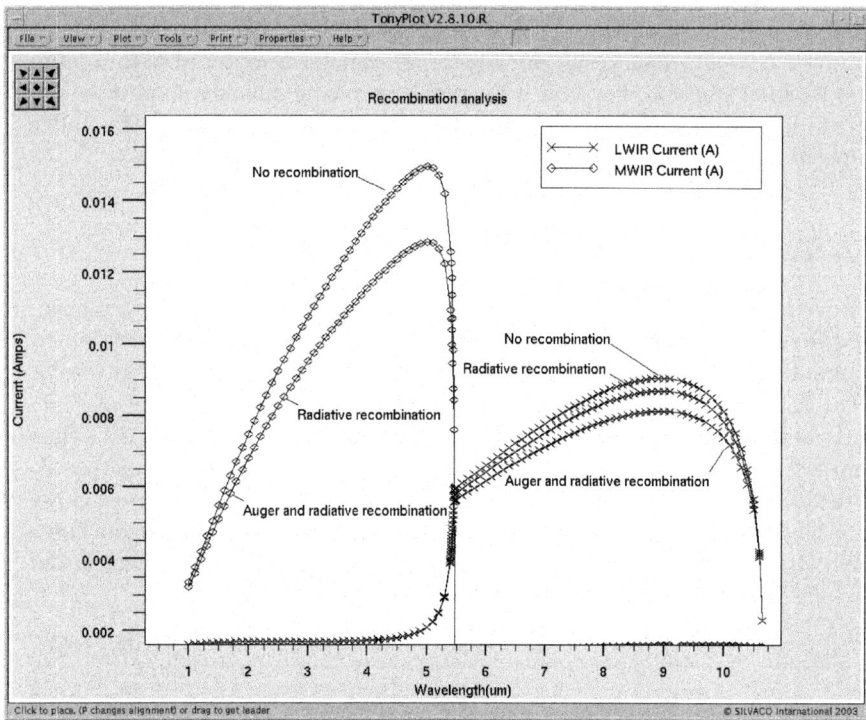

Fig. 13.8. Spectral response of LWIR and MWIR current as a function of wavelength with varying combinations of recombination mechanisms. Shown here are no recombination included, radiative recombination only, and Auger combined with radiative recombination.

of current reduction and hence recombination is greater in this MWIR device region. This is due to the dependence of the radiative recombination on compositional fraction x and in particular the material band gap. Upon inspection of (13.34), the dependence on the band gap squared term is clearly dominant in this situation compared with a material with a smaller band gap such as that associated with the LWIR region. When Auger recombination is introduced with radiative recombination, this has little effect on the MWIR response, which stays almost identical to the situation when radiative recombination was introduced independently. However, upon inspection of the LWIR response, there is a marked reduction in the device current. This is again due to the recombination factor being influenced by the fractional composition. Equation (13.32) states that the Auger coefficient for electrons is inversely proportional to the static dielectric constant ε_s squared and the material band gap E_g in addition to other parameters. As ε_s and E_g are smaller in the LWIR region, the Auger recombination will therefore be larger compared with that in the MWIR region. As such there will be a greater effect of

Auger recombination in the LWIR region, resulting in a greater reduction in the device current. This result demonstrates that for a material with a large compositional fraction, i.e., $x > 0.29$, radiative recombination dominates the electron loss and limits the device current. However, in materials of smaller compositional fraction, Auger recombination has the greater effect.

13.6 Temperature Studies

In order to investigate the effect of temperature on the device performance, two different temperatures have been used in the simulation in addition to the simulations performed at 77 K. Figure 13.9 shows the simulation results for the MWIR and LWIR spectral response for varying temperatures of 77 K, 100 K, and 120 K. It is clearly evident that operation away from 77 K has a marked affect on the device performance. As the temperature is increased, the device current is reduced in both the MWIR and LWIR responses. However, the device current is severely limited in the LWIR material region compared with the MWIR material region. The predominant factor here is the

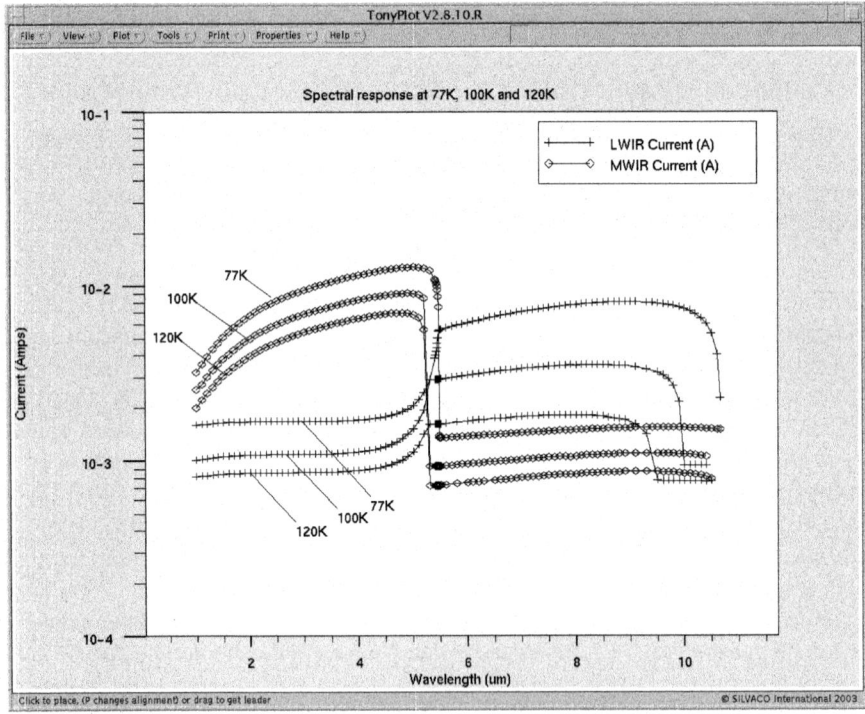

Fig. 13.9. Spectral response of LWIR and MWIR current as a function of wavelength with varying combinations of temperature. Simulations are performed at 77 K, 100 K, and 120 K.

electron mobility. It is clear from (13.29) that the electron mobility will be reduced by the factor $T^{-2(0.2/x)^{0.6}}$, and this is illustrated in Fig. 13.9. However, within this temperature-dependent factor is an inverse relationship with the compositional factor x. As a consequence, the exponent term is increased greatly when $x = 0.21$ compared with $x = 0.29$. The electron mobility will therefore be greatly reduced in the LWIR material region compared with the MWIR material region. Not only is the mobility affected with respect to temperature but also the electron and hole Auger and radiative recombination factor. It is clearly seen that in addition to a reduction in electron and hole mobility, there will also be an increase in the electron loss due to recombination that will continue to increase as the temperature is increased.

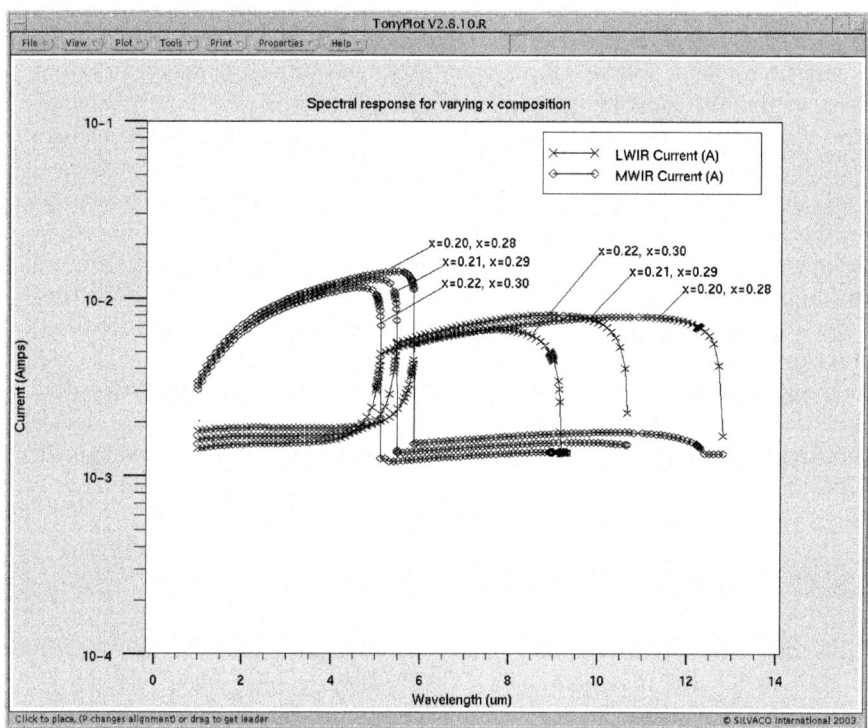

Fig. 13.10. Spectral response of LWIR and MWIR current as a function of wavelength with varying fraction of CdTe to HgTe. The simulation is performed at 77 K. Three different material compositions are shown. Case 1 is when layer 1 is Hg(0.72)Cd(0.28)Te, and layer 3 is Hg(0.8)Cd(0.2)Te. This is shown as $x = 0.2$, $x = 0.28$. Case 2 is when layer 1 is Hg(0.71)Cd(0.29)Te, and layer 3 is Hg(0.79)Cd(0.21)Te. This is shown as $x = 0.21$, $x = 0.29$. Case 3 is when layer 1 is Hg(0.70)Cd(0.30)Te, and layer 3 is Hg(0.78)Cd(0.22)Te. This is shown as $x = 0.22$, $x = 0.30$.

13.7 Variation of Composition

For a final investigation, the compositional fraction x has been changed to illustrate the effect of changing the window of spectral response for both the MWIR and LWIR regions. The results for three different cases with simulations performed at a constant temperature of 77 K are shown in Fig. 13.10.

1. LWIR $= Hg_{0.80}Cd_{0.20}Te$, MWIR $= Hg_{0.72}Cd_{0.28}Te$
2. LWIR $= Hg_{0.79}Cd_{0.21}Te$, MWIR $= Hg_{0.71}Cd_{0.29}Te$
3. LWIR $= g_{0.78}Cd_{0.22}Te$, MWIR $= g_{0.70}Cd_{0.3}Te$

In all cases, there is a change in the spectral response for both the LWIR and MWIR detectors. As the fractional composition is reduced, the width of the spectral response increases. Similarly as the fractional composition x is increased, the width of the spectral response is reduced. The magnitude of the current remains similar throughout. Also the rate of change of current is seen to remain constant throughout. The obvious reason for the change in the wavelength over which the device performs is linked to the change in the material band gap as a function of compositional fraction x. Consequently, for larger band gap materials, the energy required to excite electrons from the valence band into the conduction is increased, which is delivered by shorter wavelengths. As the wavelength increases, its energy is reduced and at some point it will have insufficient energy to cause this excitation, thus the device current reduces. Similarly for smaller x, the band gap is reduced and so less energy is required to cause such excitation and so the device will have a response in terms of deliverable device current at longer wavelengths. This is shown here. In addition to the direct band gap effect on the device responsivity, altering the fractional composition x also changes the values for (13.27–13.35).

13.8 Conclusion

In this chapter, optical detection has been discussed and the specific capabilities of Silvaco International's TCAD software have been introduced. A dual band monolithic photodetector was simulated using Atlas, S-Pieces, and Blaze. Nonstandard expressions have been incorporated into the simulation domain through the use of a C-interpreter, which has a seamless link with Atlas. Effective simulations have been performed to investigate several electrical effects by changing different paramters for each device. In all cases, it has been shown that HgCdTe detectors can be effectively simulated and investigated over a wide range of experimental factors.

References

1. W. D. Lawson, S. Nielson, E. H. Putley, and A. S. Young: J. Phys. Chem. Solids **9**, 325 (1959)
2. P. Norton: Optoelectronics Review **10**, 159 (2002)
3. P. Papadopoulou, N. Georgoulas, and A. Thanailakis: Thin Solid Films **415**, 276 (2002)
4. H. Vainola, J. Storgards, M. Yii-Koski, and J. Sinkkonen: Mater. Sci. Eng. B **91–92**, 421 (2002)
5. S. Selberherr: *Analysis and Simulation of Semiconductor Devices* (Springer, Vienna 1984)
6. R. Stratton: IEEE Trans. Electron Devices **19**, 1288 (1972)
7. W. Schockley and W.T. Read: Phys. Rev. **87**, 835 (1952)
8. D. J. Roulston, N. D. Arora, and S. G. Chamberlain: IEEE Trans. Electron Devices **29**, 284 (1982)
9. E. R. Blazejewski, J. M. Arias, G. M. Williams, W. McLerige, M. Zandian, and J. Pasko: J. Vac. Sci. Technol., B **10**, 1626 (1992)
10. K. Konnma, Y. Asano, K. Masubuchi, H. Utsumi, S. Tohyama, T. Eudo, H. Azuma, and N. Teranishi: IEEE Trans. Electron Devices **43**, 282 (1996)
11. M. B. Reinie, P. W. Norton, R. Starr, M. H. Weiler, M. Kestigian, B. L. Musicant, P. Mitra, T. Schimert, F. C. Case, I. B. Bhat, H. Ehsani, and V. Rao: J. Electron. Mater. **24**, 669 (1995)
12. G. Parish, C. A. Musca, J. F. Siliquini, J. M. Dell, B. D. Nener, L. Faraone, and G. J. Gouws: IEEE Electron Device Lett. **18**, 352 (1997)
13. C. A. Musca, J. F. Siliquini, K. A. Fynn, B. B. Nener, L. Faraone, and S. J. C. Trvine: Semicond. Sci. Technol. **11**, 1912 (1996)
14. J. Wenus, J. Rutkowski, and A. Rogalski: IEEE Trans. Elecron Devices **48**, 1326 (2001)
15. G. L. Hansen, J. L. Schmit, and T. M. Casselman: J. Appl. Phys. **53**, 7099 (1982)
16. T. J. Sanders and G. Hess: Government Microcircuit Applications Conf., Digest of Papers, Washington D.C. (1998)
17. G. T. Hess, T. J. Sanders, G. Newsome, and T. Fischer: *Proc. 4th Int. Conf. Modeling and Simulation of Microsystems*, 542 (2001)
18. M. W. Scott: J. Appl. Phys. **40**, 4077 (1969)
19. S. E. Schacham and E. Finkman: J. Appl. Phys. **57**, 2001 (1985)

14 Monolithic Wavelength Converters: Many-Body Effects and Saturation Analysis

J. Piprek[1], S. Li[2], P. Mensz[2], and J. Hader[3]

[1] University of California, Santa Barbara, CA 93106-9560, U.S.A., piprek@ieee.org
[2] Crosslight Software, 202-3855 Henning Dr., Burnaby, BC V5C 6N3 Canada, simon@crosslight.com
[3] Nonlinear Control Strategies, 1001 East Rudasill Rd., Tucson, AZ 85718, U.S.A., jhader@acms.arizona.edu

14.1 Introduction

Wavelength converters are a novel class of photonic integrated circuits that is crucial for multiwavelength fiber-optic communication networks [1]. Such converters switch the flow of information from one wavelength to another. We present here the simulation and analysis of an optoelectronic InP-based tunable wavelength converter (Fig. 14.1) that monolithically combines a preamplified receiver with a postamplified sampled-grating distributed Bragg reflector (SG-DBR) laser diode [2]. We employ the commercial software PICS3D [3], which was modified for the purpose of this investigation. Our self-consistent physical model takes into account many-body gain and absorption in the quantum wells, carrier drift and diffusion, and optical waveguiding. The time-consuming calculation of many-body spectra is performed externally [4], based on the theory outlined in Chap. 1 and in [5]. Tabulated spectra of gain, spontaneous emission, and index change are then imported into PICS3D. Performance limitations by saturation effects are the main target of this investigation.

The next section outlines the device structure. Section 14.3 describes physical device models and material parameters, including their experimental calibration. The following Sect. 14.4 investigates each component of the wavelength converter by three-dimensional (3D) steady-state simulation. Time-domain simulations of a similar SG-DBR laser can be found in Chap. 6.

14.2 Device Structure

Figure 14.1 gives the schematic design of the wavelength converter [6]. It electrically couples an optical receiver for any input wavelength of the C-band, e.g., $\lambda_\text{in} = 1525$ nm, with an optical transmitter for any output wavelength of the C-band, e.g., $\lambda_\text{out} = 1565$ nm. The receiver integrates signal preamplification by a semiconductor optical amplifier (SOA) and signal detection by a waveguide photodiode (WPD). The optical signal is converted into an

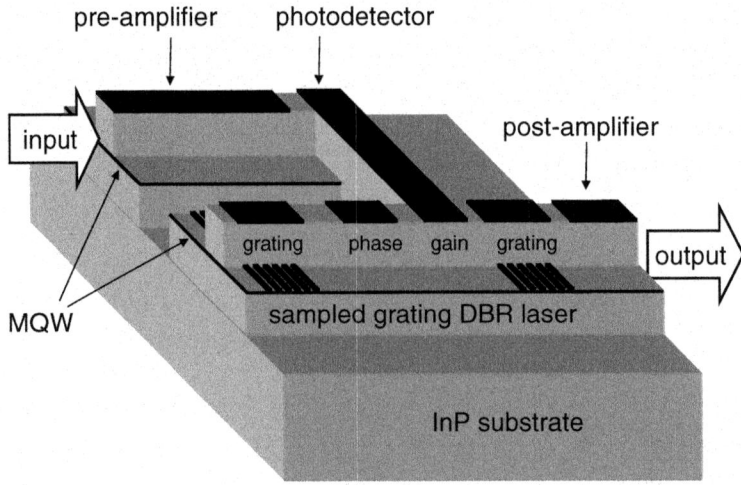

Fig. 14.1. Schematic view of the monolithic wavelength converter.

electrical signal that directly modulates a SG-DBR laser diode that is integrated with a semiconductor optical amplifier for signal enhancement [7]. The SG-DBR laser can be widely tuned to emit at any wavelength of the C-band [8].

The device is grown on InP by metal organic chemical vapor deposition; details of the fabrication process are given in [2]. The layer structure of the different components is very similar as all are based on the same epitaxial growth (Table 14.1). An offset-multi-quantum-well (MQW) active region is grown on top of the waveguide layer. For lateral confinement of optical field and current flow, a narrow ridge-waveguide structure is etched down to the MQW region. Passive device sections are formed by etching off the MQW.

14.3 General Device Physics

The different device types integrated in the wavelength converter (laser, amplifier, detector) exhibit the same epitaxial layer structure as well as some common device physics, which is discussed in this section. Material parameters are listed in Table 14.2 for binary compounds. They are linearly interpolated to obtain InGaAsP parameters, unless noted otherwise in the following.

14.3.1 Optical Waveguiding

The software solves the scalar Helmholtz equation:

$$\frac{\partial^2 \Phi}{\partial x^2} + \frac{\partial^2 \Phi}{\partial y^2} + (k^2 - \beta^2)\Phi = 0, \qquad (14.1)$$

Table 14.1. Epitaxial Layer Structure [9] (intrinsic i-regions exhibit about 10^{16}cm^{-3} n-type background doping).

	Thickness (nm)	Doping (1/cm^3)	Mobility (n/p) (cm^2/Vs)
p-InGaAs contact	100	3×10^{19}	100/20
p-InP cladding	1600	1×10^{18}	2200/70
p-InP cladding	200	3×10^{17}	2800/100
i-InP doping setback	50	—	4300/160
i-In$_{0.735}$Ga$_{0.265}$As$_{0.513}$P$_{0.487}$ barrier (7)	8	—	4300/160
i-In$_{0.735}$Ga$_{0.265}$As$_{0.845}$P$_{0.155}$ well (7)	6.5	—	4300/160
i-In$_{0.735}$Ga$_{0.265}$As$_{0.513}$P$_{0.487}$ barrier	8	—	4300/160
i-InP etch stop	10	—	4300/160
n-In$_{0.612}$Ga$_{0.338}$As$_{0.728}$P$_{0.272}$ waveguide	350	1×10^{17}	3300/130
n-InP cladding	1400	1×10^{18}	2200/70
n-InGaAs contact	100	1×10^{18}	2200/70
i-InP buffer	1000	—	4300/160

Table 14.2. Binary Material Parameters.

Parameter	Symbol	Unit	GaAs	GaP	InAs	InP
Direct band gap (unstrained)	E_g	eV	1.423	2.773	0.356	1.35
Spin–orbit split energy	Δ_0	eV	0.341	0.080	0.410	0.11
Electron eff. mass	m_c	m_0	0.0665	0.131	0.027	0.064
Luttinger parameter	γ_1	—	6.85	4.20	19.67	6.35
Luttinger parameter	γ_2	—	2.10	0.98	8.37	2.08
Luttinger parameter	γ_3	—	2.90	1.66	9.29	2.76
Lattice constant	a_0	Å	5.65325	5.451	6.0583	5.869
Elastic constant	C_{11}	10^{11}dyn/cm^2	11.81	14.12	8.329	10.22
Elastic constant	C_{12}	10^{11}dyn/cm^2	5.32	6.253	4.526	5.76
Elastic constant	C_{44}	10^{11}dyn/cm^2	5.94	7.047	3.959	4.60
Hydrost. deformation potential	a_1	eV	-7.1	-5.54	-5.9	-6.35
Shear deformation potential	a_2	eV	-1.7	-1.6	-1.8	-2.0
Dipole matrix energy	E_p	eV	22.80	21.51	14.72	15.22
LO phonon energy	$\hbar\omega$	eV	0.0354	0.046	0.0296	0.0426
opt. dielectric constant	ε_∞	—	10.9	9.075	12.25	9.61
stat. dielectric constant	ε_{st}	—	12.91	11.1	15.15	12.61
refractive index parameter	A_r	—	6.30	22.25	5.14	8.40
refractive index parameter	B_r	—	9.40	0.90	10.15	6.60

where $\Phi(x,y)$ represents any transverse component of the optical field, k is the absolute value of the wavevector, and β is the longitudinal propagation constant. The calculated vertical intensity profile is plotted in Fig. 14.2 together with the refractive index profile. The nonsymmetric index profile results in a reduced optical confinement factor for the quantum wells of $\Gamma = 0.06$. The 2D profile of the fundamental optical mode is given in Fig. 14.3 for half the device cross section. It is well confined by the p-InP ridge. The narrow ridge width $W = 3\mu$m prevents the appearance of higher-order lateral modes, as confirmed by measurements.

In propagation direction of the optical wave (z axis), the modal optical intensity $P_m(z)$ changes according to:

$$\frac{dP_m}{dz} = (g_m - \alpha_m)P_m \,, \tag{14.2}$$

with $g_m(z)$ giving the modal optical gain or absorption due to band-to-band transitions within the quantum well (see next section) and $\alpha_m(z)$ giving the modal optical loss caused by other processes.

Near 1.55-μm wavelength, optical losses are mainly attributed to intervalence band absorption (IVBA). The IVBA coefficient is considered proportional to the local hole density, i.e., it is only relevant within the quantum wells and within p-doped regions. The local loss coefficient is calculated as:

$$\alpha_{\text{opt}} = \alpha_b + k_n n + k_p p \,, \tag{14.3}$$

with the background absorption α_b, the electron density n, and the hole density p. The hole coefficient $k_p = 25 \times 10^{-18}\text{cm}^2$ [10], the electron coefficient $k_n = 1 \times 10^{-18}\text{cm}^2$, and $\alpha_b = 9$ cm^{-1} are employed in our simulations. The background loss is mainly related to photon scattering. The modal loss coefficient $\alpha_m(z)$ is obtained by 2D integration over $\alpha_{\text{opt}}(x,y,z)$ within each xy cross section, weighted by the optical intensity $P_{\text{opt}}(x,y,z)$.

The InGaAsP refractive index n_r is calculated as function of the photon energy $h\nu$ using:

$$n_r^2(h\nu) = A_r\left[f(x_1) + 0.5\left(\frac{E_g}{E_g + \Delta_0}\right)^{1.5} f(x_2)\right] + B_r \,, \tag{14.4}$$

with:

$$f(x_1) = \frac{1}{x_1^2}\left(2 - \sqrt{1+x_1} - \sqrt{1-x_1}\right), \quad x_1 = \frac{h\nu}{E_g} \tag{14.5}$$

$$f(x_2) = \frac{1}{x_2^2}\left(2 - \sqrt{1+x_2} - \sqrt{1-x_2}\right), \quad x_2 = \frac{h\nu}{E_g + \Delta_0}, \tag{14.6}$$

which was shown to give good agreement with measurements on InGaAsP by linear interpolation of the binary material parameters A_r and B_r (Table 14.2) [11].

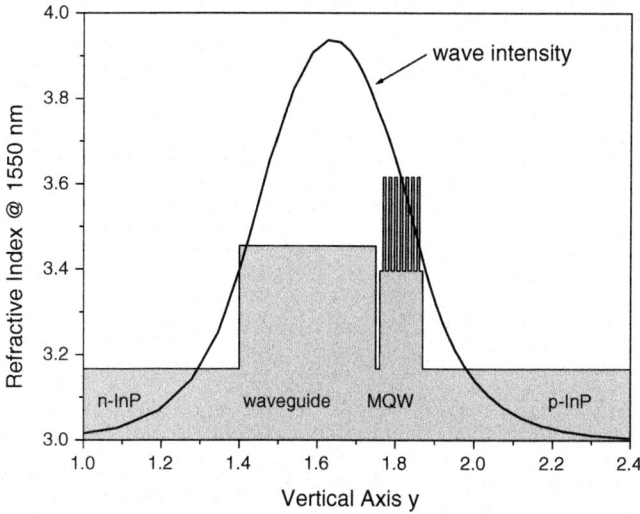

Fig. 14.2. Vertical profile of refractive index and wave intensity in the center of the device (y in microns).

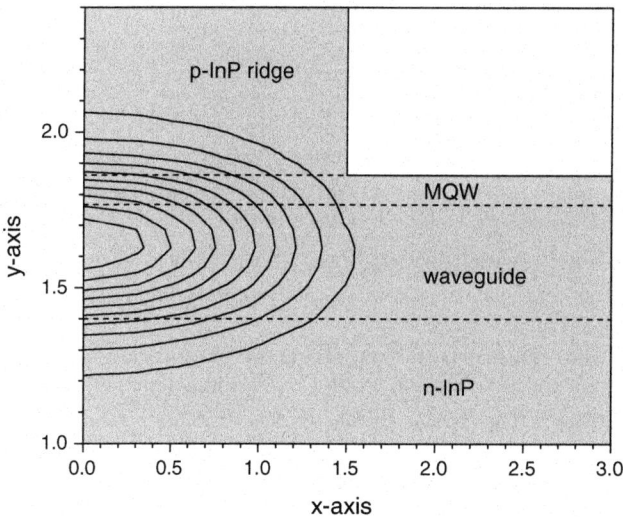

Fig. 14.3. 2D profile of the optical mode for half the device (x, y in microns).

14.3.2 Quantum Well Active Region

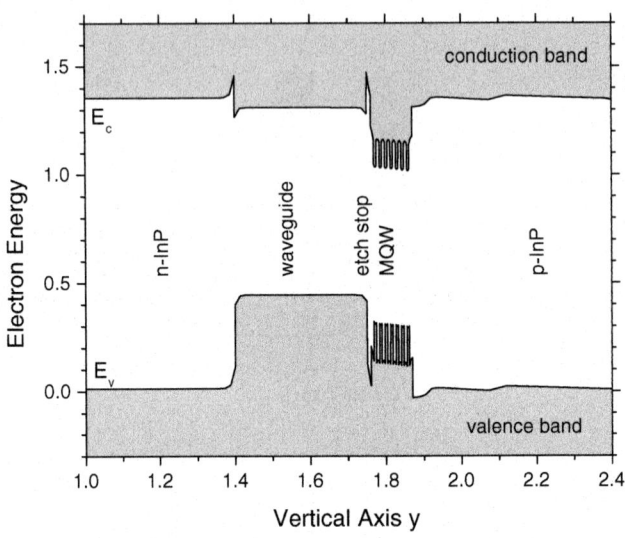

Fig. 14.4. Energy band diagram (electronvolts) for the active region at forward bias (E_c — conduction band edge, E_v — valence band edge, y in microns).

Figure 14.4 shows the energy band diagram of the MQW active region. The unstrained $In_{1-x}Ga_xAs_yP_{1-y}$ band gap is given in electronvolts by [12]:

$$E_g = 1.35 - 0.775\, y + 0.149\, y^2. \tag{14.7}$$

The conduction band edge offset ratio $\Delta E_c / \Delta E_g = 0.4$ is employed at all interfaces, which was demonstrated to give good agreement with measurements on similar devices [13].

PICS3D calculates the energy band structure of quantum wells, including strain effects using the 4×4 $\boldsymbol{k} \cdot \boldsymbol{p}$ model as published by Chuang [12]. By default, it then uses a free-carrier model to account for radiative carrier recombination within the quantum wells. However, previous investigations have shown that such free-carrier model results in poor agreement with measurements [9]. The many-body model outlined in the first chapter of this book is expected to more accurately describe optical spectra of quantum wells. The correct representation of the entire gain spectrum is of particular importance in our multiwavelength device. Therefore, we here use many-body spectra calculated for our quantum wells and imported into PICS3D. These computations are based on 8 × 8 $\mathbf{k} \cdot \mathbf{p}$ bandstructure calculations using the parameters listed in Table 14.2.

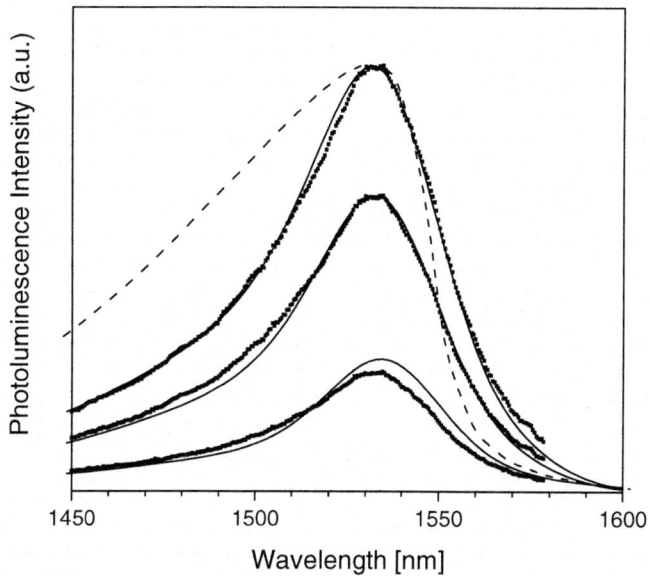

Fig. 14.5. Photoluminescence spectra of our MQW active region (dots — measured at 1×, 2×, and 3× pump intensity [14], solid lines — many-body calculations at carrier densities of 3.8, 6.2, and 7.7 $\times 10^{17} \text{cm}^{-3}$, dashed line — normalized free-carrier spectrum).

The dots in Fig. 14.5 show measured photoluminescence (PL) spectra. The emission peak at 1532 nm indicates a slight growth deviation from the intended MQW composition (cf. Table 14.1). The PL peak wavelength can be matched in the simulation by adjustment of the MQW composition, maintaining the measured biaxial MQW strain: 0.6% compressive strain in the quantum wells and 0.3% tensile strain in the barriers. This way, we obtain $\text{In}_{0.685}\text{Ga}_{0.315}\text{As}_{0.864}\text{P}_{0.136}$ for the quantum well composition and $\text{In}_{0.685}\text{Ga}_{0.315}\text{As}_{0.595}\text{P}_{0.405}$ for the barrier composition. The adjustment may also reflect inaccuracies in the calculation of the strained quaternary energy band gap. The only unknown parameter of the many-body calculation is the inhomogeneous energy broadening due to structural imperfections of the quantum well. A full-width half-maximum (FWHM) value of 14 meV gives good agreement with the PL measurement (solid lines in Fig. 14.5). The quantum well carrier density is varied in the simulation to fit the measurement. For comparison, the dashed line shows the normalized free-carrier spectrum with homogeneous broadening of 13.6 meV (scattering time = 0.1 ps). This parameter can be adjusted to find a better fit on the long-wavelength side, however, the PL intensity remains overestimated on the short wavelength side [9].

The corresponding many-body gain spectra are plotted in Fig. 14.6, and the gain at fixed wavelengths is given in Fig. 14.7 as a function of carrier density. The wavelength range of interest (1525 – 1565 nm) is completely covered by positive gain for carrier densities above $3 \times 10^{18} \text{cm}^{-3}$. The gain spectrum flattens at high densities. At very low carrier density, the many-body absorption spectrum shows a characteristic exciton peak. For comparison, free-carrier gain and absorption spectra are given by dashed lines as calculated with the 4×4 $\boldsymbol{k} \cdot \boldsymbol{p}$ band structure model, including valence band mixing. There are significant differences between many-body and free-carrier spectra, which are discussed in the following. At low carrier density, the exciton absorption peak is missing in the free-carrier spectrum because Coulomb attraction between electrons and holes is not considered. At high densities, the free-carrier gain peaks at shorter wavelength due to the missing band gap renormalization caused by many-body interaction. The larger width of the free-carrier gain spectrum is attributed to the smaller density of states resulting from the 4×4 $\boldsymbol{k} \cdot \boldsymbol{p}$ model as compared with the 8×8 $\boldsymbol{k} \cdot \boldsymbol{p}$ model used in the many-body calculation. Thus, the separation ΔE_F of the quasi-Fermi levels needs to be wider in the free-carrier model to accommodate the same number of carriers in the quantum well. The larger magnitude of the free-carrier gain is unexpected because Coulomb interaction is known to cause gain enhancement [15]. The gain magnitude is proportional to the bulk matrix element:

$$M_\text{b}^2 = \frac{m_0}{6} E_p , \qquad (14.8)$$

with the electron rest mass m_0 and the material parameter E_p, which is interpolated between the binary data given in Table 14.2. Linear interpolation results in $E_p = 17.3$ eV for our quantum well as used in both the free-carrier and the many-body model. The recent compilation of band structure parameters in [16] suggests a larger number of $E_p = 24$ eV for our case; however, validation of either number requires absorption measurements that are not available for our quantum well material. Thus, some uncertainty remains with the magnitude of gain and absorption in our quantum wells.

The quantum well gain at given carrier density and wavelength tends to decrease at very high photon densities S. Such nonlinear behavior (gain compression) is attributed to the depletion of carriers at the transition energy level. This spectral hole burning is due to the finite time required for electrons and holes to fill up the energy levels emptied by stimulated recombination

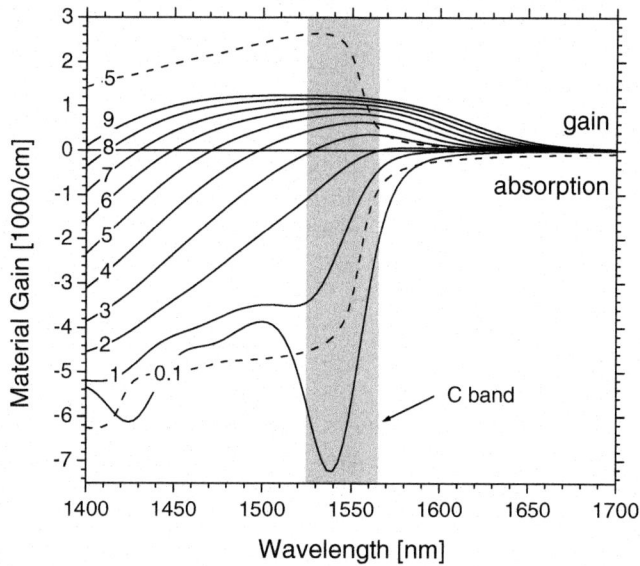

Fig. 14.6. Many-body gain spectra with the quantum well carrier density $n = p$ given as parameter ($\times 10^{18}$cm^{-3}). The dashed lines show the free-carrier spectrum at two densities.

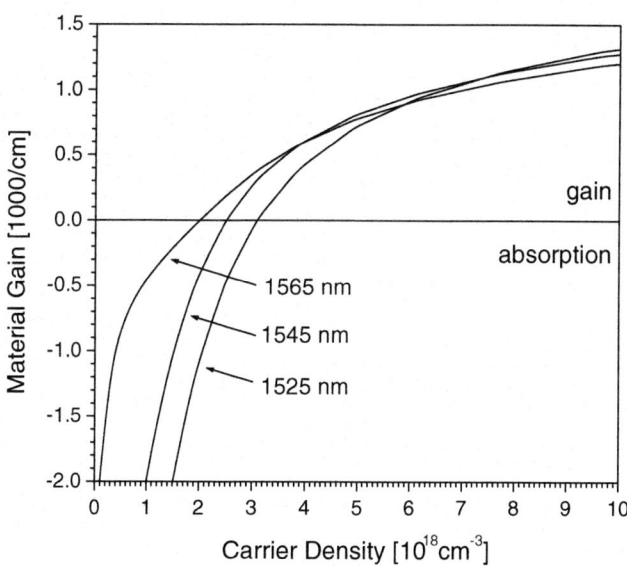

Fig. 14.7. Many-body gain vs. quantum well carrier density at three fixed wavelengths of the C-band.

and to reestablish a Fermi distribution. The nonlinear gain is commonly approximated by the phenomenological equation:

$$g_{\text{opt}} = \frac{g_0}{1 + \epsilon S}, \tag{14.9}$$

with the linear gain g_0 and the gain compression coefficient $\epsilon \approx 10^{-17}$ cm^3. This nonlinear gain formula is included in our model; however, the effect on our steady-state results is negligibly small because the photon density remains below 10^{15}cm^{-3}. Note that a different form of gain saturation is due to the reduction of the total carrier density by stimulated recombination, which is included self-consistently in our model and which shows strong effects on our results.

14.3.3 Carrier Transport

Drift and diffusion of electrons and holes is calculated by solving the semiconductor transport equations. The current densities are given by:

$$\boldsymbol{j}_n = q\mu_n n \boldsymbol{F} + qD_n \nabla n \tag{14.10}$$

$$\boldsymbol{j}_p = q\mu_p p \boldsymbol{F} - qD_p \nabla p, \tag{14.11}$$

with the elementary charge q and the mobilities μ_n and μ_p. The doping-dependent low-field mobilities listed in Table 14.1 are obtained from [17] using InP values also for InGaAsP due to the lack of experimental data for our quaternary layers. The diffusion constants D_n and D_p are replaced by mobilities using the Einstein relation $D = \mu k_B T/q$ with the Boltzmann constant k_B and the temperature T. The electrostatic field $\boldsymbol{F}$ is calculated from the charge distribution by the Poisson equation:

$$\nabla \cdot (\varepsilon_{\text{st}}\varepsilon_0 \boldsymbol{F}) = q(p - n + p_D - n_A) \tag{14.12}$$

($\varepsilon_{\text{st}}\varepsilon_0$ — electrical permittivity, p_D, n_A — donor and acceptor concentration). In our steady-state simulation, the continuity equations:

$$\nabla \cdot \boldsymbol{j}_n - qR = 0 \tag{14.13}$$

$$-\nabla \cdot \boldsymbol{j}_p - qR = 0 \tag{14.14}$$

describe the influence of carrier generation ($R < 0$) and recombination ($R > 0$). Emission and absorption of photons in the quantum wells was discussed in the previous section. The total local recombination rate is:

$$R = R_{\text{stim}} + R_{\text{spon}} + R_{\text{Auger}} + R_{\text{SRH}}, \tag{14.15}$$

with the stimulated recombination rate:

$$R_{\text{stim}} = g_{\text{opt}} \frac{P_{\text{opt}}}{h\nu}, \tag{14.16}$$

the spontaneous emission rate outside the quantum wells:[4]

$$R_{\text{spon}} = B(np - n_i^2), \qquad (14.17)$$

the Auger recombination rate:

$$R_{\text{Auger}} = (C_n n + C_p p)(np - n_i^2), \qquad (14.18)$$

and the defect-related Shockley–Read–Hall (SRH) recombination rate:

$$R_{\text{SRH}} = \frac{np - n_i^2}{\tau_p^{\text{SRH}}\left(n + N_c \exp\left[\frac{E_t - E_c}{k_B T}\right]\right) + \tau_n^{\text{SRH}}\left(p + N_v \exp\left[\frac{E_v - E_t}{k_B T}\right]\right)} \qquad (14.19)$$

(n_i — intrinsic carrier density, $N_{c,v}$ — density of states of conduction, valence band; E_t — mid-gap defect energy). For the radiative coefficient, we assume a typical value of $B = 10^{-10} \text{cm}^3 \text{s}^{-1}$ and for the SRH lifetimes, $\tau_n^{\text{SRH}} = \tau_p^{\text{SRH}} = 20$ ns. There is some uncertainty with these recombination parameters; however, Auger recombination is known to have the strongest impact on InP-based devices and we use the Auger parameter C_n as fit parameter to find agreement with the measured threshold current. Auger recombination within the valence band is believed to be negligible in our device ($C_p = 0$) [18]. Figure 14.8 compares simulated light-current characteristics with the measurement on broad-area Fabry–Perot lasers [9]. The fit results in $C_n = 10^{-29} \text{cm}^6 \text{s}^{-1}$, which is one order of magnitude smaller than the value obtained in [9] using a free-carrier gain model. This difference can be easily understood. The lower many-body gain requires a higher quantum well carrier density to reach lasing threshold. In order to obtain the same threshold current, the Auger recombination needs to be scaled down. However, our Auger coefficient is still within the rather wide range of values reported in the literature [19].

PICS3D obtains quasi-3D solutions to the transport equations by slicing the device into many transversal xy sections. Within each 2D section, the equations are solved self-consistently and the solutions are iteratively adjusted to the longitudinal photon density $S(z)$. Longitudinal variations of, e.g., the quantum well carrier density are accounted for this way; however, carrier flow in z direction is not considered.

14.4 Simulation Results

This section investigates device physics and performance of each converter component. We mainly focus on saturation effects in amplifier and photodetector, which limit the device performance. Experimental device characteristics from the wafer used for the PL calibration in Fig. 14.5 are not yet available for comparison.

---

[4] Inside the quantum wells, the many-body luminescence spectrum is integrated over all photon energies.

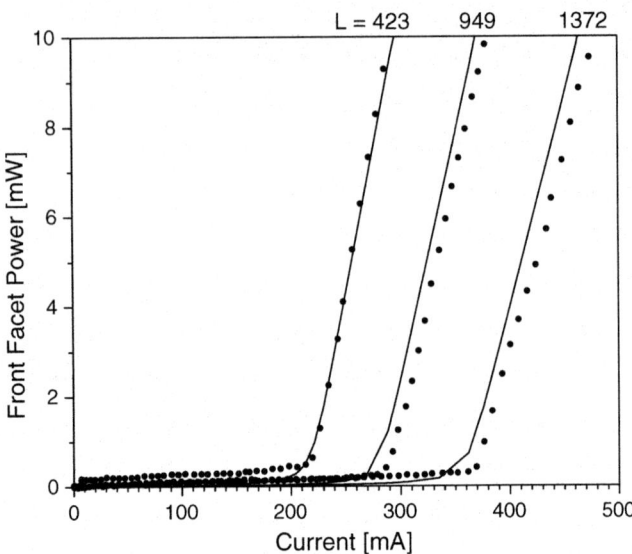

Fig. 14.8. Light-current (LI) curves for 50-μm-wide Fabry–Perot lasers of different length L/μm (dots — measurement [9], lines — simulation with $C_n = 10^{-29} \text{cm}^6 \text{s}^{-1}$).

14.4.1 Amplifier

Amplification enhances the optical signal, and it compensates for any losses during wavelength conversion. The SOA should provide maximum signal gain while maintaining linearity between input and output signal. Amplifier saturation results in a sublinear SOA response, and it restricts the maximum output signal.

We consider the ideal case of zero facet reflectance and maximum fiber-to-waveguide power coupling efficiency of 0.25, as estimated from the modal overlap between fiber and waveguide with perfect alignment. The coupling efficiency may be lower in the actual measurement [9]. The amplifier length is 600 μm in our simulation.

At low input power, the material properties of the SOA are not affected by the amplified light. For such a case, Fig. 14.9 shows the SOA output power as a function of the injection current at different wavelengths. The wavelength dependence can be understood from the gain plots in Figs. 14.6 and 14.7. Power saturation at higher currents is caused by the sublinear dependence of the quantum well gain on current and carrier density. The red-shift of the gain spectrum with higher carrier density gives maximum gain at the longest wavelength for low currents and at the shortest wavelength for high

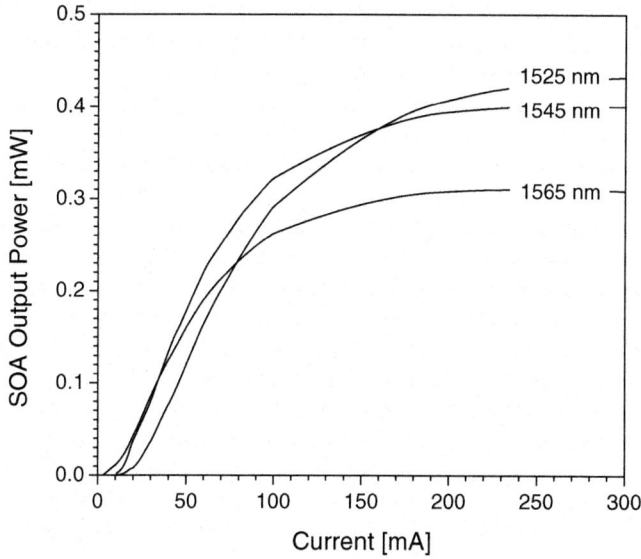

Fig. 14.9. SOA output power vs. injection current with the signal wavelength given as parameter (fiber input power = 0.1 mW).

currents. Self-heating of the SOA would cause an additional red-shift of the gain spectrum, and it would lower the gain magnitude, thereby increasing gain saturation. However, the temperature of the wavelength converter is typically stabilized by a thermoelectric cooler [2]. The gain saturation with higher current restricts the maximum SOA gain, but it hardly affects the optical linearity of the SOA, which is investigated in the following.

With higher input power, the increasing modal intensity $P_{\mathrm{m}}(z)$ leads to increasing stimulated carrier recombination, which reduces the carrier density in the quantum wells as well as the modal gain toward the output facet. Figure 14.10 visualizes longitudinal variations of electron density, modal gain, and modal power for different injection currents. At low current (15 mA), the electron density is not high enough for amplification because the active region still exhibits net absorption (including optical losses), leading to a decay of the traveling optical wave. At higher current, positive net modal gain causes the light power to increase with travel distance. If the input power is high enough, the photon-triggered stimulated recombination reduces the carrier density with travel distance, thereby lowering the local gain. High modal intensity may cause spatial hole burning near the output facet as illustrated in Fig. 14.11. The reduction of the quantum well carrier density

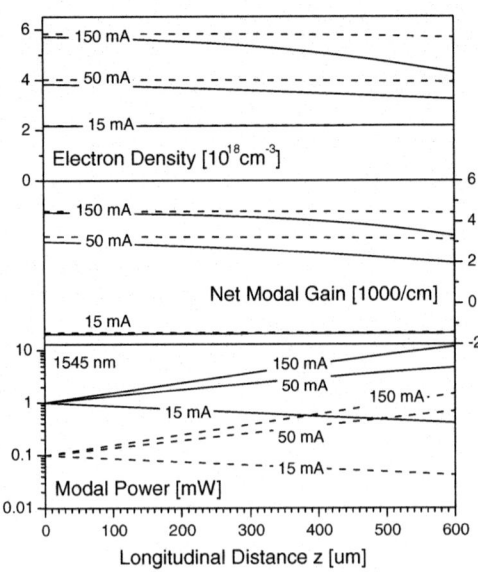

Fig. 14.10. Internal SOA parameters vs. longitudinal distance (μm) at 1545 nm wavelength. Top: quantum well electron density; middle: net modal gain; bottom: light power. The SOA current is given as parameter and the fiber input power is 0.4 mW (dashed) and 4 mW (solid), respectively (at 15-mA injection current, the modal gain is negative and the SOA partially absorbs the input signal).

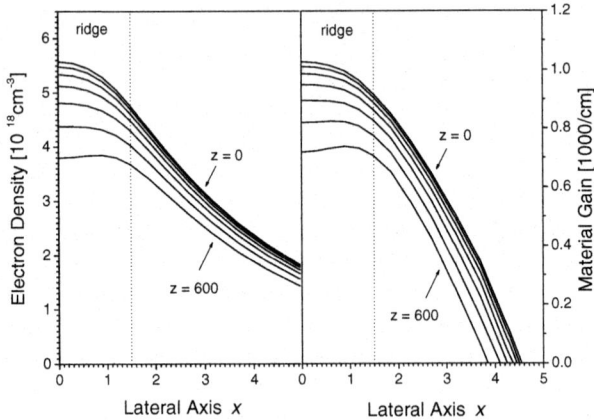

Fig. 14.11. Internal SOA parameters vs. lateral distance (μm) with the longitudinal position $z/\mu m$ in steps of 100 μm as parameter (current = 150 mA, fiber power = 4 mW, wavelength = 1545 nm). Left: quantum well electron density; right: material gain.

14 Monolithic Wavelength Converter   419

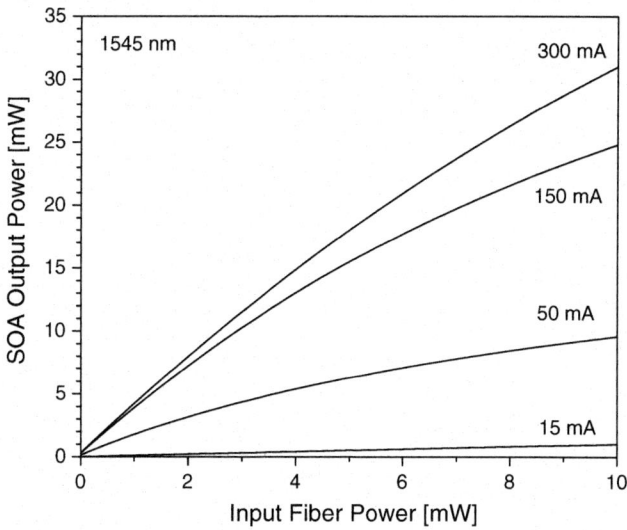

Fig. 14.12. Output power vs. input power for different SOA currents.

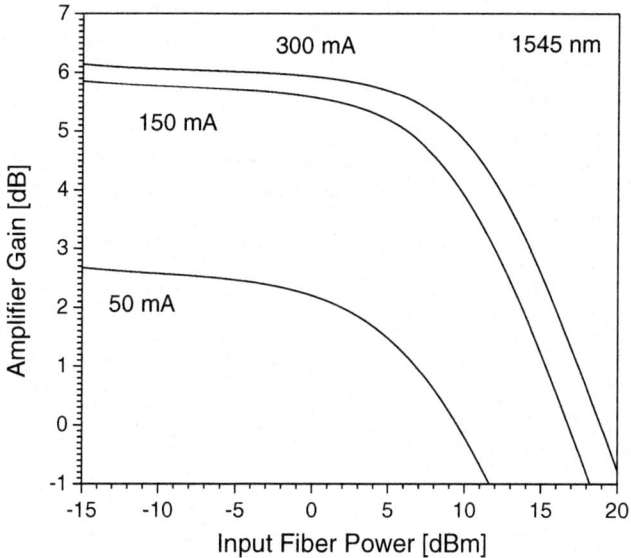

Fig. 14.13. Amplifier gain vs. input power for different SOA currents.

with higher power is the main reason for the nonlinear amplifier response shown in Fig. 14.12. The ratio of output to input power gives the amplifier gain, which is plotted in Fig. 14.13, including fiber coupling losses. At maximum material gain (300 mA), the maximum amplifier gain is about 4 (6 dB) and the -3 dB saturation power is about 25 mW (14 dBm).

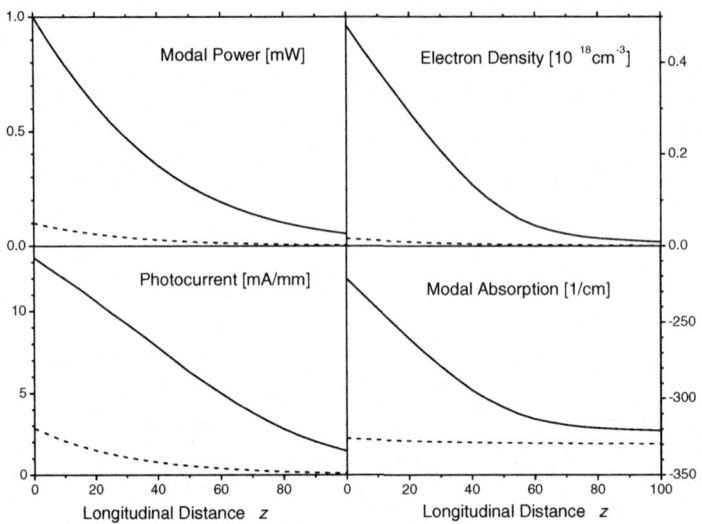

Fig. 14.14. Longitudinal variations (z/μm) within the waveguide photodetector as calculated at 1545-nm wavelength and 1-V reverse bias for 1-mW (solid) and 0.1-mW (dashed) input power.

14.4.2 Photodetector

The waveguide photodetector (WPD) is coupled monolithically to the preamplifier, and we neglect any optical coupling loss here. The WPD length is 100 μm in our simulation. Detector saturation effects are mainly related to the accumulation of photogenerated carriers within the quantum wells. Such accumulation depends on the local light power and therefore on the longitudinal position z within the detector. Figures 14.14 and 14.15, respectively, show longitudinal and vertical variations within the detector at two different input power levels. At low input power (0.1 mW, dashed lines), the quantum well carrier density remains low and the band-to-band absorption is almost constant in longitudinal direction. The photocurrent $I_{\rm ph}(z)$ decays with light penetration depth due to the reduced modal power. At higher input power (1 mW, solid lines), the photogenerated carriers pile up near the front and lead to a reduction of the band-to-band absorption. Therefore, the light penetrates deeper into the WPD and about 5% of the light power remains undetected (light reflection at the rear facet is neglected). The amount of undetected light increases with higher input light, leading to a nonlinear detector response.

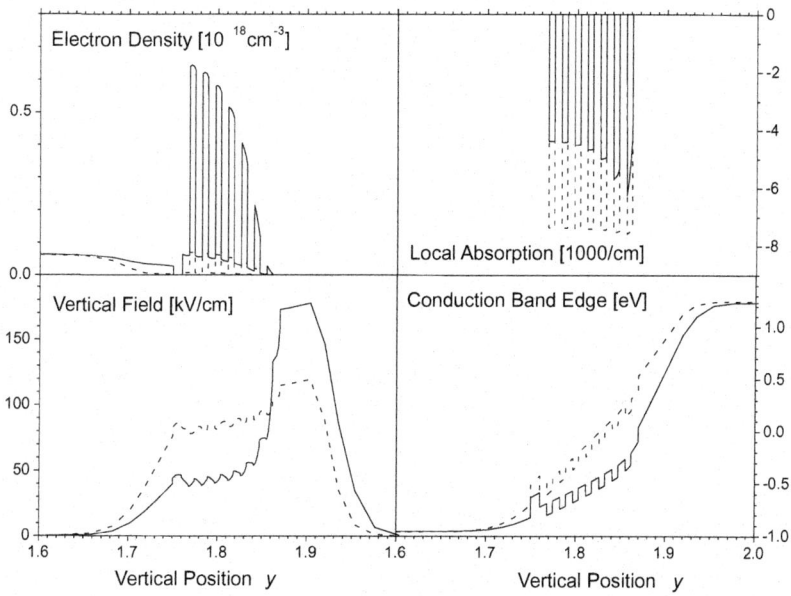

Fig. 14.15. Vertical variations in the center of the detector (y/μm, $x = 0$, $z = 0$) for 1-mW (solid) and 0.1-mW (dashed) input power (1545 nm, reverse bias = 1 V).

Figure 14.16 plots calculated detector response characteristics. There is a significant dependence on the input wavelength, which corresponds to the strong wavelength dependence of the absorption spectra shown in Fig. 14.6. At 1545 nm, 1-V reverse bias, and 1-mW input power, the total quantum efficiency is 75%, which indicates that about 20% of the photogenerated carriers are lost in recombination mechanisms without contributing to the photocurrent. The differential quantum efficiency is 60%, and it further decreases with increasing input power.

The accumulation of quantum well carriers leads to a partial screening of the electrostatic field within the active region (Fig. 14.15), hindering the carrier transport to the contacts. Such screening can be reduced by applying a larger reverse electric field. The dashed line in Fig. 14.16 is calculated for 3-V reverse bias, and it shows less nonlinearity than the low-bias characteristic.

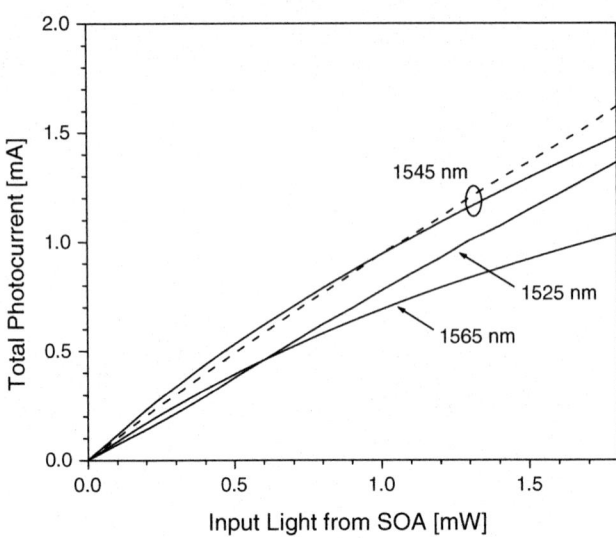

Fig. 14.16. Photodetector response for different wavelengths and for a reverse bias of 1 V (solid) and 3 V (dashed).

This leads to an improved differential quantum efficiency of 70% at 1-mW input power. However, the significant saturation calculated for our quantum well detector at relatively low input power may require the use of Franz–Keldysh-type detectors in the wavelength converter. Such photon absorption by the reverse-biased waveguide layer shows less saturation effects due to the missing carrier confinement [2].

14.4.3 Sampled-Grating DBR Laser

A sampled grating is a conventional DBR grating with grating elements removed in a periodic fashion [8]. Such sampling leads to reflection spectra with periodic peaks (Fig. 14.17). As the figure shows, the periods of the two reflectivity spectra are slightly mismatched. Lasing occurs at that pair of maxima that is aligned. Each spectrum can be shifted by inducing small refractive index changes in the mirror, thereby changing the emission wavelength by aligning a different pair of reflection peaks. This Vernier tuning mechanism allows for six to eight times the tuning range that is achieved from the index change alone [20]. An overview of the different wavelength tuning mechanisms employed in semiconductor lasers can be found in [21].

Our example device contains five sections (Table 14.3) [9]. The rear mirror employs 12 periods of 6-μm-long grating bursts with 46-μm period length.

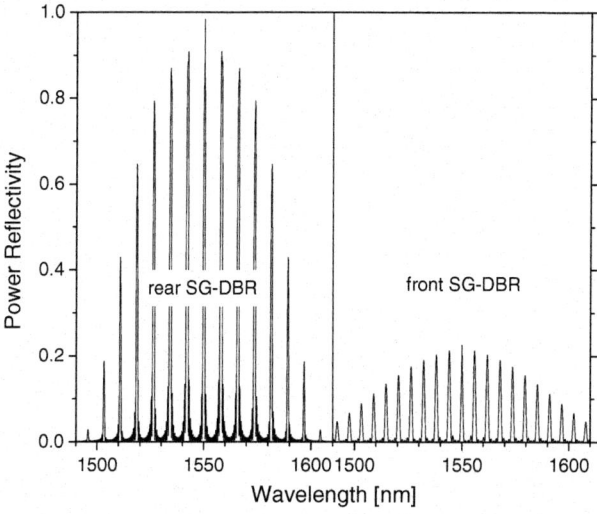

Fig. 14.17. Reflectivity spectra of the two SG-DBR mirrors.

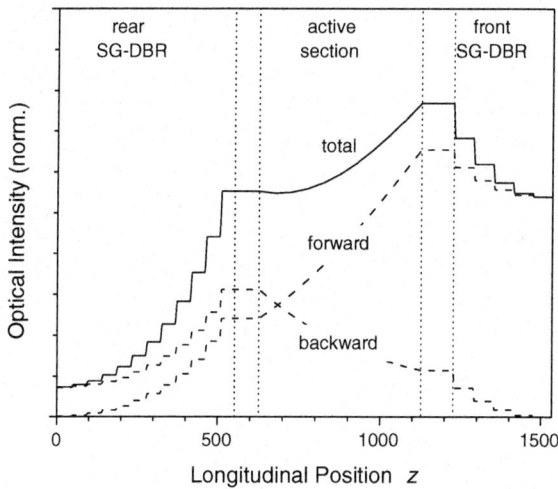

Fig. 14.18. Longitudinal intensity profile (z/μm) within the SG-DBR laser for forward and backward traveling wave (dashed) and total intensity (solid).

Table 14.3. Longitudinal Sections of the SG-DBR Laser (L — section length, κ — optical coupling coefficient of each grating burst).

	L (μm)	κ (1/cm)
Rear SG-DBR	552	250
Phase tuning	75	0
Active	500	0
Gain lever	100	0
Front SG-DBR	307.5	250

The front mirror has 5 periods of 4-μm-long grating bursts with 61.5-μm period length. Both the SG-DBR reflectivity spectra are plotted in Fig. 14.17. The phase section is used to fine-tune the wavelength. The active section is used for DC pumping of the laser. The gain lever section is connected to the photodiode for direct laser modulation. Figure 14.18 shows the optical intensity along the longitudinal axis without photocurrent. Both the forward and the backward traveling wave gain intensity in the active section according to (14.2). Within the SG-DBR sections, each grating burst reflects part of the optical wave and causes a stepwise change in intensity.

Figure 14.19 gives the simulated light-current (LI) characteristics of the SG-DBR laser for two different emission wavelengths. At 1550 nm, the threshold current is 26 mA and the slope efficiency 39%. Due to the lower reflectivity at 1526 nm, threshold gain and quantum well carrier density rise, giving an increased threshold current of 42 mA for this wavelength. The slope efficiency drops to 33%, which is partially caused by the stronger differential change of the Auger recombination. This phenomenon is related to the nonuniform carrier distribution among the quantum wells, which leads to increasing recombination with increasing current, even if the total quantum well carrier density is constant [22]. Current adjustment is required to maintain the same DC power at different wavelengths. Postamplification helps to enhance both the slope efficiency and the signal power. The rear emission is also shown for comparison (dashed lines), it can be used to monitor the output signal. The laser emission spectrum is plotted in the insets of Fig. 14.19 for 13-mW output power. The side-mode suppression ratio (SMSR) is almost 40 dB in both cases, showing that the large tuning range of SG-DBR lasers can be combined with high SMSR.

Here, we cover SG-DBR lasers only briefly because a more detailed discussion and simulation of SG-DBR lasers is given in Chapt. 6.

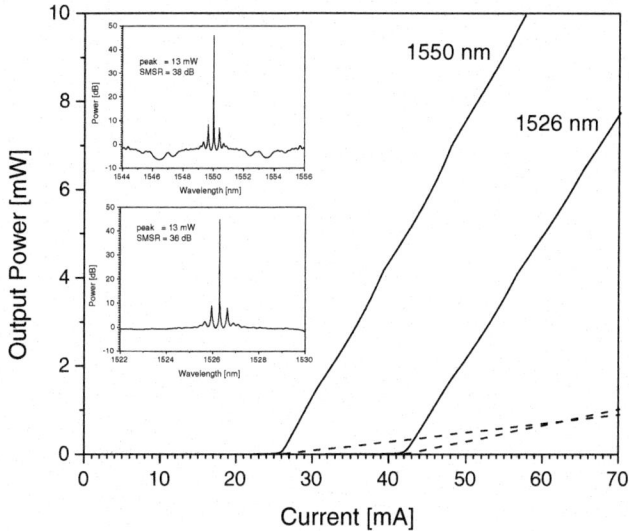

Fig. 14.19. Light vs. current characteristics for two laser wavelengths (solid — front emission, dashed — rear emission); the insets give the emission spectrum for both cases, including the SMSR at 13-mW front output power.

14.5 Summary

We have demonstrated the inclusion of many-body theory into the full device simulation of all components comprising the monolithic wavelength converter. This allows for a more realistic prediction of the device performance at different wavelength. The detailed analysis of microscopic saturation mechanisms helps to improve the design of future converter generations.

Acknowledgment

This work was partially supported by the Semiconductor Research Corporation and by the Intel Corporation. The authors are grateful to Dr. John Hutchinson, Jeff Henness, Anna Tauke Pedretti, and Prof. Larry Coldren for valuable discussions and measurements. We also acknowledge fruitful discussions with Prof. Stephan W. Koch.

References

1. J. S. B. Yoo: IEEE J. Lightwave Technol. **14**, 955 (1996)

2. J. Hutchinson, J. Barton, M. Masanovic, M. Sysak, J. Henness, L. Johansson, D. Blumenthal, and L. A. Coldren: Monolithically integrated InP-based tunable wavelength conversion. In: *Physics and Simulation of Optoelectronic Devices XII* SPIE Proceedings **5349** (The International Society for Optical Engineering, Bellingham 2004)
3. PICS3D by Crosslight Software Inc., Burnaby, Canada (http://www.crosslight.com)
4. Nonlinear Control Strategies Inc., Tucson, AZ (http://www.nlcstr.com)
5. J. Hader, J. V. Moloney, S. W. Koch, and W. W. Chow: IEEE J. Select. Topics Quantum Electron. **9**, 688 (2003)
6. J. Hutchinson, J. Henness, L. Johansson, J. Barton, M. Masanovic, and L. A. Coldren: 2.5 Gb/sec wavelength conversion using monolithically-integrated photodetector and directly modulated widely-tunable SGDBR laser. In: *Proc. 16th LEOS Annual Meeting* Lasers and Electro-Optics Society, Piscataway, NJ (2003)
7. B. Mason, J. Barton, G. A. Fish, L. A. Coldren, and S. P. DenBaars: IEEE Photon. Technol. Lett. **12**, 762 (2000)
8. V. Jayaraman, Z. M. Chuang, and L. A. Coldren: IEEE J. Quantum Electron. **29**, 1824 (1993)
9. J. Piprek, N. Trenado, J. Hutchinson, J. Henness, and L. A. Coldren: 3D Simulation of an Integrated Wavelength Converter. In: *Physics and Simulation of Optoelectronic Devices XII* SPIE Proceedings **5349** (The International Society for Optical Engineering, Bellingham 2004)
10. C. Henry, R. Logan, F. R. Merritt, and J. P. Luongo: IEEE J. Quantum Electron. **19**, 947 (1983)
11. S. Adachi: *Physical Properties of III-V Semiconductor Compounds* (Wiley, New York 1992)
12. S. L. Chuang: *Physics of Optoelectronic Devices* (Wiley, New York 1995)
13. J. Piprek, P. Abraham, and J. E. Bowers: IEEE J. Quantum Electron. **36**, 366 (2000)
14. J. Henness, *unpublished*
15. W. W. Chow and S. W. Koch: *Semiconductor-Laser Fundamentals* (Springer, Berlin 1999)
16. I. Vurgaftman, J. R. Meyer, and L. R. Ram-Mohan: J. Appl. Phys. **89**, 5815 (2001)
17. M. Sotoodeh, A. H. Khalid, and A. A. Rezazadeh: J. Appl. Phys. **87**, 2890 (2000)
18. S. Seki, W. W. Lui, and K. Yokoyama: Appl. Phys. Lett. **66**, 3093 (1995)
19. J. Piprek: *Semiconductor Optoelectronic Devices: Introduction to Physics and Simulation* (Academic Press, San Diego, 2003)
20. L. A. Coldren: IEEE J. Select. Topics Quantum Electron. **6**, 988 (2000)
21. L. A. Coldren, G. A. Fish, Y. Akulova, J. S. Barton, L. Johansson, and C.W. Coldren: IEEE J. Lightwave Technol. **22**, 193 (2004)
22. J.Piprek, P. Abraham, and J. E. Bowers: Appl. Phys. Lett. **74**, 489 (1999)

15 Active Photonic Integrated Circuits

A. J. Lowery

VPIphotonics (a division of VPIsystems), Level 2, 35 Cotham Road, Kew, Victoria, 3101, Australia, www.vpiphotonics.com

15.1 Introduction

This chapter investigates the application of computer simulation to active photonic circuit design, from a topology point of view. This is because the topology of a circuit often holds the key intellectual property, rather than the implementation [1]. Once the topology has been optimized, then there are many materials technologies in which to implement the photonic circuit. In fact, many circuits can be implemented in bulk optics or fiber optics, although thermal fluctuations can make the stability of circuits using interferometers problematic.

To illustrate a simulation-based design process for active photonic circuits, a 40-Gbit/s to 10-Gbit/s photonic time-division demultiplexer design was developed especially for this book. Starting with a published idea for an optical phase discriminator [2], and some ready-made applications examples in VPIcomponentMaker^{TM} Active Photonics, a circuit containing an optical phase-locked loop (including mode-locked laser) and an optical gate was developed and tuned. This could be integrated as shown in Fig. 15.1. During this process, the simulations suggested inventive steps, and allowed sub-components to be tuned rapidly before being simulated together.

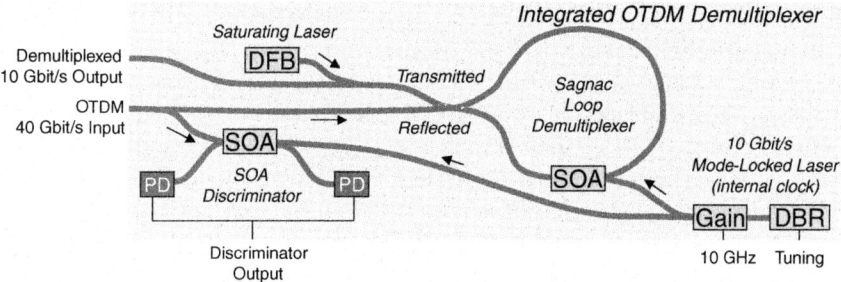

Fig. 15.1. All-optical (no 40-Gbit/s electronics) 40-Gbit/s to 10-Gbit/s demultiplexer. Arrows indicate the positions of optical isolators (SOA — Semiconductor optical amplifier. DFB — Distributed feedback laser, DBR — Distributed Bragg reflector, Gain — Active gain region, PD — Photodiode).

The simulations in this chapter are available as downloads for the free simulator VPIplayer™, available from www.VPIphotonics.com. VPIplayer™ allows simulations created with VPIcomponentMaker™ Active Photonics to be run by a wide audience on any PC.

At this point, I wish to stress the importance of the word "active" in this chapter's title. Passive circuits have many uses, especially in wavelength division multiplexed (WDM) systems, because networks of tunable filters can be used to multiplex, demultiplex, and route data encoded onto separate wavelengths. Also, optical filters require improved performance as WDM systems become "denser" (closer channel spacing, with higher modulation bandwidths). However, it is the addition of active devices that can drive innovation, particularly if the inherent nonlinear characteristics of active devices can be exploited.

15.2 Fundamental Requirements of a Simulator

Simulators should support innovation by providing a platform to study devices and concepts that go beyond the state of the art [3–5]. This is a challenging task, because it could be argued that a model cannot be developed until a device is understood. However, a modular approach helps provide a solution. If each module is based on a fundamental understanding of a section of waveguide, and the interfaces between the modules pass sufficient information to represent all interactions between the modules, then novel arrangements of the modules should accurately model novel circuits and devices [5].

The optimization of topologies requires that the simulation of circuits with tens of components be reasonably fast, as there are many permutations of components, and many free parameters. Thus, some approximations have to be made for the solution to be tractable. The following sections discuss these approximations.

15.2.1 Single-Mode Interfaces

A common approximation in microwave circuit design is that the waveguides support only a single mode with a well-defined field distribution at the ports of devices. Thus, the field patterns need not be described, and the waves can be simply represented by their amplitudes. This is a useful approximation when designing the topology of a photonic integrated circuit. It is assumed that all multiport devices (such as couplers) can be constructed to provide a standard field pattern at their interfaces (by tapering waveguides, for example). The waves at the ports are then simply described by their amplitudes.

15.2.2 Backward-Propagating Waves

The design of topologies is vastly simplified if the waves are assumed to propagate in one direction only: that is, no backward waves. This means that the simulation scheduling can progress from source (e.g., transmitter) to sink (receiver), one component at a time. Each component's model calculates its effect on the wave, and then shuts down. Unfortunately, backward waves are important because:

- They are critical for the design of Fabry–Perot resonators and gratings, which form the basis of lasers
- They are critical for circuits relying on the interaction of multiple resonators, such as clock recovery circuits
- They are critical for circuits where the control signal propagates backward, to isolate it from the controlled signal
- They severely degrade the performance of optical amplifiers, causing gain ripple and saturation due to backward-propagating ASE
- They can severely degrade the performance of transmitters, causing chaos, linewidth broadening, and intensity noise increase
- Optical isolators are the "missing component" in photonic integrated circuits, so they cannot be relied on to remove unwanted backward waves

For these reasons, a photonic integrated circuit simulator should consider backward waves.

15.2.3 Nonlinearities

Nonlinearities in optical fibers are well understood, because the lengths of transmission fibers and the high power levels along the system make nonlinearities a critical issue in long-haul transmission systems [6]. In passive optical circuits, the nonlinearities are usually disregarded, except in specialist materials or "what if" studies. However, active materials have significant nonlinearities, leading to useful or unwanted phenomena:

- A semiconductor optical amplifier has a low gain-compression power. That is, the gain is reduced even by small optical input powers. This is useful for wavelength conversion, power limiting, and switching, but unwanted for simple amplification. This is known as cross-gain modulation (XGM).
- The index of active material is often dependent on carrier density and, hence, gain. When the gain is compressed, the index changes and, hence, the optical phase of the output of the active section. This is useful for wavelength conversion, by cross-phase modulation (XPM).
- The gain at a given carrier density is dependent on the photon density, due to nonlinear material gain (which describes a number of physical effects, such as carrier heating). This is an extremely fast process, but relatively weak compared with the simple reduction of carrier density due to stimulated emission. This can be used for wavelength conversion and is also known as XGM.
- Any modulation of the gain or phase of an active medium with amplitude and phase modulate the wave passing through the medium, leading to

linewidth broadening and sidebands. In the presence of two input carriers, the output will contain modulated tones displaced from the original carriers by the difference frequency of the original carriers. This is the basis of wavelength conversion by four-wave mixing (FWM).

15.2.4 Optical Time Delays

Lumped models, in which all effects are considered to happen at one point in space at one instant in time, can represent devices much shorter than the distance propagated during the shortest optical pulse. However, this approximation rarely holds for data rates of 40 Gbit/s and above, particularly if return-to-zero pulses are used. For example, a typical wavelength-converting SOA cavity is 1-mm long, with a group index of 3.6, leading to a delay of $1e{-}3 \times 3.6/3e8 = 12\,\text{ps}$, which is comparable with a return-to-zero (RZ) pulse at 40 Gbit/s.

Furthermore, many devices are actually designed to resonate at the bit rate, particularly mode-locked lasers and clock-recovery circuits. Thus, the accurate modeling of time delays is critical to their performance.

15.2.5 Time Domain versus Frequency Domain

The modeling of nonlinearities generally requires a solution in the time domain. However, early laser models used frequency-domain approaches, where each resonant mode of the cavity was represented by a rate equation for the number of photons within the mode. The cross-coupling between the modes was via the carrier density, which provided gain to all modes (some more than others) and was reduced by all modes (by stimulated recombination) [1]. However, this approach breaks down for mode-locked lasers, because the modes are also coupled by the sidebands of adjacent modes that are generated by the strong amplitude and phase modulation imparted by the material.

A solution is to represent the optical field within the cavity as time-domain samples traveling waves between calculation nodes along the cavity [7]. By sampling the envelope of the optical field at a sufficiently high rate, all modes of a laser can be represented within the one sampled signal. The spectrum can be revealed using Fourier transforms. Because all modes are within the one field, they will be appropriately coupled by imposed amplitude and phase modulation.

The calculation nodes should represent the gain and phase shifts imposed on the traveling waves by the material. The boundaries between the nodes can then represent coupling between the forward- and backward-traveling waves, such as by gratings and facets [8]. In this way, the cavity resonances and grating characteristics are included. Because the grating can be within an active medium, its characteristics can be dependent on the carrier density, leading to laser chirping, modal instabilities, and functions based on these effects.

One issue with using time-domain signal representations is that it is more difficult to impose the spectral dependence of material gain than to use a frequency-domain representation. This is because the gain spectrum has to be approximated to that can be represented by a digital filter. The simplest form is an infinite-impulse response filter that leads to a Lorentzian function in the limit of a small timestep.

15.2.6 Transmission Line Laser Models

This time-domain optical-field method of modeling lasers was pioneered by me in the mid-1980s and is known as the transmission-line laser model (TLLM) [7]. The transmission-line name was inherited from transmission-line modeling (TLM). In essence, the calculation nodes (scattering nodes, in TLM) are interconnected by lossless transmission lines (connections in TLM). As shown in Fig. 15.2, the numerical algorithm simply calculates the scattering at all nodes (one by one), and then the connection between all nodes (one by one). This scattering-connection process is repeated for every timestep in the simulation, until a sufficiently long waveform has built up.

The transmission lines have a delay of one model timestep, and they serve to communicate results from adjacent nodes calculated during the previous iteration to the node in question. The delay is equal to the delay expected from the group velocity of the optical wave between the nodes. The transmission lines serve a useful numerical purpose without adding any approximations; they allow each node's calculations to be performed independently using only information coming from previous iterations of adjacent nodes. An added advantage of this decoupling is that multisection lasers, external cavity

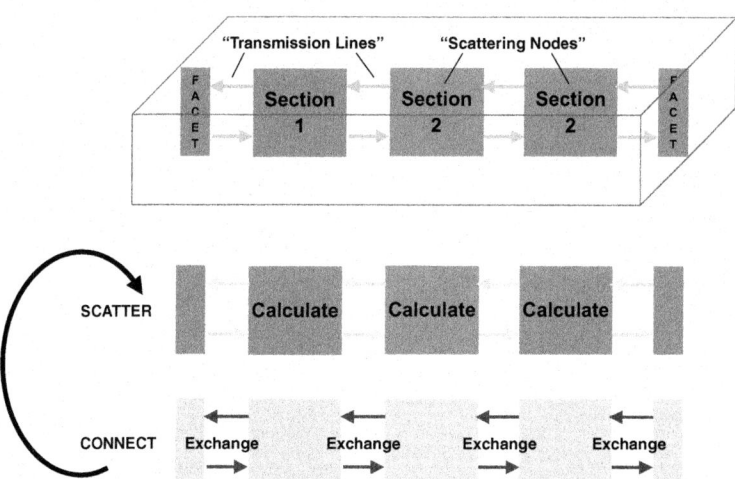

Fig. 15.2. A transmission-line laser model and the scatter-connect iteration process.

Table 15.1. Parameters Used in the TLLM (values for the SOA model).

Parameter Name	Symbol	Value	Unit
Optical Coupling Efficiency	η_{opt}	1	
Material Parameters			
Nominal Wavelength	λ	1.55252e−06	m
Linear Material Gain Coefficient	a	3.0e−20	M^2
Transparency Carrier Density	N_0	1.5e+24	m^{-3}
Linear Recombination Coefficient	A	0.00	s^{-1}
Bimolecular Recombination Coefficient	B	1.0e−16	$m^3 s^{-1}$
Auger Recombination Coefficient	C	1.3e−41	$m^6 s^{-1}$
Material Linewidth Enhancement Factor	α	3.0	
Nonlinear Gain Coefficient	ε	1.0e−23	m^3
Nonlinear Gain Timeconstant	τ_ε	0.0	s
Gain Peak Frequency	f_{st}	193.1e12	Hz
Gain Peak Frequency Carrier Dependence	df_{st}/dN	0	Hz m^3
Gain Coefficient Spectral Width	δf	1e15	Hz
Gain Coefficient Spectral Width Carrier Dependence,	$d\delta f/dN$	0	Hz m^3
Grating Stopband Frequency	f_{Bragg}	193.1e12	Hz
Chirp Reference Carrier Density	N_{ch0}	2.0e+24	m^{-3}
Population Inversion Parameter	n_{sp}	2.0	
Spontaneous Emission Coupling Factor	β	Not used	
Spontaneous Emission Peak Frequency	f_{sp}	193.1e12	Hz
Spontaneous Spectral Width	δf_{sp}	1e15	
Spontaneous Spectral Width Carrier Dependence	$d\delta f_{sp}/dN$	0	Hz m^3
Cavity Parameters			
Laser Chip Length	L	350.0e−06	m
Active Region Width	w	2.5e−06	m
(MQW) Active Region Thickness	d	0.2e−06	m
Waveguide Group Effective Index	n_g	3.7	
Internal Loss (MQWs)	α_i	3000.0	m^{-1}
Confinement Factor	Γ	0.3	
Index Grating Coupling Coefficient	κ_i	0	m^{-1}
Gain Grating Coupling Coefficient	κ_i	0	m^{-1}
Grating Facet Phase Left/Right	Θ	0	degrees
Output Signals			
Carrier Density	N		m^{-3}
Field	E		$W^{0.5}$
Photon Density	S		m^{-3}

lasers, and photonic circuits can be represented by separate numerical models, with interfaces that simply exchange samples of the forward/backward optical waveforms at each iteration. Each numerical model can be represented by a separate icon. The icons can be placed on a schematic and then linked together in an almost limitless number of topologies.

TLM has evolved to use complex-envelope field representation, which allows optical phase shifts to be implemented without approximation [9]. VPIphotonics' simulators use this representation. Many research groups have adopted similar ideas to develop time-domain laser models, sometimes known as finite-difference time-domain models. Applications include mode-locked lasers [10], gain-coupled lasers [11], external-cavity lasers [12], semiconductor optical amplifiers [13, 14], self-pulsating lasers [15], multisection lasers [16], lasers exposed to unwanted feedback [17], injection-locked lasers, and clock recovery circuits.

The parameters used in the TLLM are given in Table 15.1, together with typical values used within the simulation examples.

15.3 The Simulation Environment

VPIcomponentMaker^{TM} Active Photonics [18] incorporates time-domain models of active and passive components (including fibers) in a flexible simulation environment, shown in Fig. 15.3. These modules can be interconnected, allowing almost limitless topologies to be investigated. Each model has a comprehensive set of physical or behavioral parameters, which can be manually tuned, swept, optimized, or dithered for yield estimation.

VPIcomponentMaker^{TM} Active Photonics is part of a family of physical-layer simulators that share a common environment. This family includes VPIcomponentMaker^{TM} Optical Amplifiers for the optimization of multistage and hybrid fiber, waveguide and Raman amplifiers; VPItransmissionMaker^{TM} Cable Access for analog or mixed access services and microwave photonic applications; VPItransmissionMaker^{TM} WDM for coarse and dense WDM and optical time-division multiplexed (OTDM) systems design.

The design environment provides a large library of modules covering: photonic devices, components, subsystems; electronic and signal processing elements; signal generation, test instrumentation and visualization tools; and simulation control and data analysis functions. These modules are placed on a schematic, linked, and then the schematic run to produce visual results and output files. Interactive simulation controls lead through a design process of manual tuning, multidimensional parameter sweeps, optimization and yield estimation. Advanced features are provided, such as remote simulation, simulation scripting to drive large numerical experiments, design assistants to help synthesize and verify systems, and cosimulation with in-house software, electronic simulators, and test equipment.

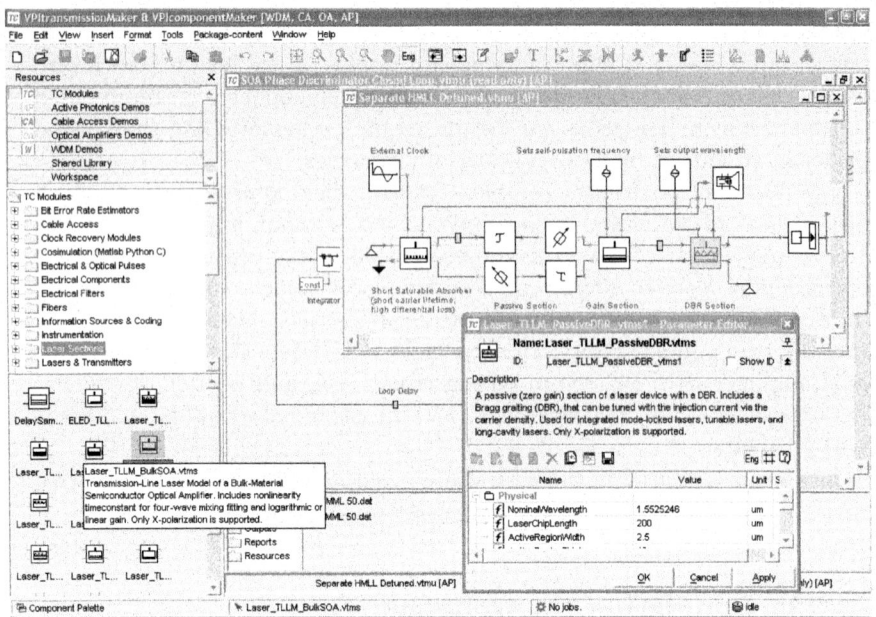

Fig. 15.3. VPIcomponentMaker^{TM} and VPItransmissionMaker^{TM} design environment.

As illustrated in Fig. 15.4, VPIcomponentMaker^{TM} Active Photonics contains a wide variety of simulation examples, including tunable lasers, mode-locked lasers, gain-coupled and index-coupled lasers, signal processing using semiconductor optical amplifiers, regenerators using multisection lasers, and systems models showing the propagation of various transmitter outputs from multimode lasers, through to near-solitons from external-cavity lasers.

To help communicate new design ideas and applications, designs can be exported to VPIplayer, which is a full simulation engine that reproduces the results of the full design environment, but is freely distributed. The applications examples presented in this book are available as VPIplayer^{TM} simulations from www.vpiphotonics.com.

15.4 Simulation Example

The example here was developed especially for this book and evolved from an interesting paper using an SOA as a phase discriminator [2] for an optical phase-locked loop. After an initial investigation of the phase discriminator concept, other components were added to the simulation to form a 40-Gbit/s to 10-Gbit/s demultiplexer for an OTDM system. This circuit could be integrated onto a single active substrate, as shown in Fig. 15.1. The integrated

15 Active Photonic Integrated Circuits 435

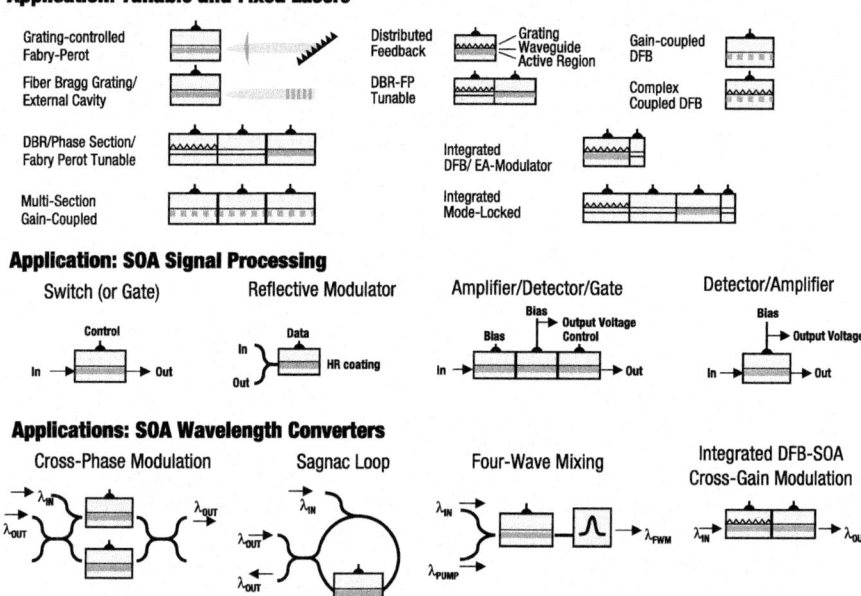

Fig. 15.4. Some applications of VPIcomponentMaker^{TM} Active Photonics (FP — Fabry–Perot, EA — Electro-Absorption).

OTDM demultiplexer has many circuit concepts within it. These are illustrated by simulation in the sections below.

15.4.1 Phase Discriminator

Central to the design is the phase discriminator, formed using an SOA, couplers, and photodiodes [2]. Its VPIcomponentMaker^{TM} schematic is shown in Fig. 15.5. From the left, a pseudo-random bit-sequence (PRBS) is used to generate logical pulses, which are then coded into electrical RZ pulses before being converted to optical pulses by an ideal laser. The signal representation is converted from a block of optical samples into individual samples before being fed into the left-hand facet of an SOA model. This is because all bidirectional laser models operate on individual time samples so that they may exchange optical data on a picosecond timescale. A similar arrangement feeds optical pulses into the right-hand facet of the SOA. The left- and right-facet output signals of the SOA are individually detected by photodiodes. A difference signal is derived using an electrical subtraction circuit.

The photodiodes detect the slowly fluctuating powers of the forward- and backward-propagating waves. The difference between their photocurrents indicates whether the forward-propagating pulses leads or lags the backward-propagating pulse.

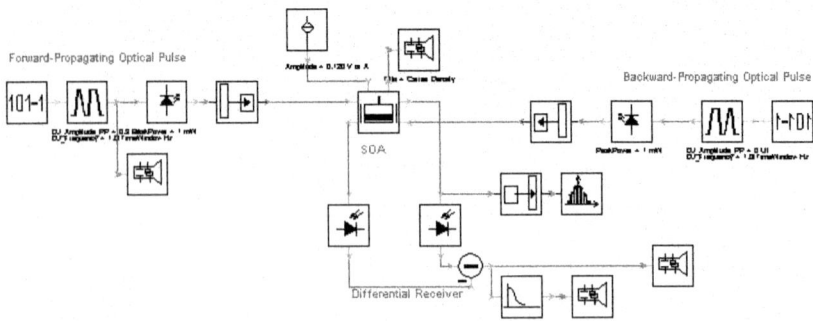

Fig. 15.5. SOA-discriminator model with idealized test pulses.

The discriminator relies on cross-gain saturation between the pulses. If a forward-propagating pulse leads a backward-propagating pulse, then the forward pulse will saturate the gain before the backward pulse arrives. Thus, the forward pulse will have a higher power than the backward pulse. Low-frequency photodetectors can be used to detect the average powers of the forward and backward pulses, and an electrical subtraction circuit can be used to find the difference in these average powers.

Figure 15.6 shows the simulated output of the discriminator without low-pass filtering. The forward pulse has a low-frequency sinusoidal jitter applied to it, so it first lags the backward pulse, then leads the backward pulse. This can be seen in Fig. 15.6, as the forward pulse produces a positive peak, and the backward pulse a negative peak. When the pulses are synchronized, the peaks cancel.

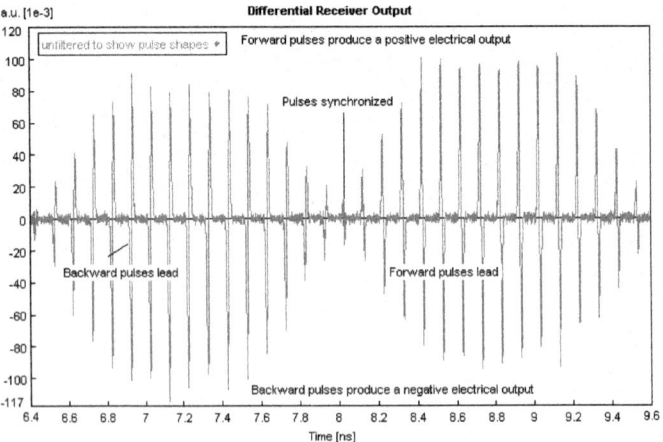

Fig. 15.6. Output of the differential photodetector (no bandlimiting).

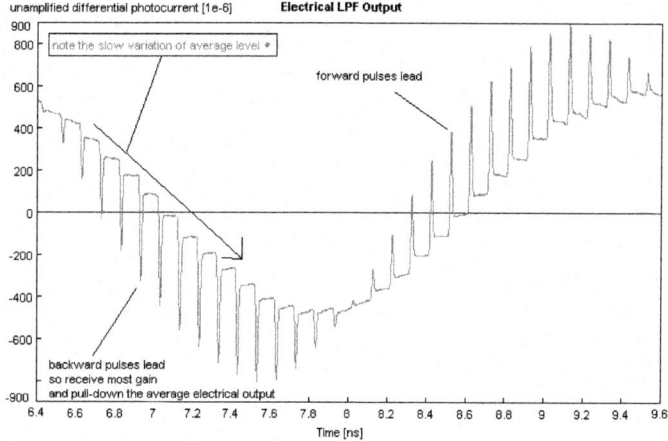

Fig. 15.7. Output of the differential photodetector after low-pass filtering.

Figure 15.7 shows the effect of the low-pass electrical filter. The filter highlights the differences in pulse energies between the forward and backward pulses. Thus, if the backward pulses lead the forward, they receive more gain from the SOA; hence, they drive the output of the discriminator downward. In a real system, the time constant of the integrator is in the order of milliseconds, so no high-frequency spikes will be seen. For the efficiency of the simulation, this time constant has been reduced to 15 ns.

15.4.2 Internal Clock Source

The purpose of the internal clock source is to generate 10 Gpulses/s, which is one-quarter of the external clock rate. These pulses can then be used to select one out of four bits from the external data stream, using an optical AND gate. The discriminator will be used to lock the internal clock to the external clock.

The internal clock source is an active-mode-locked laser, modeled as a multisection laser with a modulated gain section, a passive waveguide delay, and a grating section to act as a reflector with a band-pass response [16]. The grating bandwidth is essential to the stability of the laser, as it balances the pulse-shortening mechanism of the modulated gain region, so that stable, defined wavelength, pulses are generated. A 5-Gpulses/s demonstration example in VPIcomponentMaker™ Active Photonics was used as the basis of this design.

The design was optimized for this application by increasing the optical bandwidth of the grating so that short-enough pulses (8 ps) were obtained. For each bandwidth, the passive cavity length was swept to tune the cavity to the radio-frequency (RF) drive frequency [19]. The length of the cavity is

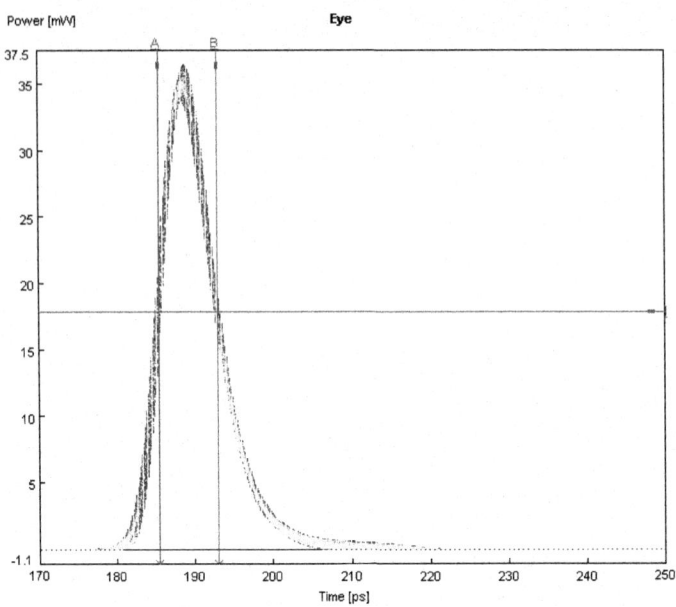

Fig. 15.8. Optical pulse from the actively mode locked laser (10 Gpulses/s).

critical to the stability of the laser: too short or too long cavities will result in cyclic instabilities, where the gain modulation prefers to amplify a new pulse, rather than the returning pulse from the cavity.

An eye diagram of the stable pulses is shown in Fig. 15.8. This comprises 160 overlapped pulses, and it shows the low timing jitter after the initial transient, which was approximately 80 pulses.

15.4.3 External Clock Source

The external clock source was a four-section integrated hybrid mode-locked laser [20]. This has saturable absorber, gain, passive and grating sections. Each laser section is represented by a separate simulation block, which is interconnected bidirectionally with its neighbors. The right-most block represents a saturable absorber. This is connected to a block representing the gain region of the laser, via optical delays and optical phase shifts, representing the passive region of the laser. The output samples at the right-hand facet of the laser are gathered into blocks (arrays) so the spectrum can be viewed using a Fourier transform. The output is also stored to file.

This laser will self-pulsate if there is no external source, but it can be locked by optical or electrical injection into its saturable absorber section. It generates very low jitter pulses. This design was available as a demo in VPIcomponentMaker[TM] Active Photonics and was used without modifica-

15 Active Photonic Integrated Circuits 439

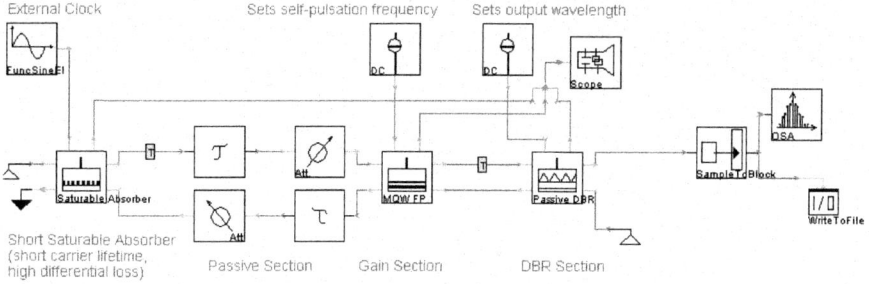

Fig. 15.9. Model of a four-section hybrid mode-locked laser.

tion, as shown in Fig. 15.9. It produces low-jitter symmetrical 4-ps pulses, shown as an eye diagram in Fig. 15.10.

The possibility of locking this source to an optical signal means that this device could be used as a clock-recovery subcircuit. Simulations could be used to investigate its resilience to input jitter, long sequences of repeated zeros and ones, and input amplitude variations.

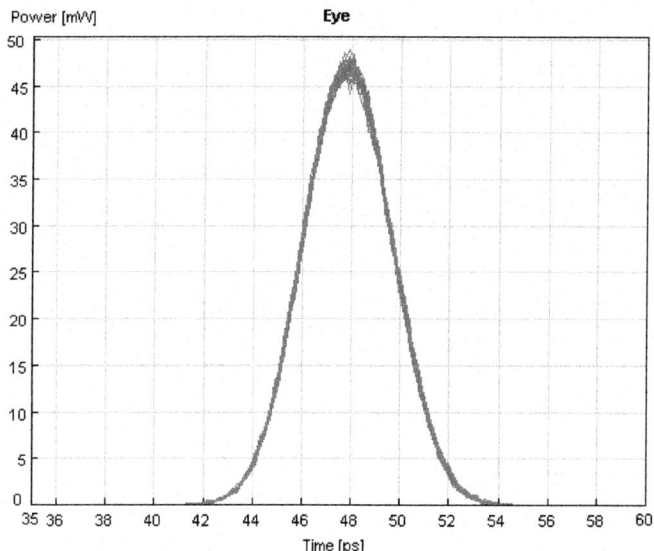

Fig. 15.10. Superimposed optical pulses from the four-section hybrid mode-locked laser (40 Gpulses/s).

15.4.4 Phase Locking the Clock Sources

A requirement of a demultiplexer is that the internal clock is phase locked to the external clock or data, so the same channel is extracted out of every four data bits. The SOA phase discriminator can be used to derive a phase error signal between an external clock (or data) and the internal clock. This phase error can be used to control the frequency of the internal clock, so that its phase will lock to the phase of the external clock. This is a phase-locked loop (PLL) [21].

In this preliminary investigation, two assumptions were made. First, the PLL was designed to work with an external clock signal, rather than a data signal. This is because every RZ pulse is present in a clock signal, whereas an unknown number is present in a data signal. This makes the design of the control circuit easier to simulate, because a much shorter time-constant can be used. In reality, data could be input into the PLL, but a control circuit with a long time constant would be required to average out short-term variations in the number of data edges. Second, the frequency variation of the internal clock was replaced with a variable time-delay to shift the clock's phase. The two are similar, as the derivative of the phase is equivalent to a frequency shift.

The first design task was to find the sensitivity of the phase discriminator, that is, the change in output voltage with phase. The sensitivity was found by sweeping the phase of the external clock using a swept time delay, and plotting the discriminator's output averaged over 800 ps for each time delay. The same segments of the clock waveforms were used for each sweep point to eliminate the effect of the clocks' jitters. However, the simulation was repeated for four different clock segments to assess the clock jitter. Figure 15.11 shows the plot of discriminator output vs. time delay. This has a skewed sinusoidal

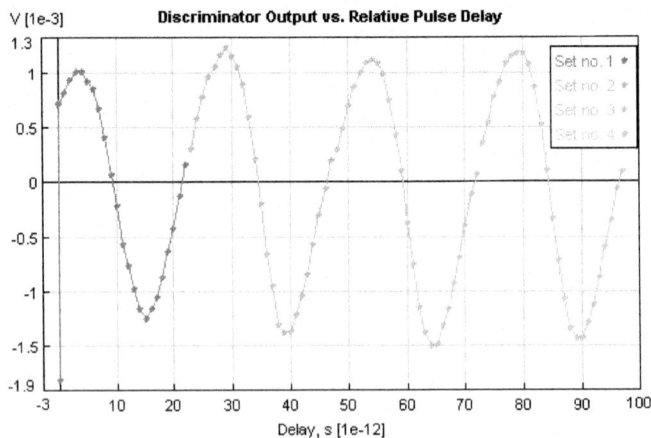

Fig. 15.11. Discriminator output vs. pulse delay using 10-G and 40-G input pulses.

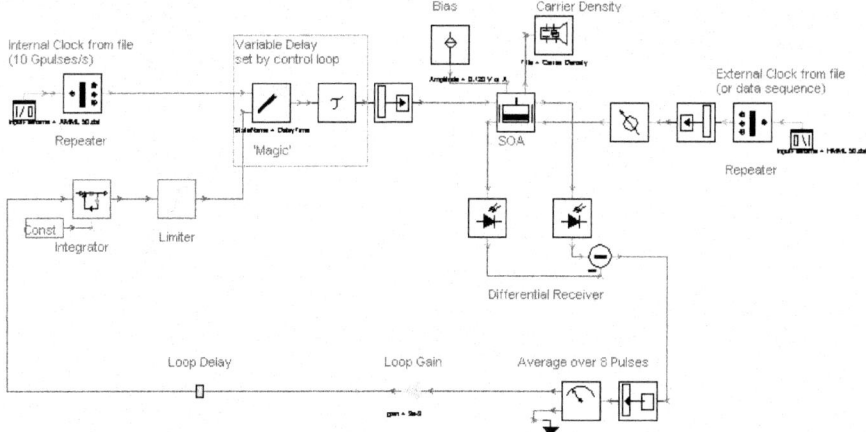

Fig. 15.12. Schematic of the phase-locked loop. The clocks are retrieved from files.

response, similar to that observed experimentally [2]. The downward slope has the greatest sensitivity (0.4 mV/ps).

The discriminator was then incorporated in a closed-loop control system, as shown in Fig. 15.12. The output of the discriminator is averaged every eight pulses using a power meter (which averages over one simulation "Block"). The power meter produces one output message per eight pulses, which is fed via a gain block, a unit delay, and through an integrator and limiter. The output of the limiter adjusts the "Time Delay" parameter of a time delay module, using a "Magic" module. A Magic module allows any module's parameter to be converted into a port, so that the port can be wired to the output of a control circuit and adjusted dynamically during a simulation. The delay ensures that the current delay is set using the previous output of the power meter (otherwise the simulation would be noncausal and would deadlock). The integrator accumulates the phase error over all previous iterations. However, it is slightly lossy, to ensure loop stability. The limiter ensures that a negative time delay is not requested.

Figure 15.13 shows the evolution of the applied time delay during the simulation of the PLL. Each point represents 800 ps of time (eight internal clock pulses). The loop gain has been set close to unity, so that a positive error will drive the time delay higher, until a stable point on the downward slope of the discriminator's curve is reached. As seen in Fig. 15.13, the PLL applies a time delay of about 8 ps in the steady state. Further tuning of the integrator and gain could improve the acquisition time and stability (jitter rejection) of the PLL.

Figure 15.14 shows an eye diagram of the discriminator output, before any bandwidth limitation. This illustrates the acquisition process. A positive pulse corresponds with the internal clock after amplification with the SOA,

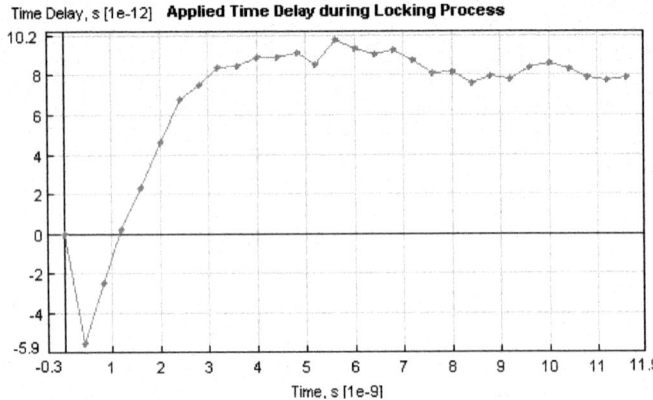

Fig. 15.13. Performance of the control loop: the applied time delay to the internal clock versus time.

minus the amplified external clock. A negative pulse is the amplified external clock only (because the internal clock is only present in a quarter of the traces, because of its lower clock rate). As the simulation progresses, greater and greater time delays are added to the internal clock, until it aligns with the external clock. Some residual jitter can be seen. The positive pulse looks asymmetric because the SOA's gain saturates during the pulse. Its peak is also missing, because the discriminator subtracts the much shorter external clock from this trace.

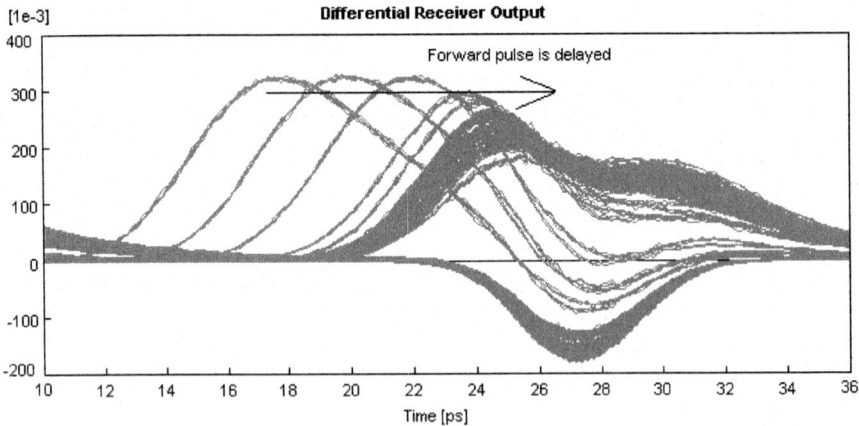

Fig. 15.14. Eye diagram of the output of the differential receiver during locking, showing the time shift of the internal clock pulses (above zero) to match the external clock pulse (below zero).

The PLL simulation illustrates the power of combining electrical and photonic simulations into one environment. The advanced simulation scheduler allows the electrical and optical signals to be simulated at completely different rates. That is, the timestep of the optical signal simulation is 0.19 ps, whereas a timestep of 800 ps is used for the control loop. Furthermore, the control loop simulations can support multiple timesteps, as can the optical parts of the simulation.

15.4.5 Optical AND Gate

An optical AND gate is required to select one in four data bits from the input 40-Gbit/s data stream. This could be achieved using FWM in an optical amplifier [22, 23], in which an FWM tone is only generated in the presence of a clock tone and a data bit, or using a Sagnac loop as shown here [24, 25].

The Sagnac loop acts as a mirror because the two counterpropagating fields experience exactly the same conditions as they travel in opposite directions through the same medium [26]. They arrive back at the coupler with the same relative phase irrespective of any changes in environmental conditions. The phase shifts within the coupler mean that any light input to the coupler is normally reflected out of the same port: It acts as a mirror. If the symmetry is broken between the phase shifts the two fields experience as they propagate around the loop, the reflection will decrease, and the output will switch to the other port of the coupler. One way to create such an asymmetry is to add a nonlinear element to the loop, such as an SOA [27, 28].

Figure 15.15 shows the simulation used to study the loop. The internal and external clock pulses are read from file. The internal clock is used to saturate the SOA, and it is called the "control" input. The external clock is the controlled input. Normally it will be reflected out of the same port as it enters (the Reflected Output). However, if a control pulse is applied to the SOA, the phase change along the SOA will change. Because the SOA is not in the middle of the loop, the counterclockwise controlled signal will see the phase change before the clockwise controlled signal, so there will be a period where the controlled signal is switched to the Transmitted Output.

Because the SOA's gain changes along with its phase, the switching of the loop is complex [24]. Also, because the SOA's gain and index do not change instantaneously (they follow the dynamics of the carrier density), the switching is not instantaneous. Figure 15.16 shows the performance of the loop when the controlled input is a continuous-wave (CW) tone. There is considerable leakage to the Transmitted Output outside the switching window, and the power of the Control Input is critical to balancing the leakage and the peak transmission. The leakage and shape of the switching window is also critically dependent on the asymmetry of the loop (the time delay) [24]. For this example, an 11-ps delay was used. Greater delays produced a squarer switching window, but more leakage.

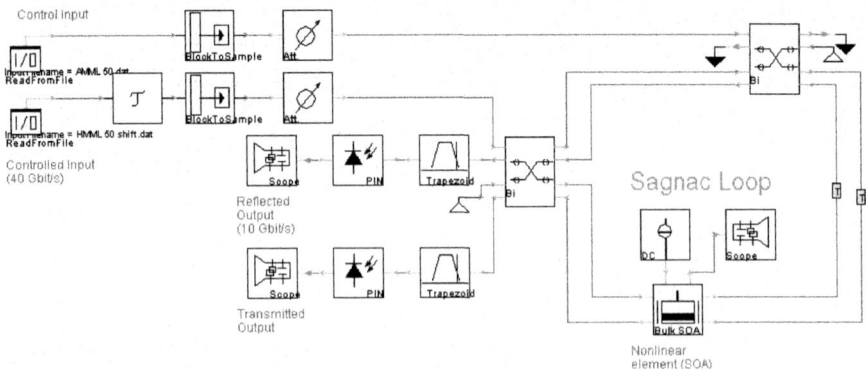

Fig. 15.15. Schematic diagram of the Sagnac loop simulation.

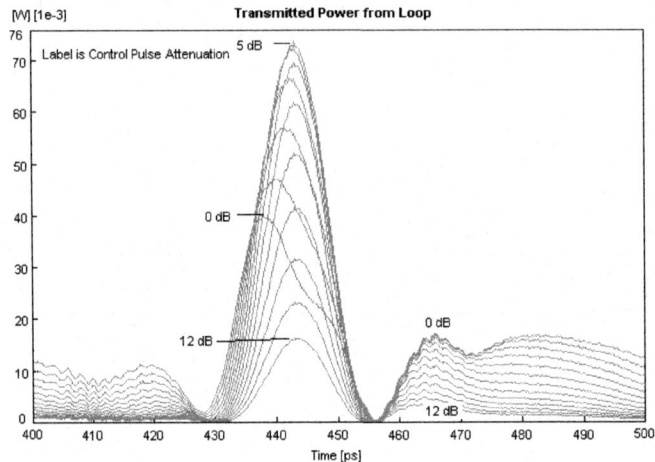

Fig. 15.16. Sagnac loop switching response for 13 levels of clock pulse.

The loop was then operated with the external clock as the controlled input. In order to separate the controlled input and the control input, the wavelength of the controlled input was tuned 500 GHz away from the internal clock. Furthermore, simulations showed that the fast switching dynamics rely on having a CW "saturating" laser injected into the controlling input (to maintain a high stimulated emission rate in the SOA at all times), so this was added at an optical frequency 500 GHz below the Control Input. The resulting spectrum is shown in Fig. 15.17. Note how the CW input has been modulated during its passage through the SOA.

Finally, the loop was operated with a modulated external clock to simulate a 40-Gbit/s RZ data sequence. Thus, the internal clock selects one in four input data pulses. The resulting eye diagram is shown in Fig. 15.18. This has

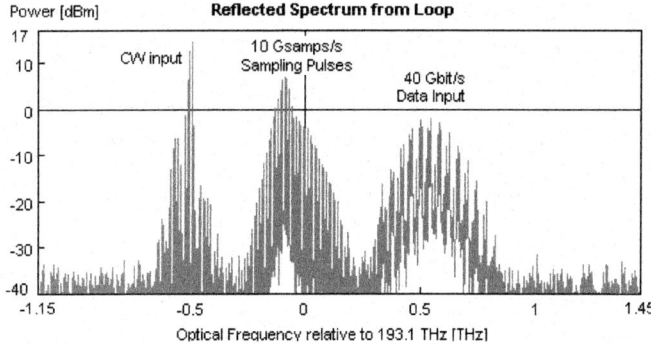

Fig. 15.17. Optical spectrum from the reflected port of the Sagnac loop.

a strong, low-jitter demultiplexed pulse, but with unwanted pulses around it, showing the finite isolation of the switch outside the switching window.

This switch is far from optimized, and simulation provides a powerful tool for parameter and bias-conditions optimization. As the introduction of the CW laser into the design showed, simulation can assess the effects of topology changes in a matter of minutes, whereas experimental design techniques would discourage such major changes in a design simply because of the effort required to resplice fibers, or redesign and manufacture a further prototype photonic integrated circuit.

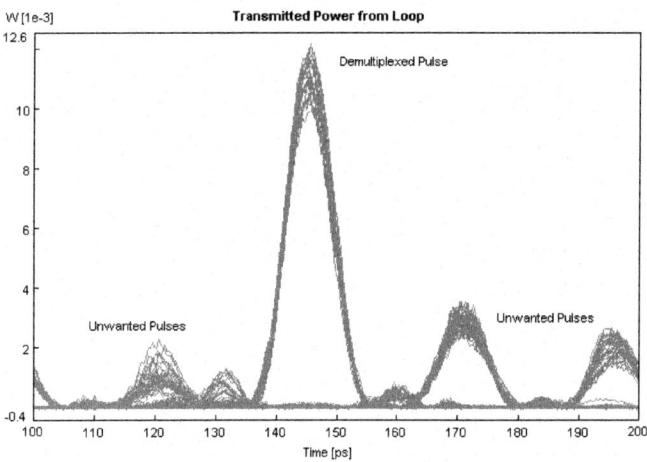

Fig. 15.18. Demultiplexed 40-Gbit/s data from the reflected port of the Sagnac Loop.

15.4.6 Open Design Issues

These design examples were produced over two days, so they are not optimized to a production standard, nor have all of the design sensitivities been explored, or the sensitivity to input and output signal properties. However, given time, simulation can address all of these issues using multidimensional sweeps, optimization, and yield functions within VPIcomponentMaker.

Some of the open design issues with this example include:

- Can the SOA discriminator and PLL operate with a modulated external data sequence rather than a clock, as shown in Fig. 15.1? This would require investigations into the effects of pattern length on the PLL's performance.
- How does the PLL respond to varying levels of input signal power and pulse length, and can a control loop be used to level the input powers?
- Could an AND gate based on FWM replace the Sagnac loop?
- What are the effects of unwanted "backward" waves on all elements of the circuit?

This last point is critical to the design, because effective optical isolators have yet to be developed for photonic integrated circuits. Fortunately, the models in VPIcomponentMaker^{TM} Active Photonics have bidirectional interfaces, so that these interactions can be studied. Preliminary simulations showed that the feedback from the loop into the internal clock mode-locked laser severely compromised its stability. With judicious use of optical filters, it may be possible to isolate the components sufficiently so that, for example, the stability of the optical clock will not be affected by the signals within the Sagnac Loop.

15.5 Conclusions

This chapter has stressed the importance of topology when designing photonic circuits. To illustrate this point, an OTDM demultiplexer design was developed, based on an SOA-based phase discriminator, combined with model-locked lasers and a Sagnac-loop switch. The design evolved from a published SOA phase discriminator, developed for equal clock rates at its ports. The simulations showed that it can also operate with unequal clock rates and unequal pulse shapes, and it is effective in a phase-locked loop. This opened up the possibility of using it to lock a 10-GHz clock to an incoming 40-GHz clock, which suggested that an OTDM demultiplexer could be developed, where one out of four pulses is selected by the 10-GHz clock. A Sagnac-loop switch was developed, with the innovation of adding a CW laser to improve its switching response.

Simulation provided the following benefits while developing this application:

- Published works were rapidly emulated, verified, and extended
- Subsystems were optimized independently combined into a larger system
- Electrical and control circuits were simulated together with the photonics, using efficient sample rates
- Subcircuits, available as demos, were rapidly tuned to new applications
- Unlimited-bandwidth instrumentation gave insight into the dynamics of the circuit operation
- Innovation was supported by allowing easy changes in topology and parameter values, including "turning-off" unwanted effects such as back-reflection

In conclusion, circuit-level simulation of photonics is critical for rapid innovation of advanced photonic subsystems, no matter what the eventual fabrication technique may be.

Acknowledgment

I would like to thank all those involved in the production and testing of VPIcomponentMaker^{TM} Active Photonics, which has its roots in the development of the OPALS device and circuit simulator, which has been successfully used around the world for over six years by many companies and universities.

References

1. A. J. Lowery, P. C. R. Gurney, X-H. Wang, L. V. T. Nguyen, Y.-C. Chan, and M. Premaratne: "Time-domain simulation of photonic devices, circuits and systems". In: *Technical Digest of Physics and Simulation of Optoelectronic Devices IV*, 29th January to 2 February 1996, San Jose, CA, vol. 2693, pp. 624–635 (1996)
2. E. S. Awad, C. J. K. Richardson, P. S. Cho, N. Moulton, and J. Goldhar: Photon. Technol. Lett. **14**, 396 (2002)
3. H. Hamster and J. Lam: Lightwave **15** (1998)
4. A. J. Lowery and P. C. R. Gurney: Appl. Opt. **37**, 6066 (1998)
5. A. J. Lowery: IEEE Spectrum **34**, 26 (1997)
6. A. J. Lowery, O. Lenzmann, I. Koltchanov, R. Moosburger, R. Freund, A. Richter, S. Georgi, D. Breuer, and H. Hamster: IEEE J. Select. Top. Quantum Electron. **6**, 282 (2000)
7. A. J. Lowery: IEE Proc. J. Optoelectron. **134**, 281 (1987)
8. A. J. Lowery: Electron. Lett. **25**, 1307 (1989)
9. J. Carroll, J. E. Whiteaway, and D. Plumb: *Distributed Feedback Semiconductor Lasers* (IEE Publishing/SPIE, London 1998)
10. A. J. Lowery, N. Onodera, and R. S. Tucker: IEEE J. Quantum Electron. **27**, 2422 (1991)
11. A. J. Lowery and D. F. Hewitt: Electron. Lett. **28**, 1959 (1992)
12. A. J. Lowery: IEE Proc. J. Optoelectron. **136**, 229 (1989)

13. A. J. Lowery and I. W. Marshall: Electron. Lett. **26**, 104 (1990)
14. A. J. Lowery: IEE Proc. J. Optoelectron. **139**, 180 (1992)
15. A. J. Lowery and H. Olesen: Electron. Lett. **30**, 965 (1994)
16. A. J. Lowery: IEE Proc. J. Optoelectron. **138**, 39 (1991)
17. A. J. Lowery: IEEE J. Quantum Electron. **28**, 82 (1992)
18. VPIcomponentMaker^{TM} is a product of VPIphotonics (www.vpiphotonics.com), which is a division of VPIsystems (Holmdel, NJ) (www.vpisystems.com)
19. Z. Ahmed, L. Zhai, A. J. Lowery, N. Onodera, and R. S. Tucker: IEEE J. Quantum Electron. **29**, 1714 (1993)
20. S. Arahira, Y. Matsui, and Y. Ogawa: IEEE J. Quantum Electron. **32**, 1211 (1996)
21. S. Kawanishi, H. Takara, K. Uchiyama, K. Kitoh, and M. Saruwatari: Electron. Lett. **29**, 1075 (1993)
22. S. Kawanishi, T. Morioka, O. Kamatani, H. Takara, and M. Saruwatari: Electron. Lett. **30**, 800 (1994)
23. I. Shake, H. Takara, K. Uchiyama, I. Ogawa, T. Kitoh, T. Kitagawa, M. Okamoto, K. Magari, Y. Suzuki, and T. Morioka: Electron Lett. **38**, 37 (2002)
24. H. Shi: J. Lightwave Technol. **20**, 682 (2002)
25. K. Chan, C.-K. Chan, W. Hung, F. Tong, and L. K. Chen: Photon. Technol. Lett. **14**, 995 (2002)
26. D. B. Mortimore: J. Lightwave Technol. **6**, 1217 (1998)
27. K. Suzuki, K. Iwatsuki, S. Nishi, and M. Saruwatari: Electron. Lett. **30**, 660 (1994)
28. K. Suzuki, K. Iwatsuki, S. Nishi, and M. Saruwatari: Electron. Lett. **30**, 1501 (1994)

Index

absorption
  band-to-band, 1
  intervalence band, 408
  spectrum, 413
advection equations, 151
amplifier, 416, 434
APSYS, 295
ATLAS, 383
Auger recombination, 35, 325, 388, 415
axial approximation, 10

backward waves, 428
band gap narrowing, 321
bifurcations, 127, 140
bit-error rate, 285
BLAZE, 383
Boltzmann statistics, 76, 385
Bragg constant, 91

carrier leakage, 307
carrier capture, 33
carrier rate equation, 123, 155, 257
carrier transport
  equations, 31, 93, 298, 384, 414
CCD, 343
  barrier region, 344
  bloomimg, 364
  burried channel, 344
  channel potential, 355
  charge capacity, 357
  charge transfer efficiency, 357
  deep depletion mode, 370
  fringing electric field, 357
  full-frame array, 346
  lateral overflow drain, 369
  storage region, 344
  traps, 373
charge-coupled device: see CCD, 343

CLADISS-2D, 158, 197
complex-envelope field representation, 433
continuity equation, 222, 299
Coulomb enhancement factor, 8
Coulomb-induced intersubband coupling, 15
cross-coupling susceptibility, 11
cross-gain saturation, 436
current conservation equation, 65

dark current, 370
DBR laser, 188
demultiplexer, 427
dephasing time, 13
DESSIS, 217, 317
DFB laser, 87
  harmonic distortion, 101
  intermodulation distortion, 101
  modulation response, 101
  small-signal analysis, 101
digital-supermode DBR laser, 178
dissipation rate, 74
distributed Bragg reflector, see DBR, 159
distributed feedback laser: see DFB laser, 87
distributed feedback section, 125
distributed resistance, 334

effective index, 92
energy bands
  wurtzite, 295
  offset, 298, 410
energy models, 63
entropy, 67
excess-carrier lifetime, 315
eye diagram, 283

Fabry–Perot laser, 27, 63
Fermi–Dirac statistics, 65, 322, 386
floating junction, 331
free energy, 67
frequency response function, 82
Fresnel equations, 304, 390

gain
  compression, 412
  dispersion, 195
  free-carrier model, 44
  linear, 3, 39
  many-body theory, 1, 44
  nonlinear, 414
  spectrum, 413
  suppression, 252, 265
  tables, 18, 405
gain-coupled laser, 434

heat flux equation, 37, 70, 97, 259, 302
heat source, 71, 223, 259, 302
Helmholtz equation, 38, 406
hot carrier effects, 252

IIR filter, 154
image sensing, 343
infrared detectors, 381
  crosstalk, 395
  recombination effects, 398
  spectral response, 396
  temperature effects, 400
intrinsic carrier density, 320

Langevin noise, 255
Laplace equation, 256
LaserMOD, 27
LDSL-tool, 121
light-emitting diode (LED), 293
linewidth enhancement factor, 4
Lorentzian gain, 154
Lorentzian lineshape function, 7, 10, 13, 40, 304
LUMINOUS, 383
lumped models, 430

many-body effects, 1, 27, 412
material parameters
  AlGaAs, 239, 278
  AlGaN, 294

AlN, 296
GaAs, 407
GaN, 296
GaP, 407
HgCdTe, 391
InAlGaAs, 109
InAs, 407
InGaAs, 241
InGaAsP, 50, 84, 105, 406
InP, 407
Si, 345
MIS structure, 330
mode beating, 134, 139
mode expansion, 131
mode locking, 185, 437
multisection laser, 121, 159, 422, 434

noise
  current supply, 290
  flicker, 290
  frequency, 289
  mode-partition, 285
  relative intensity, 287
  spectrum, 155
  stochastic, 152
  VCSEL, 254

optical AND gate, 443
optical mode
  amplitude, 133
  analysis, 139
  axial distribution, 137
  fundamental, 408
  instantaneous, 129
  longitudinal, 121
  VCSEL, 222, 261
optical polarization, 3
optical susceptibility, 3

PERL cell, 327
phase discriminator, 435
phase tuning, 126, 139, 141
phase-locked loop, 440
PhaseCOMB laser, 126, 134
photodetector
  CdTe-based, 381
  InP-based, 420
  saturation, 420
photogeneration, 351, 390
photoluminescence, 3

measured, 20, 309, 411
  spectrum, 411
photon rate equation, 38, 66, 221
photonic integrated circuit, 405, 428
PICS3D, 405
piezoelectric effect, 300
pixel, 343
Poisson equation, 33, 64, 222, 299, 414

Q-switching, 196
quantum efficiency, 309
quasi-Fermi level method, 352

ray tracing, 304, 389
refractive index
  carrier-induced, 3
  model, 305, 408
relaxation process, 24

Sagnac loop, 443
sampled grating, 159, 422
saturable absorber, 185
scattering
  electron–electron, 5
  electron–phonon, 5
self-heating
  laser, 53, 63, 97
  LED, 302
self-pulsation, 127, 135, 137, 201, 438
semiconductor Bloch equations, 5
SG-DBR laser, 160, 422
Shockley–Read–Hall recombination, 35, 299, 387, 415
side-mode suppression, 168, 424
solar cell, 313
  nongeneration loss, 337
  recombination losses, 327
  resistive loss, 334
spatial hole burning
  longitudinal, 93
  photodetector, 417
  VCSEL, 245, 250, 252
spectral broadening, 9
  due to scattering, 29
  inhomogeneous, 20, 30, 411
  parameter, 47
spectral equation, 129
spectral hole burning, 252
spontaneous emission, 36
  rate, 303, 415
  spectrum, 303
spontaneous inversion factor, 153
stimulated emission, 36
  rate, 414
surface recombination, 324

TDTW model, 151, 192
thermodynamics
  modeling, 63
  second law, 74
time-domain model, 151, 430
  distributed: see TDTW model, 192
time-evolving spectrum, 169
timing jitter, 283
transfer matrix, 145
transition matrix element, 303, 412
transmission-line laser model, 431
traveling wave model, 122, 151
  see also TDTW model, 151

ultraviolet light source, 293

VCSEL, 217, 249
  anti-guiding, 262
  anti-resonant structure, 217
  distributed Bragg reflector, 219
  dynamics, 250
  effective index method, 261
  etched mesa, 217
  harmonic distortion, 253
  intermodulation distortion, 253
  linewidth, 289
  metallic absorber, 217
  mode competition, 250, 264
  mode discrimination, 235
  modulation, 251, 280
  multimode, 253
  noise, 254, 265
  oxide aperture, 217
  power rollover, 238, 277
  quality factor, 232
  radiative losses, 228
  self-heating, 222
  single mode control, 217, 250
  spatial hole burning, 281
  surface relief, 220, 271
  thermal lensing, 250, 262
Vernier tuning, 422

vertical-cavity surface-emitting laser: see VCSEL, 217
VPIcomponentMaker^{TM}, 427
VPIplayer^{TM}, 428
VPItransmissionMaker^{TM}, 433

wave equation, 89

vectorial, 221
waveguide equation, 66, 406
wavelength converter, 405
wavelength-tunable laser, 159, 422, 434
WIAS-TeSCA, 64
wurtzite semiconductor, 295

Printed by Printforce, United Kingdom